Signal Processing Methods
for Music Transcription

Signal Processing Methods for Music Transcription

With 124 illustrations

Anssi Klapuri
Manuel Davy

Editors

 Springer

Anssi Klapuri
Tampere University of Technology
Institute of Signal Processing
Korkeakoulunkatu 1
33720 Tampere, Finland
Anssi.Klapuri@tut.fi

Manuel Davy
LAGIS/CNRS
Ecole Centrale de Lille
Cité Scientifique
BP 48
59651 Villeneuve d'Ascq
Cedex, France
Manuel.Davy@ec-lille.fr

Cover design by Alex Gerasev.

Mathematics Subject Classification (2000): 68T10, 62F15, 68P20

e-ISBN-10: 0-387-32845-9

ISBN-13: 978-1-4419-4035-3

e-ISBN-13: 978-0-387-32845-4

Printed on acid-free paper.

Printed in the United States of America.   (KeS/MP)

9 8 7 6 5 4 3 2 1

springer.com

Contents

Preface ix

List of Contributors xi

Part I Foundations

1 Introduction to Music Transcription
Anssi Klapuri 3
1.1 Terminology and Concepts 7
1.2 Perspectives on Music Transcription 11
1.3 Outline 17

**2 An Introduction to Statistical Signal Processing and
Spectrum Estimation**
Manuel Davy 21
2.1 Frequency, Time-Frequency, and Cepstral Representations 21
2.2 Basic Statistical Methods 28
2.3 Bayesian Statistical Methods 39
2.4 Pattern Recognition Methods 52

3 Sparse Adaptive Representations for Musical Signals
Laurent Daudet, Bruno Torrésani 65
3.1 Introduction 65
3.2 Parametric Representations 68
3.3 Waveform Representations 70
3.4 Conclusion 97

Part II Rhythm and Timbre Analysis

4 Beat Tracking and Musical Metre Analysis
Stephen Hainsworth 101
4.1 Introduction 101
4.2 Summary of Beat-Tracking Approaches 102
4.3 Musical Background to Rhythmic Structure 105
4.4 Onset Detection 107
4.5 Rule-Based Approaches 111
4.6 Autocorrelation Methods 112
4.7 Oscillating Filter Approaches 113
4.8 Histogramming Methods 115
4.9 Multiple Agent Approaches 116
4.10 Probabilistic Models 117
4.11 Comparison of Algorithms 124
4.12 Conclusions 127

5 Unpitched Percussion Transcription
Derry FitzGerald, Jouni Paulus 131
5.1 Introduction 131
5.2 Pattern Recognition Approaches 133
5.3 Separation-Based Approaches 142
5.4 Musicological Modelling 153
5.5 Conclusions 160
5.6 Acknowledgements 162

6 Automatic Classification of Pitched Musical Instrument Sounds
Perfecto Herrera-Boyer, Anssi Klapuri, and Manuel Davy 163
6.1 Introduction 163
6.2 Methodology 169
6.3 Features and Their Selection 171
6.4 Classification Techniques 184
6.5 Classification of Isolated Sounds 188
6.6 Classification of Sounds from Music Files 193
6.7 Conclusions 199

Part III Multiple Fundamental Frequency Analysis

7 Multiple Fundamental Frequency Estimation Based on Generative Models
Manuel Davy 203
7.1 Noisy Sum-of-Sines Models 204
7.2 Off-line Approaches 210

7.3 On-Line Approaches 217
7.4 Other On-Line Bayesian Approaches 225
7.5 Conclusions 227

8 Auditory Model-Based Methods for Multiple Fundamental Frequency Estimation
Anssi Klapuri 229
8.1 Introduction 229
8.2 Musical Sounds and F0 Estimation............................. 231
8.3 Pitch Perception Models 234
8.4 Using an Auditory Model as a Front End 244
8.5 Computational Multiple F0 Estimation Methods................ 248
8.6 Conclusions 264

9 Unsupervised Learning Methods for Source Separation in Monaural Music Signals
Tuomas Virtanen 267
9.1 Introduction 267
9.2 Signal Model 268
9.3 Independent Component Analysis............................... 274
9.4 Sparse Coding... 278
9.5 Non-Negative Matrix Factorization............................ 282
9.6 Prior Information about Sources 284
9.7 Further Processing of the Components......................... 286
9.8 Time-Varying Components...................................... 289
9.9 Evaluation of the Separation Quality 294
9.10 Summary and Discussion 295

Part IV Entire Systems, Acoustic and Musicological Modelling

10 Auditory Scene Analysis in Music Signals
Kunio Kashino 299
10.1 Introduction 299
10.2 Strategy for Music Scene Analysis 304
10.3 Probabilistic Models for Music Scene Analysis 313
10.4 Conclusion: From Grouping to Generative Estimation 324

11 Music Scene Description
Masataka Goto 327
11.1 Introduction 327
11.2 Estimating Melody and Bass Lines............................. 330
11.3 Estimating Beat Structure 341
11.4 Estimating Drums 342
11.5 Estimating Chorus Sections and Repeated Sections 342

11.6 Evaluation Issues 355
11.7 Applications of Music Scene Description 355
11.8 Conclusion 358

12 Singing Transcription
Matti Ryynänen 361
12.1 Introduction 361
12.2 Singing Signals 364
12.3 Feature Extraction 368
12.4 Converting Features into Note Sequences 375
12.5 Summary and Discussion 390

References 391

Index 429

Preface

Signal processing techniques, and information technology in general, have undergone several scientific advances which permit us to address the very complex problem of automatic music transcription (AMT). During the last ten years, the interest in AMT has increased rapidly, and the time has come for a book-length overview of this subject.

The purpose of this book is to present signal processing algorithms dedicated to the various aspects of music transcription. AMT is a multifaceted problem, comprising several subtasks: rhythm analysis, multiple fundamental frequency analysis, sound source separation, musical instrument classification, and integration of all these into entire systems. AMT is, in addition, deeply rooted in fundamental signal processing, which this book also covers. As the field is quite wide, we have focused mainly on signal processing methods and Western polyphonic music. An extensive presentation of the work in musicology and music perception is beyond the scope of this book.

This book is mainly intended for researchers and graduate students in signal processing, computer science, acoustics, and music. We hope that the book will make the field easier to approach, providing a good starting point for newcomers, but also a comprehensive reference source for those already working in the field. The book is also suitable for use as a textbook for advanced courses in music signal processing. The chapters are mostly self-contained, and readers may want to read them in any order or jump from one to another at will. Whenever an element from another chapter is needed, an explicit reference is made to the relevant chapter. Chapters 1 and 2 provide some background of AMT and signal processing for the entire book, respectively. Otherwise, only a basic knowledge of signal processing is assumed.

Editing a book is a great deal of work. This volume was made possible by those who provided us support and help. We would like to thank Vaishali Damle and Ana Bozicevic at Springer for their help and support, and for their quick replies to our e-mails. Also thanks to Teemu Karjalainen for his practical assistance with LaTeX.

Early versions of individual chapters were reviewed by the following people, whose valuable comments and suggestions are gratefully acknowledged:
- Michael Casey, Goldsmiths College, University of London, UK
- A. Taylan Cemgil, University of Cambridge, UK
- Alain de Cheveigné, Ecole Normale Supérieure, Paris, France
- Simon Dixon, Austrian Research Institute for Artificial Intelligence, Vienna
- Dan Ellis, Columbia University, New York
- Olivier Gillet, Télécom-Paris (ENST), Paris, France
- Aki Härmä, Philips Research Laboratories, Eindhoven, The Netherlands
- Marc Leman, Ghent University, Belgium
- Emanuele Pollastri, Erazero S.r.l., Milan, Italy

Thanks go also to the chapter authors, many of whom proofread another, related chapter in the book and provided helpful comments.

Tampere, Finland   *Anssi Klapuri*
Lille, France   *Manuel Davy*
December 2005

List of Contributors

Laurent Daudet
Laboratoire d'Acoustique Musicale
11 rue de Lourmel
75015 Paris, France
daudet@lam.jussieu.fr

Manuel Davy
LAGIS/CNRS
BP 48, Cité Scientifique
59651 Villeneuve d'Ascq Cedex
France
Manuel.Davy@ec-lille.fr

Derry FitzGerald
Cork Institute of Technology
Rossa Avenue
Bishopstown
Cork, Ireland
derry.fitzgerald@cit.ie

Masataka Goto
National Institute of Advanced
Industrial Science and Technology
(AIST)
1-1-1 Umezono, Tsukuba
Ibaraki 305-8568, Japan
m.goto@aist.go.jp

Stephen Hainsworth
Tillinghast-Towers Perrin
71 High Holborn
London WC1V 6TH, UK
swh21@cantab.net

Perfecto Herrera-Boyer
Institut Universitari de l'Audiovisual
Universitat Pompeu Fabra
Pg. Circumval·lació 8
08003 Barcelona, Spain
pherrera@iua.upf.es

Kunio Kashino
NTT Communication Science
Laboratories
Nippon Telegraph and Telephone
Corporation 3–1
Morinosato-Wakamiya
Atsugi, 243-0198, Japan
kunio@eye.brl.ntt.co.jp

Anssi Klapuri
Institute of Signal Processing
Tampere University of Technology
Korkeakoulunkatu 1
33720 Tampere, Finland
Anssi.Klapuri@tut.fi

Jouni Paulus
Institute of Signal Processing
Tampere University of Technology
Korkeakoulunkatu 1
33720 Tampere, Finland
Jouni.Paulus@tut.fi

Bruno Torrésani
Laboratoire d'Analyse
Topologie et Probabilités
CMI, Université de Provence
39 rue F. Joliot-Curie
13453 Marseille cedex 13, France
Bruno.Torresani@cmi.univ-mrs.fr

Matti Ryynänen
Institute of Signal Processing
Tampere University of Technology

Korkeakoulunkatu 1
33720 Tampere, Finland
Matti.Ryynanen@tut.fi

Tuomas Virtanen
Institute of Signal Processing
Tampere University of Technology
Korkeakoulunkatu 1
33720 Tampere, Finland
Tuomas.Virtanen@tut.fi

Part I

Foundations

FOUNDATIONS

1

Introduction to Music Transcription

Anssi Klapuri

Institute of Signal Processing, Tampere University of Technology
Korkeakoulunkatu 1, 33720 Tampere, Finland
Anssi.Klapuri@tut.fi

Music transcription refers to the analysis of an acoustic musical signal so as to write down the pitch, onset time, duration, and source of each sound that occurs in it. In Western tradition, written music uses *note symbols* to indicate these parameters in a piece of music. Figures 1.1 and 1.2 show the notation of an example music signal. Omitting the details, the main conventions are that time flows from left to right and the pitch of the notes is indicated by their vertical position on the staff lines. In the case of drums and percussions, the vertical position indicates the instrument and the stroke type. The loudness (and the applied instrument in the case of pitched instruments) is normally not specified for individual notes but is determined for larger parts.

Besides the common musical notation, the transcription can take many other forms, too. For example, a guitar player may find it convenient to read *chord symbols* which characterize the note combinations to be played in a more general manner. In a computational transcription system, a MIDI file[1] is often an appropriate format for musical notations (Fig. 1.3). Common to all these representations is that they capture musically meaningful parameters that can be used in performing or synthesizing the piece of music in question. From this point of view, music transcription can be seen as discovering the 'recipe', or reverse-engineering the 'source code' of a music signal.

A complete transcription would require that the pitch, timing, and instrument of all the sound events be resolved. As this can be very hard or even theoretically impossible in some cases, the goal is usually redefined as being either to notate as many of the constituent sounds as possible (complete transcription) or to transcribe only some well-defined part of the music signal, for example the dominant melody or the most prominent drum sounds (partial transcription). Both of these goals are relevant and are discussed in this book.

Music transcription is closely related to *structured audio coding*. A musical notation or a MIDI file is an extremely compact representation that retains

---

[1]Musical Instrument Digital Interface (MIDI) is a standard for exchanging performance data and parameters between electronic musical devices [462], [571].

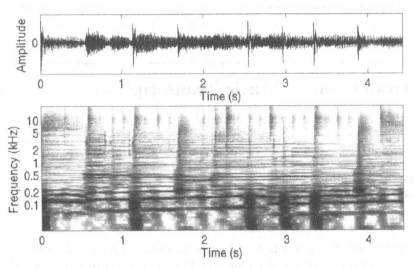

Fig. 1.1. An acoustic musical signal (top) and its time-frequency domain representation (bottom). The excerpt is from Song G034 in the RWC database [230].

Fig. 1.2. Musical notation corresponding to the signal in Fig. 1.1. The upper staff lines show the notation for pitched musical instruments and the lower staff lines show the notation for percussion instruments.

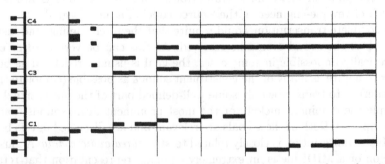

Fig. 1.3. A 'piano-roll' illustration of a MIDI file which corresponds to the pitched instruments in the signal in Fig. 1.1. Different notes are arranged on the vertical axis and time flows from left to right.

the characteristics of a piece of music to an important degree. Another related area of study is that of *music perception* [144]. Detecting and recognizing individual sounds in music is a big part of its perception, although it should be emphasized that musical notation is primarily designed to serve sound production and not to model hearing. We do not hear music in terms of note symbols but, as described by Bregman [49, pp. 457–460], music often 'fools' the auditory system so that we perceive simultaneous sounds as a single entity.

In addition to audio coding, applications of music transcription comprise

- *Music information retrieval* based on the melody of a piece, for example.
- *Music processing*, such as changing the instrumentation, arrangement, or the loudness of different parts before resynthesizing a piece from its score.
- *Human-computer interaction* in various applications, including score type-setting programs and musically oriented computer games. Singing transcription is of particular importance here.
- *Music-related equipment*, ranging from music-synchronous light effects to highly sophisticated interactive music systems which generate an accompaniment for a soloist.
- *Musicological analysis* of improvised and ethnic music for which musical notations do not exist.
- *Transcription tools* for amateur musicians who wish to play along with their favorite music.

The purpose of this book is to describe algorithms and models for the different subtopics of music transcription, including pitch analysis, metre analysis (see Section 1.1 for term definitions), percussion transcription, musical instrument classification, and music structure analysis. The main emphasis is laid on the low-level signal analysis where sound events are detected and their parameters are estimated, and not so much on the subsequent processing of the note data to obtain larger musical structures. The theoretical background of different signal analysis methods is presented and their application to the transcription problem is discussed.

The primary target material considered in this book is complex music signals where several sounds are played simultaneously. These are referred to as *polyphonic* signals, in contrast to *monophonic* signals where at most one note is sounding at a time. For practical reasons, the scope is limited to Western music, although not to any particular genre. Many of the analysis methods make no assumptions about the larger-scale structure of the signal and are thus applicable to the analysis of music from other cultures as well.

To give a reasonable estimate of the achievable goals in automatic music transcription, it is instructive to study what human listeners are able to do in this task. An average listener perceives a lot of musically relevant information in complex audio signals. He or she can tap along with the rhythm, hum the melody (more or less correctly), recognize musical instruments, and locate structural parts of the piece, such as the chorus and the verse in popular music. Harmonic changes and various details are perceived less consciously. Similarly to natural language, however, reading and writing music requires education. Not only the notation needs to be studied, but recognizing different pitch intervals and timing relationships is an ability that has to be learned – these have to be encoded into a symbolic form in one's mind before writing them down. Moreover, an untrained listener is typically not able to hear the inner lines in music (sub-melodies other than the dominant one), so musical ear training is needed to develop an analytic mode of listening where these can be distinguished. The richer the polyphonic complexity of a musical composition, the more its transcription requires musical ear training and knowledge of the particular musical style and of the playing techniques of the instruments involved.

First attempts towards the automatic transcription of polyphonic music were made in the 1970s, when Moorer proposed a system for transcribing two-voice compositions [477], [478]. His work was followed by that of Chafe et al. [81], Piszczalski [520], and Maher [426], [427] in the 1980s. In all these early systems, the number of concurrent voices was limited to two and the pitch relationships of simultaneous sounds were restricted in various ways. On the rhythm analysis side, the first algorithm for beat tracking[2] in general audio signals was proposed by Goto and Muraoka in the 1990s [235], although this was preceded by a considerable amount of work for tracking the beat in parametric note data (see [398] for a summary) and by the beat-tracking algorithm of Schloss for percussive audio tracks [567]. First attempts to transcribe percussive instruments were made in the mid-1980s by Schloss [567] and later by Bilmes [37], both of whom classified different types of conga strikes in continuous recordings. Transcription of polyphonic percussion tracks was later addressed by Goto and Muraoka [236]. A more extensive description of the early stages of music transcription has been given by Tanguiane in [619, pp. 3–6].

Since the beginning of 1990s, the interest in music transcription has grown rapidly and it is not possible to make a complete account of the work here. However, certain general trends and successful approaches can be discerned. One of these has been the use of *statistical methods*. To mention a few examples, Kashino [332], Goto [223], Davy and Godsill [122], and Ryynänen [559] proposed statistical methods for the pitch analysis of polyphonic music; in beat tracking, statistical methods were employed by Cemgil and Kappen [77], Hainsworth and MacLeod [266], and Klapuri et al. [349]; and in percussive instrument transcription by Gillet and Richard [209] and Paulus et al. [506]. In musical instrument classification, statistical pattern recognition methods prevail [286]. Another trend has been the increasing utilization of *computational models of the human auditory system*. These were first used for music transcription by Martin [439], and auditorily motivated methods have since been proposed for polyphonic pitch analysis by Karjalainen and Tolonen [627] and

---

[2] *Beat tracking* refers to the estimation of a rhythmic pulse which corresponds to the tempo of a piece and (loosely) to the foot-tapping rate of human listeners.

Klapuri [354], and for beat tracking by Scheirer [564], for example. Another prominent approach has been to model the human *auditory scene analysis* (ASA) ability. The term ASA refers to the way in which humans organize spectral components to their respective sounds sources and recognize simultaneously occurring sounds [49]. The principles of ASA were brought to the pitch analysis of polyphonic music signals by Mellinger [460] and Kashino [333], and later by Godsmark and Brown [215] and Sterian [609]. Most recently, several *unsupervised learning* methods have been proposed where a minimal number of prior assumptions are made about the analysed signal. Methods based on independent component analysis [304] were introduced to music transcription by Casey [70], [73], and various other methods were later proposed by Lepain [403], Smaragdis [598], [600], Abdallah [2], [5], Virtanen (see Chapter 9), FitzGerald [186], [188], and Paulus [505]. Of course, there are also methods that do not represent any of the above-mentioned trends, and a more comprehensive review of the literature is presented in the coming chapters.

The state-of-the-art music transcription systems are still clearly inferior to skilled human musicians in accuracy and flexibility. That is, a reliable general-purpose transcription system does not exist at the present time. However, some degree of success has been achieved for polyphonic music of limited complexity. In the transcription of pitched instruments, typical restrictions are that the number of concurrent sounds is limited [627], [122], interference of drums and percussive sounds is not allowed [324], or only a specific instrument is considered [434]. Some promising results for the transcription of real-world music on CD recordings has been demonstrated by Goto [223] and Ryynänen and Klapuri [559]. In percussion transcription, quite good accuracy has been achieved in the transcription of percussive tracks which comprise a limited number of instruments (typically bass drum, snare, and hi-hat) and no pitched instruments [209], [505]. Also promising results have been reported for the transcription of the bass and snare drums on real-world recordings, but this is a more open problem (see e.g. Zils et al. [693], FitzGerald et al. [189], Yoshii et al. [683]). Beat tracking of complex real-world audio signals can be performed quite reliably with the state-of-the-art methods, but difficulties remain especially in the analysis of classical music and rhythmically complex material. Comparative evaluations of beat-tracking systems can be found in [266], [349], [248]. Research on musical instrument classification has mostly concentrated on working with isolated sounds, although more recently this has been attempted in polyphonic audio signals, too [331], [33], [170], [647].

1.1 Terminology and Concepts

Before turning to a more general discussion of the music transcription problem and the contents of this book, it is necessary to introduce some basic terminology of auditory perception and music. To discuss music signals, we first have to discuss the perceptual attributes of sounds of which they consist. There are four subjective qualities that are particularly useful in characterizing sound events: pitch, loudness, duration, and timbre [550].

Pitch is a perceptual attribute which allows the ordering of sounds on a frequency-related scale extending from low to high. More exactly, pitch is defined as the frequency of a sine wave that is matched to the target sound by human listeners [275]. *Fundamental frequency* (F0) is the corresponding physical term and is defined for periodic or nearly periodic sounds only. For these classes of sounds, F0 is defined as the inverse of the period and is closely related to pitch. In ambiguous situations, the period corresponding to the perceived pitch is chosen.

The perceived *loudness* of an acoustic signal has a non-trivial connection to its physical properties, and computational models of loudness perception constitute a fundamental part of psychoacoustics [523].[3] In music processing, however, it is often more convenient to express the level of sounds with their mean-square power and to apply a logarithmic (decibel) scale to deal with the wide dynamic range involved. The perceived *duration* of a sound has more or less one-to-one mapping to its physical duration in cases where this can be unambiguously determined.

Timbre is sometimes referred to as sound 'colour' and is closely related to the recognition of sound sources [271]. For example, the sounds of the violin and the flute may be identical in their pitch, loudness, and duration, but are still easily distinguished by their timbre. The concept is not explained by any simple acoustic property but depends mainly on the coarse spectral energy distribution of a sound, and the time evolution of this. Whereas pitch, loudness, and duration can be quite naturally encoded into a single scalar value, timbre is essentially a multidimensional concept and is typically represented with a feature *vector* in musical signal analysis tasks.

Musical information is generally encoded into the *relationships* between individual sound events and between larger entities composed of these. Pitch relationships are utilized to make up melodies and chords. Timbre and loudness relationships are used to create musical form especially in percussive music, where pitched musical instruments are not necessarily employed at all. Inter-onset interval (IOI) relationships, in turn, largely define the rhythmic characteristics of a melody or a percussive sound sequence (the term IOI refers to the time interval between the beginnings of two sound events). Although durations of the sounds play a role too, the IOIs are more crucial in determining the perceived rhythm [93]. Indeed, many rhythmically important instruments, such as drums and percussions, produce exponentially decaying wave shapes that do not even have a uniquely defined duration. In the case of

---

[3]Psychoacoustics is the science that deals with the perception of sound. In a psychoacoustic experiment, the relationships between an acoustic stimulus and the resulting subjective sensation are studied by presenting specific tasks or questions to human listeners [550].

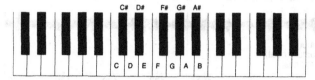

Fig. 1.4. Illustration of the piano keyboard (only three octaves are shown here).

sustained musical sounds, however, the durations are used to control *articulation*. The two extremes here are 'staccato', where notes are cut very short, and 'legato', where no perceptible gaps are left between successive notes.

A *melody* is a series of pitched sounds with musically meaningful pitch and IOI relationships. In written music, this corresponds to a sequence of single notes. A *chord* is a combination of two or more simultaneous notes. A chord can be harmonious or dissonant, subjective attributes related to the specific relationships between the component pitches and their overtone partials. *Harmony* refers to the part of music theory which studies the formation and relationships of chords.

Western music arranges notes on a quantized logarithmic scale, with 12 notes in each octave range. The nominal fundamental frequency of note n can be calculated as 440 Hz $\times 2^{n/12}$, where 440 Hz is an agreed-upon anchor point for the tuning and n varies from -48 to 39 on a standard piano keyboard, for example. According to a musical convention, the notes in each octave are lettered as C, C#, D, D#, E, F, ... (see Fig. 1.4) and the octave is indicated with a number following this, for example A4 and A3 referring to the notes with fundamental frequencies 440 Hz and 220 Hz, respectively.

There are of course instruments which produce arbitrary pitch values and not just discrete notes like the piano. When playing the violin or singing, for example, both intentional and unintentional deviations take place from the nominal note pitches. In order to write down the music in a symbolic form, it is necessary to perform *quantization*, or perceptual categorization [60]: a track of pitch values is segmented into notes with discrete pitch labels, note timings are quantized to quarter notes, whole notes, and so forth, and timbral information is 'quantized' by naming the sound sources involved. In some cases this is not necessary but a parametric or semi-symbolic[4] representation suffices.

An important property of basically all musical cultures is that corresponding notes in different octaves are perceived as having a special kind of similarity, independent of their separation in frequency. The notes C3, C4, and C5, for example, play largely the same harmonic role although they are not interchangeable in a melody. Therefore the set of all notes can be described as representing only 12 *pitch classes*. An individual musical piece usually recruits only a subset of the 12 pitch classes, depending on the *musical key* of

---

[4]In a MIDI file, for example, the time values are not quantized.

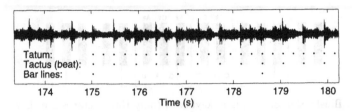

Fig. 1.5. A music signal with three metrical levels illustrated.

the piece. For example, a piece in the C major key tends to employ the white keys of the piano, whereas a piece in B major typically employs all the black keys but only two white keys in each octave. Usually there are seven pitch classes that 'belong' to a given key. These are called *scale tones* and they possess a varying degree of importance or stability in the key context. The most important is the *tonic* note (for example C in the C major key) and often a musical piece starts or ends on the tonic. Perception of pitch along musical scales and in relation to the musical key of the piece is characteristic to *tonal music*, to which most of Western music belongs [377].

The term *musical metre* has to do with the rhythmic aspects of music: it refers to the regular pattern of strong and weak beats in a piece. Perceiving the metre consists of detecting moments of musical emphasis in an acoustic signal and filtering them so that the underlying periodicities are discovered [404], [93]. The perceived periodicities, *pulses*, at different time scales (or levels) together constitute the metre, as illustrated in Fig. 1.5. Perceptually the most salient metrical level is the *tactus*, which is often referred to as the foot-tapping rate or the *beat*. The tactus can be viewed as the temporal 'backbone' of a piece of music, making beat tracking an important subtask of music transcription. Further metrical analysis aims at identifying the other pulse levels, the periods of which are generally integer multiples or submultiples of the tactus pulse. For example, detecting the *musical measure* pulse consists of determining the number of tactus beats that elapse within one musical measure (usually 2 to 8) and aligning the boundaries of the musical measures (bar lines) to the music signal.

Another element of musical rhythms is *grouping*, which refers to the way in which individual sounds are perceived as being grouped into melodic phrases; these are further grouped into larger musical entities in a hierarchical manner [404]. Important to the rhythmic characteristics of a piece of music is how these groups are aligned in time with respect to the metrical system.

The *structure* of a musical work refers to the way in which it can be subdivided into parts and sections at the largest time-scale. In popular music, for example, it is usually possible to identify parts that we label as the chorus, the verse, an introductory section, and so forth. Structural parts can be detected by finding relatively long repeated pitch structures or by observing considerable changes in the instrumentation at section boundaries.

The forthcoming chapters of this book address the extraction and analysis of the above elements in musical audio signals. Fundamental frequency estimation is considered in Parts III and IV of this book, with a separate treatise on melody transcription in Chapters 11 and 12. Metre analysis is discussed in Chapter 4 and percussion transcription in Chapter 5. Chapter 6 discusses the measurement of timbre and musical instrument classification. Structure analysis is addressed in Chapter 11, and the quantization of time and pitch in Chapters 4 and 12, respectively. Before going to a more detailed outline of each chapter, however, let us have a look at some general aspects of the transcription problem.

1.2 Perspectives on Music Transcription

When starting to design a transcription system, certain decisions have to be made before the actual algorithm development. Among the questions involved are: How should the transcription system be structured into smaller submodules or tasks? What kind of data representations would be the most suitable? Should musical information be used as an aid in the analysis? Would it be advantageous to analyse larger musical structures before going into note-by-note transcription? These general and quite 'philosophical' issues are discussed from various perspectives in the following.

1.2.1 Neurophysiological Perspective

First, let us consider a neurophysiological argument about how the music transcription problem should be decomposed into smaller subtasks. In human auditory cognition, *modularity* of a certain kind has been observed, meaning that certain parts can be functionally and neuro-anatomically isolated from the rest [517], [687], [623]. One source of evidence for this are studies with brain-damaged patients: an accidental brain damage may selectively affect musical abilities but not speech-related abilities, and vice versa [518]. Moreover, there are patients who suffer from difficulties dealing with pitch variations in music but not with temporal variations. In music performance or in perception, either of the two can be selectively lost [29], [517].

Peretz has studied brain-damaged patients who suffer from specific music impairments and she proposes that the music cognition system comprises at least four discernable 'modules' [517], [518]. An acoustic analysis module segregates a mixture signal into distinct sound sources and extracts the perceptual parameters of these (including pitch) in some raw form. This is followed by two parallel modules which carry out pitch organization (melodic contour analysis and tonal encoding of pitch) and temporal organization (rhythm and metre analysis). The fourth module, musical lexicon, contains representations of the musical phrases a subject has previously heard.

Neuroimaging experiments in healthy subjects are another way of local-
izing the cognitive functions in the brain. Speech sounds and higher-level
speech information are known to be preferentially processed in the left auditory cortex, whereas musical sounds are preferentially processed in the right
auditory cortex [623]. Interestingly, however, when musical tasks specifically
involve processing of temporal information (temporal synchrony or duration),
the processing is weighted towards the left hemisphere [687], [517]. The relative (not complete) asymmetry between the two hemispheres seems to be
related to the acoustic characteristics of the signals: rapid temporal information is characteristic for speech, whereas accurate processing of spectral and
pitch information is more important in music [581], [687], [623]. Zatorre et al.
proposed that the left auditory cortex is relatively specialized to a better time
resolution and the right auditory cortex to a better frequency resolution [687].

In computational transcription systems, rhythm and pitch have often been
analysed separately and using different data representations (see e.g. [332],
[439], [237], [223], [350], [122]). Typically, a better time resolution is applied
in rhythm analysis and a better frequency resolution in pitch analysis. Based
on the above studies, this seems to be justified to some extent. However, it
should be kept in mind that studying the human brain is very difficult and the
reported results are therefore a subject of controversy. Also, the structure of
transcription systems is often determined by merely pragmatic considerations.
For example, temporal segmentation is performed prior to pitch analysis in
order to allow an appropriate positioning of analysis frames in pitch analysis,
which is typically the most demanding stage computationally.

1.2.2 Human Transcription

Another viewpoint to the transcription problem is obtained by studying the
conscious transcription process of human musicians and by inquiring about
their transcription strategies. The aim of this is to determine the sequence of
actions or processing steps that leads to the transcription result.

As already mentioned above, reading and writing music is an acquired
ability and therefore the practice of music transcription is of course affected by
its teaching at musical institutions. In this context, the term *musical dictation*
is used to refer to an exercise where a musical excerpt is played and it has to
be written down as notes [206]. An excellent study on the practice of musical
dictation and ear training pedagogy can be found in [280].

Characteristic to ear training is that the emphasis is not on trying to *hear*
more but to *recognize* what is being heard; to hear relationships accurately
and with understanding. Students are presented with different pitch intervals,
rhythms, and chords, and they are trained to name these. Simple examples
are first presented in isolation and when these become familiar, increasingly
complex material is considered. Melodies are typically viewed as a synthesis of
pitch and rhythm. For example, Ghezzo instructs the student first to memorize
the fragment of music that is to be written down, then to write the pitch of

the notes, and finally to apply the rhythm [206, p. 6]. Obviously, ear training presumes a normally hearing subject who is able to detect distinct sounds and their pitch and timing in the played excerpts—aspects which are very difficult to model computationally.

Recently, Hainsworth conducted a study where he asked trained musicians to describe how they transcribe realistic musical material [263]. The subjects (19 in total) had transcribed music from various genres and with varying goals, but Hainsworth reports that a consistent pattern emerged in the responses. Most musicians first write down the structure of the piece, possibly with some key phrases marked in an approximate way. Next, the chords of the piece or the bass line are notated, and this is followed by the melody. As the last step, the inner lines are studied. Many reported that they heard these by repeated listening, by using an instrument as an aid, or by making musically educated guesses based on the context.

Hainsworth points out certain characteristics of the above-described process. First, it is sequential rather than concurrent; quoting the author, 'no-one transcribes anything but the most simple music in a single pass'. In this respect, the process differs from most computational transcription systems. Secondly, the process relies on the human ability to attend to certain parts of a polyphonic signal while selectively ignoring others.[5] Thirdly, some early analysis steps appear to be so trivial for humans that they are not even mentioned. Among these are style detection (causing prior expectations regarding the content), instrument identification, and beat tracking.

1.2.3 Mid-Level Data Representations

The concept of *mid-level data representations* provides a convenient way to characterize certain aspects of signal analysis systems. The analysis process can be viewed as a sequence of representations from an acoustic signal towards the analysis result [173], [438]. Usually intermediate abstraction levels are needed between these two since musical notes, for example, are not readily visible in the raw acoustic signal. An appropriate mid-level representation functions as an 'interface' for further analysis and facilitates the design of efficient algorithms for this purpose.

The most-often used representation in acoustic signal analysis is the short-time Fourier transform of a signal in successive time frames. Time-frequency decompositions in general are of fundamental importance in signal processing and are introduced in Chapter 2. Chapter 3 discusses these in a more general framework of *waveform representations* where a music signal is represented as a linear combination of elementary waveforms from a given dictionary. Time-frequency plane representations have been used in many transcription systems (see e.g. [223], [351], [5], and Chapter 9), and especially in percussive

---

[5]We may add also that the *limitations* of human memory and attention affect the way in which large amounts of data are written down [602].

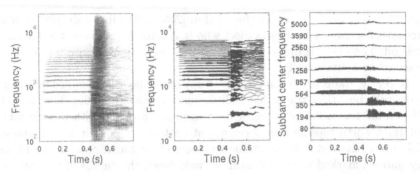

Fig. 1.6. Three different mid-level representations for a short trumpet sound (F0 260 Hz), followed by a snare drum hit. The left panel shows the time-frequency spectrogram with a logarithmic frequency scale. The middle panel shows the sinusoidal model for the same signal, line width indicating the amplitude of each sinusoid. The right panel shows the output of a simple peripheral auditory model for the same signal.

transcription where both linear [188] and logarithmic [505], [209] frequency resolution has been used.

Another common choice for a mid-level representation in music transcription has been the one based on *sinusoid tracks* [332], [440], [609], [652]. In this parametric representation, an acoustic signal is modelled as a sum of sinusoids with time-varying frequencies and amplitudes [449], [575], as illustrated in Fig. 1.6. Pitched musical instruments can be modelled effectively with relatively few sinusoids and, ideally, the representation supports sound source separation by classifying the sinusoids to their sources. However, this is complicated by the fact that frequency components of co-occurring sounds in music often overlap in time and frequency. Also, reliable extraction of the components in real-world complex music signals can be hard. Sinusoidal models are described in Chapter 3 and applied in Chapters 7 and 10.

In the human auditory system, the signal travelling from the inner ear to the brain can be viewed as a mid-level representation. A nice thing about this is that the peripheral parts of hearing are quite well known and computational models exist which are capable of approximating the signal in the auditory nerve to a high accuracy. The right panel of Fig. 1.6 illustrates this representation. Auditory models have been used for music transcription by several authors [439], [627], [434], [354] and these are further discussed in Chapter 8.

It is natural to ask if a certain mid-level representation is better than others in a given task. Ellis and Rosenthal have discussed this question in the light of several example representations commonly used in acoustic signal analysis [173]. The authors list several desirable qualities for a mid-level representation. Among these are *component reduction*, meaning that the number of objects in the representation is smaller and the meaningfulness of each is higher compared to the individual samples of the input signal. At the same

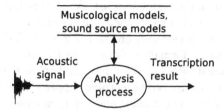

Fig. 1.7. The two main sources of information in music transcription: an acoustic input signal and pre-stored musicological and sound source models.

time, the sound should be decomposed into sufficiently fine-grained elements so as to support *sound source separation* by grouping the elements to their sound sources. Other requirements included *invertibility*, meaning that the original sound can be resynthesized from its representation in a perceptually accurate way, and *psychoacoustic plausibility* of the representation.

1.2.4 Internal Models

Large-vocabulary speech recognition systems are critically dependent on *language models* which represent linguistic knowledge about speech signals [536], [316], [321]. The models can be very primitive in nature, for example merely tabulating the occurrence frequencies of different three-word sequences (N-gram models), or more complex, implementing part-of-speech tagging of words and syntactic inference within sentences.

Musicological information is likely to be equally important for the automatic transcription of polyphonically rich musical material. The probabilities of different notes occurring concurrently or in sequence can be straightforwardly estimated, since large databases of written music exist in an electronic format. Also, there are a lot of musical conventions concerning the arrangement of notes for a certain instrument within a given genre. In principle, these musical constructs can be modelled and learned from data.

In addition to musicological constraints, internal models may contain information about the physics of musical instruments [193], and heuristic rules, for example that a human musician has only ten fingers with limited dimensions. These function as a source of information in the transcription process, along with the input waveform (see Fig. 1.7). Contrary to an individual music signal, however, these characterize musical tradition at large: its compositional conventions, selection of musical instruments, and so forth. Although these are generally bound to a certain musical tradition, there are also more universal constraints that stem from the human perception (see Bregman [49, Ch. 5]). For example, perceptually coherent melodies usually advance in relatively small pitch transitions and employ a consistent timbre.

Some transcription systems have applied musicological models or sound source models in the analysis [332], [440], [215], [559]. The principles of doing this are discussed in more detail in Part IV of this book. The term *top-down processing* is often used to characterize systems where models at a high abstraction level impose constraints on the lower levels [592], [172]. In *bottom-up* processing, in turn, information flows from the acoustic signal: features are extracted, combined into sound sources, and these are further processed at higher levels. The 'unsupervised-learning' approach mentioned on p. 7 is characterized by bottom-up processing and a minimal use of pre-stored models and assumptions. This approach has a certain appeal too, since music signals are redundant at many levels and, in theory, it might be possible to resolve this 'puzzle' in a completely data-driven manner by analysing a huge collection of musical pieces in connection and by constructing models automatically from the data. For further discussion of this approach, see Chapter 9.

Utilizing diverse sources of knowledge in the analysis raises the issue of integrating the information meaningfully. In automatic speech recognition, statistical methods have been very successful in this respect: they allow representing uncertain knowledge, learning from examples, and combining diverse types of information.

1.2.5 A Comparison with Speech Recognition

Music transcription is in many ways comparable to automatic speech recognition, although the latter has received greater academic and commercial interest and has been studied longer. Characteristic to both music and speech is that they are *generative* in nature: a limited number of discrete elements are combined to yield larger structures. In speech, phonemes are used to construct words and sentences and, in music, individual sounds are combined to build up melodies, rhythms, and songs. An important difference between the two is that speech is essentially monophonic (one speaker), whereas music is usually polyphonic. On the other hand, speech signals vary more rapidly and the acoustic features that carry speech information are inherently multi-dimensional, whereas pitch and timing in music are one-dimensional quantities.

A central problem in the development of speech recognition systems is the high dynamic variability of speech sounds in different acoustic and linguistic contexts – even in the case of a single speaker. To model this variability adequately, large databases of carefully annotated speech are collected and used to train statistical models which represent the acoustic characteristics of phonemes and words.

In music transcription, the principal difficulties stem from combinatorics: the sounds of different instruments occur in varying combinations and make up musical pieces. On the other hand, the dynamic variability and complexity of a single sound event is not as high as that of speech sounds. This has the consequence that, to some extent, synthetic music signals can be used in developing and training a music transcriber. Large amounts of training data can be generated since acoustic measurements for isolated musical sounds are available and combinations of these can be generated by mixing. However, it should be emphasized that this does not remove the need for more realistic acoustic material, too. The issue of obtaining and annotating such databases is discussed in [405] and in Chapters 6 and 11. Realistic data are also needed for the objective evaluation of music analysis systems [465].

1.3 Outline

This section discusses the different subtopics of music transcription and summarizes the contents of each chapter of this book. All the chapters are intended to be self-contained entities, and in principle nothing prevents one from jumping directly to the beginning of a chapter that is of special interest to the reader. Whenever some element from the other parts of the book is needed, an explicit reference is made to the chapter in question.

Part I Foundations

The first part of this book is dedicated to topics that are more or less related to all areas of music trancription discussed in this book.

Chapter 2 introduces *statistical and signal processing techniques* that are applied to music transcription in the subsequent chapters. First, the Fourier transform and concepts related to time-frequency representations are described. This is followed by a discussion of statistical methods, including random variables, probability density functions, probabilistic models, and elements of estimation theory. Bayesian estimation methods are separately discussed and numerical computation techniques are described, including Monte Carlo methods. The last section introduces the reader to pattern recognition methods and various concepts related to these. Widely used techniques such as support vector machines and hidden Markov models are included.

Chapter 3 discusses *sparse adaptive representations for musical signals.* The issue of data representations was already briefly touched in Section 1.2.3 above. This chapter describes parametric representations (for example the sinusoidal model) and 'waveform' representations in which a signal is modelled as a linear sum of elementary waveforms chosen from a well-defined dictionary. In particular, signal-adaptive algorithms are discussed which aim at *sparse* representations, meaning that a small subset of waveforms is chosen from a large dictionary so that the sound is represented effectively. This is advantageous from the viewpoint of signal analysis and imposes an implicit structure to the analysed signal.

Part II Rhythm and Timbre Analysis

The second part of this book describes methods for metre analysis, percussion transcription, and pitched musical instrument classification.

Chapter 4 discusses *beat tracking and musical metre analysis*, which constitute an important subtask of music transcription. As mentioned on p. 10, metre perception consists of detecting moments of musical stress in an audio signal, and processing these so that the underlying periodicities are discovered. These two steps can also be discerned in the computational methods. Measuring the degree of musical emphasis as a function of time is closely related to *onset detection*, that is, to the detection of the beginnings of discrete sound events in an acoustic signal, a problem which is separately discussed. For the estimation of the underlying metrical pulses, a number of different approaches are described, putting particular emphasis on statistical methods.

Chapter 5 discusses *unpitched percussion transcription*,[6] where the aim is to write down the *timbre class*, or the sound source, of each constituent sound along with its timing (see Fig. 1.2 above). The methods discussed in this chapter represent two main approaches. In one, a percussive track is assumed to be performed using a conventional set of drums, such bass drums, snares, hi-hats, cymbals, tom-toms, and so forth, and the transcription proceeds by detecting distinct sound events and by classifying them into these pre-defined categories. In another approach, no assumptions are made about the employed instrumental sounds, but these are learned from the input signal in an unsupervised manner, along with their occurrence times and gains. This is accomplished by processing a longer portion of the signal in connection and by trying to find such source signals that the percussive track can be effectively represented as a linear mixture of them. Percussion transcription both in the presence and absence of pitched instruments is discussed.

Chapter 6 is concerned with the *classification of pitched musical instrument sounds*. This is useful for music information retrieval purposes, and in music transcription, it is often desirable to assign individual note events into 'streams' that can be attributed to a certain instrument. The chapter looks at the acoustics of musical instruments, timbre perception in humans, and basic concepts related to classification in general. A number of acoustic descriptors, or features, are described that have been found useful in musical instrument classification. Then, different classification methods are described and compared, complementing those described in Chapter 2. Classifying individual musical sounds in polyphonic music usually requires that they are separated from the mixture signal to some degree. Although this is usually seen as a separate task from the actual instrument classification, some methods for instrument classification in complex music signals are reviewed, too.

Part III Multiple Fundamental Frequency Analysis

The term *multiple F0 estimation* refers to the estimation of the F0s of several concurrent sounds in an acoustic signal. The third part of this book describes

---

[6]Many drum instruments can be tuned and their sound evokes a perception of pitch. Here 'unpitched' means that the instruments are not used to play melodies.

different ways to do this. *Harmonic analysis* (writing down the chords of a piece) can be performed based on the results of the multiple F0 analysis, but this is beyond the scope of this book and an interested reader is referred to [621, Ch. 6] and [553, Ch. 2]. Harmonic analysis can also be attempted directly, without note-by-note F0 estimation [402], [69], [583].

Chapter 7 discusses *multiple F0 estimation based on generative models*. Here, the multiple F0 estimation problem is expressed in terms of a signal model, the parameters of which are being estimated. A particular emphasis of this chapter is statistical methods where the F0s and other relevant parameters are estimated using the acoustic data and possible prior knowledge about the parameter distributions. Various algorithms for on-line (causal) and off-line (non-causal) parameter estimation are described and the computational aspects of the methods are discussed.

Chapter 8 describes *auditory model-based methods for multiple F0 estimation*. The reader is first introduced with computational models of human pitch perception. Then, transcription systems are described that use an auditory model as a pre-processing step, and the advantages and disadvantages of auditorily motivated data representations are discussed. The second part of the chapter describes multiple F0 estimators that are based on an auditory model but make significant modifications to it in order to perform robust F0 estimation in polyphonic music signals. Two different methods are described in more detail and evaluated.

Chapter 9 discusses *unsupervised learning methods for source separation in monaural music signals*. Here the aim is to separate and learn sound sources from polyphonic data without sophisticated modelling of the characteristics of the sources, or detailed modelling of the human auditory perception. Instead, the methods utilize general principles, such as statistical independency between sources, to perform the separation. Various methods are described that are based on independent component analysis, sparse coding, and non-negative matrix factorization.

Part IV Entire Systems, Acoustic and Musicological Modelling

The fourth part of the book discusses entire music content analysis systems and the use of musicological and sound source models in these.

Chapter 10 is concerned with *auditory scene analysis (ASA) in music signals*. As already mentioned above, ASA refers to the perception of distinct sources in polyphonic signals. In music, ASA aims at extracting entities like notes and chords from an audio signal. The chapter reviews psychophysical findings regarding the acoustic 'clues' that humans use to assign spectral components to their respective sources, and the role of internal models and top-down processing in this. Various computational approaches to ASA are described, with a special emphasis on statistical methods and inference in Bayesian networks.

Chapter 11 discusses a research approach called *music scene description*, where the aim is to obtain descriptions that are intuitively meaningful to an untrained listener, without trying to extract every musical note from musical audio. Concretely, this includes the analysis of the melody, bass lines, metrical structure, rhythm, and chorus and phrase repetition. In particular, two research problems are discussed in more detail. *Predominant F0 estimation* refers to the estimation of the F0 of only the most prominent sound in a polyphonic mixture. This closely resembles the experience of an average listener who catches the melody or the 'theme' of a piece of music even though he or she would not be able to distinguish the inner lines. Here, methods for extracting the melody and the bass line in music recordings are introduced. The other problem addressed is *music structure analysis*, especially locating the chorus section in popular music.

Chapter 12 addresses *singing transcription*, which means converting a recorded singing performance into a sequence of discrete note pitch labels and their starting and ending points in time. The process can be broken into two stages, where first a continuous track of pitch estimates (and possibly other acoustic features) is extracted from an acoustic signal, and these are then converted into a symbolic musical notation. The latter stage involves the segmentation of the pitch track into discrete note events and quantizing their pitch values—tasks which are particularly difficult for singing signals. The chapter reviews state-of-the-art singing transcription methods and discusses the use of acoustic and musicological models to tackle the problem.

2

An Introduction to Statistical Signal Processing and Spectrum Estimation

Manuel Davy

LAGIS/CNRS, BP 48, Cité Scientifique, 59651 Villeneuve d'Ascq Cedex, France
Manuel.Davy@ec-lille.fr

This chapter presents an overview of current signal processing techniques, most of which are applied to music transcription in the following chapters. The elements provided will hopefully help the reader. Some signal processing tools presented here are well known, and readers already familiar with these concepts may wish to skip ahead. As we only present an overview of various methods, readers interested in more depth may refer to the bibliographical references provided throughout the chapter.

This chapter is organized as follows. Section 2.1 presents the Fourier transform and some related tools: time-frequency representations and cepstral coefficients. Section 2.2 introduces basic statistical tools such as random variables, probability density functions, and likelihood functions. It also introduces estimation theory. Section 2.3 is about Bayesian estimation methods, including Monte Carlo techniques for numerical computations. Finally, Section 2.4 introduces pattern recognition methods, including support vector machines and hidden Markov models.

2.1 Frequency, Time-Frequency, and Cepstral Representations

As explained in Chapter 1, the two main domains where musical signals can be represented and understood are *time* and *frequency*. If time is the domain where the music signals are played and recorded (air pressure vs. time data), frequency is the domain where they can be represented and understood: in Western music, the height of a note in a score represents its fundamental frequency. From a mathematics viewpoint, frequency is defined via the *Fourier transform* (FT), which is introduced in Section 2.1.1. However, the Fourier definition of frequency is not practical: frequency is well defined *only* for infinite length, stationary, continuous, pure sine waves—these objects do not belong to the real world, and non-stationarity is everywhere. The FT nevertheless provides a lot of useful information.

If time alone or frequency alone are not enough to represent music, then we need to think in terms of joint time and frequency representations (TFRs). Western musical scores are actually TFRs with a specific encoding. Such representations are introduced in Section 2.1.2, and rely on the concept of *frame*, that is, a well time-localized part of the signal. From these frames, we can define a simple TFR, the spectrogram, as well as a time and time-lag representation called cepstral representation; see Section 2.1.3. The following section discusses some basic properties of Fourier Transforms.

2.1.1 The Fourier Transform

The frequency representation of signals is given by the Fourier transform[1] (FT). In practical situations we generally have to deal with sampled, real-valued discrete time signals, denoted $x(n)$, where n denotes discrete time. The corresponding discrete FT is defined in (2.1) where k denotes the discrete frequency. It is sometimes more convenient to define music transcription methods in terms of continuous time signals, denoted $x(t)$, where t denotes continuous time. The continuous FT is defined in (2.2), and f denotes continuous frequency.

$$\mathsf{DFT}_x(k) = X(k) = \sum_{n=-\infty}^{\infty} x(n)e^{-j2\pi kn}, \tag{2.1}$$

$$\mathsf{CFT}_x(f) = X(f) = \int_{-\infty}^{\infty} x(t)e^{-j2\pi ft}dt. \tag{2.2}$$

The continuous and discrete FTs map the signal from the time domain to the frequency domain; $X(f)$ and $X(k)$ are generally complex valued. The inverse Fourier transforms (IFTs) are also quite useful for music processing; they are defined in (2.3) and (2.4) below.

$$\mathsf{IDFT}_X(n) = \sum_{k=-\infty}^{\infty} X(k)e^{j2\pi kn} = x(n), \tag{2.3}$$

$$\mathsf{ICFT}_X(t) = \int_{-\infty}^{\infty} X(f)e^{j2\pi ft}df = x(t). \tag{2.4}$$

An efficient approach to computing DFTs (2.1) and IDFTs (2.3) is the *fast Fourier transform* (FFT) algorithm; see [531].

Some properties of the FT/IFT are of importance in music transcription applications. In particular, the FT is a linear operation. Moreover, it maps the *convolution* operation into a simple product.[2] In other words, considering

---

[1]Readers interested in more advanced topics may refer to a dedicated book; see [48] for example.

[2]The convolution operation is used for signal filtering: applying a filter with time impulse response $h(n)$ to a signal $x(n)$ is done by convolving x with h or, equivalently, by multiplying their FTs.

for example the discrete time case, $\mathsf{DFT}_{x_1 * x_2}(k) = \mathsf{DFT}_{x_1}(k) \times \mathsf{DFT}_{x_2}(k)$, where $*$ denotes the convolution defined as follows:

$$x_1(n) * x_2(n) = \sum_{i=-\infty}^{\infty} x_1(i)x_2(n-i) \qquad \text{(discrete time signals)}, \qquad (2.5)$$

$$x_1(t) * x_2(t) = \int_{-\infty}^{\infty} x_1(\tau)x_2(t-\tau)d\tau \qquad \text{(continuous time signals)}. \qquad (2.6)$$

This result is also true for the continuous-time FT, as well as for the inverse transforms. Another important property, the *Shannon theorem*, states that the range of frequencies where the discrete FT is meaningful has an upper limit given by $k_s/2$ (this is the *Nyquist frequency*, where k_s is the sampling frequency). A straightforward consequence of the Shannon theorem is that digital signals sampled at the CD rate $k_s = 44100\,\text{Hz}$ can be analysed up to the maximum frequency $22050\,\text{Hz}$.

2.1.2 Time-Frequency Representations

There exists only *one* Fourier transform of a given signal, as defined above. However, there is an infinite number of time-frequency representations (TFRs). The most popular one is the *spectrogram*, defined as the Fourier transform of successive signal frames.[3]

Frames are widely used in audio processing algorithms. They are portions of the signal with given time localizations. More precisely, the frame localized at time t_0, computed with window w and denoted $s_{t_0}^{\text{w}}(\tau)$, is

$$s_{t_0}^{\text{w}}(t) = x(t)\text{w}(t_0 - t), \qquad (2.7)$$

where w is a *window*; see Fig. 2.1. Windows are generally positive and symmetric (that is, $\text{w}(-t) = \text{w}(t)$, for all t) with a limited support (i.e., there exists a time t_1 such that $\text{w}(t) = 0$, for all $t \geq t_1$). Standard window shapes are Gaussian, Hamming, Hanning, or rectangular[4] and typical frame durations are from 20 ms to 100 ms in audio processing. As a rule of thumb, rectangular windows should not be used in practice, except under special circumstances. From frames, it is easy to build short time Fourier transforms (STFTs) as the FT of successive frames:

$$\mathsf{STFT}_x^{\text{w}}(t, f) = \mathsf{CFT}_{s_t^{\text{w}}}(f) = \int_{-\infty}^{\infty} x(\tau)\text{w}(t-\tau)e^{-j2\pi f\tau}d\tau. \qquad (2.8)$$

---

[3]We restrict this discussion to continuous time signals, for the sake of simplicity. Discrete time-frequency representations of discrete signals are, in general, more complex to write in closed form, and they are obtained by discretizing the continuous representations; see [191].

[4]A list of windows and reasons for choosing for their shapes can be found in [244], [61].

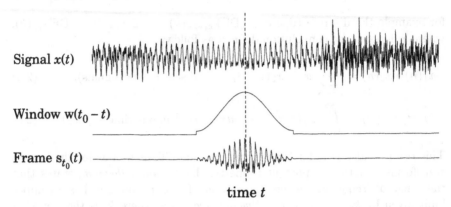

Fig. 2.1. The frame $s_{t_0}^{\mathrm{w}}$ is obtained by multiplying the signal x by a sliding window w centred at time t.

Spectrograms are energy representations and they are defined as the squared modulus of the STFT:

$$\mathsf{SP}_x^{\mathrm{w}}(t, f) \;=\; |\mathsf{STFT}_x^{\mathrm{w}}(t, f)|^2 . \tag{2.9}$$

Changing the window w defines a different STFT and thus a different spectrogram. Any spectrogram can also be interpreted as the output of a bank of filters: considering a given frequency f, $\mathsf{SP}_x^{\mathrm{w}}(t, f)$ is the instantaneous energy at time t of the output of the filter with frequency response given by $\mathsf{CFT}_{\mathrm{w}}(f - \nu)$, $\nu \in [-\infty, \infty]$, applied to the signal x. As a consequence, a short duration window leads to a spectrogram with good time resolution and bad frequency resolution, whereas a longer window leads to the opposite situation.[5] Figure 2.2 represents two spectrograms of a piano excerpt, illustrating the influence of the window length on the representation.

Another interpretation of the STFT arises when considering (2.8) as the dot product between $x(\tau)$ and the windowed complex sinusoid $\mathrm{w}(t-\tau)e^{-j2\pi f\tau}$: under some conditions, the family of elementary time-frequency atoms $\{\mathrm{w}(t-\tau)e^{-j2\pi f\tau}\}_{t,f}$ forms a basis called a *Gabor basis*. In such cases, $\mathsf{STFT}_x^{\mathrm{w}}$ is a decomposition of x on this basis, yielding a *Gabor representation* of x; see Chapter 3 and [191], [184].

Spectrograms being energy representations, they are quadratic in the signal x. They also have the *time-frequency covariance* property: Let us shift x in time and frequency, defining $x_1(t) = x(t-t_0)\exp\{j2\pi f_0\}$. The time-frequency covariance property ensures that $\mathsf{SP}_{x_1}^{\mathrm{w}}(t, f) \;=\; \mathsf{SP}_x^{\mathrm{w}}(t-t_0, f-f_0)$. Many other

---

[5]A TFR is said to have good time (frequency) resolution if the signal energy is displayed around its true location with small spread along the time (frequency) axis. The product of time resolution and frequency resolution is lower bounded via the Heisenberg–Gabor inequality [191].

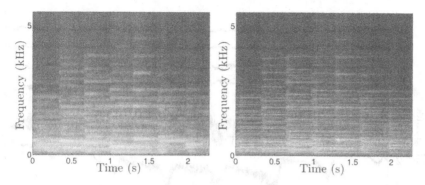

Fig. 2.2. Spectrograms of a piano excerpt computed with Hamming windows with length 20 ms (left) and 80 ms (right). The time-frequency resolution is highly dependent on the window length.

quadratic and covariant time-frequency representations can be defined. They are all derived from the *Wigner–Ville representation*:

$$\mathsf{WV}_x(t, f) = \int_{-\infty}^{\infty} x\left(t + \frac{\tau}{2}\right) x^\star\left(t - \frac{\tau}{2}\right) e^{-j2\pi f\tau} d\tau, \qquad (2.10)$$

where $x^\star$ is the complex conjugate of x, via a time and frequency convolution with a time-frequency kernel $\phi(t, f)$ [191]. In particular, the spectrogram is obtained by using $\phi(t, f) = \mathsf{WV}_w(t, f)$; that is, $\mathsf{SP}_x^w(t, f) = \left[\mathsf{WV}_x \underset{t,f}{*} \mathsf{WV}_w\right](t, f)$, where $\underset{t,f}{*}$ denotes the time-frequency convolution and WV_w is the Wigner–Ville representation of the window.

2.1.3 The Cepstrum

Filtering a signal is actually a convolution operation in the time domain, and a product operation in the frequency domain. The aim of cepstral representations was initially to turn filtering into an addition operation by using the logarithm of the FT. More precisely, the cepstrum of a continuous signal $x(t)$, with FT denoted $X(f)$, is defined as

$$\mathsf{Cep}_x(\tau) = \mathsf{ICFT}_{\log(|X|)}(\tau) = \int_{-\infty}^{\infty} \log\left(|X(f)|\right) e^{j2\pi f\tau} df, \qquad (2.11)$$

and we see that filtering a signal in the frequency domain through the product $X(f)H(f)$ becomes, after taking the log, the addition $\log[X(f)] + \log[H(f)]$ where $H(f)$ is the filter frequency response. When dealing with discrete time signals, the cepstrum is also discrete and is generally referred to in terms of *cepstral coefficients*. Cepstral coefficients can be computed from the discrete

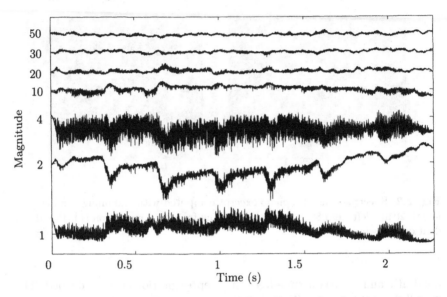

Fig. 2.3. Time evolution of the first cepstral coefficients for the piano signal of Fig. 2.2. The coefficient number is written along the vertical axis. For the sake of figure clarity, the magnitude of the first cepstral coefficient has been divided by five, and an offset has been added to each coefficient to avoid overlap.

equivalent of (2.11). Figure 2.3 displays the time evolution of some of the first cepstral coefficients for a music excerpt.

In practice, one uses *linear prediction cepstral coefficients* (LPCCs) or *mel frequency cepstral coefficients* (MFCCs) [536]. LPCCs are computed by applying a linear prediction model (also called an *autoregressive* model; see Section 2.2.9, p. 39) to the signal and transforming the resulting coefficients into LPCCs. MFCCs are used much more in music processing, and the steps of their computation are plotted in Fig. 2.5. The input signal $x(n)$ is first windowed, resulting in frames $s_{n_0}^w(n)$ centred at time n_0. The magnitude spectrum of each frame, denoted $|S_{n_0}^w(k)|$, is sent into a bank of K_{mel} mel frequency filters (typically $K_{\text{mel}} = 40$). Mel frequency filters have a triangular shape in the frequency domain; see Fig. 2.4. The central frequencies of the filters are equally spaced in terms of *mel frequencies*, which relate to the standard, linear frequency through the relation [297]

$$k_{\text{mel}} = 2595 \log_{10}\left[\frac{k}{700} + 1\right]. \tag{2.12}$$

For each mel filter, frequency components within its passband are weighted by the magnitude response of the filter, and then squared and summed. The resulting filter-related coefficient is denoted $\chi_{n_0}(k_{\text{mel}})$. For the full set of filters, the $\chi_{n_0}(k_{\text{mel}})$'s are stacked into a vector of size K_{mel}, whose logarithm

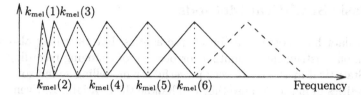

Fig. 2.4. Mel filter bank. Each filter has a triangular shape, and unity response at its centre. Its edges coincide with the adjacent filters' central frequencies. The central frequencies are linearly spaced on the mel frequency scale, which results in an exponential interval between the filter centres onto the linear scale, through (2.12).

is transformed back into the time lag domain using the discrete cosine transform (DCT), where the DCT of a discrete signal x with length T is defined as follows:

$$\mathsf{DCT}_x(i) = \sum_{n=1}^{T} x(n) \cos\left[\frac{\pi}{T}i\left(n - \frac{1}{2}\right)\right]. \tag{2.13}$$

In addition to yielding time-domain coefficients, the DCT also decomposes the coefficients in a way similar to principal component analysis [417] (see p. 54), so that each mel cepstral coefficient carries different information from the other MFCCs. In many applications, one is also interested in the time evolution of MFCCs, which can be given as derivatives. The first and second derivatives for each MFCC $\mathsf{Cep}_{n_0}(i)$ ($i = 1, \dots, K_{\mathrm{mel}}$) at each time n_0 are denoted $\Delta_{n_0}(i)$ and $\Delta\Delta_{n_0}(i)$.

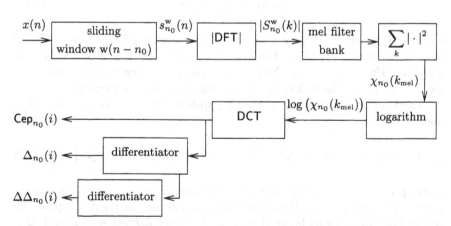

Fig. 2.5. MFCC computation steps. The magnitude spectrum $|S_{n_0}^{\mathrm{w}}(k)|$ of each frame $s_{n_0}^{\mathrm{w}}(n)$ is filtered through the mel filter bank. The squared outputs of each filter are summed over the filter frequency range, yielding a coefficient $\chi_{n_0}(k_{\mathrm{mel}})$ for each filter. The vector made of the logarithm of these coefficients is mapped back to the time domain using the discrete cosine transform.

2.2 Basic Statistical Methods

Musical data have some common characteristics, but they may also vary a lot from one instrument to another, or because of changing recording conditions. Statistical methods are very powerful for modelling sound sources and other entities in music because they can handle their variability. Even a single sound source produces a great variety of sounds, and it cannot be modelled deterministically. The main concept in statistical methods is that of *random variables* which are characterized by a *probability density function* (pdf).

2.2.1 Probability Distributions and Density Functions

Roughly speaking,[6] a random variable **a** is a scalar or a vector in some space $\mathcal{A}$ which can be continuous (e.g., $\mathcal{A} = \mathbb{R}$) or discrete (e.g., $\mathcal{A} = \mathbb{Z}$). It is called *random* because it can take several values in $\mathcal{A}$ which are impossible to predict without any errors. If random variables cannot be predicted, they can be characterized, since some values of **a** may be more likely to appear than others. In the case where **a** is a discrete variable, it is characterized by the probabilities that each value of **a** will appear, denoted $P(\mathbf{a})$. For example, in the coin tossing problem, $\mathcal{A} = \{\text{tail}, \text{head}\}$ and the probabilities are $P(\mathbf{a} = \text{head}) = P(\mathbf{a} = \text{tail}) = 1/2$. Standard discrete random variables distributions are uniform (like in coin tossing), Poisson, binomial, etc. [544].

In the continuous case, it may happen that all values of **a** are equally likely in a part $\mathcal{B}$ of the space $\mathcal{A}$, and never appear out of $\mathcal{B}$. In that case, **a** is a continuous random variable with a *uniform pdf* denoted $\mathcal{U}_{\mathcal{B}}$, with

$$\mathcal{U}_{\mathcal{B}}(\mathbf{a}) = \frac{1}{\mu(\mathcal{B})} \mathbb{I}_{\mathcal{B}}(\mathbf{a}), \tag{2.14}$$

where $\mu(\mathcal{B})$ is called *Lebesgue measure* of $\mathcal{B}$ (it measures the volume of $\mathcal{B}$) and $\mathbb{I}_{\mathcal{B}}$ is the indicator function which satisfies $\mathbb{I}_{\mathcal{B}}(\mathbf{a}) = 1$ if $\mathbf{a} \in \mathcal{B}$ and $\mathbb{I}_{\mathcal{B}}(\mathbf{a}) = 0$ otherwise. Uniform pdfs are often met in audio processing problems, and they are used to model the lack of precise information about a parameter.

Another important pdf in engineering problems is the *Gaussian*, also called *normal pdf*:

$$\mathcal{N}(\mathbf{a}; \boldsymbol{\mu}, \boldsymbol{\Sigma}) = [\det(2\pi\boldsymbol{\Sigma})]^{-1/2} \exp\left\{-\frac{1}{2}(\mathbf{a} - \boldsymbol{\mu})^{\mathsf{T}} \boldsymbol{\Sigma}^{-1}(\mathbf{a} - \boldsymbol{\mu})\right\}, \tag{2.15}$$

where $\boldsymbol{\mu}$ is the *mean vector*, and $\boldsymbol{\Sigma}$ is the *covariance matrix*, which is symmetric. The vector $\boldsymbol{\mu}$ has the same size as **a**, and the matrix $\boldsymbol{\Sigma}$ is square, its size in both dimensions being that of **a**. An exact definition of $\boldsymbol{\mu}$ and $\boldsymbol{\Sigma}$ is given in (2.19) and (2.20). An important special case is when **a** is scalar. In this case, the Gaussian pdf is written as

---

[6]The precise definition of random variables is beyond the scope of this chapter; see [548] for a more precise introduction.

$$N(a; m, \sigma^2) = \frac{1}{\sqrt{2\pi}\sigma} \exp\left[-\frac{1}{2\sigma^2}(a-m)^2\right], \tag{2.16}$$

where m is the mean and σ^2 is the variance. There exist many other standard continuous pdf shapes such as the gamma, inverse gamma, Cauchy, Laplace, Dirichlet, etc. [544]. When the random variable $\mathbf{a}$ is a vector with dimension $d_{\mathbf{a}}$, $\mathbf{a} = [a_1, \ldots, a_{d_{\mathbf{a}}}]^\mathsf{T}$, then the pdf $\mathsf{p}(\mathbf{a})$ is in fact a joint pdf $\mathsf{p}(a_1, \ldots, a_{d_{\mathbf{a}}})$. This is a function of dimension $d_{\mathbf{a}}$ and it gives us the probability of observing jointly $a_1, \ldots, a_{d_{\mathbf{a}}}$.

Two very important properties of pdfs and discrete variable distributions are that they are *always positive* and their *sum equals one*:

$$\sum_{\mathbf{a} \in \mathcal{A}} \mathsf{P}(\mathbf{a}) = 1 \qquad \text{(discrete case)}, \tag{2.17}$$

$$\int_{\mathcal{A}} \mathsf{p}(\mathbf{a}) \, d\mathbf{a} = 1 \qquad \text{(continuous case)}. \tag{2.18}$$

2.2.2 Mean, Covariance, Expectation, Moments, and Modes

For any continuous pdf $\mathsf{p}(\mathbf{a})$, it is generally possible to define the mean vector $\boldsymbol{\mu}$ and covariance matrix $\boldsymbol{\Sigma}$ as follows:

$$\boldsymbol{\mu} = \int_{\mathcal{A}} \mathbf{a} \, \mathsf{p}(\mathbf{a}) \, d\mathbf{a}, \tag{2.19}$$

$$\boldsymbol{\Sigma} = \int_{\mathcal{A}} (\mathbf{a} - \boldsymbol{\mu})(\mathbf{a} - \boldsymbol{\mu})^\mathsf{T} \, \mathsf{p}(\mathbf{a}) \, d\mathbf{a}. \tag{2.20}$$

Note that both $\boldsymbol{\mu}$ and $\boldsymbol{\Sigma}$ cannot be defined whenever the integrals in (2.19) and (2.20) do not converge.

For both discrete and continuous random variables, the mean and covariance matrix can be expressed in terms of the *expectation* $\mathbb{E}$ of a function h over $\mathcal{A}$, defined as follows:

$$\mathbb{E}_{\mathsf{P}(\mathbf{a})}[h(\mathbf{a})] = \sum_{\mathbf{a} \in \mathcal{A}} h(\mathbf{a})\mathsf{P}(\mathbf{a}) \qquad \text{(discrete case)}, \tag{2.21}$$

$$\mathbb{E}_{\mathsf{p}(\mathbf{a})}[h(\mathbf{a})] = \int_{\mathcal{A}} h(\mathbf{a}) \, \mathsf{p}(\mathbf{a}) \, d\mathbf{a} \qquad \text{(continuous case)}. \tag{2.22}$$

The expectation can be interpreted as follows: consider for example the continuous case, and assume that a set of random variables $\{\mathbf{a}_1, \ldots, \mathbf{a}_N\}$ are identically distributed according to some pdf $\mathsf{p}(\mathbf{a})$; then the so-called *empirical average*

$$\hat{\boldsymbol{\mu}}_N\big(h(\mathbf{a}_1), \ldots, h(\mathbf{a}_N)\big) = \frac{1}{N} \sum_{i=1}^{N} h(\mathbf{a}_i) \tag{2.23}$$

is an approximation of $\mathbb{E}_{\mathsf{p}(\mathbf{a})}[h(\mathbf{a})]$ in the sense that $\widehat{\boldsymbol{\mu}}_N(h(\mathbf{a}_1), \ldots, h(\mathbf{a}_N))$ converges to $\mathbb{E}_{\mathsf{p}(\mathbf{a})}[h(\mathbf{a})]$ when the number of random variables N becomes infinite (this is the *law of large numbers*). It is easy to understand the expectation as the average of many samples $h(\mathbf{a})$ where $\mathbf{a}$ is distributed according to $\mathsf{p}(\mathbf{a})$ (this is also true for discrete random variables).

Using the expectation operator for any continuous or discrete random variable, the mean vector and covariance matrix can be written as follows:

$$\boldsymbol{\mu} = \mathbb{E}_{\mathsf{P}(\mathbf{a})}[\mathbf{a}] \text{ and } \boldsymbol{\Sigma} = \mathbb{E}_{\mathsf{P}(\mathbf{a})}[(\mathbf{a} - \boldsymbol{\mu})(\mathbf{a} - \boldsymbol{\mu})^\mathsf{T}] \text{ (discrete case),} \qquad (2.24)$$

$$\boldsymbol{\mu} = \mathbb{E}_{\mathsf{p}(\mathbf{a})}[\mathbf{a}] \text{ and } \boldsymbol{\Sigma} = \mathbb{E}_{\mathsf{p}(\mathbf{a})}[(\mathbf{a} - \boldsymbol{\mu})(\mathbf{a} - \boldsymbol{\mu})^\mathsf{T}] \text{ (continuous case).} \qquad (2.25)$$

More generally, the *order r moment* $\mathcal{M}_r$ of a scalar random variable a with mean m is given by the expectation

$$\mathcal{M}_r[\mathsf{P}(a)] = \mathbb{E}_{\mathsf{P}(a)}[(a - m)^r] \quad \text{(discrete case),} \qquad (2.26)$$

$$\mathcal{M}_r[\mathsf{p}(a)] = \mathbb{E}_{\mathsf{p}(a)}[(a - m)^r] \quad \text{(continuous case).} \qquad (2.27)$$

In the case where the random variable $\mathbf{a}$ is a vector of dimension $d_\mathbf{a}$, its order r moment $\mathcal{M}_r$ is an r-dimensional tensor with size $d_\mathbf{a} \times d_\mathbf{a} \times \ldots \times d_\mathbf{a}$.

Finally, distributions generally have *modes*, that is, local maxima: their locations indicate in which regions of the space $\mathcal{A}$ the random variable $\mathbf{a}$ is more likely to appear. Figure 2.6 summarizes graphically the concepts introduced in this subsection.

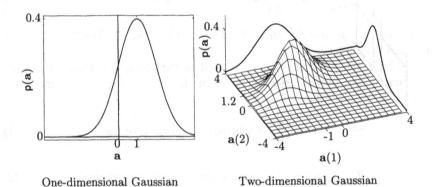

One-dimensional Gaussian   Two-dimensional Gaussian

Fig. 2.6. Graphical representation of the Gaussian pdf of a scalar (continuous) random variable (left) and a two-dimensional Gaussian random variable (right). In the scalar case (left), the mean and standard deviation are equal to one. In the two-dimensional case (right), the mean vector is $[-1\ 1.2]^\mathsf{T}$ and the covariance matrix is $\Sigma(1,1) = 1$, $\Sigma(1,2) = \Sigma(2,1) = 0.6$ and $\Sigma(2,2) = 1.2$. The diagonal terms of the covariance matrix being non-zero, the pdf is tilted around the mean point $(-1, 1.2)$. At the back of the right figure, the marginal pdfs (see Section 2.2.3) have been plotted.

2.2.3 Marginal and Conditional pdfs

Considering two random variables $a_1 \in \mathcal{A}_1$ and $a_2 \in \mathcal{A}_2$, we can define their joint pdf $p(a_1, a_2)$, which informs us about the probability of observing jointly (a_1, a_2). Then we can define the *marginal pdfs*,

$$p(a_1) = \int_{\mathcal{A}_2} p(a_1, a_2)\, da_2 \quad \text{and} \quad p(a_2) = \int_{\mathcal{A}_1} p(a_1, a_2)\, da_1, \quad (2.28)$$

which can be interpreted as the pdf of one of the random variable irrespective of the value of the other one; see Fig. 2.6, right. Using marginal pdfs, it is possible to decompose the joint pdf $p(a_1, a_2)$ as follows:

$$p(a_1, a_2) = p(a_1|a_2)p(a_2) = p(a_2|a_1)p(a_1), \quad (2.29)$$

where $p(a_1|a_2)$ is the conditional pdf of a_1, and should be read as 'the pdf of a_1 conditional on a_2'. Its interpretation is simple: Imagine that a_2 has some fixed, non-random value denoted a_2^{fixed}. Then a_1 is still a random variable, and its pdf is $p(a_1, a_2 = a_2^{\text{fixed}})$ up to a normalizing constant. This constant is necessary because the integral of $p(a_1, a_2 = a_2^{\text{fixed}})$ with respect to (w.r.t.) a_1 does not equal one anymore. The notation $p(a_1|a_2)$ should be understood as $p(a_1|a_2^{\text{fixed}})$, that is, $(1/C)p(a_1, a_2 = a_2^{\text{fixed}})$ where the normalizing constant C is $p(a_2^{\text{fixed}})$ from (2.29).

Finally, two random variables a_1 and a_2 are independent if and only if $p(a_1, a_2) = p(a_1)\, p(a_2)$ or, equivalently, $p(a_1|a_2) = p(a_1)$ or $p(a_2|a_1) = p(a_2)$. This means that the knowledge of a_1 provides no information about a_2, and vice versa.

2.2.4 Probabilistic Models

Random variables are extremely useful in signal processing because they can model the lack of certainty about a physical phenomenon. Imagine we have a good model for some process: for example, a 'pure sine' acoustic waveform generated by an electronic instrument. A model for the pressure signal recorded is given by the following discrete time sine model:

$$x(n) = \alpha \sin(2\pi k_0 n + \phi_0) \text{ for } n = 1, \ldots, T, \quad (2.30)$$

where k_0 is the unknown sine waveform frequency, ϕ_0 is the initial phase, and α is the signal amplitude. It is clear that the recorded pressure signal will not fit *exactly* the model (2.30), and that it will deviate from it. As these deviations may have various causes (air temperature/pressure inhomogeneity, non purely sinusoidal loudspeaker behaviour when emitting the sound, digital to analog conversion artifacts, etc.), it is unrealistic to model them deterministically, and a random model can be used. A possible such model is

$$x(n) = \alpha \sin(2\pi k_0 n + \phi_0) + \epsilon(n) \text{ for } n = 1, \ldots, T, \quad (2.31)$$

where $\epsilon(n)$ is a so-called *additive random noise* with a given pdf. Including this noise in the model is aimed at modelling the deviations of the recorded data from the model in (2.30). In the general case, it is assumed that $\epsilon(n)$ is a *stationary white noise*, or *independent identically distributed (i.i.d.) noise*; that is, the noise at any time n_1 is statistically independent of the noise at any time n_2, and their pdfs are equal. More precisely, the joint pdf of the noise samples equals the product of the pdfs of each sample: Writing the noise samples as a vector $\epsilon = [\epsilon(1), \ldots, \epsilon(T)]^\mathsf{T}$ for $n = 1, \ldots, T$, we have $\mathsf{p}(\epsilon) = p(\epsilon(1))p(\epsilon(2)) \ldots p(\epsilon(T))$, with $p(\epsilon(1)) = p(\epsilon(2)) = \ldots = p(\epsilon(T))$. Finally, since the noise is aimed at modelling small deviations around the sine model, it is assumed to be zero-mean. Equation (2.31) defines a probabilistic signal model and directly yields the *likelihood function*.

2.2.5 Likelihood Functions

In (2.31), it is assumed that the recorded signal x follows a sine model with additive zero-mean white noise. In practice, this model is interesting in the sense that it relates the recorded signal $x(n)$, $n = 1, \ldots, T$ to the parameters k_0, ϕ_0, and α. In the following, we denote by $\boldsymbol{\theta}$ the set of unknown parameters, i.e., $\boldsymbol{\theta} = [k_0, \phi_0, \alpha]$. From the probabilistic model defined above, and given a recorded signal x, we see that some values of $\boldsymbol{\theta}$ are more likely than others: for example, if the sine wave is generated with frequency 440 Hz, finding $k_0 = 440$ Hz is very likely. It is also likely that the loudspeaker which emits the sound adds partials (that is, additional sine waves with lower amplitudes) at frequencies 880 Hz, 1320 Hz, etc. These are also likely, to a lower extent than 440 Hz, though. Conversely, assume $\boldsymbol{\theta}$ is given; then the signal $x(n)$, $n = 1, \ldots, T$ can be seen as a random vector denoted by $\mathbf{x} = [x(1), \ldots, x(T)]^\mathsf{T}$. It admits a joint pdf $\mathsf{p}(\mathbf{x}|\boldsymbol{\theta}) = p(x(1), x(2), \ldots, x(T)|\boldsymbol{\theta})$, conditional on $\boldsymbol{\theta}$: by changing $\boldsymbol{\theta}$, the signal pdf $\mathsf{p}(\mathbf{x}|\boldsymbol{\theta})$ is changed. In the sine example presented above, assuming the noise is Gaussian, the covariance matrix of ϵ is diagonal of size T, with the variance σ^2 of each $\epsilon(n)$, $n = 1, \ldots, T$ on its diagonal. This pdf is

$$
\begin{aligned}
\mathsf{p}(\mathbf{x}|\boldsymbol{\theta}) &= [2\pi\sigma^2]^{-N/2} \exp\left\{ -\frac{1}{2\sigma^2}[\mathbf{x} - \mathbf{f}(\boldsymbol{\theta})]^\mathsf{T}[\mathbf{x} - \mathbf{f}(\boldsymbol{\theta})] \right\} \\
&= [2\pi\sigma^2]^{-N/2} \exp\left\{ -\|\mathbf{x} - \mathbf{f}(\boldsymbol{\theta})\|^2 / 2\sigma^2 \right\},
\end{aligned} \tag{2.32}
$$

where $\mathbf{f}(\boldsymbol{\theta})$ is the model given in (2.30) written in vector form, i.e., $\mathbf{f}(\boldsymbol{\theta}) = [\alpha\sin(2\pi k_0 1 + \phi_0), \ldots, \alpha\sin(2\pi k_0 T + \phi_0)]^\mathsf{T}$.

The mathematical object $\mathsf{p}(\mathbf{x}|\boldsymbol{\theta})$ admits two interpretations: First, when read as a pdf of $\mathbf{x}$ for a given $\boldsymbol{\theta}$, $\mathsf{p}(\mathbf{x}|\boldsymbol{\theta})$ is called the *conditional pdf of* $\mathbf{x}$, *conditioned on* $\boldsymbol{\theta}$. Second, when seen as a function of $\boldsymbol{\theta}$ for given $\mathbf{x}$, $\mathsf{p}(\mathbf{x}|\boldsymbol{\theta})$ is called the *parameter likelihood function* and it is defined over the space of all possible values of $\boldsymbol{\theta}$ denoted Θ. For example $\Theta = [0, 1/2] \times [0, 2\pi] \times [0, \infty)$ in the sine example (2.30). Note that the function $\mathsf{p}(\mathbf{x}|\boldsymbol{\theta})$ is not a pdf of $\boldsymbol{\theta}$, because its integral w.r.t. $\boldsymbol{\theta}$ over Θ may not equal one.

In practice, we always refer to $p(\mathbf{x}|\boldsymbol{\theta})$ as the likelihood function. It should be interpreted as a measure of similarity between the model $\mathbf{f}(\boldsymbol{\theta})$ tuned by $\boldsymbol{\theta}$ and the signal $\mathbf{x}$: the larger the likelihood, the more likely the parameter vector $\boldsymbol{\theta}$. In some problems, it can happen that the likelihood function takes infinite values for some values of $\boldsymbol{\theta}$. In this case, it is called *degenerate*.

2.2.6 Maximum Likelihood Estimation

As outlined above, the likelihood can be seen as a similarity measure between the signal and the model, via its parameters. In the sine example given in (2.30), the parameter vector $\boldsymbol{\theta} = [k_0, \phi_0, \alpha]$ is unknown and it may be useful to estimate it. It is quite natural to use as an estimate the value of $\boldsymbol{\theta}$ that maximizes the likelihood (assumed non-degenerate):

$$\widehat{\boldsymbol{\theta}}_{\mathrm{ML}} = \underset{\boldsymbol{\theta} \in \Theta}{\mathrm{argmax}}\ p(\mathbf{x}|\boldsymbol{\theta}), \qquad (2.33)$$

where the subscript ML stands for *maximum likelihood* (ML). Figure 2.7 shows the likelihood function corresponding to the model (2.31), and the maximum can be seen: it is located at point $\widehat{\boldsymbol{\theta}}_{\mathrm{ML}}$. The optimization problem (2.33) is often transformed into the equivalent problem of maximizing the log-likelihood[7] defined as follows:

$$\mathsf{L}_{\mathbf{x}}(\boldsymbol{\theta}) = \log\left[p(\mathbf{x}|\boldsymbol{\theta})\right]. \qquad (2.34)$$

Thanks to their excellent properties (see subsection below), ML estimators are often used in signal processing applications.

2.2.7 Elements of Estimation Theory

There exist many estimation techniques with possible application to music transcription: Aside from maximum likelihood, one can quote *minimum mean square error* (MMSE) estimators;[8] they consist of estimating the expectation $\mathbb{E}_{p(\boldsymbol{\theta})}\left[h(\boldsymbol{\theta})\right]$ of some function h of the parameter by using an empirical average (2.23). Estimators are characterized by two important quantities: the bias and the estimator covariance. In the sine wave example presented above, the sound produced by the electronic instrument through a loudspeaker can be recorded several times, resulting in several discrete signal vectors $\mathbf{x}_i$, $i = 1, \ldots, I$, all described by the model (2.31) (this is because the signal is random). When applying some estimation technique on each of these signals, such as ML estimation, we obtain estimates $\widehat{\boldsymbol{\theta}}_i$, $i = 1, \ldots, I$, which all have a different value. In other words, the estimate $\widehat{\boldsymbol{\theta}}$ is a random variable with pdf $p(\widehat{\boldsymbol{\theta}})$.

---

[7]The logarithm function is order preserving, and thus maximizing a function or maximizing its logarithm yield the same result.

[8]MMSE estimates are further considered in Section 2.3.4, p. 41.

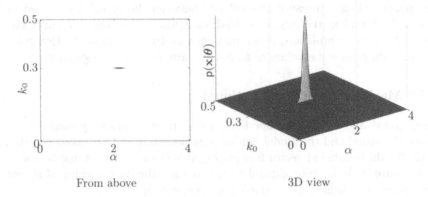

Fig. 2.7. Likelihood $p(\mathbf{x}|k_0, \phi_0, \alpha)$ in (2.32) as a function of the frequency k_0 and the amplitude α for a sine wave with frequency $k_0 = 0.3$ and amplitude $\alpha = 2$. The sine wave is corrupted with a white Gaussian zero-mean additive noise with variance $\sigma^2 = 4$. Both this variance and the initial phase ϕ_0 are assumed to be known. The likelihood shows a very sharp peak, which leaves no doubt about the parameter value, but which may be hard to localize.

Putting aside estimation issues, and assuming the model (2.31) is correct, the recorded signals $\mathbf{x}_i$ are all generated by one value of $\boldsymbol{\theta}$, denoted $\boldsymbol{\theta}_{\text{gt}}$, where the subscript 'gt' stands for 'ground truth': this is the value we want to estimate.

In the general case, we define the bias of the estimator which produces the estimate $\widehat{\boldsymbol{\theta}}$ of $\boldsymbol{\theta}$ as follows:

$$\text{Bias}(\widehat{\boldsymbol{\theta}}) \;=\; \mathbb{E}_{p(\widehat{\boldsymbol{\theta}})}(\widehat{\boldsymbol{\theta}}) - \boldsymbol{\theta}_{\text{gt}}. \tag{2.35}$$

Note that the bias characterizes the estimator (e.g., the ML estimator applied to the model (2.31) in the sine wave example) and not an estimate $\widehat{\boldsymbol{\theta}}$. An estimator is said to be *unbiased* whenever $\text{Bias}(\widehat{\boldsymbol{\theta}}) = 0$, and this is, of course, a very important property.

The estimator covariance matrix $\text{Var}(\widehat{\boldsymbol{\theta}})$ is defined for both biased and unbiased estimators (in the latter case, it is also referred to as the *mean square error*):

$$\text{Var}(\widehat{\boldsymbol{\theta}}) \;=\; \mathbb{E}\big[(\widehat{\boldsymbol{\theta}} - \mathbb{E}[\widehat{\boldsymbol{\theta}}])(\widehat{\boldsymbol{\theta}} - \mathbb{E}[\widehat{\boldsymbol{\theta}}])^{\mathsf{T}}\big]. \tag{2.36}$$

For unbiased estimators, $\text{Var}(\widehat{\boldsymbol{\theta}})$ is lower bounded by the so-called Cramér–Rao bound.[9] When the information about the signal becomes stronger,

---

[9]The definition of a *lower bound* for scalar numbers has to be adapted to matrices: A matrix A is a lower bound for the covariance matrix $\boldsymbol{\Sigma}$ if and only if the matrix $\boldsymbol{\Sigma} - A$ is a definite non-negative matrix (where A and $\boldsymbol{\Sigma}$ have the same size); see [295].

typically because we have longer records $(T \rightarrow \infty)$, we can study the estimator asymptotic properties. In particular, the estimator is said to be *consistent* whenever the variance $\mathrm{Var}(\widehat{\boldsymbol{\theta}})$ tends to zero.

ML estimators are asymptotically unbiased (in some cases, they are even unbiased for any number T of data); they are consistent; and they reach asymptotically the Cramér–Rao bound. Because of these excellent properties, maximum likelihood approaches are widely used in signal processing. When a good model is defined, the implementation difficulty consists mainly of the optimization problem (2.33), which can usually be solved by the expectation-maximization algorithm.

2.2.8 The Expectation-Maximization Algorithm

The expectation-maximization (EM) algorithm is an optimization algorithm specially designed for ML parameter estimation. Before describing it, we emphasize two of its important characteristics: First, it is 'gradient-based'; that is, it does not find the global maximum in the general case, unless conveniently initialized. Second, it is only applicable to problems with *latent variables*. There are a great number of problems which can be addressed using EM. As an illustration, we consider here the important problem of fitting a mixture of Gaussians to some data.[10]

An Example: Gaussian Mixture Models (GMMs)

Gaussian mixture models are quite important in audio processing. For example in speech processing the data considered are generally sets of cepstral coefficients. However, this algorithm is not restricted to cepstral data. In general, GMMs are used to estimate the pdf of a set of data. This is because they have two key properties: 1) given that there are enough Gaussians in (2.37), GMMs can approximate any pdf (versatility), and 2) finding their parameters is easy thanks to the EM algorithm.

Consider the set of data $\mathsf{X} = \{\mathbf{x}_1, \ldots, \mathbf{x}_m\}$, where each individual datum is a vector in $\mathbb{R}^{d_\mathbf{x}}$. In the speech processing example mentioned above, each datum $\mathbf{x}_i$ $(i = 1, \ldots, m)$ is made of the first $d_\mathbf{x}$ cepstral coefficients of an audio signal frame. Gaussian mixture modelling consists of fitting a mixture pdf made of J Gaussians on each datum in X. The mixture pdf is

$$p\big(\mathbf{x}_i | \{\beta_j, \boldsymbol{\mu}_j, \boldsymbol{\Sigma}_j\}_{j=1,\ldots,J}\big) \;=\; \sum_{j=1}^{J} \beta_j \mathcal{N}\big(\mathbf{x}_i; \boldsymbol{\mu}_j, \boldsymbol{\Sigma}_j\big), \qquad (2.37)$$

where the mixture coefficients β_j, $j = 1, \ldots, J$ are such that $\sum_{j=1}^{J} \beta_j = 1$. The likelihood of the mixture for the complete set X is written as

---

[10]Further details about the EM algorithm for general purposes may be found in [471] and the special case of Gaussian mixtures is addressed in [536], [38].

$$p\left(\mathsf{X}|\{\beta_j,\boldsymbol{\mu}_j,\boldsymbol{\Sigma}_j\}_{j=1,\ldots,J}\right) \;=\; \prod_{i=1}^{m}\left[\sum_{j=1}^{J}\beta_j\mathcal{N}\left(\mathbf{x}_i;\boldsymbol{\mu}_j,\boldsymbol{\Sigma}_j\right)\right], \qquad (2.38)$$

because the data in X are assumed to be independent and identically distributed (thus, the joint dataset pdf is the product of the individual datum pdfs).

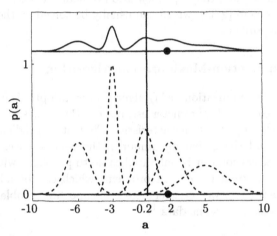

Fig. 2.8. Mixture of five Gaussians. The five Gaussians composing the mixture are represented in dotted lines, whereas the full mixture pdf is represented in solid line (with offset $+1.1$ for better visibility). The mixture coefficients are $\beta_1 = \ldots = \beta_5 = 1/5$. The means of the Gaussians are $[-6, -3, -0.2, 2, 5]$ and the variances are $[1, 0.16, 0.64, 1, 3.24]$. Any random variable distributed according to this mixture pdf may be generated either directly from the mixture pdf, or by first selecting randomly one of the five Gaussians, and then generating the variable from it. For example, the dot at abscissa 1.8 can either result from direct sampling form the mixture in solid line, or by first selecting randomly one of the Gaussians (here, the Gaussian centred which mean equals 2), and sampling from this Gaussian.

In Gaussian mixture approaches, X is described statistically through the mixture parameters $\{\beta_j,\boldsymbol{\mu}_j,\boldsymbol{\Sigma}_j\}_{j=1,\ldots,J}$. This modelling approach is also interesting because it compresses the information of m data into J parameter sets $\{\beta_j,\boldsymbol{\mu}_j,\boldsymbol{\Sigma}_j\}$, with $J \ll m$. These parameters need to be estimated, however. This is implemented here by the EM algorithm for ML estimation of the unknown parameters, denoted for convenience by $\boldsymbol{\theta} = \{\boldsymbol{\theta}_1,\ldots,\boldsymbol{\theta}_J\}$, where $\boldsymbol{\theta}_j = (\beta_j,\boldsymbol{\mu}_j,\boldsymbol{\Sigma}_j)$, $j = 1,\ldots,J$.

In order to apply the EM algorithm, we note that the mixture (2.37) admits two interpretations: First, the pdf of $\mathbf{x}_i$ is seen as a sum of weighted Gaussians, as depicted in Fig. 2.8. Second, with probability β_j $(j = 1,\ldots,J)$, the datum $\mathbf{x}_i$ is distributed according to the pdf $\mathcal{N}\left(\mathbf{x}_i;\boldsymbol{\mu}_j,\boldsymbol{\Sigma}_j\right)$. In the second interpretation, we can associate the latent variables z_i, $i = 1, \ldots, m$ to each datum $\mathbf{x}_i$, where z_i takes its values in $\{1, 2, \ldots, J\}$. The latent variable z_i indicates the component $\mathcal{N}(\mathbf{x}_i; \boldsymbol{\mu}_j, \boldsymbol{\Sigma}_j)$ of the mixture (2.37) that corresponds to $\mathbf{x}_i$. In other words, when z_i is known, the distribution of $\mathbf{x}_i$ is

$$p(\mathbf{x}_i | z_i = j, \beta_j, \boldsymbol{\mu}_j, \boldsymbol{\Sigma}_j)_, = \mathcal{N}(\mathbf{x}_i; \boldsymbol{\mu}_j, \boldsymbol{\Sigma}_j). \qquad (2.39)$$

The latent variable z_i probability is $\mathsf{P}(z_i = j) = \beta_j$.

Principle of the EM Algorithm

The general structure of the EM algorithm is presented in Algorithm 2.1 below. It is assumed that the model involves the data set X and latent variables $\mathsf{Z} = \{z_1, \ldots, z_J\}$.

Algorithm 2.1: General EM Algorithm

1. <u>Initialization:</u> Select the starting point $\boldsymbol{\theta}^{(0)}$ inside Θ.
2. <u>Iterations:</u> For $j = 1, 2, \ldots$, do (until convergence is reached)
   - Set

$$\boldsymbol{\theta}^{(j+1)} = \operatorname*{argmax}_{\boldsymbol{\theta}} Q_{\mathsf{X}}(\boldsymbol{\theta} | \boldsymbol{\theta}^{(j)}), \qquad (2.40)$$

   where

$$Q_{\mathsf{X}}(\boldsymbol{\theta} | \boldsymbol{\theta}^{(j)}) = \mathbb{E}_{p(\mathsf{Z} | \mathsf{X}, \boldsymbol{\theta}^{(j)})} [\log \mathsf{p}(\mathsf{X}, \mathsf{Z} | \boldsymbol{\theta})]. \qquad (2.41)$$

Algorithm 2.1 is quite simple. The two steps used to compute $\boldsymbol{\theta}^{(j+1)}$ from $\boldsymbol{\theta}^{(j)}$ are

- The *expectation* step: compute $Q_{\mathsf{X}}(\boldsymbol{\theta} | \boldsymbol{\theta}^{(j)})$ using (2.41).
- The *maximization* step: maximize $Q_{\mathsf{X}}(\boldsymbol{\theta} | \boldsymbol{\theta}^{(j)})$ with respect to $\boldsymbol{\theta}$. This yields $\boldsymbol{\theta}^{(j+1)}$.

The principle of this algorithm [471] is that we want to maximize the log-likelihood $\log \mathsf{p}(\mathsf{X}, \mathsf{Z} | \boldsymbol{\theta})$ with respect to $\boldsymbol{\theta}$ without knowing the latent variables Z. The expectation in (2.41) permits us to get rid of the latent variables, so as to perform the maximization. However, since this expectation is computed for the parameter value $\boldsymbol{\theta}^{(j)}$, which is not the 'true' value, the expectation in (2.41) is not the 'true' one. The iterations in Algorithm 2.1 ensure that $\boldsymbol{\theta}^{(j)}$ becomes closer to the 'true' value, yielding a better approximation of the true expectation in (2.41), and thus a better ML estimate at each iteration.

Application to the Gaussian Mixture Model

In the Gaussian mixture case, (2.41) and (2.40) can be calculated analytically.

$$\log \mathsf{p}(\mathsf{X}, \mathsf{Z} | \boldsymbol{\theta}) = \sum_{i=1}^{m} \log \mathsf{p}(\mathbf{x}_i | z_i, \boldsymbol{\theta}) + \log \mathsf{P}(z_i | \boldsymbol{\theta}), \qquad (2.42)$$

and, using (2.41) (see [38] for details),

$$Q_X(\boldsymbol{\theta}|\boldsymbol{\theta}^{(j)}) = \sum_{j=1}^{J}\sum_{i=1}^{m} \log\left[\beta_j \mathsf{p}(\mathbf{x}_i|z_i = j, \boldsymbol{\theta}_j)\right]\mathsf{P}(z_i = j|\mathbf{x}_i, \boldsymbol{\theta}_j^{(j)}), \qquad (2.43)$$

where $\mathsf{P}(z_i = j|\mathbf{x}_i, \boldsymbol{\theta}_j^{(j)})$ is computed using Bayes's rule; see (2.51) in Section 2.2.3, p. 31. Moreover, $Q_X(\boldsymbol{\theta}|\boldsymbol{\theta}^{(j)})$ in (2.43) can be maximized analytically. Defining $S_{j,\boldsymbol{\theta}_j^{(j)}} = \sum_{i=1}^{m} \mathsf{P}(z_i = j|\mathbf{x}_i, \boldsymbol{\theta}_j^{(j)})$, the EM update equations are [38]

$$\beta_j^{(j+1)} = \frac{1}{m}S_{j,\boldsymbol{\theta}_j^{(j)}}, \qquad (2.44)$$

$$\boldsymbol{\mu}_j^{(j+1)} = S_{j,\boldsymbol{\theta}_j^{(j)}}^{-1}\sum_{i=1}^{m}\mathbf{x}_i\mathsf{P}(z_i = j|\mathbf{x}_i, \boldsymbol{\theta}_j^{(j)}), \qquad (2.45)$$

$$\boldsymbol{\Sigma}_j^{(j+1)} = S_{j,\boldsymbol{\theta}_j^{(j)}}^{-1}\sum_{i=1}^{m}\left[\mathbf{x}_i - \boldsymbol{\mu}_j^{(j+1)}\right]\left[\mathbf{x}_i - \boldsymbol{\mu}_j^{(j+1)}\right]^{\mathsf{T}}\mathsf{P}(z_i = j|\mathbf{x}_i, \boldsymbol{\theta}_j^{(j)}). \qquad (2.46)$$

An interesting remark is that the likelihood of Gaussian mixture models is degenerate. In particular, when there are more components than data points (i.e., when $J \geq m$), the ML estimate includes variances equal to zero. In this case, the Gaussians degenerate into Dirac delta functions, and the model is exactly equivalent to the dataset X itself: this is a useless solution. A solution to avoid this problem consists of maximizing a *penalized likelihood* instead of the likelihood itself; see [91].

As mentioned earlier, the EM algorithm is gradient based. In other words, its convergence to the global maximum is not ensured in the general case. In particular, assume that the likelihood admits several local maxima (that is, points where $\partial \mathsf{p}(X|\boldsymbol{\theta})/\partial\boldsymbol{\theta} = 0$); then the solution found by the EM algorithm is one local maximum, determined by the initialization point. Solutions to overcome this problem consist of implementing other optimization methods such as simulated annealing or stochastic EM [544], but none of them is simple and/or computationally cheap.

2.2.9 Penalized Likelihood Approaches

Penalized likelihood approaches are really useful when a parameter estimation problem admits many possible solutions, including trivial ones (problems which lead to such trivial or useless solutions are said to be *ill posed*). In this case, the ML estimate found is generally not the one sought. This can however be avoided by modifying the likelihood into a *penalized likelihood*, defined as

$$\mathsf{L}_X^{\text{pen}}(\boldsymbol{\theta}) = \mathsf{L}_X(\boldsymbol{\theta}) + \lambda\Omega(\boldsymbol{\theta}), \qquad (2.47)$$

where $L_x(\boldsymbol{\theta})$ is defined in (2.34) and $\Omega(\boldsymbol{\theta})$ is aimed at lowering the likelihood in parts of the space Θ which are to be avoided.[11] In the Gaussian mixture example where $J \geq m$, the penalty term can be set so as to disable variance parameters that are too small; see [91]. The EM algorithm can be adapted in order to address penalized likelihood problems [251].

An important application of penalized likelihood concerns the *model selection problem*, where one tries to estimate the probabilistic model that best fits some data. A typical example is that of autoregressive models.

Autoregressive Models

Autoregressive models form a major signal processing tool. They are used for spectrum estimation, coding, or noise reduction. An autoregressive (AR) model (also called a *linear prediction model*) expresses a signal $x(n)$ at time n as a linear combination of its previous values:

$$x(n) = a_1 x(n-1) + a_2 x(n-2) + \cdots + a_p x(n-p) + \epsilon(n), \qquad (2.48)$$

where $\boldsymbol{a} = [a_1, \ldots, a_p]^\mathsf{T}$ is the vector made of the AR coefficients. The order of this model is the number p of coefficients. Assuming the noise $\epsilon(n)$ is zero-mean, white and Gaussian with known variance σ^2, the log-likelihood of the AR coefficients and the model order is

$$L_x(\boldsymbol{a}, p) = -\frac{T}{2} \log\left(2\pi\sigma^2\right) - \frac{1}{2\sigma^2} \sum_{n=1}^{T} \|x(n) - \boldsymbol{a}^\mathsf{T} \mathbf{x}_{n-1:n-p}\|^2, \qquad (2.49)$$

where $\mathbf{x}_{n-1:n-p} = [x(n-1), \ldots x(n-p)]^\mathsf{T}$ and assuming, e.g., $\mathbf{x}_{n-1:n-p} = [0, 0, \ldots, 0]^\mathsf{T}$ for $n < p$. Here, we denote by $\mathbf{x}$ the vector made of signal samples $x(n)$, $n = 1, \ldots, T$. Maximizing $L_x(\boldsymbol{a}, p)$ yields uninteresting solutions, typically p becomes very large. This problem can be overcome by maximizing instead the penalized log-likelihood (2.47). Choosing $\Omega(\boldsymbol{a}, p) = -p$ and $\lambda = 1$ leads to Akaike's penalized log-likelihood, which leads to reliable model order estimation [252]:

$$L_x^{\text{pen}}(\boldsymbol{a}, p) = -\frac{T}{2} \log\left(2\pi\sigma\right) - \frac{1}{2\sigma^2} \sum_{n=1}^{T} \|x(n) - \boldsymbol{a}^\mathsf{T} \mathbf{x}_n^p\|^2 - p. \qquad (2.50)$$

The next section presents an estimation framework different from likelihood approaches: *Bayesian statistical methods*.

2.3 Bayesian Statistical Methods

In maximum likelihood parameter estimation, it is implicitly assumed that there exists a fixed, ground truth parameter value. It can be argued, however,

---

[11] See [252] for a review of penalized likelihood approaches.

that the parameter is well defined insofar as the model is correct, but in real cases, the model is always an approximation of the real world. In Bayesian approaches, it is assumed instead that the unknown parameter is a random variable, characterized by a pdf. This pdf is stated before any data are collected and is called the parameter *prior distribution*.

2.3.1 Bayes's Rule and Posterior Distributions

The most important concept of Bayesian statistical theory is the so-called Bayes's rule. Consider two random variables $\mathbf{a}_1 \in \mathcal{A}_1$ and $\mathbf{a}_2 \in \mathcal{A}_2$, with joint pdf $p(\mathbf{a}_1, \mathbf{a}_2)$. Recall (2.29): $p(\mathbf{a}_1, \mathbf{a}_2) = p(\mathbf{a}_1|\mathbf{a}_2)p(\mathbf{a}_2) = p(\mathbf{a}_2|\mathbf{a}_1)p(\mathbf{a}_1)$. We can deduce Bayes's rule

$$p(\mathbf{a}_2|\mathbf{a}_1) = \frac{p(\mathbf{a}_1|\mathbf{a}_2)p(\mathbf{a}_2)}{\int_{\mathcal{A}_2} p(\mathbf{a}_1|\mathbf{a}_2)p(\mathbf{a}_2)\, d\mathbf{a}_2}, \tag{2.51}$$

which enables us to 'reverse the conditioning'. Note that the denominator in (2.51) equals $\int_{\mathcal{A}_2} p(\mathbf{a}_1, \mathbf{a}_2), d\mathbf{a}_2 = p(\mathbf{a}_1)$, but we keep the integral form in order not to be confused with a possible prior over $\mathbf{a}_1$ which would also be denoted $p(\mathbf{a}_1)$. Bayes's rule can be applied straightforwardly to Bayesian parameter estimation. Assume we want to learn the value of some parameter $\boldsymbol{\theta} \in \Theta$ from a set of data X. We already have the likelihood $p(\mathsf{X}|\boldsymbol{\theta})$, and, being Bayesian, a parameter prior $p(\boldsymbol{\theta})$ is selected. Using Bayes's rule,

$$p(\boldsymbol{\theta}|\mathsf{X}) = \frac{p(\mathsf{X}|\boldsymbol{\theta})p(\boldsymbol{\theta})}{\int_{\Theta} p(\mathsf{X}|\boldsymbol{\theta})p(\boldsymbol{\theta})\, d\boldsymbol{\theta}}, \tag{2.52}$$

where $p(\boldsymbol{\theta}|\mathsf{X})$ is called the parameter *posterior distribution*, which incorporates information from both the data and the parameter prior. Roughly speaking, the posterior provides information about the probability of each possible value of $\boldsymbol{\theta}$ given the data and some prior knowledge. In practice, it suffices to write

$$p(\boldsymbol{\theta}|\mathsf{X}) \propto p(\mathsf{X}|\boldsymbol{\theta})p(\boldsymbol{\theta}), \tag{2.53}$$

where $\propto$ stands for 'proportional to'. The proportionality constant only depends on the data X and it does not give additional information about $\boldsymbol{\theta}$. Bayesian statistical methods are concerned with posterior distributions.

2.3.2 Bayesian Estimation

Similar to ML estimation, where estimation is performed from the likelihood, it is possible to perform Bayesian estimation from the posterior. In particular, two estimation methods are generally considered:

- *Maximum a posteriori* (MAP). The estimated value maximizes the posterior pdf

$$\widehat{\boldsymbol{\theta}}_{\mathrm{MAP}} = \underset{\boldsymbol{\theta} \in \Theta}{\operatorname{argmax}}\ p(\boldsymbol{\theta}|\mathsf{X}). \tag{2.54}$$

- *Minimum mean square error* (MMSE). The estimate is the expectation of the posterior pdf

$$\widehat{\theta}_{\mathrm{MMSE}} = \mathbb{E}_{\mathsf{p}(\theta|\mathsf{X})}[\theta]. \tag{2.55}$$

An important remark is that Bayesian estimators can be 'biased': for example, if the likelihood $\mathsf{p}(\mathsf{X}|\theta)$ and the prior $\mathsf{p}(\theta)$ are Gaussian, with the prior mean μ being different from the 'ground truth' parameter value θ_{gt}, then the above Bayes estimates are equal[12] $(\widehat{\theta}_{\mathrm{MAP}} = \widehat{\theta}_{\mathrm{MMSE}})$ and their expectation differs from θ_{gt}. However, in the Bayesian philosophy, θ_{gt} is meaningless as there is no true value of θ explicitly or implicitly assumed: all values θ are possible true values.

2.3.3 Bayesian Interpretation of Penalized Likelihood

In (2.47), the penalty term is introduced in the log likelihood. Going back to the likelihood via the exponential function, we can rewrite it as

$$\exp\left(\mathsf{L}_{\mathsf{X}}^{\mathrm{pen}}\left(\theta\right)\right) = \mathsf{p}(\mathsf{X}|\theta) \times \exp\left(\lambda\Omega\left(\theta\right)\right) \tag{2.56}$$

and it appears that the likelihood has been multiplied by a term that only depends on θ. Assuming that the following integral is finite,

$$\int_{\Theta} \exp\left(\lambda\Omega\left(\theta\right)\right) d\theta = \mathrm{C}, \tag{2.57}$$

then we can define the pdf of θ:

$$\mathsf{p}(\theta) = \frac{1}{\mathrm{C}} \exp\left(\lambda\Omega\left(\theta\right)\right). \tag{2.58}$$

We recognize in (2.58) the prior pdf of θ, and (2.56) can be interpreted as the product of the likelihood with the prior, that is, the posterior $\mathsf{p}(\theta|\mathsf{X})$ (up to a normalizing constant). Maximizing the penalized likelihood becomes equivalent to performing MAP estimation.

2.3.4 Monte Carlo Methods

Bayes's theory provides a general framework for statistical inference. The main concept is that of posterior distributions, which result from both the likelihood and the prior. However, computing estimates such as (2.54) and (2.55) is a difficult problem in the general case. For example, MMSE estimates are obtained by the following integral:

$$\widehat{\theta}_{\mathrm{MMSE}} = \int_{\Theta} \theta \mathsf{p}(\theta|\mathsf{X}) d\theta. \tag{2.59}$$

---

[12]It is worth mentioning that the product of two Gaussians is Gaussian; thus the posterior pdf is also Gaussian in this example, which explains why its mean value coincides with its maximum.

In statistical inference problems, the dimension of Θ may be very large (sometimes, more that 1000). Whenever the closed-form expression of (2.59) cannot be found, a numerical integration technique should be implemented. Consider the slightly more general integral computation case

$$I[h] = \int_\Theta h(\boldsymbol{\theta})\pi(\boldsymbol{\theta}) \, d\boldsymbol{\theta}, \qquad (2.60)$$

where $\pi(\boldsymbol{\theta})$ denotes any pdf of $\boldsymbol{\theta}$, such as the posterior $p(\boldsymbol{\theta}|X)$. When the dimension of Θ is small (typically, smaller than 3), $I[h]$ can be computed numerically à la Riemann using a regular grid (provided Θ is a compact set). For example, if $\boldsymbol{\theta}$ is one dimensional, and $\Theta = [0,1]$, Riemann integration consists of computing

$$I[h] \approx \frac{1}{N} \sum_{i=1}^{N} h(i/N)\pi(i/N). \qquad (2.61)$$

The limit of this approach is that the grid size increases exponentially with the dimension d_Θ of Θ: assuming 100 grid points are used in each dimension of Θ of dimension $d_\Theta = 50$, the grid size is $100^{50} = 10^{100}$; this is out of reach of today's computers.

However, another numerical computation technique can be implemented. Assume random samples $\widetilde{\boldsymbol{\theta}}^{(i)}$, $i = 1, \ldots, N$ are available, where each sample $\widetilde{\boldsymbol{\theta}}^{(i)}$ is distributed according to $\pi(\boldsymbol{\theta})$ (this is denoted $\widetilde{\boldsymbol{\theta}}^{(i)} \sim \pi(\boldsymbol{\theta})$ in the following). Then, from the law of large numbers,

$$\widehat{I}_N[h] = \frac{1}{N} \sum_{i=1}^{N} h(\widetilde{\boldsymbol{\theta}}^{(i)}) \approx I[h]. \qquad (2.62)$$

The random samples $\widetilde{\boldsymbol{\theta}}^{(i)}$, $i = 1, \ldots, N$ are generally referred to as *Monte Carlo samples* and $\widehat{I}_N[h]$ is the *Monte Carlo estimate* of $I[h]$.

The estimate $\widehat{I}_N[h]$ is unbiased for any N, i.e., $\mathbb{E}_{\pi(\widetilde{\boldsymbol{\theta}}^{(1)},\ldots,\widetilde{\boldsymbol{\theta}}^{(N)})}\left(\widehat{I}_N[h]\right) = I[h]$ and consistent. Finally, the *empirical mean square error* $\widehat{\sigma}_N^2[h]$ provides a reliable indication of the variance of $\widehat{I}_N[h]$ (in the sense that, asymptotically as $N \to \infty$, $\widehat{\sigma}_N^2[h]$ tends to the true estimation variance):

$$\widehat{\sigma}_N^2[h] = \frac{1}{N} \sum_{i=1}^{N} \left[h(\widetilde{\boldsymbol{\theta}}^{(i)}) - \widehat{I}_N[h] \right]^2. \qquad (2.63)$$

Monte Carlo methods may also be used to compute other kinds of estimates, inside or outside of the Bayesian framework. For example, Monte Carlo optimization methods may be used to compute maximum likelihood of MAP estimates; see [544].

We have assumed so far that Monte Carlo samples are available. The real difficulty is actually in generating these samples. When $\pi(\boldsymbol{\theta})$ is a standard pdf (e.g., uniform or Gaussian), it is possible to implement direct sampling using a random variable generator. In other cases, typically when $\pi(\boldsymbol{\theta})$ is a posterior of large dimension with an unknown normalizing constant, the problem can be much harder. Several techniques have been developed for random variable generation from any pdf. We present here Monte Carlo Markov chain (MCMC) methods and importance sampling.[13]

2.3.5 Monte Carlo Markov Chain (MCMC) Methods

The principle of MCMC algorithms is as follows: Given some pdf $\pi(\boldsymbol{\theta})$ we want to sample from, a chain of samples is generated iteratively; see Algorithm 2.2. The chain is statistically fully determined by the pdf of the initial sample $\widetilde{\boldsymbol{\theta}}^{(0)}$ denoted $\pi_0(\boldsymbol{\theta})$ and a so-called *Markov kernel* $\mathcal{K}(\boldsymbol{\theta}|\boldsymbol{\theta}')$, which is a pdf w.r.t. $\boldsymbol{\theta}$ (for fixed $\boldsymbol{\theta}'$). Provided the kernel $\mathcal{K}(\boldsymbol{\theta}|\boldsymbol{\theta}')$ satisfies some properties [544], the pdf of each sample $\widetilde{\boldsymbol{\theta}}^{(i)}$ slowly converges to the target pdf $\pi(\boldsymbol{\theta})$ as i increases. In particular, the kernel must be built so that $\pi(\boldsymbol{\theta})$ is the invariance distribution of $\mathcal{K}(\boldsymbol{\theta}|\boldsymbol{\theta}')$, namely,

$$\int_{\Theta} \mathcal{K}(\boldsymbol{\theta}|\boldsymbol{\theta}')\pi(\boldsymbol{\theta}')\,d\boldsymbol{\theta}' = \pi(\boldsymbol{\theta}) \quad \text{for all } \boldsymbol{\theta} \in \Theta. \qquad (2.64)$$

Algorithm 2.2: Generic MCMC Algorithm

1. <u>Initialization:</u> Sample $\widetilde{\boldsymbol{\theta}}^{(0)} \sim \pi_0(\boldsymbol{\theta})$.
2. <u>Iterations:</u> For $i = 1, 2, \ldots, N$, do
   - Sample $\widetilde{\boldsymbol{\theta}}^{(i)} \sim \mathcal{K}(\boldsymbol{\theta}|\widetilde{\boldsymbol{\theta}}^{(i-1)})$.

Markov chain algorithms provide a series of samples which are asymptotically distributed according to $\pi(\boldsymbol{\theta})$. In practice, it is necessary to run 'burn-in' iterations before reaching convergence. Burn-in samples are not kept for the computation of Monte Carlo estimates (2.62). Figure 2.9, p. 46, shows typical random chains, including the burn-in iterations: the curves show more fluctuation of the current value in the beginning (during the burn-in) and then the value converges.

In can be hard to design such a kernel from scratch. Fortunately, two simple algorithms enable easy building of such kernels: the *Metropolis–Hastings algorithm*, based on accept-reject moves, and the *Gibbs sampler*, based on conditional sampling.

The Gibbs Sampler

Assume we want to generate Monte Carlo samples from $p(\boldsymbol{\theta})$, with $\boldsymbol{\theta} = [\theta_1, \ldots, \theta_{d_\Theta}]^{\mathsf{T}}$, where d_Θ is the dimension of the space Θ. Assume moreover

---

[13]See for example [544] for a complete overview of these techniques.

that we can sample easily from each of the conditional pdfs $p(\theta_1|\theta_2,\ldots,\theta_{d_\Theta})$, $p(\theta_2|\theta_1,\theta_3,\ldots,\theta_{d_\Theta})$, $\ldots$, $p(\theta_{d_\Theta}|\theta_1,\ldots,\theta_{d_\Theta-1})$. Typically, this situation happens when some of the conditionals are Gaussian and the others are, for example, gamma distributions. The Gibbs sampler consists of sampling one component of θ at a time from the conditional posteriors, as presented in Algorithm 2.3.

Algorithm 2.3: The Gibbs Sampler

1. <u>Initialization:</u> Sample $\widetilde{\theta}^{(0)} \sim \pi_0(\theta)$.
2. <u>Iterations:</u> For $i = 1, 2, \ldots, N$, and for $j = 1, \ldots, d_\Theta$, do

   - Sample $\widetilde{\theta}_1^{(i)} \sim p(\theta_1|\widetilde{\theta}_2^{(i-1)},\ldots,\widetilde{\theta}_{d_\Theta}^{(i-1)})$.
   - Sample $\widetilde{\theta}_2^{(i)} \sim p(\theta_2|\widetilde{\theta}_1^{(i)},\widetilde{\theta}_3^{(i-1)},\ldots,\widetilde{\theta}_{d_\Theta}^{(i-1)})$.
   - $\ldots$
   - Sample $\widetilde{\theta}_{d_\Theta}^{(i)} \sim p(\theta_{d_\Theta}|\widetilde{\theta}_1^{(i)},\ldots,\widetilde{\theta}_{d_\Theta-1}^{(i)})$.

The Gibbs sampler is quite simple; however, it requires the ability to sample from the conditional pdfs. This is sometimes not possible to implement, and we can use instead the Metropolis–Hastings (MH) algorithm.

The Metropolis–Hastings Algorithm

In addition to the target pdf $\pi(\theta)$, the MH algorithm requires a *proposal pdf* $q(\theta|\theta')$ that we can directly sample from, such that $q(\theta|\theta') \neq 0$ whenever $\pi(\theta) \neq 0$. Algorithm 2.4 presents its structure.

Algorithm 2.4: The Metropolis–Hastings Algorithm

1. <u>Initialization:</u> Sample $\widetilde{\theta}^{(0)} \sim \pi_0(\theta)$.
2. <u>Iterations:</u> For $i = 1, 2, \ldots, N$ do

   - Sample a candidate parameter value $\theta^\star \sim q(\theta|\widetilde{\theta}^{(i-1)})$.
   - Compute

$$\alpha_{\text{MH}}(\theta^\star, \widetilde{\theta}^{(i-1)}) = \min\left[1, \frac{\pi(\theta^\star)}{\pi(\widetilde{\theta}^{(i-1)})}\frac{q(\widetilde{\theta}^{(i-1)}|\theta^\star)}{q(\theta^\star|\widetilde{\theta}^{(i-1)})}\right]. \quad (2.65)$$

   - With probability $\alpha_{\text{MH}}(\theta^\star, \widetilde{\theta}^{(i-1)})$, accept the candidate, i.e., set $\widetilde{\theta}^{(i)} \leftarrow \theta^\star$.
   - Otherwise (that is, with probability $1 - \alpha_{\text{MH}}(\theta^\star, \widetilde{\theta}^{(i-1)})$, reject the candidate, i.e., set $\widetilde{\theta}^{(i)} \leftarrow \widetilde{\theta}^{(i-1)}$.

From Algorithm 2.4, it is clear than the MH kernel $\mathcal{K}(\theta|\theta')$ is determined by the proposal pdf $q(\theta|\theta')$. Note that the pdf $\pi(\theta)$ appears in a ratio: it can be known only up to a normalizing constant (which is often the case for a posterior pdf due to Bayes's theorem) because it cancels out. Various choices of proposals lead to various MH algorithms. Three important subcases are presented below.

- Set $q(\boldsymbol{\theta}|\boldsymbol{\theta}') = \mathcal{N}(\boldsymbol{\theta}; \boldsymbol{\theta}', \boldsymbol{\Sigma})$. This corresponds to a *random walk proposal*. It also is a symmetric proposal in the sense that $q(\boldsymbol{\theta}|\boldsymbol{\theta}') = q(\boldsymbol{\theta}'|\boldsymbol{\theta})$, and the ratio in (2.65) does not depend on q. This kind of proposal is often called a *local proposal* because the chain evolves locally around the last accepted point. Such chains have the advantage of not remaining at the same point for many iterations: the fraction of accepted candidates can be set to about 50% by tuning the covariance matrix $\boldsymbol{\Sigma}$. However, local proposals cannot explore the whole space Θ very quickly.

- Set $q(\boldsymbol{\theta}|\boldsymbol{\theta}') = q(\boldsymbol{\theta})$, independent of $\boldsymbol{\theta}'$. This proposal is called *independent* or *global*. As opposed to local proposals, it does not consider the last accepted value when building the candidate. This permits large jumps over the space Θ, and quick convergence. However, it also causes many steady points: the chain may keep at the same value for many iterations (i.e., $\widetilde{\boldsymbol{\theta}}^{(i)} = \widetilde{\boldsymbol{\theta}}^{(i+1)} = \cdots = \widetilde{\boldsymbol{\theta}}^{(i+I)}$ where I typically equals 100 or 1000) because all the proposed candidates are rejected. It is thus quite important to build global proposals upon heuristics about the regions of Θ where $\pi(\boldsymbol{\theta})$ is likely to be large.

- For each component θ_j $(j = 1, \ldots, d_\Theta)$, use a proposal $q(\theta_j|\theta_j')$ while not touching the other components. Overall, this proposal can be written $q_j(\boldsymbol{\theta}|\boldsymbol{\theta}') = q(\theta_j|\theta_j') \prod_{i=1, i\neq j}^{d_\Theta} \delta_{\theta_i'}(\theta_i)$ for any $j = 1, \ldots, d_\Theta$, where $\delta_u(v)$ is the Dirac delta function.[14] This is a *one-at-a-time proposal pdf* because it only updates one component of $\boldsymbol{\theta}$ at each iteration, using a local or a global proposal.

Figure 2.9 displays example outcomes for the proposal distributions listed above. These MH kernels can be mixed, yielding another admissible kernel: Given a family of MH kernels $\{\mathcal{K}_j(\boldsymbol{\theta}|\boldsymbol{\theta}'), j = 1, \ldots, J\}$ that have the same invariant pdf $\pi(\boldsymbol{\theta})$, and positive coefficients $\{\beta_j, j = 1, \ldots, J\}$ such that $\sum_{j=1}^{J} \beta_j = 1$, then the mixture kernel

$$\mathcal{K}(\boldsymbol{\theta}|\boldsymbol{\theta}') = \sum_{j=1}^{J} \beta_j \mathcal{K}_j(\boldsymbol{\theta}|\boldsymbol{\theta}') \qquad (2.66)$$

also has $\pi(\boldsymbol{\theta})$ as invariant pdf. In practice, this kernel is implemented as follows: At each iteration, sample a discrete variable u in $\{1, 2, \ldots, J\}$ with discrete distribution $P(u = j) = \beta_j$, and use the kernel with the index $j = u$ drawn. This leads to, e.g., kernels that update all components one at a time (using a mixture of kernels with one-at-a-time proposals for each component), and mix local and global moves for faster convergence; see Fig. 2.9.

---

[14]The Dirac delta function can be viewed as the derivative of the step function at point 0, where the step function equals zero over $]-\infty, 0[$ and one over $]0, \infty[$. An important property is that $\delta_u(v) = 0$ whenever $u \neq v$ and for a function $h(v)$, we have $\int h(v)\delta_u(v)dv = h(u)$. When used in a probabilistic context, writing **a** *has distribution* $\delta_u(\mathbf{a})$ means that $\mathbf{a} = u$ deterministically.

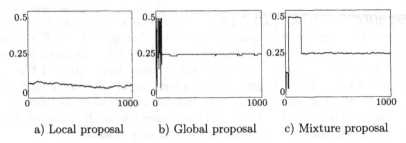

a) Local proposal   b) Global proposal   c) Mixture proposal

Fig. 2.9. Typical Markov chains produced by different Markov kernels for the Metropolis–Hastings algorithm. The parameter sampled is a frequency having the true value 0.25. a) Local, Gaussian random walk proposal pdf. b) Global, independent proposal pdf. c) Mixture of a local and a global proposal. The local proposal fails to explore the frequencies in a short time, whereas the global proposal keeps at the same frequency for many iterations. The mixed proposal performs well.

An Example: Bayesian Estimation of Sinusoids in Noise

In order to examplify the use of MCMC methods, we present here a simplified version of the MCMC algorithm of Andrieu and Doucet [21] which is aimed at estimating the parameters of an unknown number of sinusoids in noise, as well as the number of sinusoids itself. Here, we consider an extension of the model (2.31) with M sinusoids in Gaussian noise, with M known. The unknown parameters are the frequencies (denoted $\mathbf{k} = [k_1, \ldots, k_M]$), the amplitudes, the initial phases, and the Gaussian noise variance σ^2. This model can equivalently be written as

$$x(n) = \sum_{m=1}^{M} \alpha_m^c \cos(2\pi k_m n) + \alpha_m^s \sin(2\pi k_m n) + \epsilon(n) \text{ for } n = 1, \ldots, T,$$

(2.67)

where the (non-linear) phase parameters have been replaced by an additional set of (linear) amplitude parameters. Let us define $\boldsymbol{\alpha} = [\alpha_1^s, \alpha_1^c, \ldots, \alpha_M^s, \alpha_M^c]^\mathsf{T}$ and

$$\mathbf{D}(\mathbf{k}) = \begin{bmatrix} \cos(2\pi k_1\,1), & \cos(2\pi k_1\,2), & \ldots, & \cos(2\pi k_1\,T) \\ \sin(2\pi k_1\,1), & \sin(2\pi k_1\,2), & \ldots, & \sin(2\pi k_1\,T) \\ \vdots & \vdots & & \vdots \\ \cos(2\pi k_M\,1), & \cos(2\pi k_M\,2), & \ldots, & \cos(2\pi k_M\,T) \\ \sin(2\pi k_M\,1), & \sin(2\pi k_M\,2), & \ldots, & \sin(2\pi k_M\,T) \end{bmatrix}^\mathsf{T}.$$

(2.68)

Then, using the notations introduced in Section 2.2.5, p. 32, (2.67) becomes

$$\mathbf{x} = \mathbf{D}(\mathbf{k})\boldsymbol{\alpha} + \boldsymbol{\epsilon},$$

(2.69)

corresponding to the Gaussian likelihood $p(\mathbf{x}|\mathbf{k}, \boldsymbol{\alpha}, \sigma^2) = \mathcal{N}(\mathbf{x}; \mathbf{D}(\mathbf{k})\boldsymbol{\alpha}, \sigma^2\mathbf{I})$. In order to estimate the unknown parameters, we can embed this model into the Bayesian framework by defining parameter prior pdfs, as follows:

- The amplitudes follow a Gaussian prior pdf with mean $\mathbf{0}$ and covariance $\sigma^2 \boldsymbol{\Sigma}_\alpha$, which is proportional to the additive noise variance (this is to adjust the noise 'amplitude' to the sinusoids' average amplitude).

$$\mathsf{p}(\boldsymbol{\alpha}) \;=\; \mathcal{N}(\boldsymbol{\alpha}; [0,0]^\mathsf{T}, \sigma^2 \boldsymbol{\Sigma}_\alpha), \tag{2.70}$$

where $\boldsymbol{\Sigma}_\alpha$ is a covariance matrix chosen as $\boldsymbol{\Sigma}_\alpha = \gamma^2 [\mathbf{D}^\mathsf{T}(\mathbf{k})\mathbf{D}(\mathbf{k})]^{-1}$ (this is the *g-prior*, see [21]). The parameter γ^2 can be interpreted as an expected signal-to-noise ratio, and it is thus quite difficult to set its value a priori. Consider now γ^2 as a unknown random *hyperparameter* to be estimated, and set its prior $\mathsf{p}(\gamma^2) = \mathcal{IG}(\gamma^2; \nu_2, \nu_3)$; see (2.72). Setting $\nu_2 = 2$ ensures that $\mathsf{p}(\gamma^2)$ has an infinite variance (it is non-informative), and the value of ν_3 has very limited influence ($\nu_3 = 20$ is the standard choice).
- The frequencies may be anywhere in the interval $[0; 0.5]$, and thus we select a uniform prior pdf

$$\mathsf{p}(\mathbf{k}) \;=\; \prod_{m=1}^{M} \mathcal{U}_{[0,0.5]}(k_m). \tag{2.71}$$

- The additive noise variance should be as small as possible, and we select its prior as an inverse Gamma distribution

$$\mathsf{p}(\sigma^2) \;=\; \mathcal{IG}\left(\sigma^2; \frac{\nu_0}{2}, \frac{\nu_1}{2}\right) \;=\; \left(\left(\frac{\nu_1}{2}\right)^{\frac{\nu_0}{2}} / \Gamma\left(\frac{\nu_0}{2}\right)\right) \exp\left(-\frac{\nu_1}{2\sigma^2}\right) / \sigma^{(\nu_0+2)} \tag{2.72}$$

where $\Gamma(\cdot)$ is the Gamma function [8]. This choice has two main justifications: First, for small ν_0 and ν_1, this density favors small values of σ^2. Second, this pdf is called a *conjugate prior* because the posterior distribution can be calculated in closed form. In practice, we may choose $\nu_0 \ll 1$ and $\nu_1 \ll 1$ (e.g., $\nu_0 = \nu_1 = 10^{-2}$); the precise selected values have little influence on the estimation results.

Using these priors and the likelihood, it is possible to write the parameters posterior. Moreover, we can express the two conditional posteriors $\mathsf{p}(\mathbf{k}, \boldsymbol{\alpha}, \sigma^2 | \gamma^2, \mathbf{x})$ and $\mathsf{p}(\gamma^2 | \mathbf{k}, \boldsymbol{\alpha}, \sigma^2, \mathbf{x})$. The former can be further decomposed into the product $\mathsf{p}(\mathbf{k} | \gamma^2, \mathbf{x}) \mathsf{p}(\sigma^2 | \mathbf{k}, \gamma^2, \mathbf{x}) \mathsf{p}(\boldsymbol{\alpha} | \mathbf{k}, \sigma^2, \gamma^2, \mathbf{x})$, with

$$\mathsf{p}(\mathbf{k} | \gamma^2, \mathbf{x}) = \frac{2}{\gamma^2 + 1} \left[\nu_0 + \mathbf{x}^\mathsf{T} \mathbf{P}(\mathbf{k}, \gamma^2) \mathbf{x}\right]^{-\frac{T+\nu_1}{2}}, \tag{2.73}$$

$$\mathsf{p}(\sigma^2 | \mathbf{k}, \gamma^2, \mathbf{x}) = \mathcal{IG}\left(\sigma^2; \frac{T+\nu_1}{2}, \nu_0 + \mathbf{x}^\mathsf{T} \mathbf{P}(\mathbf{k}, \gamma^2) \mathbf{x}\right), \tag{2.74}$$

$$\mathsf{p}(\boldsymbol{\alpha} | \mathbf{k}, \sigma^2, \gamma^2, \mathbf{x}) = \mathcal{N}(\boldsymbol{\alpha}; \mathbf{S}(\mathbf{k}, \gamma^2) \mathbf{D}(\mathbf{k})^\mathsf{T} \mathbf{x}, \sigma^2 \mathbf{S}(\mathbf{k}, \gamma^2)), \tag{2.75}$$

where $\mathbf{P}(\mathbf{k}, \gamma^2) = \mathbf{I}_T - \mathbf{D}(\mathbf{k}) \mathbf{S}(\mathbf{k}, \gamma^2) \mathbf{D}(\mathbf{k})^\mathsf{T}$ and $\mathbf{S}(\mathbf{k}, \gamma^2) = \frac{\gamma^2}{1+\gamma^2} [\mathbf{D}(\mathbf{k})^\mathsf{T} \mathbf{D}(\mathbf{k})]^{-1}$. The hyperparameter conditional posterior is the inverted gamma pdf

$$\mathsf{p}(\gamma^2 | \mathbf{k}, \boldsymbol{\alpha}, \sigma^2, \mathbf{x}) \;=\; \mathcal{IG}\left(\gamma^2; \nu_3 + 1, \frac{\boldsymbol{\alpha}^\mathsf{T} \mathbf{D}(\mathbf{k})^\mathsf{T} \mathbf{D}(\mathbf{k}) \boldsymbol{\alpha}}{2\sigma^2} + \nu_4\right), \tag{2.76}$$

where the parameters are chosen as $\nu_3 = 2$ and $\nu_4 = 20$ to ensure that the prior over γ has an infinite variance and a vague shape. Following [21], we can now write the MCMC algorithm that mixes a Gibbs sampler with one local and one global MH kernel.

Algorithm 2.5: An MCMC Algorithm for Sinusoids in Noise Estimation

1. Initialization:
   - Sample $\widetilde{\gamma}^{2(0)} \sim \mathcal{IG}(\gamma^2; \nu_3, \nu_4)$.
   - For $m = 1, \ldots, M$, sample $\widetilde{k}_m^{(0)} \sim q_g(k_m|\mathbf{x})$.
2. Iterations: For $i = 1, 2, \ldots, N$ do

   - For $m = 1, \ldots, M$ do
     - Frequency MH step
       · With probability β, sample a candidate using a local proposal $k_m^\star \sim q_l(k_m|\widetilde{k}_m^{(i-1)})$.
       · Otherwise, sample a candidate frequency using a global proposal $k_m^\star \sim q_g(k_m|\mathbf{x})$.
       · Compute $\alpha_{\text{MH}}(\boldsymbol{\theta}^\star, \widetilde{\boldsymbol{\theta}}^{(i-1)})$ using (2.65) with target pdf $p(k_m|\widetilde{\gamma}^{2(i-1)}, \mathbf{x})$ and with the proposal selected randomly above, then perform the accept/reject step (see Algorithm 2.4). This yields $\widetilde{k}_m^{(i)}$.
     - Noise variance and amplitude direct sampling
   - Sample the noise variance $\widetilde{\sigma}^{2(i)}$ using $p(\sigma^2|\widetilde{\mathbf{k}}^{(i)}, \widetilde{\gamma}^{2(i-1)}, \mathbf{x})$ given in (2.74).
   - Sample the amplitudes $\widetilde{\boldsymbol{\alpha}}^{(i)}$ using $p(\alpha|\widetilde{\mathbf{k}}^{(i)}, \widetilde{\sigma}^{2(i)}, \widetilde{\gamma}^{2(i-1)}, \mathbf{x})$ given in (2.75).
     - Hyperparameter sampling
   - Sample $\widetilde{\gamma}^{2(i)}$ using $p(\gamma^2|\widetilde{\mathbf{k}}^{(i)}, \widetilde{\boldsymbol{\alpha}}^{(i)}, \widetilde{\sigma}^{2(i)}, \mathbf{x})$ given in (2.76).

In Algorithm 2.5, the frequency parameter is sampled using a mixture of two MH kernels that rely on either a local proposal, chosen as $q_l(k_m|\widetilde{k}_m^{(i-1)}) = \mathcal{N}(k_m; \widetilde{k}_m^{(i-1)}, \sigma_{\text{RW}}^2)$, where σ_{RW}^2 is a user-defined random walk variance, or a global pdf chosen proportional to the frequency spectrum[15] of $\mathbf{x}$, i.e., $q_g(k_m|\mathbf{x}) \propto |X(k_m)|^2$. The mixture coefficient is selected as $\beta = 0.75$ (standard choice). After the frequency is sampled, the noise variance parameter and amplitudes are sampled directly from the conditional posterior pdfs. These three sampling operations produce a sample $(\widetilde{\mathbf{k}}^{(i)}, \widetilde{\boldsymbol{\alpha}}^{(i)}, \widetilde{\sigma}^{2(i)})$ at each iteration i. This is the first step of the overall Gibbs sampler; the second step consists of sampling the hyperparameter $\widetilde{\gamma}^{2(i)}$. Figure 2.10 displays an example chain built using this algorithm. This algorithm may be complex for estimating one or two sinusoids; however, its settings remain the same for large number of sinusoids, which makes it an attractive solution in such cases; see Chapter 7.

---

[15]It is easy to build a random variable generator from a stepwise approximation of this pdf (onto $k(j)$, $j = 1, \ldots, J$) by first sampling a continuous uniform random variable u on $[0, 1]$, then choosing j_0 such that $(1/C) \sum_{j=1}^{j_0-1} |X(k(j))|^2 \leq u < (1/C) \sum_{j=1}^{j_0} |X(k(j))|^2$, where $C = \sum_{j=1}^{J} |X(k)|^2$, and finally sampling $\mathbf{k}$ uniformly on $[k(j_0 - 1), k(j_0)]$.

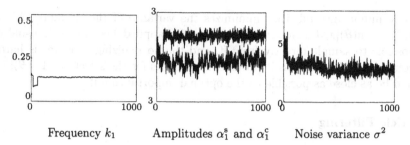

Frequency k_1 Amplitudes α_1^s and α_1^c Noise variance σ^2

Fig. 2.10. The frequency $\mathbf{k}$ and amplitudes $\boldsymbol{\alpha}$ vectors and the noise variance σ^2 sampled by Algorithm 2.5 for $m = 1$ and $M = 12$. As can be seen, convergence is reached after 200 iterations. The true parameter values are $m_1 = 0.12$, $\alpha_1^s = 1$, $\alpha_1^c = 0$, and $\sigma^2 = 2$.

Algorithm 2.5 can be extended to an unknown number of sinusoids in noise; see [21]. However, considering the number of sinusoids as a parameter to be estimated makes the implementation a bit more complicated, though still feasible. In particular, the parameter space dimension changes whenever the number of sinusoids changes, and a special algorithmic structure must be used, such as the *reversible jump MCMC* algorithm, which includes sinusoid *birth* and *death* moves [20].

2.3.6 Importance Sampling and Sequential Importance Sampling

Another interesting technique for the computation of Monte Carlo integrals is importance sampling. Let $\pi(\boldsymbol{\theta})$ be some pdf, for example a posterior pdf in the Bayesian setting. Let $\mathsf{q}(\boldsymbol{\theta})$ be another pdf (this is a proposal pdf, called here an *importance pdf*) such that $\mathsf{q}(\boldsymbol{\theta}) \neq 0$ whenever $\pi(\boldsymbol{\theta}) \neq 0$. Moreover, $\mathsf{q}(\boldsymbol{\theta})$ is selected so that it is possible to sample directly from it. Then (2.60) can be written

$$I[\mathsf{h}] = \int_{\Theta} \mathsf{h}(\boldsymbol{\theta})\pi(\boldsymbol{\theta})\,d\boldsymbol{\theta} = \int_{\Theta} \mathsf{h}(\boldsymbol{\theta})\frac{\pi(\boldsymbol{\theta})}{\mathsf{q}(\boldsymbol{\theta})}\mathsf{q}(\boldsymbol{\theta})\,d\boldsymbol{\theta}. \qquad (2.77)$$

Now, assume a set of Monte Carlo samples $\widetilde{\boldsymbol{\theta}}^{(i)}$, $i = 1, \ldots, N$ are generated using $\mathsf{q}(\boldsymbol{\theta})$. Then the following Monte Carlo estimate holds:

$$\widehat{I}_N[\mathsf{h}] = \sum_{i=1}^{N} \widetilde{\omega}^{(i)}\mathsf{h}\big(\widetilde{\boldsymbol{\theta}}^{(i)}\big) \approx I[\mathsf{h}], \quad \text{with } \widetilde{\omega}^{(i)} = \frac{\pi\big(\widetilde{\boldsymbol{\theta}}^{(i)}\big)}{\mathsf{q}\big(\widetilde{\boldsymbol{\theta}}^{(i)}\big)}, \qquad (2.78)$$

where $\widetilde{\omega}^{(i)}$ is the *importance weight* of sample $\widetilde{\boldsymbol{\theta}}^{(i)}$, and is aimed at correcting the discrepancy between $\mathsf{q}(\boldsymbol{\theta})$ and $\pi(\boldsymbol{\theta})$. A key remark is that the variance of the estimator (2.78) strongly depends on the importance pdf $\mathsf{q}(\boldsymbol{\theta})$ selected. It can be demonstrated that, for a given number of Monte Carlo samples, the optimal importance pdf that minimizes the variance of the estimate (2.78) is $q(\boldsymbol{\theta}) \propto |h(\boldsymbol{\theta})|\pi(\boldsymbol{\theta})$. However, it cannot be applied because it is usually impossible to sample directly from it, and the normalizing constant is hard to compute.[16] Nevertheless it provides some hints for the selection of $q(\boldsymbol{\theta})$: it should be as close as possible to the optimal importance pdf.

Particle Filtering

The main application of importance sampling is *particle filtering*. Assume a dynamical system with *hidden state parameter* $\boldsymbol{\theta}_n$ and *observation* $\mathbf{x}_n$ for $n = 1, 2, \ldots$ is governed by the following equations[17] (see also the graphical representation in Fig. 2.11):

$$\boldsymbol{\theta}_n = f_n[\boldsymbol{\theta}_{n-1}] + \mathbf{v}_n \quad \text{(transition equation)}, \quad (2.79)$$

$$\mathbf{x}_n = g_n[\boldsymbol{\theta}_n] + \boldsymbol{\epsilon}_n \quad \text{(observation equation)}. \quad (2.80)$$

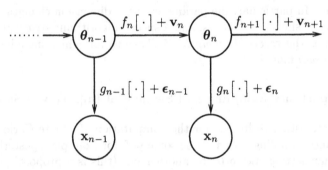

Fig. 2.11. Graphical illustration of the dynamic model in (2.79) and (2.80). In this model, only $\mathbf{x}_n$ is observed. The state parameter vector $\boldsymbol{\theta}_n$ is hidden for $n = 1, 2, \ldots$. $\boldsymbol{\theta}_n$ follows a Markov evolution, hence the name *hidden Markov model*.

Assuming the pdfs of $\mathbf{v}_n$ and $\boldsymbol{\epsilon}_n$ are known, we can express (2.79)–(2.80) as the state transition prior $p(\boldsymbol{\theta}_n|\boldsymbol{\theta}_{n-1})$ and the observation likelihood $p(\mathbf{x}_n|\boldsymbol{\theta}_n)$. Together with the initial state distribution $p_0(\boldsymbol{\theta}_0)$, this defines a sequential Bayesian model. Such dynamic systems are often met in practice: for example, $\boldsymbol{\theta}_n$ is a vector composed of the position, speed, and acceleration of a plane at time n, and $\mathbf{x}_n$ is the observation received by radar. The problem is that $\boldsymbol{\theta}_n$ cannot be observed directly; it has to be estimated.

---

[16]Whenever h is always positive, the normalizing constant is $I[h]$ itself!

[17]For consistency with most works in the field, and for notational simplicity, we use subscript time indices in this subsection, e.g., $\boldsymbol{\theta}_n$ is the state at time n, instead of the usual notation $\boldsymbol{\theta}(n)$.

The overall objective of filtering is the sequential estimation of the state trajectory for times between 0 and n, denoted $\boldsymbol{\theta}_{0:n} = \{\boldsymbol{\theta}_0, \boldsymbol{\theta}_1, \ldots, \boldsymbol{\theta}_n\}$ via the posterior $\mathsf{p}(\boldsymbol{\theta}_{0:n}|\mathbf{x}_{1:n})$ at each time n. In the special case where the functions $f_n[\,\cdot\,]$ and $g_n[\,\cdot\,]$ are linear, and the noises $\mathbf{v}_n$ and $\boldsymbol{\epsilon}_n$ are Gaussian, this can be solved analytically: the posterior is Gaussian, and its mean and covariance matrix are provided by the *Kalman filter* [18]. In the general case, the estimation of $\boldsymbol{\theta}_{0:n}$ relies on intractable integrals computation (such as (2.55) for MMSE estimates) at each time n. Particle filtering provides a Monte Carlo approximation via sequential importance sampling, where $\mathsf{p}(\boldsymbol{\theta}_{0:n-1}|\mathbf{x}_{1:n-1})$ plays the role of $\pi(\boldsymbol{\theta})$ in (2.77). In order to make the problem sequential, we first write the posterior $\mathsf{p}(\boldsymbol{\theta}_{0:n}|\mathbf{x}_{1:n})$ at time n as a function of $\mathsf{p}(\boldsymbol{\theta}_{0:n-1}|\mathbf{x}_{1:n-1})$ at time $n-1$:

$$\mathsf{p}(\boldsymbol{\theta}_{0:n}|\mathbf{x}_{1:n}) \;=\; \mathsf{p}(\boldsymbol{\theta}_{0:n-1}|\mathbf{x}_{1:n-1}) \frac{\mathsf{p}(\boldsymbol{\theta}_n|\boldsymbol{\theta}_{n-1})\mathsf{p}(\mathbf{x}_n|\boldsymbol{\theta}_n)}{\mathsf{p}(\mathbf{x}_n|\mathbf{x}_{1:n-1})}, \qquad (2.81)$$

where $\mathsf{p}(\mathbf{x}_n|\mathbf{x}_{1:n-1})$ is a normalization term which need not be computed. Second, we select an importance pdf that can be written sequentially, namely

$$\mathsf{q}_n(\boldsymbol{\theta}_{0:n}) \;=\; \mathsf{q}_{n-1}(\boldsymbol{\theta}_{0:n-1})\mathsf{q}_n(\boldsymbol{\theta}_n|\boldsymbol{\theta}_n - 1) \;=\; \mathsf{q}_0(\boldsymbol{\theta}_0)\prod_{l=1}^{n}\mathsf{q}_l(\boldsymbol{\theta}_l|\boldsymbol{\theta}_{l-1}). \quad (2.82)$$

It is now possible to sample the state at time n using $\mathsf{q}_n(\boldsymbol{\theta}_n|\boldsymbol{\theta}_n - 1)$ and compute sequentially the weight as the ratio $\mathsf{p}(\boldsymbol{\theta}_{0:n}|\mathbf{x}_{1:n})/\mathsf{q}_n(\boldsymbol{\theta}_{0:n})$. These elements lead to the particle filter presented in Algorithm 2.6 below.

Algorithm 2.6: Particle Filtering Algorithm

1. Initialization: For each particle $i = 1, \ldots, N$, sample independently $\widetilde{\boldsymbol{\theta}}_0^{(i)} \sim \mathsf{q}_0(\boldsymbol{\theta}_0)$ and compute the initial weight $\widetilde{\omega}_0^{(i)} = \mathsf{p}_0(\widetilde{\boldsymbol{\theta}}_0^{(i)})/\mathsf{q}_0(\widetilde{\boldsymbol{\theta}}_0^{(i)})$.
2. Iterations: For $n = 1, 2, \ldots, N$ do

   - The particle trajectories are updated using the sequential importance pdf
     - For $i = 1, \ldots, N$, sample the new state at time n for particle i,

     $$\widetilde{\boldsymbol{\theta}}_n^{(i)} \sim \mathsf{q}_n(\boldsymbol{\theta}_n|\widetilde{\boldsymbol{\theta}}_{n-1}^{(i)}). \qquad (2.83)$$

   - The weights are computed and normalized
     - For $i = 1, \ldots, N$, compute the sequential importance weight $\widetilde{\omega}_n^{(i)}$ as follows:

     $$\widetilde{\omega}_n^{(i)} \;\propto\; \widetilde{\omega}_{n-1}^{(i)} \frac{\mathsf{p}(\widetilde{\boldsymbol{\theta}}_n^{(i)}|\widetilde{\boldsymbol{\theta}}_{n-1}^{(i)})\mathsf{p}(\mathbf{x}_n|\boldsymbol{\theta}_n)}{\mathsf{q}_n(\widetilde{\boldsymbol{\theta}}_n^{(i)}|\widetilde{\boldsymbol{\theta}}_{n-1}^{(i)})}. \qquad (2.84)$$

     - Compute the weight normalization constant $W_n = \sum_{i=1}^{N}\widetilde{\omega}_n^{(i)}$ (this normalizes the weights, and avoids computing the term $\mathsf{p}(\mathbf{x}_n|\mathbf{x}_{1:n-1})$ in (2.77)).
     - For particles $i = 1, \ldots, N$, set $\widetilde{\omega}_n^{(i)} \leftarrow \widetilde{\omega}_n^{(i)}/W_n$.
   - The current state in estimated
     - Estimate the state from the weighted particles, e.g., $\widehat{\boldsymbol{\theta}}_{n\,\mathrm{MMSE}} = \sum_{i=1}^{N}\widetilde{\omega}_n^{(i)}\widetilde{\boldsymbol{\theta}}_n^{(i)}$.

- The particles are resampled
  - Compute the efficiency number $N_{\text{eff}} = \left[\sum_{i=1}^{N} (\widetilde{\omega}_n^{(i)})^2 \right]^{-1}$.
  - If $N_{\text{eff}} \leq N_{\text{Threshold}}$, then resample the particles: duplicate particles with large weights, suppress particles with small weights, and set $\widetilde{\omega}_n^{(i)} \leftarrow 1/N$ for $i = 1, \ldots, N$.

At each time n, Algorithm 2.6 produces Monte Carlo samples $\widetilde{\boldsymbol{\theta}}_{0:n}^{(i)}$, $i = 1, \ldots, N$ (called *particles* in this context) with weights $\widetilde{\omega}_n^{(i)}$ which approximate the posterior $\mathsf{p}(\boldsymbol{\theta}_{0:n}|\mathbf{x}_{1:n})$. However, this simple strategy produces samples which degenerate: after a few iterations, the weights of most particles are close to zero, whereas the weights of a few particles are significantly non-zero. This can be monitored by N_{eff}: if it becomes too small, then degeneracy has occurred and particles are resampled; particles with low weights are moved to more accurate points (i.e., to the location of particles with large weights). This can be implemented by first setting randomly the number of copies of each particles using, e.g., stratified sampling [344], where the expected number of copies of each particles equals its weight times N. Particles with 0 copies are suppressed.

Different importance pdfs lead to different algorithms. A simple choice consists of using the transition pdf $\mathsf{q}_n(\boldsymbol{\theta}_n|\boldsymbol{\theta}_n - 1) = \mathsf{p}(\boldsymbol{\theta}_n|\boldsymbol{\theta}_n - 1)$. The corresponding algorithm is simple but the variance of estimators based on the particles generated is large in general. The importance pdf that minimizes the variance of the weights (and thus, the variance of estimators) uses the new observation. It is written $\mathsf{q}_n(\boldsymbol{\theta}_n|\boldsymbol{\theta}_n - 1) = \mathsf{p}(\boldsymbol{\theta}_n|\mathbf{x}_n, \boldsymbol{\theta}_{n-1})$, but it is often impossible to sample from it. However, it can be approximated locally by Gaussians: $\mathsf{q}_n(\boldsymbol{\theta}_n|\boldsymbol{\theta}_n - 1) = \mathcal{N}(\boldsymbol{\theta}_n; \boldsymbol{\theta}_{n|n}, \boldsymbol{\Sigma}_{n|n})$ where $\boldsymbol{\theta}_{n|n}$ and $\boldsymbol{\Sigma}_{n|n}$ are the state and state covariance estimates given by the extended Kalman filter or by the unscented Kalman filter.[18]

2.4 Pattern Recognition Methods

Pattern recognition algorithms are mainly concerned with data classification.[19] There are generally two frameworks considered. In the *supervised classification* framework, a set of labelled training data is provided. A set of pairs

---

[18]The Kalman filter is devoted to linear and Gaussian dynamic models. When the model is non-linear and/or non-Gaussian, the posterior is non-Gaussian but it is still possible to approximate it by a Gaussian. This can be implemented by the extended Kalman filter, which operates on linearized state equations (2.79)–(2.80) with Gaussian noises, or by the unscented Kalman filter, which approximates the posterior as a Gaussian, without linearizing (2.79)–(2.80).

[19]Detection is also a pattern recognition problem. However, it can be considered, to some extent, as a subcase of classification. For the sake of simplicity, here we mainly cover classification. Some elements about detection theory will also be provided when appropriate.

$(X, Y) = \{(x_1, y_1), \ldots, (x_m, y_m)\}$ is given, where $x_i \in \mathcal{X}$ and $y_i \in \mathcal{Y}$ for $i = 1, \ldots, m$. Example data and label spaces are $\mathcal{X} = \mathbb{R}^{d_x}$ and $\mathcal{Y} = \{1, 2, \ldots, \}$. Labels indicate to which class each data belongs. The aim of supervised classification is to predict the class label $y \in \mathcal{Y}$ of a new datum $x \in \mathcal{X}$ which is not in (X, Y). This prediction must be elaborated from the sole knowledge of x and of the training set (X, Y). In the *unsupervised classification* framework (which is often called *clustering*), a set of unlabelled data $X = \{x_1, \ldots, x_m\}$ is provided, and each datum in X must be assigned a label from $\mathcal{Y}$ (i.e, it must be assigned to a class).

In this section, we review classification techniques. One unsupervised classification method has already been presented in Section 2.2.8, p. 35. Indeed, Gaussian mixtures models assign a latent variable z_i to each datum x_i, which indicates to which Gaussian x_i is related, that is, to which Gaussian class it is related. The number of classes J need not be known, because a penalized likelihood approach [91] or a Bayesian approach [541] could be implemented to estimate both the labels and the number J of mixture components (that is, the number of classes) found. Another unsupervised classification technique is presented in the next subsection, while the remaining subsections are devoted to supervised classification.

2.4.1 K-Means

The K-means algorithm is an unsupervised classification method. The set of data $X = \{x_1, \ldots, x_m\}$ in $\mathcal{X}$ is provided. Given the number of classes, denoted K, the algorithm assigns a label in $\mathcal{Y} = \{1, \ldots, K\}$ to each x_i. The algorithm is summarized below, where $d(\cdot, \cdot)$ is a distance measure in $\mathcal{X}$.

Algorithm 2.7: K-Means Algorithm

1. Initialization: Set the K initial means μ_j in $\mathcal{X}$, $j = 1, \ldots, K$. For example, an easy and robust choice consists of choosing any K (not equal) data in X. This ensures that no cluster is empty in the beginning.
2. Iterations: While the means keep evolving, do
   - For each datum x_i, $i = 1, \ldots, m$,
     - Compute its distance to each means $d(x_i, \mu_j)$, for $j = 1, \ldots, K$.
     - Assign to x_i $(i = 1, \ldots, m)$ the label $y_i = \underset{j=1,\ldots,K}{\operatorname{argmin}} \, d(x_i, \mu_j)$.
   - Update the means by setting

$$\mu_j \leftarrow \operatorname{mean}\{x_i | y_i = j\}, \quad j = 1, \ldots, K. \tag{2.85}$$

This algorithm converges to useful solutions insofar as the data in X form separated classes. Of course, the classification results are closely related to the distance measure used in Algorithm 2.7. Typical distance measures are the Euclidean distance $d(x_1, x_2) = (x_1 - x_2)^{\mathsf{T}}(x_1 - x_2) = \|x_1 - x_2\|_{\mathcal{X}}^2$, the L_1 distance $d(x_1, x_2) = \|x_1 - x_2\|_{\mathcal{X}}$ or the Mahalanobis distance $d(x_1, x_2) = (x_1 - x_2)^{\mathsf{T}} \Sigma^{-1}(x_1 - x_2)$, where Σ is the empirical covariance matrix of the set X, computed as in (2.86) below.

2.4.2 Principal Component Analysis

In front of an unlabelled dataset X in the space $\mathcal{X}$, one may want to reduce
the data dimension. This may be because this dimension is too large to enable the use of a classification technique (e.g., Gaussian mixture model). It
may also be to enable 2D data visualization of larger dimensional data. Principal component analysis (PCA) is aimed at reducing the dimension of data
without losing too much information. The amount of information carried by
a dimension is summarized in the *empirical covariance matrix*

$$\Sigma_X^{\text{emp}} = \frac{1}{m}\sum_{i=1}^{m}(\mathbf{x}_i-\boldsymbol{\mu}_X^{\text{emp}})(\mathbf{x}_i-\boldsymbol{\mu}_X^{\text{emp}})^{\mathsf{T}} \quad \text{with } \boldsymbol{\mu}_X^{\text{emp}} = \frac{1}{m}\sum_{i=1}^{m}\mathbf{x}_i, \quad (2.86)$$

where $\boldsymbol{\mu}_X^{\text{emp}}$ is the empirical mean of X.[20] In addition, Σ_X^{emp} also informs us
about the redundancy between the dimensions. In order to reduce the number
of dimensions of the data, PCA provides us with the most informative *linear
combinations* of the dimensions of the data [161]. The steps of PCA are

1. Compute the empirical covariance matrix Σ_X^{emp} as in (2.86);
2. Apply matrix diagonilization to Σ_X^{emp}, i.e., compute

$$\Sigma_X^{\text{emp}} = \Xi\Lambda\Xi^{-1}, \quad (2.87)$$

   where Λ is a diagonal matrix whose diagonal terms are called the *eigenvalues* of Σ_X^{emp}. The eigenvalues are positive, and they are assumed sorted
   in decreasing order along the diagonal. The columns of Ξ are called the
   eigenvectors[21] of Σ_X^{emp}, and the matrix Ξ is orthogonal, meaning that
   $\Xi^{-1} = \Xi^{\mathsf{T}}$.
3. Keep the d_{pca} columns of Ξ that correspond to the d_{pca} largest eigenvalues
   in Λ. They are stored in a matrix Ξ_{pca} with size $d_{\mathcal{X}} \times d_{\text{pca}}$ and the
   corresponding eigenvalues are in the squared matrix Λ_{pca} with size d_{pca}.
4. Compute the lower dimensional data for $i = 1,\ldots,m$:

$$\zeta_i = \Xi_{\text{pca}}(\mathbf{x}_i - \boldsymbol{\mu}_X^{\text{emp}}). \quad (2.88)$$

The matrix diagonalization in (2.87) is almost always possible[22] whenever
m is larger than $d_{\mathcal{X}}$. The transform in (2.88) reduces the dimension of $\mathbf{x}_i$

---

[20]The empirical covariance matrix in (2.86) is computed with normalization term
$1/m$. The corresponding empirical covariance matrix is known to be a biased estimate of the true covariance matrix. The unbiased empirical covariance matrix should
be computed with normalization term $1/(m-1)$.

[21]Details about eigenvalues, eigenvectors, and matrix diagonalization may be
found in any textbook about matrices; see, e.g., [295].

[22]A property is said to be *almost true* (here, the property is 'diagonalizing the
covariance matrix is possible') if it is true with probability one. Surprisingly, even
though it is true with probability one, it is still possible that the property is false,
but this is quite unlikely; see [548] for further details. Here, the property would be
false if at least one datum was a linear combination of others.

$(i = 1, \ldots, m)$, i.e., the dimension of ζ_i equals d_{pca}. The variables ζ_i have covariance matrix Λ_{pca}.

The PCA approach has two major advantages. First, it reduces the dimension of the data to the most informative dimensions. Second, it removes the correlations between the initial data dimensions, which removes redundancy.

2.4.3 Supervised Classification: Cost Function and Risk

Supervised classification is concerned with the design of a *classification function* $\mathsf{F}_{(\mathsf{X},\mathsf{Y})} \colon \mathcal{X} \to \mathcal{Y}$ which associates a class label $\mathsf{y} \in \mathcal{Y}$ to each datum $\mathbf{x} \in \mathcal{X}$, and which is designed from the training set (X, Y). In this section, we restrict to two-class classification and set $\mathcal{Y} = \{-1; 1\}$. Multi-class classification algorithms can be derived from two-class algorithms; see [568].

Loosely speaking, the function $\mathsf{F}_{(\mathsf{X},\mathsf{Y})}$ is relevant only if it has the smallest possible risk of performing classification errors. Let F be any classification function. Errors occur whenever $\mathsf{F}(\mathbf{x}) = \mathsf{y}'$, whereas the true class label of $\mathbf{x}$ is y, with $\mathsf{y}' \neq \mathsf{y}$. In practice, the seriousness of errors needs be measured in order to design $\mathsf{F}_{(\mathsf{X},\mathsf{Y})}$; this is the role of loss functions. Let $\mathsf{c}\,(\mathbf{x}, \mathsf{y}; \mathsf{F}(\mathbf{x}))$ be such a loss function for a classification function F. A standard example is the 0-1 loss $\mathsf{c}\,(\mathbf{x}, \mathsf{y}; \mathsf{F}(\mathbf{x})) = 1 - \delta_{\mathsf{y}}(\mathsf{F}(\mathbf{x}))$. In the two-class classification case, it is helpful to assume that F delivers values over $\mathbb{R}$ (and not just $\mathcal{Y}$), and set $\mathsf{y} = \mathrm{sign}\,(\mathsf{F}(\mathbf{x}))$. This enables the use of, e.g., the quadratic loss, the hinge loss, etc. (see Fig. 2.12).

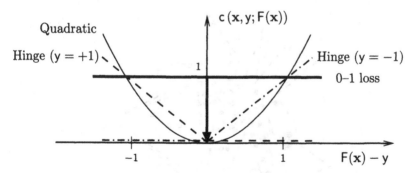

Fig. 2.12. Standard loss functions to be used in learning problems. The quadratic loss $\mathsf{c}\,(\mathbf{x}, \mathsf{y}; \mathsf{F}(\mathbf{x})) = (\mathsf{y} - \mathsf{F}(\mathbf{x}))^2$ is used in so-called *least square methods*, whereas the hinge losses are used in support vector machines. The 0-1 loss $\mathsf{c}\,(\mathbf{x}, \mathsf{y}; \mathsf{F}(\mathbf{x})) = 1 - \delta_{\mathsf{y}}(\mathsf{F}(\mathbf{x}))$ is often used in Bayesian classification problems.

From loss functions, we can define the *risk* $\mathsf{R}\,[\mathsf{F}]$ of a classification function F as the expected loss over all possible pairs $(\mathbf{x}, \mathsf{y})$, namely

$$\mathsf{R}\,[\mathsf{F}] = \mathbb{E}_{\mathsf{p}(\mathbf{x},\mathsf{y})}\big[\mathsf{c}\,(\mathbf{x}, \mathsf{y}; \mathsf{F}(\mathbf{x}))\big], \tag{2.89}$$

where $p(\mathbf{x}, y)$ is the 'pdf' of the data and labels. Of course, this 'pdf' is unknown in practice, and the risk cannot be computed directly. Assume however that $R\,[F]$ of any classification function F could be computed: the optimal classification function could be found by minimizing the risk w.r.t. F over a set of functions denoted $\mathcal{F}$. This approach is not really possible, but the risk can still be estimated from the training set. Define the *empirical risk*

$$R^{\mathrm{emp}}_{(\mathsf{X},\mathsf{Y})}\,[F] \;=\; \frac{1}{m}\sum_{i=1}^{m} c\,(\mathbf{x}_i, y_i; F(\mathbf{x}_i)) \qquad (2.90)$$

as an estimate of the true risk $R\,[F]$, based on the information provided by the training set. A well-known problem is that $R^{\mathrm{emp}}_{(\mathsf{X},\mathsf{Y})}\,[F]$ is a poor estimator of $R\,[F]$, and there usually exist many functions F such that $R^{\mathrm{emp}}_{(\mathsf{X},\mathsf{Y})}\,[F] = 0$, including trivial ones. In Fig. 2.13, a classification function which achieves a zero empirical risk is depicted. It fits the training data perfectly, but its shape is so complicated that it is likely to make many errors on data which are not in the training set. In other words, this function fails at generalizing its correct classification ability to new data. Such a solution is said to *overfit* the training set, and over-fitting is often related to generalization issues. As explained in the next section, it is necessary to prefer solutions which are simple in some sense, so as to limit over-fitting and favour good generalization.

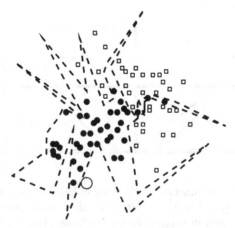

Fig. 2.13. Supervised classification into two classes with 2-dimensional data. In the training set (X, Y), data with label $y = -1$ are represented with dots, whereas data with label $y = 1$ are represented with squares. The dotted line is a classification function F such that $R^{\mathrm{emp}}_{(\mathsf{X},\mathsf{Y})}\,[F] = 0$. Though it achieves zero empirical risk, F is not a good classification function, as it makes an error for a new datum which is not in the training set (circle at the bottom, with the true label $y = -1$).

2.4.4 Regularization

Minimizing the empirical risk leads to useless functions F. This problem can be overcome by minimizing an objective function that includes both the empirical risk $R^{emp}_{(X,Y)}$ [F] and a term Ω (F) that penalizes unwanted solutions. Define the *regularized risk*

$$R^{reg} [F] = R^{emp}_{(X,Y)} [F] + \lambda \Omega (F), \tag{2.91}$$

where λ is the regularization tuning parameter. The optimal classification function $F_{(X,Y)}$ is found by solving

$$F_{(X,Y)} = \underset{F \in \mathcal{F}}{\text{argmin }} R^{reg} [F]. \tag{2.92}$$

Given a training set, various supervised classification methods employ different loss functions, penalty terms and sets of functions $\mathcal{F}$ so as to design $F_{(X,Y)}$. Examples are penalized likelihood approaches (see Section 2.2.9, p. 38) and Bayesian approaches (see Section 2.3, p. 39). The following subsection presents another one: the support vector machine.

2.4.5 Support Vector Machines

Support vector machines (SVMs) are specific instances of the above regularization scheme. In SVMs,

$$\mathcal{F} = \{F(\cdot) = f(\cdot) + b \text{ such that } f \in \mathcal{H} \text{ and } b \in \mathbb{R}\}, \tag{2.93}$$

where $\mathcal{H}$ is a space of functions called a *reproducing kernel Hilbert space* with kernel $k(\cdot, \cdot)$.[23] In practice, the dot product defined over $\mathcal{H}$, denoted $\langle \cdot, \cdot \rangle_{\mathcal{H}}$, can generally not be written in an explicit form—all we need to know is that it has the properties of a dot product, and that it enables the reproducing kernel property for some kernel. In (2.93), the set $\mathcal{F}$ of possible classification functions is specified by selecting the kernel $k(\cdot, \cdot)$. Provided the kernel is positive definite,[24] it gives rise to a RKHS. A common choice is the Gaussian kernel (for any $\mathbf{x}$ and $\mathbf{x}'$ in $\mathcal{X}$)

---

[23] A Hilbert Space $\mathcal{H}$ is a nonempty space such that 1) $\mathcal{H}$ is a vector (linear) space, that is, any linear combinations of elements of $\mathcal{H}$ are also elements of $\mathcal{H}$; and 2) A dot product $\langle \cdot, \cdot \rangle_{\mathcal{H}}$ is defined over $\mathcal{H}$. This dot product induces a norm for any $f \in \mathcal{H}$ such that $\|f(\cdot)\|^2_{\mathcal{H}} = \langle f(\cdot), f(\cdot) \rangle_{\mathcal{H}}$ 3) $\mathcal{H}$ is complete for this norm. Moreover, assuming $\mathcal{H}$ is a Hilbert space of functions $f(\cdot) : \mathcal{X} \to \mathbb{R}$, it is a reproducing kernel Hilbert space (RKHS) if and only if there exists a kernel function $k(\cdot, \cdot) : \mathcal{X} \times \mathcal{X} \to \mathbb{R}$ with the following reproducing property: For any $\mathbf{x} \in \mathcal{X}$, $k(\mathbf{x}, \cdot)$ is in $\mathcal{H}$ and for any function $f \in \mathcal{H}$, we have $\langle k(\mathbf{x}, \cdot), f(\cdot) \rangle_{\mathcal{H}} = f(\mathbf{x})$.

[24] A kernel $k(\cdot, \cdot)$ is positive definite if and only if for any set $\{\mathbf{x}_1, \ldots, \mathbf{x}_m\}$ and any $m > 0$, the matrix with entries $k(\mathbf{x}_i, \mathbf{x}_j)$, $(i, j) = 1, \ldots, m$ is a positive definite matrix.

$$k(\mathbf{x}, \mathbf{x}') = \exp\left\{-\frac{1}{2\sigma^2}\|\mathbf{x} - \mathbf{x}'\|^2\right\}. \tag{2.94}$$

In SVMs, the hinge loss $c_{\text{hinge}}(\mathbf{x}, y; F(\mathbf{x}))$ (see Fig. 2.12) is chosen and the penalty term $\Omega(F)$ is the squared norm induced by the dot product in $\mathcal{H}$, i.e., $\Omega(F) = \|f\|_{\mathcal{H}}^2$. For any $\mathbf{x}$, from the reproducing property, $F(\mathbf{x}) = \langle k(\mathbf{x}, \cdot), f(\cdot)\rangle_{\mathcal{H}} + b$, which means that classifying a datum $\mathbf{x}$ is an affine operation (that is, a linear + constant operation) in terms of elements of $\mathcal{H}$, because the dot product $\langle f(\cdot), f'(\cdot)\rangle_{\mathcal{H}}$ is linear w.r.t. both $f(\cdot)$ and $f'(\cdot)$, and a non-linear operation in terms of elements of $\mathcal{X}$. In SVMs, the regularized risk (2.92) becomes

$$R^{\text{reg}}[F] = \frac{1}{m}\sum_{i=1}^{m} c_{\text{hinge}}\left(\mathbf{x}_i, y_i; \langle k(\mathbf{x}_i, \cdot), f(\cdot)\rangle_{\mathcal{H}} + b\right) + \lambda\|f\|_{\mathcal{H}}^2. \tag{2.95}$$

It is possible to modify this regularized risk slightly so as to replace the regularization tradeoff parameter λ by another parameter $\nu \in [0, 1]$ that can be interpreted more easily. Minimizing this modified regularized risk can be equivalently written as the *soft margin SVM* optimization: Minimize

$$\frac{1}{2}\|f\|_{\mathcal{H}}^2 - \nu\rho + \frac{1}{m}\sum_{i=1}^{m}\xi_i \quad \text{with respect to } f, \rho, \xi_i, b, \tag{2.96a}$$

with

$$y_i\left(\langle k(\mathbf{x}_i, \cdot), f(\cdot)\rangle_{\mathcal{H}} + b\right) \geq \rho - \xi_i \quad \text{for all } i = 1, \ldots, m, \tag{2.96b}$$

and

$$\xi_i \geq 0, \quad \text{for all } i = 1, \ldots, m, \ \rho \geq 0, \tag{2.96c}$$

where the *slack variables* ξ_i $(i = 1, \ldots, m)$ are used to implement the hinge loss [568]: they are non-zero only if $y_i\left(\langle k(\mathbf{x}_i, \cdot), f(\cdot)\rangle_{\mathcal{H}} + b\right) < \rho$ and, in this case, they induce a linear cost in the objective function $\frac{1}{2}\|f\|_{\mathcal{H}}^2 - \nu\rho + \frac{1}{m}\sum_{i=1}^{m}\xi_i$. Introducing Lagrange multipliers $\alpha_1, \ldots, \alpha_m$, the optimization in (2.96) can be turned into the equivalent quadratic convex problem with linear constraints (dual problem): Maximize

$$-\frac{1}{2}\sum_{i=1}^{m}\sum_{j=1}^{m}\alpha_i\alpha_j y_i y_j k(\mathbf{x}_i, \mathbf{x}_j) \quad \text{with respect to } \alpha_i \ (i = 1, \ldots, m), \tag{2.97a}$$

with

$$0 \leq \alpha_i \leq 1/m \quad (i = 1, \ldots, m), \quad \sum_{i=1}^{m}\alpha_i y_i = 0 \quad \text{and} \quad \sum_{i=1}^{m}\alpha_i \geq \nu. \tag{2.97b}$$

The set of Lagrange multipliers α_i, $i = 1, \ldots, m$ that solve (2.97) leads to the optimal classification function $F_{(X,Y)}$, which is written for all $\mathbf{x} \in \mathcal{X}$

$$F_{(X,Y)}(\mathbf{x}) = \sum_{i=1}^{m} y_i \alpha_i k(\mathbf{x}_i, \mathbf{x}) + b. \tag{2.98}$$

The classification of a new datum $\mathbf{x}$ into one of the two classes $\{-1; +1\}$ is performed by assigning to $\mathbf{x}$ the class label $y = \text{sign}(F_{(X,Y)}(\mathbf{x}))$ where $F_{(X,Y)}$ is given in (2.98). The support vector machine admits a gemoetrical interpretation; see Fig. 2.14.

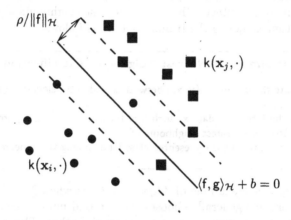

Fig. 2.14. Geometrical interpretation of the ν-soft margin support vector machine in the reproducing kernel Hilbert space $\mathcal{H}$. Each element of $\mathcal{H}$ is a function which is plotted here as a point in an infinite-dimensional space (here, $\mathcal{H}$ is represented as a two-dimensional space). The classification function f is the vector othogonal to the hyperplane with equation $\langle f, g \rangle_{\mathcal{H}} + b = 0$. This hyperplane separates the data from the two classes with maximum margin, where the dots represent the functions $k(\mathbf{x}_i, \cdot)$ with label $y_i = 1$, and the squares represent the functions $k(\mathbf{x}_j, \cdot)$ with label $y_j = -1$. The margin width is $\rho/\|f\|_{\mathcal{H}}$. As this is the soft margin SVM, some training vectors are allowed to be located inside the margin. The vectors located on the margin hyperplanes (dotted lines) and inside the margin are called the support vectors.

An important property of this SVM is its *sparseness*: ν is an upper bound on the fraction of non-zero Lagrange multipliers. In other words, assume for example $\nu = 0.2$; then at least 80% of the Lagrange multipliers $\{\alpha_i, i = 1, \ldots, m\}$ equal zero, and computing $F_{(X,Y)}(\mathbf{x})$ requires us to compute $k(\mathbf{x}_i, \mathbf{x})$ for only a small fraction (at most 20%) of the training data $\mathbf{x}_i$, $i = 1, \ldots, m$. Training data with non-zero Lagrange multipliers are called *support vectors*. The support vectors such that $0 < \alpha_i < 1/m$ are called the *margin support vectors*, and we denote them by $\mathbf{x}_{i^*}$. The set of margin support vectors is denoted X^* and it includes m^* vectors. In (2.98), b is computed as the average over all the support vectors:

$$b = -\frac{1}{m^*} \sum_{x_i^* \in X^*} \sum_{j=1}^{m} \alpha_j y_j k(x_j, x_{i^*}). \qquad (2.99)$$

2.4.6 Nearest Neighbours Classification

Nearest neighbours classification is a very simple supervised classification technique. Similar to the K-means algorithm, nearest neighbours classification requires a distance measure $d(\cdot, \cdot)$ to be defined over the data space $\mathcal{X}$. Let k be some non-zero integer; then the k-nearest neighbours algorithm classifies a new datum x (not in the training set X) as follows:

Algorithm 2.8: k-Nearest Neighbours Classification Algorithm

1. Compute the distance $d(x, x_i)$ between x and every element in the training set x_i, $i = 1, \ldots, m$.
2. Select the k training data x_i such that the distances $d(x, x_i)$ are the smallest (they are called the k-nearest neighbours of x).
3. Assign to x the most represented class label among the k-nearest neighbours.

An important example of Algorithm 2.8 is when $\mathcal{Y} = \{-1; 1\}$ (two-class case), where one generally chooses k as an odd number so as to ensure that one class label is more represented than the other. This case is illustrated in Fig. 2.15 with $k = 5$. Further insight into this method can be found in [161].

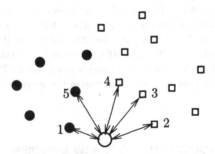

Fig. 2.15. k-nearest neighbours classification of two-dimensional data in the two-class case, with $k = 5$. The new datum x is represented by a non-filled circle. Elements of the training set (X, Y) are represented with dots (those with label -1) and squares (those with label $+1$). The arrow lengths represent the Euclidean distance between x and its 5 nearest neighbours. Three of them are squares, which makes x have the label $y = +1$.

2.4.7 Linear Discriminant Analysis

Linear Discriminant Analysis (LDA) is a supervised classification approach, also called *Fisher linear discriminant*. In its simplest form, it is devoted to the two-class case where $\mathcal{Y} = \{-1; 1\}$. This classification technique has some connections with principal component analysis (see Section 2.4.2, p. 54) in the sense that it reduces the dimension of the data. In two-class linear discriminant analysis, the dimensionality of the data is reduced to one: the direction of the space that best separates the classes. More formally, we want to transform the dataset $(\mathsf{X}, \mathsf{Y}) = \{(\mathbf{x}_1, y_1), \ldots, (\mathbf{x}_m, y_m)\}$ into a one-dimensional dataset by setting

$$\zeta_i = \mathbf{w}^\mathsf{T} \mathbf{x}_i, \qquad i = 1, \ldots, m. \tag{2.100}$$

This consists of projecting the data onto a line with direction given by $\mathbf{w}$. In order for $\mathbf{w}$ to be effectively the best projection line for class separation, we want to make sure that it maximizes the Fisher ratio (also known as the *Rayleigh quotient*)

$$J(\mathbf{w}) = \frac{|\mathbf{w}^\mathsf{T} \boldsymbol{\mu}_{\mathsf{X}|y=+1}^{\mathrm{emp}} - \mathbf{w}^\mathsf{T} \boldsymbol{\mu}_{\mathsf{X}|y=-1}^{\mathrm{emp}}|^2}{\mathbf{w}^\mathsf{T} \big(\boldsymbol{\Sigma}_{\mathsf{X}|y=+1}^{\mathrm{emp}} + \boldsymbol{\Sigma}_{\mathsf{X}|y=-1}^{\mathrm{emp}} \big) \mathbf{w}}, \tag{2.101}$$

where $\boldsymbol{\Sigma}_{\mathsf{X}|y=+1}^{\mathrm{emp}}$ (respectively $\boldsymbol{\Sigma}_{\mathsf{X}|y=-1}^{\mathrm{emp}}$) is defined as in (2.86), by only considering the training data with label $y = +1$ (respectively $y = -1$), and similarly for $\boldsymbol{\mu}_{\mathsf{X}|y=+1}^{\mathrm{emp}}$ and $\boldsymbol{\mu}_{\mathsf{X}|y=-1}^{\mathrm{emp}}$. The Fisher ratio is computed as the squared distance between the means of the projected data ζ_i $(i = 1, \ldots, m)$ divided by their variances, and can be understood as a scaled distance between the training data classes. The vector that minimizes $J(\mathbf{w})$ can be computed analytically as

$$\mathbf{w}_{\mathrm{LDA}} = \big(\boldsymbol{\Sigma}_{\mathsf{X}|y=+1}^{\mathrm{emp}} + \boldsymbol{\Sigma}_{\mathsf{X}|y=-1}^{\mathrm{emp}} \big)^{-1} \big(\boldsymbol{\mu}_{\mathsf{X}|y=+1}^{\mathrm{emp}} - \boldsymbol{\mu}_{\mathsf{X}|y=-1}^{\mathrm{emp}} \big). \tag{2.102}$$

In practice, two-class LDA classification is implemented as in Algorithm 2.9 below. See [161] for additional details. In multi-class classification, LDA can be also used to reduce data to $C - 1$ dimensions, where C is the number of classes.

Algorithm 2.9: Linear Discriminant Analysis

1. Compute the empirical means $\boldsymbol{\mu}_{\mathsf{X}|y=+1}^{\mathrm{emp}}$ and $\boldsymbol{\mu}_{\mathsf{X}|y=-1}^{\mathrm{emp}}$ from the training data, as in (2.86).
2. Compute the empirical covariance matrics $\boldsymbol{\Sigma}_{\mathsf{X}|y=+1}^{\mathrm{emp}}$ and $\boldsymbol{\Sigma}_{\mathsf{X}|y=-1}^{\mathrm{emp}}$ from the training data, as in (2.86).
3. Compute the optimal projection direction $\mathbf{w}_{\mathrm{LDA}}$ according to (2.102).
4. Compute the projected data from the training set by $\zeta_i = \mathbf{w}_{\mathrm{LDA}}^\mathsf{T} \mathbf{x}_i$, $i = 1, \ldots, m$
5. Implement any one-dimensional classification technique in order to classify the projection $\zeta = \mathbf{w}_{\mathrm{LDA}}^\mathsf{T} \mathbf{x}$ of a new datum $\mathbf{x}$ by using the mapped training set $\{(\zeta_1, y_1), \ldots, (\zeta_m, y_m)\}$.

2.4.8 Bayesian Supervised Classification Using Generative Models

In this subsection, we assume the set of data $X = \{x_1, \ldots, x_m\}$, as well as unobserved data, to be classified following a known probabilistic model. In the context of classification, such models are said to be *generative* because they can be used to generate data by changing the parameters and the noise. Assume for example that data are sinusoidal signals in noise; see (2.31). Signals with label $y = 1$ have frequency $k_{(1)}$, and signals with label $y = -1$ have frequency $k_{(-1)}$ with $k_{(-1)} \neq k_{(1)}$. These models can be turned into likelihood functions of the form (2.32) for each class. More generally, to each class with label $y \in \mathcal{Y}$ corresponds a likelihood function $p(x|\theta_y, y)$, denoted $p(x|\theta_y)$ for brevity.

Standard Bayesian classification consists of first learning the class parameters θ_y for each class label y using (2.103), and then computing the class likelihood of a new datum x using (2.104). In other words, given class-priors $p(\theta_y|y)$, the posterior $p(\theta_y|X_y)$ is computed using class-training sets $X_y = \{x_i \in X \text{ such that } y_i = y\}$, as follows:

$$p(\theta_y|X_y) \propto p(\theta_y|y) \prod_{x_i \in X_y} p(x_i|\theta_y). \tag{2.103}$$

In the sine signals example, the training set is cut into $X_{(1)}$ and $X_{(-1)}$ containing respectively the signals with frequency $k_{(1)}$ and $k_{(-1)}$. The posterior of $k_{(1)}$ (respectively $k_{(-1)}$) is computed independently for each class.

Whenever the posterior is computed, it is possible to implement the classification of x via the computation of the *class likelihood*

$$p(x|X_y) = \int_\Theta p(x|\theta_y, y) \, p(\theta_y|X_y) \, d\theta_y. \tag{2.104}$$

The integral (2.104) can be computed in analytic closed form in some cases; in all other cases, MCMC enable its computation. Finally, assuming the 0-1 loss function, classification is performed by assigning x to the class for which $p(x|X_y)$ is maximum.

This technique requires some refinements for practical implementation because the parameter to learn may vary from one datum to another inside a given class. In this case, it is not possible to learn the *value* of the parameter of each class; we need to learn instead its *density function*. In this framework, θ_y becomes a set of pdf parameters, which is now denoted ψ_y to distinguish it from θ_y used in the likelihood function. ψ_y is called the *class hyperparameter* with prior $p(\psi_y|y)$. Consider again the sine signals example and, e.g., the class $y = 1$. This class might include signals with frequencies *close to* k_1. Assuming known amplitudes and noise variance, the signal unknown parameter is $\theta_y = k_1$. Choosing a Gaussian distribution $\mathcal{N}(\theta_y; \mu_y, \Sigma_y)$ for the parameter θ_y in each class, the hyperparameter ψ_y is composed of μ_y and Σ_y (see [121] for further details).

2.4.9 Hidden Markov Models

Hidden Markov models (HMMs) are widely used in speech recognition mainly because they offer a robust pattern recognition dynamic scheme. HMMs include continuous state space systems such as (2.79)–(2.80): The parameter vector $\boldsymbol{\theta}_n$ lives in a continuous state space Θ; its evolution follows a so-called Markov process (in the sense that the pdf of $\boldsymbol{\theta}_n$ only depends on the pdf of $\boldsymbol{\theta}_{n-1}$ at each time $n = 1, 2, \ldots$); and the state cannot be observed directly (it is *hidden*). In the speech and audio processing literature, HMMs mostly refer to *finite, discrete state space dynamic models*; that is, $\boldsymbol{\theta}_n$ is a discrete random variable in a finite space with E possible state values, namely $\Theta = \{e_1, \ldots, e_E\}$; see Fig. 2.11. A finite, discrete state space HMM is governed by

- the state transition probabilities $\mathsf{P}(\boldsymbol{\theta}_n = e_i | \boldsymbol{\theta}_{n-1} = e_j)$ for $(i, j) = 1, \ldots, E$—this is the discrete equivalent of (2.79);
- the state likelihoods $\mathsf{p}(\mathbf{x}_n | \boldsymbol{\theta}_n = e_i)$, $i = 1, \ldots, E$—equivalent to (2.80);
- and the initial probabilities $\mathsf{P}(\boldsymbol{\theta}_1)$.

Similar to the particle filtering problem, the issue is here to estimate the sequence of states over time. Typically, in speech recognition the observations are MFCCs extracted from frames over the speech signal, and the state is the phoneme pronounced by the speaker. Likelihood functions are typically Gaussian mixtures over the MFCCs (there is one GMM for each possible state). Transition probabilities and GMM parameters are learned from a large database and from the speaker's voice.[25]

The Viterbi Algorithm

The aim of the Viterbi algorithm is the estimation of the sequence of states $\boldsymbol{\theta}_{1:T}$ from time 1 to time T by maximum a posteriori. In other words, given a sequence of observations $\mathbf{x}_{1:T}$, the Viterbi algorithm finds the sequence of states $\widehat{\boldsymbol{\theta}}_{1:T}$ such that

$$\widehat{\boldsymbol{\theta}}_{1:T} = \underset{\boldsymbol{\theta}_{1:T}}{\operatorname{argmax}} \; \mathsf{P}(\boldsymbol{\theta}_{1:T} | \mathbf{x}_{1:T}), \tag{2.105}$$

where the posterior probability of a sequence of state $\boldsymbol{\theta}_{1:T}$ is

$$\mathsf{P}(\boldsymbol{\theta}_{1:T} | \mathbf{x}_{1:T}) \propto \mathsf{P}(\boldsymbol{\theta}_1) \prod_{n=2}^{T} \mathsf{p}(\mathbf{x}_n | \boldsymbol{\theta}_n) \mathsf{P}(\boldsymbol{\theta}_n | \boldsymbol{\theta}_{n-1}). \tag{2.106}$$

In this context, estimation of the state is also called *sequence decoding*. The Viterbi algorithm is presented below.

---

[25] Additional details may be found in [534], which proposes a tutorial on HMMs and describes several basic algorithms such as the Baum–Welch algorithm for parameter learning.

Algorithm 2.10: Viterbi Algorithm

1. Initialization:
   - For $i = 1, \ldots, E$, set $w_1(e_i) = p(\mathbf{x}_1|\boldsymbol{\theta}_1 = e_i)P(\boldsymbol{\theta}_1 = e_i)$.

2. Iterations: For $n = 2, \ldots, T$,
   - For $i = 1, \ldots, E$, compute

$$w_n(e_i) = p(\mathbf{x}_n|\boldsymbol{\theta}_n = e_i) \left[\max_{j=1,\ldots,E} w_{n-1}(e_j)P(\boldsymbol{\theta}_n = e_i|\boldsymbol{\theta}_{n-1} = e_j) \right]. \quad (2.107)$$

   - For $i = 1, \ldots, E$, set $\psi_n(e_i) = \operatorname*{argmax}_{e_j, j=1,\ldots,E} w_{n-1}(e_j)P(\boldsymbol{\theta}_n = e_i|\boldsymbol{\theta}_{n-1} = e_j)$.

3. Termination:
   - Compute $\widehat{\boldsymbol{\theta}}_T = \max_{e_i, i=1,\ldots,E} w_T(e_i)$.

4. State sequence backtracking
   - For $n = T - 1, \ldots, 2, 1$, extract the estimate at time n by $\widehat{\boldsymbol{\theta}}_n = \psi_n(\widehat{\boldsymbol{\theta}}_{n+1})$.

The Viterbi algorithm implements an exhaustive search method along all possible paths. Its computational complexity is $O(E^2T)$, which makes it a very efficient decoding algorithm, where the notation $O(E^2T)$ means that the computation time is proportional to T and to the square of E when other factors are kept fixed. Note that Algorithm 2.10 has some similarities with particle filtering, in particular in the way the weight $w_n(e_i)$ is computed; compare (2.107) with (2.81).

3

Sparse Adaptive Representations for Musical Signals

Laurent Daudet[1] and Bruno Torrésani[2]

[1] Laboratoire d'Acoustique Musicale, 11 rue de Lourmel 75015 Paris, France
daudet@lam.jussieu.fr
[2] Laboratoire d'Analyse, Topologie et Probabilités, CMI, Université de Provence,
39 rue F. Joliot-Curie, 13453 Marseille cedex 13, France
Bruno.Torresani@cmi.univ-mrs.fr

3.1 Introduction

Musical signals are, strictly speaking, acoustic signals where some aesthetically relevant information is conveyed through propagating pressure waves. Although the human auditory system exhibits a remarkable ability to interpret and understand these sound waves, these types of signals cannot be processed as such by computers. Obviously, the signals have to be converted into digital form, and this first implies sampling and quantization. In time-domain digital formats, such as the Pulse Code Modulation (PCM)—or newer formats such as one-bit oversampled bitstreams used in the Super Audio CD—audio signals can be stored, edited, and played back. However, many current signal processing techniques aim at extracting some musically relevant high-level information in (optimally) an unsupervised manner, and most of these are not directly applicable in the above-mentioned time domain. Among such semantic analysis tasks, let us mention segmentation, where ones wants to break down a complex sound into coherent sound objects; classification, where one wants to relate these sound objects to putative sound sources; and transcription, where one wants to retrieve the individual notes and their timings from the audio signals. For such algorithms, it is often desirable to transform the time-domain signals into other, better suited representations. Indeed, according to the Merrian-Webster dictionary,[3] to 'represent' primarily means 'to bring clearly before the mind'.

Among alternate representations, the most popular is undoubtedly the time-frequency representation, whose visual counterpart (usually without phase information) is the widely used 'spectrogram' (see Chapter 2). Here, at least visually, higher-level features such as note onset, fundamental frequency or formants can be distinguished and estimated. However, a major

---

[3]See http:www.m-w.com.

consideration for time-frequency representation is that we have to decide on a size for the analysis window, and it can be seen (Fig. 3.1) that this choice can allow radically different features to be emphasized. Among all possible choices for the analysis window, it is generally not possible to find the one that is simultaneously optimal for all the features we want to extract.

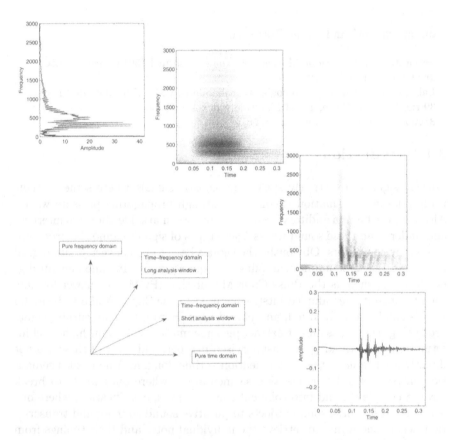

Fig. 3.1. Although equivalent, these four representations of the same signal (impulse response of an open tube, kindly provided by M. Castellengo) highlight different features of the signal (for the sake of clarity, only the magnitude of complex values has been displayed).

The aim of this chapter is to introduce more advanced techniques, where we put in strong assumptions about our signals: better performance is obtained, but of course this is at the cost of a loss of generality. However, in many applications mentioned above, there is a *model* for our class of analysed signals.

Within this framework, the *representation* of a sound is simply the set of *parameters* for the model. For a given information extraction task, 'good' representations are the ones where the parameters are simply related to the desired high-level feature.

Here, representation methods can be divided into two large classes:

- Analysis/synthesis methods: from the set of parameters (or coefficients), it is possible to reconstruct the original time-domain signal, either in an exact manner or approximately.
- Pure analysis methods: by focusing on only one type of feature that we want to estimate, we allow some (irrelevant in this context) information to be lost, and sound reconstruction is not possible anymore. For instance, estimation of the fundamental frequency F0 can usually be performed without phase information.

In this chapter, we will focus on the former, in a typical analysis-by-synthesis framework. The next section describes parametric representations of signals, from the original McAulay and Quatieri sinusoidal modelling by additive synthesis, towards most recent advances that also take into account models for transients and noise. These parametric approaches are said to be 'deterministic', in the way that the parameters are extracted locally in the signal through deterministic algorithms. Alternate approaches tentatively embed these parametric expansions into statistical models, where each sound is seen as a particular realization of a random process having the hyper-parameters we want to estimate. Section 3.3 is devoted to so-called *waveform* representations, where the signal is seen as a linear combination of elementary waveforms, chosen within a well-defined *dictionary*. This dictionary is usually structured; i.e., its elements can be derived from a few elementary waveforms through a number of simple operations such as modulations, translations, and dilations. From a practical viewpoint, they are often chosen in such a way that fast algorithms are available in order to get the expansion of a given signal. After describing the general framework for waveform expansions, we will focus on so-called *sparse* representations, in which the representation is compact: few large coefficients accurately represent the signal, while the vast majority of the coefficients are small (hence neglected). Different models for sparsity will be discussed, as well as measures for the distortion caused by neglecting a large number of (small) coefficients. Within this framework, it is also possible to enforce sound models such as { tones + transients + noise }. In *hybrid* representations, we decompose a sound as a union of orthonormal bases, each of them being well suited for a particular type of feature (for instance a basis of lapped local cosines for the tonal part, a basis of dyadic wavelets for the transients). In such models, sparsity can be further exploited by taking into account structures of significant coefficients, locally in the parameter (time-frequency/time-scale) space.

Throughout this chapter, we will treat in parallel the discrete time and continuous time settings. However, we found it necessary to be more explicit in a few specific situations, and use either the discrete or the continuous setting. We use the following notation: we reserve the letter n to denote discrete time variables (i.e., we denote discrete time signals as $x(n)$, $n \in \mathbb{Z}$), and the letter t for continuous time signals $x(t)$, $t \in \mathbb{R}$.

3.2 Parametric Representations

In this section we discuss some of the most widely used models for parametric representations of sounds. Most of these originate from the additive synthesis models, originally developed by McAulay and Quatieri [449] for speech analysis and synthesis.

3.2.1 Sinusoidal Models and Relatives

The sinusoidal additive synthesis model [449] approximates a sound by a sum of M sinusoids. In a discrete time setting,

$$x(n) = \sum_{m=1}^{M} \alpha_m(n) \sin\left(\varphi_m(n)\right) + \epsilon(n), \tag{3.1}$$

where the phase φ_m of the m-th partial is sometimes written as a primitive of a smooth time-dependent frequency $f_m(t)$:

$$\varphi_m(n) = \int_0^{n/k_s} 2\pi f_m(t) dt + \varphi_m(0),$$

k_s being the sampling frequency and $\epsilon(n)$ representing the error of the model. Here, it is implicitly assumed that the amplitude and frequency parameters (α_m, f_m) for each partial sinusoid evolve slowly over time, in such a way that their values can be estimated frame by frame (a typical frame size is 23 ms, representing 1024 samples at a 44.1 kHz sampling rate).

The selection of the partials is made first by peak-picking the magnitude of the short-time Fourier transform of the signal. Chaining the obtained peaks into partials (i.e., curves $n \to f_m(n)$ for all m in a joint time-frequency domain) is a non-trivial task which has received significant attention since the early contribution of [449], including the hidden Markov chain approach of [139], or linear prediction [383]. Whatever the chosen approach and algorithm for partial chaining, the underlying idea is to exploit the supposed smoothness of the time-dependent frequencies $f_m(t)$, and chain the peaks that are close to each other in the frequency domain. Note that, for resynthesis using (3.1), the requirement of phase continuity at frame boundaries implies some non-trivial interpolation scheme for the partials frequencies $f_m(t)$ (for instance a cubic spline interpolation).

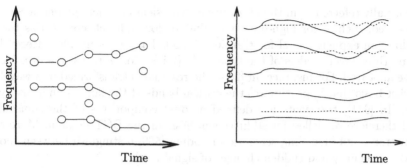

Fig. 3.2. Local spectrum peaks, chained into partials (left). Partials grouped into locally harmonic structures (right): partials corresponding to two locally harmonic sources are represented by full lines and dashed lines respectively.

The partials may in turn be used for various purposes, including harmonic source separation and possibly transcription, as shown in [652]. There, distances between time-dependent amplitudes $\alpha_m(t)$ and frequencies $f_m(t)$ are proposed, together with a measure of harmonic concordance between partials. A *perceptual distance* between partials involving these three distances is then built, and numerically optimized for grouping partials into *locally harmonic sources*. An elementary example of such grouping is presented in Fig. 3.2.

This model has been refined by Serra in the spectral model synthesis (SMS) model [576], where the residual of a sinusoidal model is also taken into account as a so-called *stochastic part*. This modification has made it possible to perform high-quality processing of general musical signals [649] (see also [408]) that have a more complex behavior than speech signals. Applications that make use of SMS range from audio effects (morphing, time-scaling, etc.) and source separation to sound analysis (transcription).

One of the main limitations of the SMS approach is the lack of an explicit model for the residual part. This results in having to keep this residual in the time domain, hence requiring a very large number of parameters. As a first-order approximation, this residual can be modelled as filtered white noise, the parameters of the filter being estimated by the energy in perceptual frequency bands [219] or through classical autoregressive (AR) or autoregressive moving average (ARMA) methods [472] (see Chapter 2). This solution is used in the Harmonics and Individual Lines plus Noise (HILN) coder [532]. With further improvements such as grouping harmonically related components, HILN achieves fair sound quality at bit rates as low as 6 kbit/s.

However, in all the above-described modelling of the residual, there is always the implicit assumption that the parameters evolve slowly over the analysis frame. Obviously, this does not always hold; for instance it does not hold at the sharp note attacks of many percussive instruments. These fast-varying features are in certain cases characterized by sudden bursts of noise, and/or fast changes of the local spectral content. These components are usually referred to in the literature as 'transients'. When further accuracy is needed, such as for high-quality coding, or for audio effects such as time-scaling, an explicit model for transient sounds has to be included. Recently, many three-layer models of the type STN { sinusoids + transients + noise } have appeared. For instance, in [269], the residual is transformed in a wavelet packet basis that approximates the critical bands of the human auditory system. In [641], transients are defined as dual components of the sinusoids, and therefore modelled by additive synthesis in the DCT domain. More recent transients modelling schemes for coding [569] distinguish between short bursts of energy and sudden changes of signal level.

3.2.2 Parametric Random Models

The deterministic parametric models briefly described above may also be seen from a probabilistic point of view. The advantage of such an approach is twofold. On the one hand, it provides a way to model the intrinsic variability of the phenomena under study. On the other hand, it also allows one to use the powerful estimation algorithms that have been developed in the context of statistical estimation. The price paid is of course an increased complexity of the models. Chapter 7 presents a thorough presentation of the latter models; see in particular (7.9) for the explicit several-notes model.

Remark 1. The model in (7.9) actually expresses the signal as a linear combination of 'elementary waveforms' of the form $\phi_i(n)\cos(2\pi f_m(n)n)$ and $\phi_i(n)\sin(2\pi f_m(n)n)$, with unknown (random) parameters. This model may be written in matrix notation; see (7.10) page 210. This creates, in some sense, the connection between parametric harmonic models and sparse waveform models, to be discussed in the next section.

A nice feature of the approach above is the fact that given estimates for parameters of the model, the latter may be used for signal synthesis. When synthesis is not necessary, it is no longer necessary to start from the signal waveform, and it may be easier to start from other representations of the signal, for example a short-time Fourier spectrum. This approach has been taken, among others, in [644], [645], where a new stochastic model for musical instruments local spectra was introduced. The model parameters are first estimated on a training set, and may then be used for transcription. Since only spectra (and not signals) are modelled in this approach, separation may only be performed by an appropriate post-processing, e.g. local Wiener filtering. We refrain from discussing this approach in more details here, and refer to Vincent [644] for a thorough description.

3.3 Waveform Representations

By *waveform representations*, we mean representations of musical signals as linear combinations of *elementary waveforms*, or *time-frequency atoms*, generated using simple construction rules. These include translations (i.e., time shifts) and modulations (frequency shifts) of a unique waveform for 'Gabor-type' waveforms, translations and dilations (rescalings) for 'wavelet-type' waveforms, or more general transformation rules. Depending on the problem at hand, and on the chosen approach, identifying the signal decomposition may be more or less simple. We describe below a number of possible approaches of various levels of difficulty. Even though it is difficult to give a sensible classification of these approaches, we attempt to class them according to two criteria:

1. *Rigid vs. adaptive:* Rigid techniques are characterized by the fact that the signals under consideration are expanded onto a fixed family of waveforms (basis or frame of the underlying signal space—these terms will be explained below). Conversely, adaptive techniques choose the waveform family as a function of the analysed signal, generally using greedy search algorithms.

2. *Redundant vs. strict:* In 'strict' expansion techniques, the goal is to minimize the redundancy between the coefficients of the expansion. Strict families generally correspond to orthonormal bases of the signal space. Being strict often imposes severe constraints on the waveforms, and it is sometimes more suitable to turn to redundant families (frames, dictionaries, etc.), for which such constraints no longer apply. Redundancy (also called *overcompleteness* since we are in finite dimension, or *oversampling* in engineering terminology) offers a way to 'customize' the waveforms to be used for the signal expansion. The price paid for redundancy is an increased number of parameters (which can be penalizing for compression purposes), and an increase in the complexity both at the analysis and the resynthesis stages.

3.3.1 Rigid Expansions: Waveform Bases and Frames

We first make the distinction between non-redundant waveform systems (i.e., bases) and redundant ones (frames, dictionaries), before moving to distinctions of more 'algorithmic' nature (rigid vs. adaptive).

Waveform representations generally make use of linear algebra language and techniques, which we briefly recall here for the sake of completeness. The reader can find more detailed descriptions in many classical mathematics textbooks. We refer to [642] for a reference in the signal processing context.

Signals are assumed to belong to some underlying vector space $\mathcal{H}$, called the *signal space*. $\mathcal{H}$ is generally assumed to be an *inner product space* which is equipped with an inner product $\langle \cdot, \cdot \rangle : x, y \in \mathcal{H} \rightarrow \langle x, y \rangle \in \mathbb{C}$. The latter defines the *norm* $\|x\|$ of the signal $x \in \mathcal{H}$ by $\|x\| = \sqrt{\langle x, x \rangle}$. Commonly used signal spaces include

- The (infinite-dimensional) L^2 spaces of square-integrable functions;

$$L^2(I) = \left\{ x : I \to \mathbb{C}, \quad \int_I |x(t)|^2 \, dt < \infty \right\},$$

  (I being either the real line $\mathbb{R}$ or some subset), with inner product

$$\langle x, y \rangle = \int_I x(t) y(t)^* \, dt,$$

  where '$*$' denotes complex conjugation;
- The (infinite-dimensional) ℓ^2 spaces of square-summable sequences

$$\ell^2(\mathbb{Z}) = \left\{ x : \mathbb{Z} \to \mathbb{C}, \quad \sum_{n=-\infty}^{\infty} |x(n)|^2 < \infty \right\},$$

  with inner product

$$\langle x, y \rangle = \sum_{n=-\infty}^{\infty} x(n) y(n)^* \ ;$$

- The (finite-dimensional) space $\mathbb{C}^N$, with inner product

$$\langle x, y \rangle = \sum_{n=0}^{N-1} x(n) y(n)^* \ .$$

Notice that in the latter case, signals are finite-dimensional vectors, whereas in the first two cases, signals are infinite sequences or functions, which may nevertheless be seen as 'vectors with infinitely many components'.

Waveform Bases

The simplest example of elementary waveform representations is provided by orthonormal bases. There exists already a considerable amount of literature dedicated to wavelet and local cosine bases (see e.g. [89], [113], [429], [642], [667]), and therefore we do not go into details on the associated theory.

An orthonormal basis of an inner product space $\mathcal{H}$ is a family $\mathcal{B} = \{\varphi_i, i \in \mathcal{I}\}$ of elements of $\mathcal{H}$ ($\mathcal{I}$ being an index set whose size equals the dimension of the space $\mathcal{H}$), such that

- The family is *complete* in $\mathcal{H}$; i.e., any signal $x \in \mathcal{H}$ admits an expansion of the form

$$x(t) = \sum_{i \in \mathcal{I}} \alpha_i \varphi_i(t) , \tag{3.2}$$

- For all $i, i' \in \mathcal{I}$,

$$\langle \varphi_i, \varphi_{i'} \rangle = \delta_{ii'} , \tag{3.3}$$

where $\delta_{ii'}$ is Kronecker's symbol, equal to 1 if $i = i'$ and 0 otherwise.

Under such assumptions, it may be proved that the expansion (3.2) is unique, and that the coefficients α_i in the expansion are nothing but the inner products

$$\alpha_i = \langle x, \varphi_i \rangle, \qquad i \in \mathcal{I}. \tag{3.4}$$

Finally, the latter satisfy Parseval's formula (this is an 'energy preservation' formula)

$$\sum_{i \in \mathcal{I}} |\alpha_i|^2 = \|x\|^2. \tag{3.5}$$

As mentioned above, the signal space may be finite dimensional (for example finite-length discrete signals or finite-length band limited continuous time signals), or infinite dimensional (for example finite or infinite support discrete or continuous time signals), but the general framework remains the same. We shall limit ourselves to spaces of finite-energy signals, i.e., L^2 spaces in the continuous-time case, and spaces of square-summable sequences in the discrete and finite cases, with norms and inner products as above.

Remark 2. The reader who is not interested in the mathematical details may simply remember that in the case of discrete time, finite-length signals (i.e., a finite-dimensional signal space), expanding a signal onto an orthonormal basis is nothing but applying a unitary matrix to the signal vector (a simple example is the DFT matrix). The inverse operation, i.e., reconstructing a signal from the coefficients of a basis expansion, is also a matrix–vector multiplication, using the Hermitian conjugate (i.e., the complex conjugate of transpose) of the transform matrix.

Among the orthonormal bases, wavelet and local cosine (with extensions such as the modified discrete cosine transform, or MDCT) bases have been particularly popular. The main difference between these two systems is the fact that the time and frequency resolution of local cosine waveforms is uniform in the time and the frequency axis, while wavelets offer finer time resolution (and thus broader frequency resolution) at high frequencies.

Wavelets

Wavelets (see for example [113] and [429] for thorough reviews) are generated from a single 'atom' ψ by regular translations and dilations. In the continuous-time settings, a deep result by Mallat and Meyer states that it is possible to construct (continuous-time) functions $\psi \in L^2(\mathbb{R})$ such that, introducing the corresponding translations and dilations ψ_{jn} of ψ defined by

$$\psi_{jn}(t) = 2^{-j/2}\, \psi\big(2^{-j}t - n\big), \quad t \in \mathbb{R},$$

the family $\{\psi_{jn},\, j, n \in \mathbb{Z}\}$ is an orthonormal basis of $L^2(\mathbb{R})$. Therefore, any signal $x \in L^2(\mathbb{R})$ may be expanded in a unique way as

$$x(t) = \sum_{j=-\infty}^{\infty} \sum_{n=-\infty}^{\infty} \langle x, \psi_{jn} \rangle \psi_{jn}(t).$$

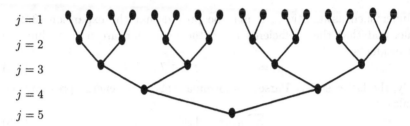

$j = 1$

$j = 2$

$j = 3$

$j = 4$

$j = 5$

Fig. 3.3. Dyadic tree of scale-time indices for a wavelet basis ψ_{jn}.

Here, j is a scale parameter, and n is a time parameter. The scale-time index (j,n) is usually represented by a *dyadic grid* (see Fig. 3.3), which illustrates the fact that the time sampling provided by wavelets depends upon the value of the scale parameter j: at large scales (bottom of the tree), the wavelets have larger support, and are sampled more coarsely than at small scales (top of the tree).

Remarkably enough, such wavelet bases may (in most situations) be generated through a mathematical algorithm called *multiresolution analysis* (MRA), which is itself intrinsically connected to the *sub-band coding* techniques of signal and image analysis. More precisely, a wavelet orthonormal basis and thus an MRA involve an auxiliary function $\phi \in L^2(\mathbb{R})$, called the *scaling function*, such that the coefficients $\langle x, \phi_{jn} \rangle$ of $x \in L^2(\mathbb{R})$ represent samples of a corresponding low-pass approximation of x at scale 2^j (the functions ϕ_{jn} are defined similarly to the wavelets ψ_{jn}). Given some (arbitrary) reference scale j_0, it is possible to show that the family $\{\phi_{j_0,n}, n \in \mathbb{Z}\} \cup \{\psi_{j,n}, j < j_0, n \in \mathbb{Z}\}$ is another orthonormal basis of $L^2(\mathbb{R})$. Therefore, the above wavelet expansion of a signal x may also be replaced with another expansion

$$x(t) = \sum_{n=-\infty}^{\infty} \langle x, \phi_{j_0 n} \rangle \phi_{j_0 n}(t) + \sum_{j=-\infty}^{j_0-1} \sum_{n=-\infty}^{\infty} \langle x, \psi_{jn} \rangle \psi_{jn}(t),$$

the basis functions $\phi_{j_0 k}, k \in \mathbb{Z}$ collecting the information contained in the (doubly labelled) family of wavelets $\psi_{jk}, k \in \mathbb{Z}, j > j_0$. Given a wavelet basis (or an MRA), the computation of the coefficients $\langle x, \psi_{jn} \rangle$ of the expansion of a signal $x \in L^2(\mathbb{R})$ on the corresponding wavelet basis may be performed through a sub-band coding algorithm, i.e., recursive filtering of the scaling function coefficients and wavelet coefficients from a fixed pair of digital filters (details on sub-band filtering may be found e.g. in [642]). The reconstruction of the signal from its wavelet coefficients $\langle x, \psi_{jn} \rangle$ also rests on sub-band coding. The case of discrete signals (the discrete wavelet transform, DWT for short) is handled by identifying signal samples $x(n)$ with the scaling function coefficients $\langle x, \phi_{j_1 n} \rangle$ of some 'underlying continuous time' signal, at some reference (finest) scale j_1.

Local cosines

Local cosine bases arose from an attempt to construct orthonormal bases that 'look like' Gabor functions, i.e., waveforms generated by regular translates and modulates of a single waveform (see (3.10) below). For simplicity we describe the case of infinite, continuous-time signals first, and then address discretization issues. Starting from a family of compactly supported, positive-valued window functions w_n, essentially supported in intervals $[a_n, a_{n+1}]$ (up to rapidly decaying tails) subject to some technical conditions (see for example [429], [667] for a detailed presentation) it was shown by Coifman and Meyer that the corresponding family of functions

$$u_{nk}(t) = \sqrt{\frac{2}{a_{n+1}-a_n}} \, w_n(t) \, \cos\left(\pi(k+1/2)\frac{t-a_n}{a_{n+1}-a_n}\right), \quad n \in \mathbb{Z}, \, k \in \mathbb{Z}^+,$$

(3.6)

forms an orthonormal basis of $L^2(\mathbb{R})$. Therefore, any $x \in L^2(\mathbb{R})$ admits a unique expansion of the form

$$x(t) = \sum_{n=-\infty}^{\infty} \sum_{k=0}^{\infty} \langle x, u_{nk}\rangle u_{nk}(t) \ .$$

Here, n is a time index and k is a frequency index. The usual practice[4] is to choose regularly spaced values of the a_n, so that the local cosine atoms u_{nk} have constant length. In such cases, one may say that the local cosine atoms are labelled by a *regular grid* in the time-frequency domain.

Local cosine bases adapted to continuous-time signals defined on finite intervals are constructed similarly, modulo a mild 'twist' in (3.6) for treating boundaries (essentially, the windows w become square windows at boundaries). Remarkably enough, a corresponding discrete theory has been developed, mainly based upon appropriate 'smoothing' of the windows used in DCT-IV transform (a particular form of discrete cosine transform (DCT); see [429] for example), yielding the *modified discrete cosine transform* (MDCT for short). Again, Remark 2 above applies, and it is not necessary to master the theory of cosine transforms and corresponding lapped transforms (see [431] for a review) to use them practically.

Remark 3. Comparing wavelets and local cosines: As mentioned above, the main difference between these two waveform systems lies in the fact that local cosine atoms are constant size, variable shape functions, whereas wavelets have constant shape and variable size. This is illustrated in Fig. 3.4, where different wavelets and local cosine atoms have been represented, with various sizes, frequencies, and locations. The two plots on the left each represent

---

[4]Although there are exceptions; for example, some audio coders use MDCT atoms with two different lengths: wide windows w for describing partials, and narrow windows in 'transient' regions.

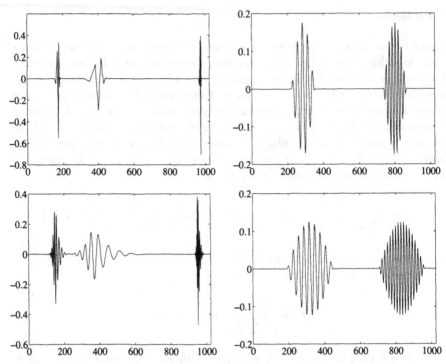

Fig. 3.4. Comparing wavelets and local cosines. Daubechies 6 (top left) and Vaidyanathan (bottom left) wavelets for three different values of the scale-time index (j, n), and narrow (top right) and wide (bottom right) local cosine atoms for two different values of the time-frequency index (n, k).

three wavelets with different scales and position, using Daubechies wavelets of order 6 (top) and Vaidyanathan wavelets (bottom).[5] This illustrates a general property of wavelet bases: the smoothness of the wavelet increases with the size of its support.

On the right are represented local cosine atoms corresponding to two different window sizes (narrow window on the top plot, wide window on the bottom plot) for two different locations and frequencies. In both cases, the window function is compactly supported, which results in a poor frequency localization for its Fourier transform.

Remark 4. Non-orthonormal bases, multi-bases, ...: Many variations around the theme of orthonormal bases in signal spaces are possible and have been considered. One of the most popular is an approach in which the orthogonality assumption is relaxed, while the completeness and non-redundancy assumptions are retained. As is often the case, relaxing one assumption yields more

---

[5]We refer to [113], [667] for detailed discussions of the common choices of wavelet bases.

freedom in the construction of the basis, and such non-orthonormal bases may be constructed from 'nicer' time-frequency atoms (for example, the atoms can be made smoother, or more symmetric, or better time localized). Another variant is obtained by assuming that the waveform basis is obtained from two or more time-frequency atoms rather than a single one. Again, one obtains in such situations 'nicer' time-frequency atoms. Examples of such a strategy are provided by the multiwavelets, or the more general multiple bases (see for example [17], [545]).

Frames and Redundant Representations

A major shortcoming of the (orthonormal or not) bases for application to audio signal representation is the fact that orthonormality is often too strong a constraint, in the sense that the corresponding waveforms do not have the nice shapes one would like to see. They often have either poor time localization or poor regularity (i.e., poor frequency localization). Even when some assumptions are relaxed (for example, requiring biorthogonality rather than orthogonality, or using multiwavelets—i.e., wavelet-like bases generated from more than one basis function), the corresponding waveforms do not look natural. Also, while bases are clearly adapted for signal compression, representing a signal by its coefficients with respect to a basis generally breaks some desirable properties such as time-frequency covariance (see Chapter 2), which is often useful for other tasks, including all signal analysis problems. *Frames*[6] provide a way of introducing waveforms that are more 'natural', by introducing extra redundancy in the waveform system. They may also offer a way to preserve the above-mentioned 'nice' properties, in particular time-frequency covariance.

By definition, a family of waveforms $\{\varphi_i, i \in \mathcal{I}\}$ is a frame of the signal space $\mathcal{H}$ if the following weaker form of Parseval's formula holds[7]: There exist positive constants $0 < A \leq B < \infty$ such that for all $x \in \mathcal{H}$,

$$A\|x\|^2 \leq \sum_{i \in \mathcal{I}} |\langle x, \varphi_i \rangle|^2 \leq B\|x\|^2 . \tag{3.7}$$

This implies in particular that the frame is complete in $\mathcal{H}$ (as defined above), and therefore that any $x \in \mathcal{H}$ may be expanded as in (3.2). The main difference is the fact that the frame is generally not *exact*, which means that the waveforms $\{\varphi_i, i \in \mathcal{I}\}$ form a *redundant* system in $\mathcal{H}$ (one sometimes speaks of

---

[6] Notice that contrary to a common usage in the signal processing literature, the term 'frame' does not represent a time interval, but rather a family of vectors in a vector space. Since this terminology is also standard in mathematics, we shall nevertheless use it here.

[7] While Parseval's formula (3.5) expresses energy conservation, such a weak Parseval's formula expresses energy equivalence, i.e., the fact that the energy of the sequence of coefficients is controlled by the energy of the signal.

an *overcomplete system*). In such situations, the expansion (3.2) is not unique, as there exists an infinity of choices for the set of coefficients $\{\alpha_i, i \in \mathcal{I}\}$ yielding signal expansions as in (3.2). The bad news is that the inner products $\langle x, \varphi_i \rangle$ can't be used as α_i coefficients. Nevertheless, a wide variety of choices is obtained by using a so-called *dual frame*, which is another family of waveforms $\{\tilde{\varphi}_i, i \in \mathcal{I}\}$ such that any $x \in \mathcal{H}$ admits the following expansion with respect to the φ_i waveforms:

$$x(t) = \sum_{i \in \mathcal{I}} \langle x, \tilde{\varphi}_i \rangle \, \varphi_i(t). \tag{3.8}$$

Notice that in such situations, one also has

$$x(t) = \sum_{i \in \mathcal{I}} \langle x, \varphi_i \rangle \, \tilde{\varphi}_i(t), \tag{3.9}$$

which may be interpreted as an *inversion* of the *transform* $x \rightarrow \{\langle x, \varphi_i \rangle, i \in \mathcal{I}\}$. In both situations, the coefficients $\langle x, \varphi_i \rangle$ or $\langle x, \tilde{\varphi}_i \rangle$ provide an alternate *representation* of the signal that often proves useful for several signal analysis tasks.

The most classical choices for waveform frames are provided by the so-called *Gabor frames* and *wavelet frames*, whose basic theory is discussed in detail in [113] (see also [184] and [261]). We shall limit our discussion here to Gabor frames and generalizations. The reader more interested in wavelet and multiresolution frames is invited to refer to the vast literature on the subject (see for example [89], [113] for tutorials).

Gabor frames, which actually correspond to sampled versions of the short-time Fourier transform described in Chapter 2, have been fairly popular in the musical signal representation community. Gabor frames are generated from a unique window function g by time and frequency translations. In the discrete time $\ell^2(\mathbb{Z})$ setting, the corresponding discrete Gabor functions g_{nk} (where n and k control time and frequency respectively) read

$$g_{nk}(l) = e^{2j\pi k k_s (l - n n_s)} g(l - n n_s). \tag{3.10}$$

Here, n_s and k_s represent time and frequency sampling rates, respectively. It may be shown that under some mild assumptions, for any window $g \in \ell^2(\mathbb{Z})$, the family $\{g_{nk}, n, k \in \mathbb{Z}\}$ is a frame of $\ell^2(\mathbb{Z})$ as soon as the product $n_s k_s$ is small enough. The latter essentially controls the *redundancy* of the Gabor frame: the smaller $n_s k_s$, the closer the atoms. For example, increasing n_s reduces time redundancy, which may be compensated for by decreasing k_s. Many examples of signal processing applications of Gabor frames may be found in the literature; see e.g. [185].

Gabor frames are often well adapted to musical signal processing, as they provide 'direct access' to time and frequency variables simultaneously. Also, using Gabor frames rather than the corresponding orthonormal bases[8] allows

---

[8]A 'no-go theorem' known as Balian–Low phenomenon states that there cannot exist 'nice' orthonormal bases of Gabor functions [113].

one to use 'nicer' (i.e., smoother and better localized) windows, and keep translation invariance, which is a very important feature. On the other hand, Gabor frames are made out of fixed resolution waveforms, which can therefore not be adapted to the features of the signal. Hence, it is difficult to find in such schemes a representation that would be well adapted to both transients and partials of sound signals.

A good illustration of this fact is provided by the two images in Fig. 3.5 below, in which are displayed the Gabor frame expansions of a short piece of guitar signal using two different windows. It clearly appears that a Gabor frame generated using a narrow window (left image) is able to capture accurately the transient parts of the signal, while a Gabor frame generated using a wide window is much more precise for capturing partials. Neither of them is able to do a good job for both types of components.

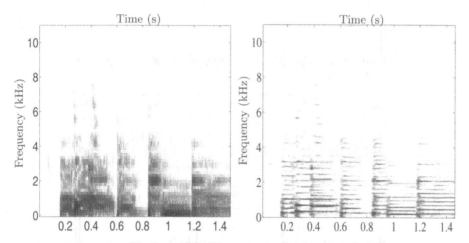

Fig. 3.5. Grey level images of two different Gabor frame representations of a short piece of guitar sound. Left: narrow Hanning window. Right: wide Hanning window.

From the above example, the following question arises naturally: Is there a way of decomposing a signal into 'layers' that could be adequately represented by an appropriate waveform system (Gabor frame or other)? A very interesting outcome of the frame theory is the fact that given a pair (or a larger, finite family) of frames in a given signal space, their union is still a frame of the same space, and thus suitable for expanding signals. Frames generated as unions of Gabor frames are called *multiple Gabor frames*. Multiple Gabor frames have been considered in the literature for various purposes, including source separation in signals and images, or musical signal processing; see for example [150], [312], [690]. Examples discussed in these references were generally based on a family of Gabor frames with identical windows at different time scales. The goal in such situations is to be able to represent transient components of sound signals using narrow Gabor functions, partials using broad Gabor functions, etc.—in a few words, find the best adapted waveform for each component of the signal. An example of such a strategy may be found in Fig. 3.6, where we displayed such a decomposition performed on a harmonic sum of sine waves to which were added spikes with random amplitude at random locations. The decomposition of this synthetic signal was performed using the 'Time–Frequency Jigsaw Puzzle' method [312], to be described below, which only exploits sparsity arguments. Even though the result is not perfect, we can see from the middle and bottom plots that the two significantly different components were succesfully separated. The main point of this example is the fact that broad Gabor functions were able to select the partials of the harmonic components, while the narrow ones estimated almost perfectly the transient components. Other illustrations of this technique on real sounds may be found in [312] and [311].

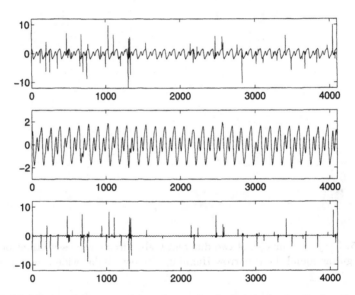

Fig. 3.6. Multilayered decomposition of a synthetic signal, obtained using the Time–Frequency Jigsaw Puzzle technique, described in Section 3.3.2 below. From top to bottom: original signal, tonal layer, and transient layer.

Another illustration of a similar strategy (using Gabor and wavelet families) may also be found in Figs. 3.11 and 3.12 below, where the tonal and transient components of a vibraphone signal were succesfully separated using structured approximation techniques (to be described below).

Remark 5. The need for adaptivity: In some sense, such multiple Gabor frames provide a way to retain the best of the Gabor and multiresolution worlds. However, achieving such a program turns out to be quite difficult for practical

purposes, mainly because one has to deal with the extra redundancy introduced by the use of several frames together. Indeed, given R Gabor frames
$\{g_{nk}^{(r)}, (n,k) \in \mathcal{I}, r = 1, \ldots, R\}$, there is an infinite number of ways to expand
any $x \in \mathcal{H}$ as

$$x(l) = \sum_{r=1}^{R} \sum_{(n,k) \in \mathcal{I}} \alpha_{n,k}^{(r)} g_{nk}^{(r)}(l) \,, \tag{3.11}$$

and one is naturally led to ask for the 'best' one, according to some given
criterion. In such a way, we are naturally led to leave the world of 'rigid
expansion' techniques, and enter the world of 'adaptive expansions'. To seek
the 'best' expansion, possible criteria are

- Optimize the sparsity of the expansion: find a 'minimal set' of non-zero
  coefficients such that (3.11) holds, at least within some prescribed accuracy. One is then naturally led to consider subsets of the multiple frame
  one started with. These subsets can be frames themselves (the so-called
  'quilted frames' introduced in [150]), or may not even be complete families
  if adaptive methods are considered.
- Organize the expansion in such a way that 'components' of the signal (for
  example, transients, partials) are represented by large coefficients with
  respect to a given waveform system that suits it.

In an ideal world, these two requirements would be equivalent, and they are
in fact nearly so. Unfortunately, coming up with an actual algorithm able to
perform such a task is far from easy. This again brings us to the problem of
sparse representations, to be discussed in Section 3.3.2 below.

Remark 6. We have only described above the multiple Gabor frame construction. Similar developments may be done starting from wavelet frames, the
union of bases, or the union of Gabor and wavelet systems (bases or frames).
In such situations, one talks of 'hybrid systems'.

Dictionaries

By 'waveform dictionary', one generally means a family of waveforms which
is more redundant than a frame. By definition, a dictionary in some signal
space $\mathcal{H}$ is a *complete* family of elements of $\mathcal{H}$, that is, a family such that
any signal $x \in \mathcal{H}$ admits an expansion as a linear combination of elements
of the dictionary. In infinite-dimensional signal spaces, dictionaries may even
not be frames, as they may contain too many elements for the right-hand side
inequality of (3.7) to be satisfied. However, when it comes to practical situations, i.e., finite-dimensional signal spaces, the dictionaries which are generally
considered in the literature are also frames, so that the distinction between
'frame methods' and 'dictionary methods' refer to the techniques that are used
to find the expansion of signals with respect to such systems rather than the
intrinsic properties of these systems. Therefore, we shall address 'dictionary
techniques' in Section 3.3.2 below.

3.3.2 Adaptive Expansions: The Quest for the Sparse Grail

It is an 'experimental fact' that in the case of audio signals, the expansions described above are generally sparse: any signal in the considered class may be represented with good accuracy by a truncated expansion of the type (3.2), in which only the largest coefficients (in absolute value) are retained:

$$x(t) \approx \sum_{j=1}^{J} \alpha_{i_j} \varphi_{i_j}(t) , \qquad (3.12)$$

where the indices $i_1, \ldots i_J$ have been chosen so that the absolute values of the corresponding coefficients α_{i_j} are sorted in decreasing order.

Remark 7. There exist different ways of evaluating the sparsity of an expansion. Essentially, any subadditive functional of the coefficients (i.e., giving less importance to large values) would do the job. Classical choices are provided by entropies, namely the Shannon or Rényi entropies[9]

$$H = -\sum_{i \in \mathcal{I}} |\alpha_i|^2 \log_2(|\alpha_i|^2) ; \qquad R_\beta = \frac{1}{\beta - 1} \log_2 \left(\sum_{i \in \mathcal{I}} |\alpha_i|^{2\beta} \right), \qquad (3.13)$$

assuming that the sequence of coefficients α_i has been normalized so that $\sum_{i \in \mathcal{I}} |\alpha_i|^2 = 1$. Rényi entropies may be interpreted as sparsity measures for $0 < \beta < 1$. Moreover, it may be shown that the Shannon entropy is in fact equal to the limit of the Rényi entropy as $\beta \to 1$.

Unfortunately, these different criteria generally yield different results, as noted for instance by Jaillet [311]. Nevertheless, they turn out to essentially agree in the case of simple signals, which motivated Jaillet to use them locally in the time-frequency domain.

The fact that a signal admits a sparse expansion with respect to a given waveform system also manifests itself by the fact that the histogram of the coefficients of the expansion is significantly peaked at the origin (meaning that a large number of coefficients are close to zero) and heavy tailed (a slower decay of large coefficients). An example is provided in Fig. 3.8, where the empirical probability density function (pdf) (computed here by an appropriate smoothing of the histogram) of the various representation coefficients (time samples, Fourier samples, wavelet and MDCT coefficients) of two significantly different audio signals are displayed. The two signals, a polyphonic organ signal which is quite 'tonal', and a castanet signal which is extremely transient, are shown in Fig. 3.7. The pdfs in Fig. 3.8 represent various degrees of sparsity. As may be expected, the frequency and MDCT representations are better suited for the organ signal: the corresponding pdfs have

---

[9]Notice that Rényi entropies R_β are essentially a logarithmic form of the $\ell^{2\beta}$ norms $\|\tilde{\alpha}\|_{2\beta} = \left(\sum_i |\tilde{\alpha}_i|^{2\beta} \right)^{1/2\beta}$ of the normalized coefficient sequences, which have also been used as sparsity measures.

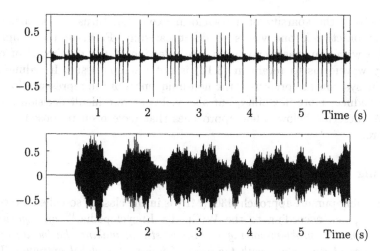

Fig. 3.7. Two sample musical signals: a castanet signal (top) and an organ signal (bottom).

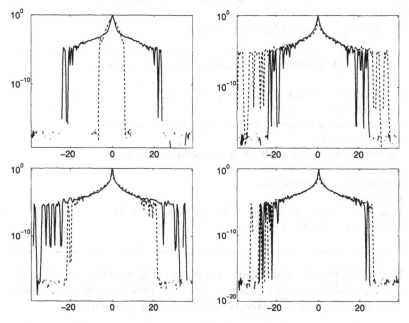

Fig. 3.8. pdf of various representations of the two sample signals of Fig. 3.7: $\ln(P(\tilde{\alpha}))$ vs $\tilde{\alpha}$. castanet (solid line) and organ (dashed line): Top: time samples and Fourier coefficients; Bottom: wavelet and MDCT coefficients.

a maximum at the centre which is more sharply peaked, meaning that one has a large number of very small values, and a smaller number of significant values.

Exploiting the sparsity of a waveform expansion turns out to become tricky in practice if one wants to do it carefully. First, in the simplest case of a waveform basis expansion, the selection of the number J of elementary waveforms to retain in (3.12) is not a simple issue. If redundant waveform systems are preferred, the non-uniqueness of the representation indeed introduces more flexibility, but also yields more difficult decision problems. We describe below a few approaches that have been proposed in the literature.

Matching Pursuit and Orthogonal Matching Pursuit

The Matching pursuit approach [430] (MP) is in the class of so-called 'greedy' algorithms, i.e., according to the Wikipedia Encyclopedia,[10] *an algorithm which follows the problem-solving meta-heuristic of making the locally optimum choice at each stage with the hope of finding the global optimum.* The MP is an iterative procedure that aims at approximating a signal through a weighted sum of atoms such as in (3.12), where the atoms belong to a given redundant dictionary $\mathcal{D}$.

The basic principle of MP is as follows:

Algorithm 3.1: Matching Pursuit Algorithm

1. Initialization: Compute all the inner products $\alpha_\lambda = \langle x, u_\lambda \rangle$.
   Let $r_0(t) = x(t)$ and $i = 0$.
2. Find maximum absolute modulus amongst all inner products:
   $\lambda_i = \arg\max_\lambda |\alpha_\lambda|$.
3. Update the residual by subtracting the corresponding atom:
   $r_{i+1}(t) = r_i(t) - \alpha_{\lambda_i} u_{\lambda_i}(t)$.
4. Update all the inner products:
   $\alpha_\lambda = \langle r_{i+1}, u_\lambda \rangle$.
5. If *stopping criteria* is satisfied, then stop; otherwise update
   $i \leftarrow i + 1$ and iterate to step 2.

Depending on the application, many stopping criteria can be used, for instance a condition on the total number I of iterations, or on the norm of the residual $\|r_{i+1}\|$. After I iterations, the signal is written as

$$x(t) = \sum_{i=0}^{I-1} \alpha_{\lambda_i} u_{\lambda_i}(t) + r_{i+1}(t), \qquad (3.14)$$

and approximated by the first term of the right-hand side of the above equation, provided the residual r_{I+1} is small enough (in norm).

---

[10]See http://en.wikipedia.org/wiki.

A few important remarks can be made:

- In very redundant dictionaries, finding the maximum inner product can be a lengthy operation, as can be updating of the inner products. Speed-enhancing tricks can in some cases be implemented, such as storing the cross inner products of all elements in the dictionary, or implementing a suboptimal search through a hierarchical organization of the parameters.
- Due to the orthogonality between the residual and the selected atom, choosing the atom that maximizes the inner product is equivalent to choosing the atom that minimizes the norm of the residual. A further refinement, called orthogonal matching pursuit [119], guarantees that the approximation at any i-th iteration is the best (in the L^2 sense) *on the span $\mathcal{D}_i$ of the i already selected atoms.* This is done by recursively orthogonalizing the family of the i previously selected atoms. It is important to keep in mind that it is in no way guaranteed that this is the best i-terms approximation, except under very stringent conditions on the dictionary and the signal, i.e., when the signal is exactly made of a small number of atoms from the dictionary, this case being of little relevance for most practical audio applications.

The classical choices for dictionaries include the union of Gabor frames with different window sizes, extensions of the latter with 'chirp' waveforms (chirps are frequency-modulated waveforms), or the union of several bases such as wavelet and MDCT bases. We shall come back to such choices later on, and refer the reader to [255], [257] for more details.

Time–Frequency Jigsaw Puzzles

The time–frequency jigsaw puzzle (TFJP) algorithm [312] is another way of obtaining 'good' multiple Gabor frame expansions (as described above) through a greedy approach, but it is simpler than matching pursuit. The idea is essentially to partition the time–frequency plane into time–frequency 'supertiles' Ω_n, and to find the optimal Gabor frame within each supertile, according to some sparsity criterion (such as the entropies described above). Supertiles are rectangular subdomains in the time–frequency plane, whose area is significantly larger than the time–frequency area of the windows (typically, 25 times larger). A supertile is said to be of size $M \times N$ if it contains M time sampling points of the wide window, and N frequency sampling points of the narrow windows. A typical value for M and N is about 5.

More precisely, start from a fixed family of Gabor frames (typically, two Gabor frames, one with a narrow window function, the other one with a wide window function). Given a supertile Ω_n of the time–frequency index set, the corresponding Gabor frame coefficients with respect to all considered Gabor frames are computed, normalized locally (so that their absolute moduli sum up to unity), and corresponding entropies R_β (as in (3.13)) are computed. The Gabor frame whose coefficient set is sparsest (within Ω_n) wins.

In such a way a subset of each frame is selected, and the signal may be iteratively projected orthogonally onto the corresponding subspace of the signal space (as in orthogonal matching pursuit). At each iteration, a residual signal is produced, and processed in a similar way. The iteration stops when the precision is considered satisfactory. So far, no known proof exists for the convergence of this method. However, the convergence is quite fast in practice: less than 20 iterations are needed to achieve 300 dB signal-to-noise ratio.

In addition, TFJP provides decompositions of signals into 'layers' as follows: Assuming that two frames $G^{(1)} = \{g_\delta^{(1)}\}$ and $G^{(2)} = \{g_\lambda^{(2)}\}$ are considered, the algorithms provides an expansion of any signal x in the form

$$x(t) = x^{(1)}(t) + x^{(2)}(t) , \quad x^{(1)}(t) = \sum_{\delta \in \Delta} \beta_\delta g_\delta^{(1)}(t) , \quad x^{(2)}(t) = \sum_{\lambda \in \Lambda} \alpha_\lambda g_\lambda^{(2)}(t) ,$$

where $x^{(1)}$ ($x^{(2)}$) is the 'component' (termed 'layer') of the signal which has been 'identified' by the frame $G^{(1)}$ ($G^{(2)}$), and Δ and Λ are (small) subsets of the global index set (in general a subset of $\mathbb{Z}^2$).

The simplest instance of this method is based on a pair of two Gabor frames, with significantly different window sizes; say, for audio signals, 5 ms and 45 ms. The corresponding Gabor atoms, when used in the framework of the TFJP method, identify nicely partials and transients. An illustration of such a strategy on a simple synthetic signal may be found in Fig. 3.6 above. As long as the signal can be correctly modelled as a superposition of partials (with slowly varying amplitude and frequency) and transients, TFJP is able to identify and separate them. In the presence of more complex phenomena, the method should be refined, and should include different types of atoms (for example chirps). However, as is well known from matching pursuit approaches, enlarging the dictionary of atoms does not necessarily improve the accuracy of the identification of signal components: the more redundant the dictionary, the larger the ambiguity of the selection.

The basic principle of TFJP (see [312, variant 2]) is presented in Algorithm 3.2 below.

Algorithm 3.2: Time–frequency Jigsaw Puzzle

Choose a value for $\beta \in (0, 1]$. Choose two windows $g^{(1)}$, $g^{(2)}$ and corresponding sampling lattices; choose supertiles. Define a maximal number of iterations I and a precision threshold ϵ.

1. Initialization: Set $r_0(t) = x(t)$ and $i = 0$.
2. Main loop: While $i \leq I$ and $\|r_i\| \geq \epsilon$,
   - Compute coefficients $\langle r_i, g_{nk}^{(j)} \rangle$ for the two windows $j = 1, 2$. Compute entropies R_β for both windows within each supertile.
   - Select supertiles for which window #1 yields the smallest entropy. Reconstruct corresponding contribution $x_i^{(1)}$ to layer 1.
   - Set $r_{i+1/2}(t) = r_i(t) - x_i^{(1)}(t)$.
   - Compute coefficients $\langle r_{i+1/2}, g_{nk}^{(j)} \rangle$ for the two windows $j = 1, 2$. Compute entropies R_β for both windows within each supertile.

- Select supertiles for which window #2 yields the smallest entropy. Reconstruct corresponding contribution $x_i^{(2)}$ to layer 2.
- Set $r_{i+1}(t) = r_{i+1/2}(t) - x_i^{(2)}(t)$.

3. Reconstruct layers ℓ_1 and ℓ_2 by summing up contributions $x_i^{(1)}$ and $x_i^{(2)}$, respectively.

3.3.3 Unstructured and Structured Hybrid Representations

In the approaches described above, the coefficients of the expansion of the signal with respect to the waveform system are treated individually, and only their magnitude is taken into account. However, in practice, it rarely happens that a signal is characterized by isolated (in the index set) large coefficients, with respect to any known waveform system. It could be so if one could use, for expanding the signal, dictionaries of waveforms that make sense as elementary sound objects, but such a goal seems quite far away for the moment (see nevertheless [43]). Therefore, sensible elementary sounds are rather made of several chained elementary time–frequency atoms, forming *time–frequency molecules*. Estimating such time–frequency molecules is the goal of what we call *structured approximation*. Notice that structured sets of waveforms already appeared in Section 3.2, when we discussed the random parametric models.

Structured Bases and Frame Representations

Structured basis approximation techniques have already been proposed in different contexts in the signal processing literature. A good and famous example is provided by the so-called embedded zero-tree wavelet (EZW) algorithm [582] for image coding, which exploits the binary tree structure of orthonormal wavelet bases (see Fig. 3.3). This algorithm is based on the following 'experimental fact': When a wavelet coefficient corresponding to a node of the coefficient tree is zero or very small, then the coefficients attached to the corresponding subtree are likely to be zero as well, yielding 'zero trees' of wavelet coefficients.

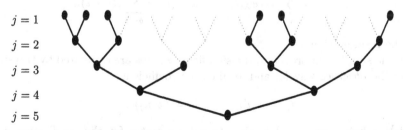

Fig. 3.9. Subtree (in black) of the dyadic tree of Fig. 3.3 (suppressed edges appear as dotted lines).

Similarly, structured sets of MDCT functions have also been considered in [114], [469] from an audio coding perspective, exploiting the fact that when an audio signal is expanded onto a local cosine basis, it often happens that large coefficients form 'time-persistent' sets, or *ridges* (in the same spirit as the chained peak sets in the Fourier domain described in Section 3.2; see Fig. 3.2). In fact, as a result of the poor frequency localization of MDCT atoms, such a localization is not so sharp. It has been shown that sine waves rather manifest themselves in the MDCT domain by 'tubes' of three (or fewer) consecutive (in frequency) ridges of significant coefficients (see Fig. 3.10).

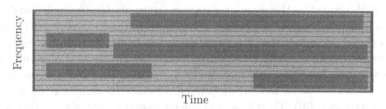

Fig. 3.10. MDCT coefficient domain, and corresponding tubes of significant MDCT coefficients.

Such 'persistent structures' may be exploited in different ways, in the framework of Hybrid waveform audio models. In such contexts, the signal is sought in the form

$$x(t) = \sum_{\delta \in \Delta} \beta_\delta u_\delta(t) + \sum_{\lambda \in \Lambda} \alpha_\lambda \psi_\lambda(t) + r(t) , \tag{3.15}$$

where the atoms ψ_λ (respectively u_δ) are wavelets (respectively MDCT atoms), the set Λ (respectively Δ) is a small subset of the wavelet (respectively MDCT) coefficient set, and r is a (hopefully small) residual. Notice that this model is again of the type 'tonal + transient + noise'.

Practically, given a signal x, the goal is to estimate the *significance maps* Λ and Δ, estimate the corresponding coefficients, and reconstruct the corresponding layers of the signal

$$x^{(\psi)}(t) = \sum_{\lambda \in \Lambda} \alpha_\lambda \psi_\lambda(t), \qquad x^{(u)}(t) = \sum_{\delta \in \Delta} \beta_\delta u_\delta(t) ,$$

and the residual signal.

In the simplest approaches, the significance maps are estimated by thresholding the *observed* wavelet and local cosine coefficients

$$a_i = \langle x, \psi_i \rangle , \qquad b_j = \langle x, u_j \rangle ,$$

the latter being processed further to get estimates for the coefficients α_λ and β_δ. This type of approach, which was taken in Berger et al. [34] and

Daudet et al. [115], does not yield structured approximations, the coefficients being processed individually. In the above-mentioned references, an iterative approach was chosen, in which the MDCT layer was estimated first and removed from the signal prior to the estimation of the wavelet layer. The difficulty is then to provide a prior estimate for the number of MDCT coefficients to retain for estimating the tonal layer. To this end, a *transientness index* was proposed, based on entropic measures [468]. The latter actually provides an estimate for the proportion of wavelet versus MDCT coefficients present in the signal. Using this ingredient, the (unstructured) hybrid model estimation procedure presented in Algorithm 3.3 is obtained.

Algorithm 3.3: Hybrid Model Estimation Procedure

Fix an overall number of coefficients J to be retained. Choose a wavelet basis and a MDCT basis.

1. Estimate the number J_M of MDCT and J_W of wavelet coefficients to retain.
2. Compute the MDCT coefficients of the signal as $b_m = \langle x, u_m \rangle$; select the J_M largest ones (in magnitude) $b_{m_1}, \ldots b_{m_{J_M}}$ and construct the tonal estimate

$$x_{\text{ton}}(t) = \sum_{j=1}^{J_M} b_{m_j} u_{m_j}(t).$$

3. Substract the tonal estimate from the signal to get the non-tonal estimate $x_{\text{nton}}(t) = x(t) - x_{\text{ton}}(t)$.
4. Compute the wavelet coefficients of the non-tonal estimate $a_n = \langle x_{\text{nton}}, \psi_n \rangle$; select the J_W largest (in magnitude) ones $a_{m_1}, \ldots a_{m_{J_W}}$ and construct the transient estimate $x_{\text{trans}}(t) = \sum_{j=1}^{J_W} a_{n_j} u_{n_j}(t).$
5. Substract the transient estimate from the signal to get the residual estimate $x_{\text{res}}(t) = x_{\text{nton}}(t) - x_{\text{trans}}(t)$.

Berger et al. [34] also suggested a greedy approach in which several passes of this two-step procedure are expected to yield more precise estimates for the layers $x^{(\psi)}$ and $x^{(u)}$. More precisely, a first estimate of the tonal layer is obtained by peaking the largest coefficients of an MDCT expansion. This estimate is substracted from the signal, and a first estimate of the transient layer is obtained from the largest wavelet coefficients of this residual. The tonal estimate is then updated by peaking the largest coefficients of the MDCT expansion of this 'second-order residual', and so on. The difficulty of such approaches is mainly in answering the question 'how many large coefficients should one keep at each step?' The transientness index alluded to above could perhaps be used at this point, but to our knowledge this has not been done up to now.

To estimate structured significance maps, coefficients have to be processed jointly rather than individually. In [115], a functional on the space of connected subtrees of the wavelet tree is proposed. Numerical optimization of this functional yields estimates for *significance trees* Δ of wavelet coefficients, and thus for the wavelet layer. Similarly, a functional on the space of time–frequency 'tubes' is exploited in [114], yielding estimates for the MDCT layer. The corresponding algorithm is again a two-step algorithm: A structured tonal estimate is first obtained from a MDCT expansion, and substracted from the signal. A structured transient estimate is then obtained from the wavelet coefficients of the non-tonal signal. Finally the residual is obtained by substracting the transient estimate from the non-tonal signal.

The latter approach has also been modified and refined in [469] to incorporate a more specific structured waveform model in the expression (3.15). The main idea is to introduce a stochastic model for the significance maps Λ and Δ that implements the desired persistence properties: time persistence for Δ and scale persistence for Λ.

Let us first study the tonal layer. For the corresponding tonal significance map, a standard first-order Markov chain model is used: For a given value of the frequency index k, the membership probability of the index (n, k) to Δ is governed by persistence (conditional) probabilities

$$\tilde{\pi}_k = \mathsf{P}\big[(n, k) \in \Delta | (n - 1, k) \in \Delta\big], \quad \tilde{\pi}'_k = \mathsf{P}\big[(n, k) \notin \Delta | (n - 1, k) \notin \Delta\big] .$$

The corresponding observed MDCT coefficients of the signal $b_{nk} = \langle x, u_{nk} \rangle$ are distributed following zero-mean normal distributions, with large variance σ_T^2 when the considered index (n, k) belongs to Δ, and small variance σ_R^2 when (n, k) does not belong to Δ. Given the parameters of the model (i.e., the variances and the persistence probabilities), the likelihood may be computed explicitly. Therefore, parameters of the model may be estimated using some training material and maximum likelihood procedures (here, EM algorithms perform quite well; see Chapter 2). Once the parameters have been estimated, the significance maps can be estimated in turn, using dynamic programming procedures or others (see [469]). The tonal layer of the signal is then constructed as

$$x_{\text{ton}}(t) = \sum_{\delta \in \Delta} \langle x, u_\delta \rangle u_\delta(t),$$

and the non-tonal layer reads

$$x_{\text{nton}}(t) = x(t) - x_{\text{ton}}(t).$$

Similarly, using the natural tree structure of wavelet coefficients (see Fig. 3.9), the transient significance map (that is, the significance tree) is modelled using a Markov tree, following the lines defined in [107]. In a similar manner, the distribution of the significance tree is then governed by persistence probabilities: at scale j, we denote by

$$\pi_j = \mathsf{P}\big[(j - 1, \ell) \in \Lambda | (j, n) \in \Lambda\big], \quad \ell = 2n, 2n + 1,$$

the probability that a node of the tree belongs to the significance tree Λ, assuming that its parent belongs to Λ. The corresponding observed wavelet coefficients of the non-tonal signal $\langle x_{\text{nton}}, \psi_{jn} \rangle$ are modelled as random variables, distributed according to a mixture of two Gaussian distributions: large variance for the indices belonging to Λ, and small variance for the other indices. Again, the parameters of the transient model may be estimated from the observed wavelet coefficients in some training material, and the significance trees may be estimated as well using the ML approach. From this, the estimated transient layer is given by

$$x_{\text{trans}}(t) = \sum_{\lambda \in \Lambda} \langle x_{\text{nton}}, \psi_\lambda \rangle \psi_\lambda(t),$$

and the residual reads

$$x_{\text{res}}(t) = x_{\text{nton}}(t) - x_{\text{trans}}(t).$$

The actual implementation of such an approach actually involves more complicated decisions; for example, parameters feature (slow) time variations, and therefore have to be estimated locally, within (large) time frames; the size of these frames is one of the important parameters to choose, which is not always an easy task (for example, for musical signals it clearly depends on the musical 'style'). We shall not go into details of these aspects, and refer the reader to [469] for a more detailed account of the model, and [114] for an application to audio coding.

Remark 8. Such two-step approaches have two major drawbacks:

- As stressed before, the relative proportion of MDCT and wavelet coefficients to be picked has to be determined prior to the estimation of the significance maps and trees. The entropy-based estimate proposed and studied in [468] yields satisfactory results as long as the tonal and transient signal representations are sparse enough (i.e., the significance maps Δ and Λ are small enough). However, a global estimation procedure would be far preferable.
- It is extremely difficult to obtain any error estimate in such schemes. For example, the errors in the estimation of proportions of the two layers should be taken into account, which does not appear to be easy.

An example of such multilayered decomposition is given in Fig. 3.11. A vibraphone signal (about 6 seconds, sampled at 44.1 kHz) has been decomposed following the lines of the Markov model described above (see [469] for details). The original signal is displayed in the top plot, and the tonal and transient layers are respectively represented in the middle and bottom plots. The spectrograms of these three signals are shown in Fig. 3.12. As appears clearly in the latter figure, the algorithm was able to capture the time persistence (scale persistence) of the waveform coefficients representing the tonal (transient) layer.

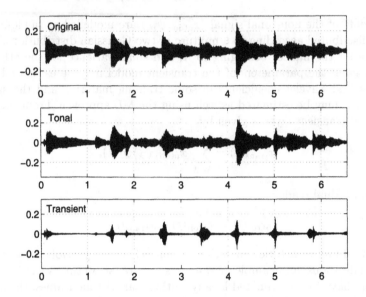

Fig. 3.11. Multilayered decomposition of a vibraphone signal, obtained using the Markov-based structured approximation techniques described in Section 3.3.3. From top to bottom: original signal, tonal layer, and transient layer.

Structured Matching Pursuits

Within the MP framework, it is also possible to design 'structured' versions of the basic algorithm presented above. The underlying principle is to select, at every iteration, a whole group of atoms that have some relation in the parameter space. In the harmonic matching pursuit (HMP) [256], this relation is a harmonic frequency relation:[11] the dictionary is made of a multiscale family of Gabor atoms (compare with (3.10))

$$g_{s,n,k}(l) = e^{2j\pi kk_s(l-nn_s)} g_s(l - nn_s),$$

the window g_s being a rescaled copy of g, at scale s, and the indices $\{s, n, k\}$ belonging to some fixed index set. A 'harmonic atom' is defined as a group of M harmonically related atoms:

$$HG_{s,n,k}(t) = \{g_{s,n,k}(t)\}_{m=1...M}.$$

The search is restricted to an interval of fundamental frequencies $k_1 \leq k \leq k_2$ such that all atoms forming a given harmonic atom can be considered to be approximately orthogonal. This requirement is crucial since in that case the signal energy carried by the harmonic atom—which is now the criterion

---

[11] The technique may easily be modified to account for any prescribed inharmonicity, as soon as the relationship between partials is known.

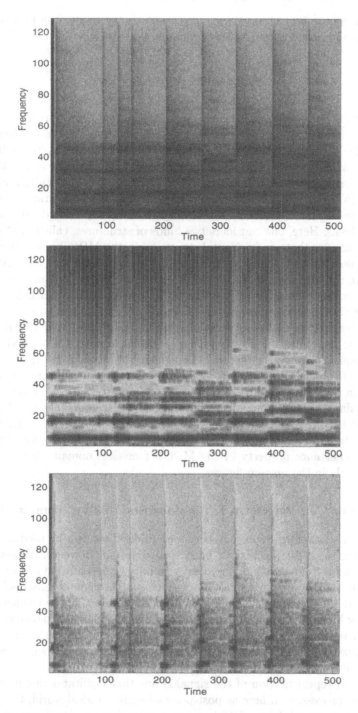

Fig. 3.12. Time–frequency representation (spectrogram) of the vibraphone signal (top) and the two layers: tonal layer (middle) and transient layer (bottom).

for choosing the best atom at every iteration—is (approximately) the sum of the squared inner products of the individual atoms, i.e.,

$$|\langle x, HG_{s,n,k}\rangle|^2 \approx \sum_{m=1}^{M} |\langle x, g_{s,n,mk}\rangle|^2 .$$

The rest of the procedure is the same as classical matching pursuit.

In a similar way, the MP can be used to construct structured decompositions for a sines + transients + noise model, in the same spirit as before. This is particularly adequate for sounds that are not harmonic and that have sharp onset transients, such as percussive sounds. In the molecular matching pursuit (MMP) [116], the dictionary is, as in the case of hybrid expansions above, the union of a local cosine (MDCT) basis and a discrete dyadic wavelet (DWT) basis. Here, one considers two kinds of structures, called 'molecules': tonal molecules that are horizontal structures in the MDCT basis, and transient molecules that are local subtrees of the dyadic wavelet tree. At each iteration, the algorithm looks for the strongest correlation among neighbouring atoms, then identifies and subtracts the corresponding molecule. Again, the procedure is iterated until some stopping criterion is met.

An example of molecular matching pursuit may be found in Fig. 3.13, where the locations of selected atoms (wavelets and MDCT atoms) in the time–frequency domain are shown. The signal is the vibraphone signal analysed in Fig. 3.12, and MP and MMP were required to select the same number of atoms. In Fig. 3.13, the first two panels (from top to bottom) correspond to the standard MP, and a significant overlap between atoms may be observed. The last two images correspond to the 'molecular' version, and the redundancy between selected atoms has been greatly reduced. In addition, the time (scale) persistence property in the MDCT (wavelet) domain appears much more clearly in the bottom figures.

3.3.4 How Can Waveform Expansions Be Used for Transcription?

Each of the waveform representations described above can be used as a pre-processing stage for the task of music transcription. However, transcription is much more than just looking at an alternate representation of the signal: proper transcription algorithms, as can be seen throughout this book, have to include a lot of high-level heuristic rules, for instance in the frequency domain, for lifting the usual transcription ambiguities due to harmonic relations between notes (octave 1:2, fifth 2:3, etc.) that have a large number of overlapping partials; or in the time domain, for making the distinction between vibrato and the start of a new note.

A good representation of the signal is one that facilitates this note identification process, seen here as post-processing. In an ideal world, this would be a rather straightforward task: when one projects a signal on a basis of tonal atoms, the notes would simply be given by the large tonal atoms, and

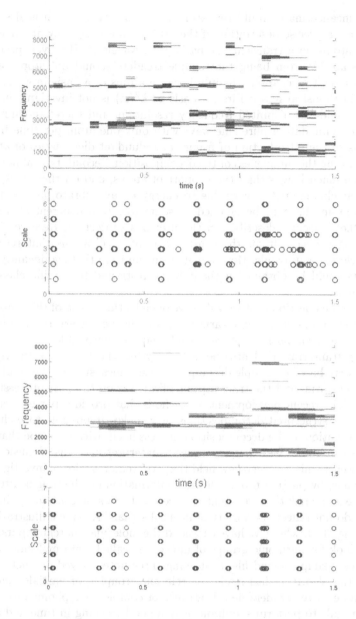

Fig. 3.13. Localization of selected coefficients in matching pursuit, in a standard implementation (MP, top two plots), and in a structured 'Molecular' implementation (MMP, bottom two plots), for the same total number of selected atoms. From top to bottom: MDCT coefficients in MP, discrete wavelets in MP, MDCT coefficients in MMP, discrete wavelets in MMP. The signal is the vibraphone excerpt analysed in Fig. 3.12.

their significance maps would represent 'piano-roll' types of musical scores. If one wants a precise description of the attack transients, they are given by the large atoms on a transient-like basis (e.g., wavelets). However, practical situations are far from being that simple (readers should not stop reading this book after this chapter), mainly because atoms of a basis in general do not look like musical notes: First, a basis of $L^2(\mathbb{R})$ is not invariant through time shifts and has a limited frequency resolution, and therefore cannot in general accommodate a pure sine wave with only one atom per time frame. This leads to the introduction of frames or redundant dictionaries of atoms, therefore losing the uniqueness of the decomposition. Second, typical note durations are much longer than the support of atoms, and on these long time scales their characteristics vary due to energy decay, vibrato, and so forth; therefore atoms have to be chained in some way into coherent structures such as the molecules described in the previous subsection. Research is actively being carried out for the construction of 'nice' atoms that resemble notes as closely as possible. However, it seems unlikely that significantly better results could be obtained without being restricted to specific classes of instruments.

Finally, the question that we will leave open is, How much of this empirical information (harmonicity, time variations, rules for octave errors, etc.) has to be put in the decomposition process itself, and how much is left for the post-processing transcription? Until the last few years, most of the transcription methods were based on simple decomposition schemes, such as the short time Fourier transform, and the transcription effort was on the post-processing of this data. The recent development of methods that provide sparse, overcomplete and/or structured decompositions now makes it possible to include a lot of information in the decomposition process itself, with the hope that this would provide some data that are easier to interpret in terms of musical objects. However, this is a very new field that still requires a lot of investigation. In particular, by putting too much prior information on the sought-after objects there is the risk that we might enforce constraints that are too stringent, and that do not reflect the variety of musical signals. A typical illustration is given in Fig. 3.14, where we have extended the molecular matching pursuit in order to look for harmonic groups of molecules, called meta-molecules [376]. On this very simple sound file, containing three notes played distinctly on a clarinet, the algorithm has made the relevant grouping of partials, and the notes can be correctly identified. However, for complex polyphonic mixes, the algorithm fails to perform simultaneously a good tracking in time and a relevant harmonic grouping. Clearly, this algorithm (which incidentally was not designed for transcription) is going too far.

Finally, it is likely that future transcription systems will go beyond the traditional one-way *transform* $\rightarrow$ *post-processing*, and try to optimize the system globally. For instance when a high-level notes hypothesis is made, some further signal-adaptive analysis can be performed to (in)validate the hypothesis.

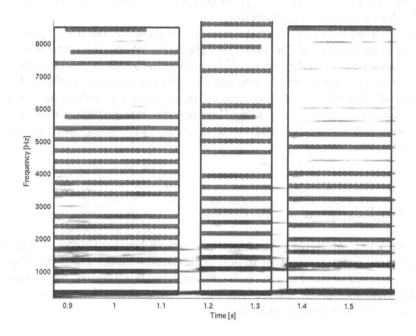

Fig. 3.14. Result of the meta-molecular matching pursuit (M3P) algorithm on three notes played by a clarinet. The molecules are shown as grey rectangles superimposed on the spectrogram. Black boxes show the detected meta-molecules, which in this very simple case correspond to the notes.

3.4 Conclusion

We have described in this chapter a number of approaches for musical signal representation, emphasizing parametric modelling and waveform expansion approaches. Parametric models are intellectually appealing, as they may sometimes be interpreted in terms of physical models for instruments, or more generally, in terms of sound production. Waveform approaches do not allow such easy interpretations, but offer other advantages, such as their flexibility and their computational efficiency.

Interestingly enough, these two competing approaches tend to somewhat converge. Indeed, as may be seen from Chapter 7 (see in particular (7.9)), expanding the amplitude functions of sinusoidal models onto bases of spline functions (for example) generate waveform systems close to Gabor systems. Similarly, the introduction of 'structured approximation' techniques in the waveform approaches may be understood as an attempt to move waveform models in the direction of more physically realistic models, and recover the possibility of identifying partials in the signal. In both cases, this convergence may be interpreted as an attempt to keep the best of the two approaches, while eliminating their shortcomings.

Another major separation between the waveform approaches we just reviewed is between rigid and adaptive approaches. While the first ones are more computationally efficient, the latter offer more flexibility, as they exploit redundant 'dictionaries' which generally contain more realistic waveforms. Unfortunately, flexibility has a price: sparse expansions in very redundant dictionaries are more difficult to identify. However, the introduction of structured expansion techniques help to reduce the ambiguity, while being closer to parametric approaches.

Acknowledgements

This work was supported in part by the European Union's Human Potential Programme, under contract HPRN-CT-2002-00285 (HASSIP), and the *Math-STIC* program of the French *Centre National de la Recherche Scientifique*, in the framework of the project 'Approximations parcimonieuses structurées pour le traitement de signaux audio'.

Part II

Rhythm and Timbre Analysis

Beat Tracking and Musical Metre Analysis

Stephen Hainsworth

Tillinghast-Towers Perrin, 71 High Holborn, London WC1V 6TH, UK
swh21@cantab.net

4.1 Introduction

Imagine you are sitting in a bar and your favourite song is played on the jukebox. It is quite possible that you might start tapping your foot in time to the music. This is the essence of beat tracking and it is a quite automatic and subconscious task for most humans. Unfortunately, the same is not true for computers; replicating this process algorithmically has been an active area of research for well over twenty years, with reasonable success achieved only recently.

Before progressing further, it would be useful to define beat tracking clearly. This involves estimating the possibly time-varying tempo and the locations of each beat. In engineering terms, this is the frequency and phase of a time-varying signal, the phase of which is zero at a beat location (i.e., where one would tap one's foot). When musical audio signals are used as an input, the aim of 'beat-tracking' algorithms is to estimate a set of beat times from this audio which would match those given by a trained human musician. In the case where a notated score of the music exists, the musician is used as a proxy for it (hopefully the musician's set of beats would align with those in the score). Where no score exists, the musician's training must be accepted to return a metre equivalent to how the music would be notated. Note that this implies that it is the intended rather than the percieved beat structure that is the focus here.

Beat tracking as just described is not the only task possible. Some algorithms attempt only tempo analysis—finding the average tempo of the sample; others attempt to find the phase of the beat process and hence produce a 'tapping signal'. Meanwhile, some methods also attempt a full rhythmic transcription and attempt to assign detected note onsets to musically relevant locations in a temporally quantized representation. This is often considered in terms of the score which a musician would be able to read in order to recreate the musical example [352]. MIDI signals are also commonly used as inputs and, assuming that the signal is of an expressive performance, all of the above tasks are again possible aims.

This chapter is organized as follows. Section 4.2 gives an overview of methods and approaches to beat tracking. However, as with any engineering system which is trying to replicate a real-world process, it is useful to examine the actual process before trying to build a model. Section 4.3 of this chapter briefly discusses some of the musical background behind beat tracking. Next, detection of onsets in musical audio signals is discussed in Section 4.4 before some of the more influential approaches to beat tracking are presented in Sections 4.5 to 4.9. Probabilistic models are examined in more detail in Section 4.10. Section 4.11 presents trials of various algorithms on a comprehensive test database and conclusions will be drawn in Section 4.12.

There are many immediate and commercial applications of a successful beat tracking program which have perhaps motivated some of the research. Some of these are: automatic accompaniment of a solo performance [538], synchronization of two music streams (e.g. for DJing [94]), correctly timed recovery from CD skipping (see [660] for a similar application), intelligent time stretching of musical samples [151], determination of good points for looping algorithms (useful for studio samplers which are heavily utilized in the creation of dance music) and adding tempo synchronous effects. Other uses include database retrieval [633] and metadata generation [566], provision of a 'rhythmic similarity' function to listeners (either in playback or for purchase recommendation) and rhythmic expressiveness transformations (e.g. adding swing to a musical example [244]). In addition, beat tracking can form a good basis for any automated transcription program (e.g. [126], [231], [353], [611]) from which to begin its analysis.

4.2 Summary of Beat-Tracking Approaches

Beat tracking with computers has been an active area of research since the early 1980s, though psychological models of human rhythmic perception predate this. The early work was undertaken in the fields of music perception and computer science, though the emphasis shifted towards engineering and statistics as computing power increased.

As a result of this paradigm shift, the aims and approaches of the methods described below vary considerably. It would hence be useful to categorize them. The first and most important distinction is by type of input; most of the earlier algorithms for beat tracking used a symbolic[1] or MIDI input while audio signals have been used more recently. This is at least partly because the signal processing required to extract rhythmic cues from the audio was beyond the power of early computers. It should be noted, however, that many of the more recent methods implicitly convert an audio stream to a set of MIDI-type inputs via the use of a pre-processing onset-detection algorithm.

---

[1]Symbolic data usually consists of a quantized set of note start times.

The second important differentiation between approaches is the intended purpose of the algorithm. Much of the early work was conducted with the music psychology goal of understanding how humans perceive music and attempting to model this. Other approaches have goals based more in engineering and attempt to capture information in the signal without direct reference to human perception. Specifically, those studies undertaken within the framework of automated transcription attempt to return to the underlying score rather than any human perception of the performance.

The next major distinction between the algorithms is the broad approach used. Categorizations here could include

- rule-based;
- autocorrelative;
- oscillating filters;
- histogramming;
- multiple agent;
- probabilistic;

though there are methods which do not fall neatly into any of these classes. Descriptions of these six broad approaches can be found later in Sections 4.5 to 4.10.

Another, more subtle method of classifying algorithms is by *causal* [572] operation. In a causal model, the estimate of the metre at a given time depends only on past and present data. A non-causal model allows the use of future data and backward decoding. Another way to consider it is that a causal algorithm attempts to mimic human tapping and uses data only up to the current time to decide whether a beat should be marked or not. Semi-causal algorithms have also been produced where the estimate is made after a short time-lag, typically around 20 ms. These can often give a 'strict' causal estimate but at the cost of optimality.

Finally, the algorithms can be grouped by their intended output; some only produce a best estimate of tempo while others evaluate phase as well, therefore giving the beat. Gouyon [242] separates these into *tempo induction*, the estimation of the most likely tempo given a segment of data, and *beat tracking*, which is the following of the beat through an extended example. Some methods also extract the super-beat and/or sub-beat structure (that is, slower and faster pulses than the beat, respectively), while some only attempt estimation of either the super- or sub-beat and not the actual beat.

Table 4.1 summarizes some methods found in the literature, indicating the type of input used and any causal nature. Others which do not fall into any particular category are Sethares and Staley [578], Smith [601], Miller et al. [464], and Bilmes [37]. Two other studies which present surveys or reviews of beat tracking are [243], [249].

Table 4.1. Summary of beat-tracking methods. Key for *Input* column: A = audio, M = MIDI, and S = symbolic.

Approach	Author and year [Ref]	Input	Causal
1) rule-based	Steedman 1977 [607]	S	
	Longuet-Higgins & Lee 1982 [418]	S	
	Povel & Essens 1985 [529]	S	
	Parncutt 1994 [497]	S	
	Temperley & Sleator 1999 [622]	M	
	Eck 2000 [165]	S	
2) autocorrelative	Brown 1993 [55]	S	
	Tzanetakis et al. 2001 [632]	A	
	Foote 2001 & Uchihashi [194]	A	
	Mayor 2001 [445]	A	
	Paulus & Klapuri 2002 [503]	A	
	Alonso et al. 2003 [15]	A	
	Davies & Plumbley 2004 [118]	A	X
3) oscillating filters	Large 1994 [390]	M	X
	McAuley 1995 [450]	M	X
	Scheirer 1998 [564]	A	X
	Toiviainen 1998 [626]	M	X
	Eck 2001 [166]	A	
4) histogramming	Gouyon et al. 2001 [245]	A	
	Seppänen 2001 [573]	A	X
	Wang & Vilermo 2001 [661]	A	
	Uhle & Herre 2003 [635]	A	
	Jensen & Andersen 2003 [318]	A	X
5) multiple agent	Allen & Dannenberg 1990 [14]	M	
	Rosenthal 1992 [546]	M	
	Goto et al. 1994 [221]	A	X
	Dixon 2001 [148]	A/M	
6) probabilistic	Laroche 2001 [392]	A	
	Cemgil et al. 2000 [75], [76]	M	
	Raphael 2001 [537]	A/M	
	Sethares et al. 2004 [577]	A	
	Hainsworth & Macleod 2003 [266]	A	
	Klapuri 2003 [349]	A	X
	Lam & Godsill 2003 [386]	A	
	Takeda et al. 2004 [617]	M	
	Lang & de Freitas 2004 [387]	A	

4.3 Musical Background to Rhythmic Structure

Typically, music consists of sounds generated concurrently by a number of different sources (usually musical instruments of varying kinds). These are organized in a temporal manner, the structure of which forms the 'rhythm' of the piece. Most music has a coherent temporal structure, as this is pleasing to most listeners. Thus the rhythm of a piece more readily lends itself to analysis than the harmonic structure, which can often be much more complex.

At the top level, the rhythm describes the timing relationships between musical events within a piece. The Oxford English Dictionary [624] gives the definition of rhythm as

a. *The aspect of musical composition concerned with periodical accent and the duration of notes.*
b. *A particular type of pattern formed by this.*

Cooper and Meyer [102] define it as the way in which one or more unaccented beats are grouped in relation to an accented one. The term *metre* is sometimes used in an equivalent manner to rhythm, though in music psychology it takes on a different meaning. Here, metre is the number of pulses between the more or less regularly recurring accents in a piece of music [102]. Thus, the metre is a constituent of the rhythm of a piece of music; however, the grouping of accents into patterns and the interaction of this process and the metre are closer to describing the rhythm of a piece.

Some further analysis can be made; Bilmes [37] breaks down musical timing into four subdivisions. The first is the hierarchical *metrical structure*, which relates the idealized timing relationships as they would exist in a musical score, i.e., quantized to a grid.[2] Next is *tempo variation*, which gives the possibly time-varying speed at which the events are sounded. Another level of abstraction gives *timing deviations*, which are individual timing discrepancies around the time-varying metrical grid (e.g. 'playing ahead of the beat'; swing[3] can also be considered a timing deviation). Finally there are *arrhythmic sections*, where there is no established rhythm. These will be ignored from now on as fundamentally impossible to analyse rhythmically, except as a collection of unrelated note start times.

The metrical structure can also be broken down into a set of three hierarchical levels. Klapuri [349] describes the *beat* or *tactus* as the preferred (trained) human tapping tempo and is what most of the beat-tracking algorithms attempt to extract at a minimum. This usually corresponds to the 1/4 note or *crotchet* when written out in common notation, though this is not always the case: in fast jazz music, the pulse is often felt at half this rate (1/2 note or *minim*), while hymns are often notated with the beat given in minims.

---

[2]Dixon [148] uses the term 'scoretime', measured in beats since the start of the sample to describe this representation.

[3]Swing is a style where the second 1/8th note of every beat is slightly delayed; it is a characteristic of jazz and some rock music.

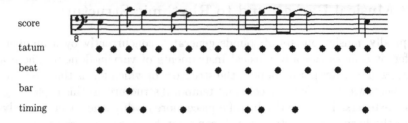

Fig. 4.1. Diagram of relationships between metrical levels.

However it is notated, the rate at which beats occur defines the tempo of the music [404].

At a lower level than the beat is the *tatum*, which is defined to be the shortest commonly occurring time interval. This is often defined by the 1/8th notes (*quavers*) or 1/16th notes (*semiquavers*). Conversely, the main metrical level above the beat is that of the *bar* or *measure*. This is related to the rate of harmonic change within the piece, usually to a repeated pattern of emphasis and also notational convention. Fig. 4.1 gives a diagrammatic representation of the above discussion. Included is a set of expressive timings for the score given. While obvious, it should also be noted that onsets do not necessarily fall on beats and that beats do not necessarily have onsets associated with them.

From here, metrical levels below the beat, including the tatum level, will be termed the *sub-beat* structure, while the converse—bar levels, etc.—will be labelled the *super-beat* structure. In between the tatum and beat, there may be intermediary levels, usually related by multiples of two or three (compound time divides the beat into three sub-beats, for instance). The same applies between the beat and bar levels. Gouyon [242] gives a comprehensive discussion of the semantics behind the words used to describe rhythm, pointing out many of the dualities and discrepancies of terminology. One point he raises is that the terms beat or pulse are commonly used to describe both an individual element in a series and the series as a whole.

An interesting point is raised by Honing [294], who discusses the duality between tempo variations and timing: the crux of the problem is that a series of expressively timed notes can be represented either as timing deviations around a fixed tempo, as a rapidly varying tempo, or as any intermediate pairing. This is a fundamental problem in rhythm perception and most algorithms arrive at an answer which lies between the extremes by applying a degree of smoothing to the processes—this usually means that estimated tempo change over an analysis segment is constrained by the algorithm and any additional error in expected timing of onsets is modelled as a timing deviation.

This leads to the concept of *quantization*, which is the process of assessing with which score location an expressively timed onset should be associated. Here, *score location* refers to the timing position the onset would take when notated upon a traditional Western musical score or other equivalent representation. However, for most purposes, it can be reduced to the number of beats (and sub-beats) since the start of the sample. Quantization is an important problem and other specific studies on this topic include Cemgil et al. [75] and Desain and Honing [142].

The phase of the beat is determined by a series of stresses or accents, termed *phenomenal* accents [404], [497] or *salience* [148], [529]. These usually correspond to note starts, though not uniquely—it is possible that note ends or changes in intensity can indicate beat, too. It is generally assumed that stresses fall on the beat more often than not and that significant chordal changes also do so. While this is not always the case, and indeed many musical styles exhibit *syncopation*, where there are off-beat stresses, Steedman notes, 'No event inconsistent with either key or metre will occur in a piece until sufficient framework (of key or time signature) has been established for it to be obvious that it is inconsistent' [607]. There are counter-examples to this statement, but it holds in the main.

There is a large body of literature in the music psychology and neuroscience fields on how humans perceive rhythm. In particular, there is some literature on human tapping processes and the behaviour of musicians versus non-musicians (e.g. [155]). However, as the aim of most audio beat trackers is to return to the underlying score or performance intentions rather than replicate the perceptions of a listener, the general psychology literature will not be discussed in detail here.

4.4 Onset Detection

While the metre and tempo of a piece of music can be thought of as a constantly evolving signals, the musical events which underpin this are the starts of notes, and these are discrete events. Many methods for beat tracking deal with symbolic or MIDI data which represent these note start (onset) times. It is highly possible, and indeed common, to simply attach an onset detector to find the note starts in an audio signal and then track the resulting set of discrete impulses. When this approach is used, the success of any beat tracker is dependent upon the reliability of the data which is provided as an input. Thus, detecting note starts in the audio can be as important as the actual beat-tracking algorithm.

Note ends, even when played exactly as written in the score, can be ignored as unreliable indications of beat due to reverberation, sustain or at the opposite extreme, staccato events, where the note is cut short.

Note sources generally fall into two categories: harmonic and percussive. The former produce sounds which would be regarded as notes, have an identifiable pitch and harmonically related partials. Percussive sounds, in comparison, are more analogous to noise clouds. Drums and cymbals are the obvious examples of this class. It should be noted that many (indeed most) pitched instruments have a transient onset which has much in common with percussive sounds. Percussive sounds are usually characterized by significant increases in signal energy (a 'transient') and methods for detecting this type of musical sound are relatively well developed. Harmonic change with little associated energy variation is much harder to reliably detect and has received less attention in the literature. Two recent studies of onset detection are Bello et al. [30] and Collins [95].

While the discussion below assumes that a hard detection decision is made as to whether an onset is present at a given location, the beat trackers discussed below which work on continuous detection functions also need to transform the raw audio into something more amenable. They also process the signal in ways similar to those described below but do not perform the step of making hard onset detection decisions, instead leaving this to the later beat-tracking process. The hard-decision onset detection method yields a set of discrete onset times, whereas the latter method results in a continuous function from which beat tracking is performed.

4.4.1 Transient Event Detection

Transient events, such as drum sounds or the start of notes with a significant energy change (e.g. piano, guitar), are easily detected by examining the signal envelope. A typical approach, which is an adaptation of methods used by a variety of other researchers [148], [392], [564], proceeds as follows: An energy envelope function $E_j(t)$ is formed by summing the power of frequency components in the spectrogram for each time slice over the range required:

$$E_j(n) = \sum_{k \in \kappa_j} |\mathsf{STFT}_x^w(n,k)|^2, \qquad (4.1)$$

where $\mathsf{STFT}_x^w(k,n)$ is the short-time Fourier transform (STFT) of the signal $x(n)$ with rectangular window w centred at time n; k is the frequency index (see Chapter 2 for details). Usually analysis frames of about 20 ms are used in computing the energy envelope, with 50–75% overlap between successive frames. Different bands j can be used; for instance, low frequency information covering the range 20–200 Hz is useful to separate. Setting κ_j to the middle range of 200 Hz to 15 kHz covers the majority of the harmonic information; meanwhile, extending over 15–22.05 kHz (assuming a sample rate of 44.1 kHz), the upper band is often generally free from harmonic content but contains a clear indication of any strong transient information [444]. This is contrary to the opinion of Duxbury [164], who claimed that there is no useful information in this range. Many other ways to split the frequency spectrum have also been proposed. One common approach is to use 5–10 sub-bands that are distributed uniformly on a logarithmic frequency scale.

$E_j(n)$ is not an ideal signal representation for detecting onsets. A potential approach for improving it uses a three-point linear regression to find $D_j(n)$, the gradient of $E_j(n)$, and peaks in this function are detected. The linear regression fits a line $Y_i = a + bX_i + e_i$ to a set of N data pairs; we are only interested in the estimate of b which is given by $\hat{b} = (\sum_{i=1}^{N} X_i Y_i - N\bar{X}\bar{Y})/(\sum_{i=1}^{N} X_i^2 - N\bar{X}^2)$, where $\bar{X}$ and $\bar{Y}$ denote the average of X and Y, respectively. In the case here, X is the equi-spaced set of time indices n in $E_j(n)$ and Y is the corresponding E_j. In the case where $N = 3$, this reduces to

$$D_j(n) = \frac{E_j(n+1) - E_j(n-1)}{3}. \tag{4.2}$$

It should be noted that the commonly used technique of differencing the signal, where $D_j(n) = E_j(n) - E_j(n-1)$, is simply linear regression with N set to 2. The linear regression approach, like that of Klapuri [347], aims to detect the start of the transient, rather than the moment it reaches its peak power.

$D_j(n)$ is often called a detection function [30] and is a transformed and reduced signal representation. Subsequent processing needs to detect the onsets contained within this. This is usually done by simply selecting maxima in $D_j(n)$ and discarding peaks which do not pass a series of tests. Low-energy peaks should be ignored (for instance by testing if they are less than two times the local 1.5-second average of E_j) and peaks can also be ignored if there is a higher-energy peak in the local vicinity[4] by using Dixon's timing criterion [148]. Thresholds and constants are usually heuristically determined and designed to give reasonable performance with a large range of styles. Figure 4.2 shows an example of a peak extraction method. When several sub-bands j are involved, the functions $D_j(n)$ can be combined by half-wave rectifying and across-band summing before the peak-picking process [37], [347].

4.4.2 Pitched Event Detection

Detection of note starts where there is no associated energy transient (e.g. violins, choral music) has received less attention than the easier problem addressed above. Notable recent exceptions are Klapuri [349], who used very narrow frequency bands to detect changes in frequency; Laurent et al. [395], who used wavelets; Davy and Godsill [123], who took a support vector machine approach; Desobry et al. [143], who furthered Davy's research and also used kernel methods; and Abdallah and Plumbley [1], who used independent component analysis (ICA) to generate a 'surprise' measure followed by an HMM to perform reliable detection. Also, Bello et al. [31] utilized phase inconsistencies in a manner very similar to time reassignment and Duxbury et al. [164] proposed a spectral change distance measure adapted from the Euclidean measure which was then applied to adjacent spectrogram frames. Recently, Duxbury, Bello et al. [162], [163] have combined the previous two approaches into a single measure for detection of harmonic changes via either

---

[4]This is similar to the psychoacoustic masking thresholds found for humans [475], [694].

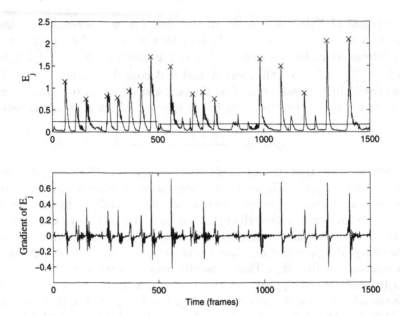

Fig. 4.2. Example of onset detection for transient events. The upper plot shows the energy-based detection function, $E_j(n)$; also shown are horizontal lines giving the 1.5 s local average of the energy function and ×'s showing the detected onsets. The lower plot shows the gradient function $D_j(n)$ from which peaks are found.

or both of phase inconsistency or spectral change. This method shows improvements over both individual approaches.

An alternative proposed by Hainsworth and Macleod [265] is the so-called modified Kullback–Leibler distance measure given by

$$d_n(k) = \log_2 \left(\frac{|\mathsf{STFT}_x^{\mathrm{w}}(n, k)|}{|\mathsf{STFT}_x^{\mathrm{w}}(n - 1, k)|} \right), \tag{4.3}$$

$$d_{\mathrm{MKL}}(n) = \sum_{k \in \mathcal{K}, \mathrm{d}(k) > 0} d_n(k), \tag{4.4}$$

where $\mathsf{STFT}_x^{\mathrm{w}}(n, k)$ is the STFT computed with window w. The measure emphasizes positive energy change between successive frames and $\mathcal{K}$ defines the spectral range over which the distance is evaluated (30 Hz to 5 kHz is suggested as it represents the majority of clear harmonic information in the spectrum). Another advantage of this method is that it also takes into account any transient energy which happens to be present as a useful aid.

A window length of about 90 ms is sufficient to give good spectral resolution. To overcome frame to frame variation, histogramming of five frames (weighted backwards and forwards with a triangular function) before and after the potential change point was used and also a very short frame hop length (namely, 87.5% overlap) was chosen to increase time resolution.

Detection of the peaks in this measure is a separate problem and is discussed more fully in [265]. Figure 4.3 gives an example of detection using this method.

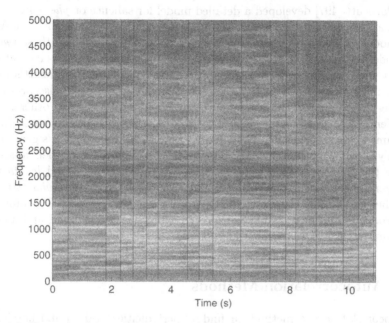

Fig. 4.3. Example of the output from the MKL harmonic change detection measure for an excerpt of Byrd's 4-Part Mass. Onsets were missed at 1 s and 5.9 s while the onset at 8.1 s is mis-estimated and should occur about 0.1 s later.

4.5 Rule-Based Approaches

We shall now discuss a number of broad methodologies for beat tracking in turn. Rule-based approaches were among the earliest used when computers were not capable of running complex algorithms. They tend to be simple and encode sensible music-theoretic rules. Tests were often done by hand and were limited to short examples. Often these did not even have expressive timing added to them and only aimed to extract the most likely pulse given the rhythmic pattern and tempo.

Steedman [607] produced one of the earliest computational models for rhythmic analysis of music. His input was symbolic and with a combination of musical structure recognition (especially melodic repetition) and psychologically motivated processing, he attempted to parse the rhythmic structure of Bach's 'Well Tempered Clavier' set of melodies. Similarly, Longuet-Higgins and Lee [418] proposed a series of psychologically motivated rules for finding the beat and higher metrical levels from lists of onset times in a monophonic melody. The rules were never implemented by the authors in the original paper for more than five-bar examples, though there have since been several papers by Lee which are summarized by Desain and Honing [141].

Parncutt [497] developed a detailed model for salience or *phenomenal accent*, as he termed it, and used this to inform a beat induction algorithm. Also, he modelled medium tempo preference explicitly and combined these two in a model to predict the tactus for a series of repeated rhythms played at different speeds. Comparison to human preferences was good. Parncutt's focus was similar to that of Povel and Essens [529], while Eck [165] also produced a rule-based model which he compared to Povel and Essens and others.

Temperley and Sleator [622] also used a series of rules to parse MIDI streams for beat structure. They quoted Lerdahl and Jackendoff's generative theory of tonal music (GTTM) [404] as the starting point of their analysis, using the GTTM *event* rule (align beats with event onsets) and *length* rule (longer notes aligned with strong beats). Other rules such as *regularity* and a number based on harmonic content were also bought into play. The aim was to produce a full beat structure from the expressive MIDI input, and a good amount of success was achieved.[5]

4.6 Autocorrelation Methods

Autocorrelation is a method for finding periodicities in data and has hence been used in several studies. Without subsequent processing, it can only find tempo and not the beat phase.

The basic approach is to define an energy function $E(n)$ to which local autocorrelation is then applied (in frames of length T_w, centred at time n):

$$r(n, i) = \sum_{u=-(T_w/2)+1}^{T_w/2} E(n + u)E(n + u - i). \qquad (4.5)$$

The value of i which maximizes $r(n, i)$ should correspond to the period-length of a metrical level. This will often be the beat, but it is possible that if the tatum is strong that autocorrelation will pick this instead.

Tzanetakis et al. [631], [633] included a series of rhythmic features in their algorithm for classification of musical genre. While not specifically extracting a beat, it performs a function similar to beat analysis. Their method was based upon the wavelet transform, followed by rectification, normalization, and summation over different bands before using autocorrelation to extract periodicity. Local autocorrelation functions were then histogrammed over the entire piece to extract a set of features for further use; these tend to show more coherence for rock pieces than for classical music.

---

[5]Source code for Temperley's method is available in [596].

Foote and Uchihashi [194] used the principle of audio self-similarity to examine rhythmic structure. The assumption was that within the space of a single sub-beat, the sound is approximately constant and therefore the spectrum will have high similarity. They therefore defined a similarity measure as the normalized scalar product (computed over the frequencies k) of the magnitude spectra of frames at times n_i and n_j

$$d_{\text{Foote}}(n_i, n_j) = \frac{\langle |\text{STFT}_x^w(n_i, k)|, |\text{STFT}_x^w(n_j, k)| \rangle}{\|\text{STFT}_x^w(n_i, k)\| \; \|\text{STFT}_x^w(n_j, k)\|}, \quad (4.6)$$

where w is some window. This produced a two-dimensional plot of similarity between any two frames of the audio signal, which was then autocorrelatively analysed for tempo hypotheses using

$$B(n_i, n_j) = \sum_{i', j'} d_{\text{Foote}}(n_{i'}, n_{j'}) d_{\text{Foote}}(n_i + n_{i'}, n_j + n_{j'}). \quad (4.7)$$

This was extended to be time varying, hence producing their 'beat spectrogram', which was a plot of the local tempo hypothesis versus time.

Other autocorrelation approaches include Mayor [445], who presented a somewhat heuristic approach to audio beat tracking: a simple multiple hypothesis algorithm was maintained which operated on his so-called BPM spectrogram, BPM referring to beats per minute. Also Paulus and Klapuri's method [503] for audio beat analysis utilized an autocorrelation-like function (based on de Cheveigné's fundamental frequency estimation algorithm [135]), which was then Fourier transformed to find the tatum. Higher-level metrical structures were inferred with probability distributions based on accent information derived using the tatum level. This was then used as part of an algorithm to measure the similarity of acoustic rhythmic patterns. Brown [55] used her narrowed autocorrelation method to examine the pulse in musical scores. Davies and Plumbley [118] and Alonso et al. [15], [16] have also produced autocorrelation-based beat trackers.

4.7 Oscillating Filter Approaches

There are two distinct approaches using oscillating filters: In the first, an adaptive oscillator is excited by an input signal and, hopefully, the oscillator will resonate at the frequency of the beat. The second method uses a bank of resonators at fixed frequencies which are exposed to the signal and the filter with the maximum response is picked for the tempo. Beat location can be calculated by examining the phase of the oscillator. This method is particularly suited to causal analysis.

The first, single-filter approach is typified by Large [389], [390], who used a single non-linear oscillator with adaptive parameters for the phase, frequency, and update rate, though these were initialized to the correct settings by hand.

The observed signal is a set of impulses $s(n) = 1$ when there is an onset event and $s(n) = 0$ otherwise. The oscillator is given by

$$o(n) = 1 + \tanh \alpha(\cos 2\pi\phi(n) - 1), \tag{4.8}$$

where $o(n)$ defines an output waveform with pulses at beat locations with width tuned by α; see Fig. 4.4. The phase is given by

$$\phi(n) = \frac{n - n_i}{p}, \tag{4.9}$$

where n_i is the location of the previous beat and p is the period of oscillation (tempo). Crucially, the single oscillator in (4.8) is assumed not to have a fixed period or phase and updates are calculated every time an onset event is observed in $s(n)$ using

$$\Delta n_i = \eta_1 s(n) \tfrac{p}{2\pi} \text{sech}^2 \{\alpha(\cos 2\pi\phi(n) - 1)\} \sin 2\pi\phi(n), \tag{4.10}$$

$$\Delta p = \eta_2 s(n) \tfrac{p}{2\pi} \text{sech}^2 \{\alpha(\cos 2\pi\phi(n) - 1)\} \sin 2\pi\phi(n), \tag{4.11}$$

where η_1 and η_2 are 'coupling strength' parameters. The update equations enable the estimation of the unknown parameters p and n_i. Marolt [433], however, points out that oscillators can be relatively slow to converge because they adapt only once per observation.

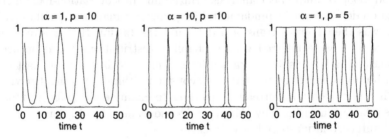

Fig. 4.4. Example output signals $o(t)$ generated using (4.8) for various values of α and p.

Large's test data was a series of impulses derived from expressive MIDI performances and the aim was to track the pulse through the example. An extra level of complexity which allowed the system to continue following the beat was to have a second oscillator 180° out of phase which could take over control from the first if confidence dropped below a certain threshold.

McAuley [450] presented a similar adaptive oscillator model to that of Large and indeed compared and contrasted the two models. Similarly, Toivi-ainen [626] extended Large's model to have short- and long-term adaption mechanisms. The former was designed to cope with local timing deviations while the latter followed tempo changes. It was tested on expressive MIDI performances. Another variation is that of Eck [166], who used Fitzhugh–Nagumo oscillators (models of neural action) linked by Heaviside coupling functions into networks. His focus was to reproduce the downbeat extraction of Povel and Essens [529] from synthetic onset data. Various authors [166], [389] have also suggested that adaptive filters have neurological plausibility and this is their motivation for its use.

The second approach is typified by Scheirer [564], who produced one of the first systems for beat tracking of musical audio. The difference compared with Large's method is that Scheirer's method implemented a bank of comb filters at different fixed feedback delays and searched for the one which resonated best with the input signal at any given time. It should be noted that the bank responds in a comb-like manner with multiples and subdivisions of the tempo also showing resonation to the signal. Scheirer implemented 150 filters logarithmically spaced between 60 bpm and 240 bpm, where bpm stands for 'beats per minute'. The input audio signal was treated in six sub-bands to find rectified power envelopes as a function of time. Each sub-band was processed by a separate comb-filter bank before the outputs were summed and the oscillator with the greatest response picked as the current tempo. Phase was also considered so as to generate a tapping signal corresponding to the tactus.

The model worked with considerable success, although there remained the problem of a 2–3 second burn-in period needed to stabilize the filters, and also a propensity for the algorithm to switch between tracking the tactus and its subdivisions/multiples since Scheirer did not explicitly address the stability of the beat estimate. Klapuri [349] (see below) capitalized on the latter observation in his method, using a bank of comb-filter resonators as the initial processing method for his system. McKinney and Moelants [452] also found a resonator method for tempo extraction to outperform histogramming and autocorrelation approaches.

4.8 Histogramming Methods

Several approaches have focused on audio beat tracking using histogramming of inter-onset intervals. First, the signal is analysed to extract onsets before the subsequent processing takes place. This was discussed above in Section 4.4. Differences between successive onsets can be used (first-order intervals), though it is more productive to also use the differences between onsets that are further apart (all-order intervals). The motivation for this is that often the successive onsets define the tatum pulse rather than the tactus, which can be better found using onsets spaced further apart. Histogramming has similarities to the autocorrelation approaches of Section 4.6, though with a discrete input rather than the continuous signal used for autocorrelation.

There are various methods of performing the histogramming operation; defining the set of calculated inter-onset intervals (IOIs), denoted o_i, $i = 1, 2, \ldots$, one can follow Seppänen [572] and divide the IOI time axis into J

bins and count the number of IOIs which fall in each: $h(j) = \text{count}(i, |o_i - u(j + 0.5)| < 0.5u)$ where u is the width of a bin. In contrast, Gouyon et al. [247] and Hainsworth [263] treat the IOI data as a set of Dirac delta functions and convolve this with a suitable shape function (e.g. a Gaussian). The resulting function generates a smoothly varying histogram. This is defined as $h(j) = \sum_i o_i * \mathcal{N}(j)$, where $*$ denotes convolution and $\mathcal{N}(j)$ is a suitable Gaussian function (low variance is desirable). Peaks can then be identified and the maximum taken as the tempo. Alternatively, Dixon [148] gives pseudocode for an IOI histogram clustering scheme.

Seppänen [572] produced an archetypal histogramming method. After an onset detection stage, he first extracted tatums via an inter-onset interval histogramming method. He then extracted a large number of features (intended to measure the musical onset salience) with the tatum signal informing the locations for analysis. These features were then used as the input to an algorithm based on pattern recognition techniques to derive higher metrical levels including the pulse and bar lines. Seppänen [573] gives further details of the tatum analysis part of the algorithm. The final thing to note is that the method was the first to be tested on a statistically significant audio database (around three hundred examples, with an average length of about one minute).

Gouyon et al. [247] applied a process of onset detection to musical audio followed by inter-onset interval histogramming to produce a beat spectrum. The highest peak (which invariably corresponded to the tatum) was then chosen as the 'tick'. This was then used to attempt drum sound labelling in audio signals consisting solely of drums [245], to modify the amount of swing in audio samples [244], and to investigate reliable measures for higher beat level discrimination (i.e., to determine whether the beat divided into groups of two or three) [246]. Other histogramming methods include Wang and Vilermo [661], Uhle and Herre [635], and Jensen and Andersen [318], all of which present variations on the general approach and use the results for different applications.

4.9 Multiple Agent Approaches

Multiple agent methods are a computer science architecture. While there is a great deal of variation in the actual implementation and often the finer details are left unreported, the basic philosophy is to have a number of agents or hypotheses which track independently; these maintain an expectation of the underlying beat process and are scored with their match to the data. Low-scoring agents are killed while high-scoring ones may be branched to cover differing local hypotheses. At the end of the signal, the agent with the highest score wins and is chosen. Older multiple agent architectures include the influential model of Allen and Dannenberg [14] and Rosenthal [547]. The two most notable multiple agent architectures are those of Goto and that of Dixon.

Goto has produced a number of papers on audio beat tracking of which [221], [238], [240] are a good summary. His first method centred on a multiple agent architecture where there were fourteen transient onset finders with slightly varying parameters, each of which fed a pair of tempo hypothesis agents (one of which was at double the tempo of the other). A manager then selected the most reliable pulse hypothesis as the tempo at that instant, thereby making the algorithm causal. Expected drum patterns as a strong prior source of information were used and tempo was tracked at one sub-beat level (twice the speed) as well as the pulse in order to increase robustness.

This method worked well for audio signals with drums but failed on other types of music. Thus, he expanded the original scheme to include chord change detection [240], each hypothesis maintaining a separate segmentation scheme and comparing chords before and after a beat boundary.

Dixon [148] has also investigated beat tracking both for MIDI and audio, with the aim of outputting a sequence of beat times. The algorithm performed well with a MIDI input, and with the addition of an energy envelope onset detection algorithm, it could also be used for audio (though with lower performance). The approach was based upon maintaining a number of hypotheses which extended themselves by predicting beat times using the past tempo trajectory, scored themselves on musical salience, and updated the (local) tempo estimate given the latest observation. The tempo update was a function of the time coherence of the onset, while the salience measure included pitch and chord functions where the MIDI data was available. Hypotheses could be branched if onsets fell inside an outer window of tolerance, the new hypothesis assuming that the onset was erroneous and maintaining an unadjusted tempo. Initialization was by analysis of the inter-onset interval histogram. Dixon has also used his beat tracker to aid the classification of ballroom dance samples by extracting rhythmic profiles [149].

4.10 Probabilistic Models

Probabilistic approaches can have similarities to multiple agent architectures in that the models underlying each can be very similar. However, while the latter use a number of discrete agents which assess themselves in isolation, probabilistic models maintain distributions of all parameters and use these to arrive at the best hypothesis. Thus, there is an explicit, underlying model specified for the rhythm process, the parameters of which are then estimated by the algorithm. This allows the use of standard estimation procedures such as the Kalman filter [41], Markov chain Monte Carlo (MCMC) methods [208], or sequential Monte Carlo (particle filtering) algorithms [22] (see Chapter 2 for an overview of these methods).

This section will concentrate on some of the models developed rather than details of the estimation procedures which are used to evaluate the final answer, as these can often be interchangeable (a point made by Cemgil, who used a variety of estimation algorithms with the same model [77]).

Again, the various methods can be broken down into two general groups: those that work with a set of MIDI onsets (or equivalently a set of onsets extracted from an audio sample) and those that work to directly model a continuous detection function[6] computed from the original signal.

4.10.1 Discrete Onset Models

Those who have worked on the problem include Cemgil et al. [77], who worked with MIDI signals, and Hainsworth [263], who used Cemgil's algorithm as a starting point for use with audio signals.

The crux of the method is to define a model for the sequential update of a tempo process. This is evaluated at discrete intervals which correspond to note onsets. The tempo process has two elements: the first defines the tempo and phase of the beat process. The second is a random process which proposes notations for the rhythm given the tempo and phase. A simple example of this is that, given a tempo, the time between onsets could either be notated as a quaver or a crotchet, one speeding the tempo up and the other requiring it to slow down. The probabilistic model will propose both and see which is more likely, given the past data (and future if allowed).

The model naturally falls into the framework for jump-Markov linear systems where the basic equations for update of the beat process are given by

$$\boldsymbol{\theta}_n = \boldsymbol{\Phi}_n(\gamma_n)\boldsymbol{\theta}_{n-1} + \mathbf{v}_n, \qquad (4.12)$$

$$s_n = \mathbf{H}_n\boldsymbol{\theta}_n + \epsilon_n. \qquad (4.13)$$

$\{s_n\}$ is the set of observed onset times, while $\boldsymbol{\theta}_n$ is the tempo process at iteration (observed onset) n and can be expanded as

$$\boldsymbol{\theta}_n = \begin{bmatrix} \rho_n \\ \Delta_n \end{bmatrix}. \qquad (4.14)$$

ρ_n is the predicted time of the nth observation s_n, and Δ_n is the beat period in seconds, i.e. $\Delta_n = 60/p_n$ where p_n is the tempo in beats per minute. $\boldsymbol{\Phi}_n(\gamma_n)$ is the state update matrix, $\mathbf{H}_n = \begin{bmatrix} 1 & 0 \end{bmatrix}$ is the observation model matrix, and $\mathbf{v}_n$ and ϵ_n are noise terms; these will be described in turn.

The principal problem is one of quantization—deciding to which beat or sub-beat in the score an onset should be assigned. To solve this, the idealized

---

[6]Strictly speaking, it will be pseudo-continuous due to sampling.

Fig. 4.5. Figure showing two identical isochronous rhythms. The top rhythm is much more likely in a musical notation context than the lower.

(quantized) number of beats between onsets is encoded as the random jump parameter, γ_n, in $\boldsymbol{\Phi}_n(\gamma_n)$,

$$\boldsymbol{\Phi}_n(\gamma_n) = \begin{bmatrix} 1 & \gamma_n \\ 0 & 1 \end{bmatrix}, \tag{4.15}$$

$$\gamma_n = c_n - c_{n-1}. \tag{4.16}$$

While the state transition matrix is dependent upon γ_n, this is a difference term between two absolute locations, c_n and c_{n-1}. c_n is the unknown quantized number of beats between the start of the sample and the nth observed onset. It is this absolute location which is important and the prior on c_n becomes critical in determining the performance characteristics. This can be elucidated by considering a simple isochronous set of onsets—if absolute score location is unimportant, then the model has no way of preferring aligning them to be on the beat over placing them on, say, the first semiquaver of each beat. This is demonstrated in Fig. 4.5. Cemgil [77] broke a single beat into subdivisions of two and used a prior related to the number of significant digits in the binary expansion of the quantized location. In MIDI signals there are no spurious onset observations and the onset times are accurate. In audio signals, however, the event detection process introduces errors both in localization accuracy and in generating completely spurious events. Thus, Cemgil's prior is not rich enough; also, it cannot cope with compound time, triplet figures, or swing. To overcome this, Hainsworth [263] broke down notated beats into 24 sub-beat locations, $c_n = \{1/24, 2/24, \dots, 24/24, 25/24, \dots\}$, and a prior was assigned to the fractional part of c_n,

$$\mathsf{p}(c_n) \propto \exp\left(-\lambda \log_2\{\underline{c}_n\}\right), \tag{4.17}$$

where $\underline{c}_n$ is the denominator of the fraction of c_n when expressed in its most reduced form; i.e., $d(3/24) = 8$, $d(36/24) = 2$, etc. λ is a scale parameter determining the sensitivity of the prior. This is shown graphically in Fig. 4.6. The prior is improper (i.e., it does not sum to unity), which is why $\mathsf{p}(c_n)$ is only expressed as a proportionality. The integer part of c_n increases as the number of beats processed increases. As a result of this, γ_n is always strictly positive; it will be less than 1 if a sub-beat interval is observed, but if there is more than one beat between observed, onsets, γ_n will be greater than 1.

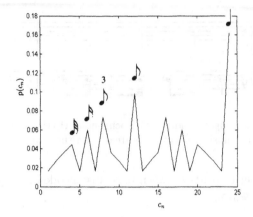

Fig. 4.6. Graphical description of the prior upon c_n. The horizontal axis is the sub-beat location from 1 to 24, while the associated probability $p(c_n)$ is shown on the vertical axis.

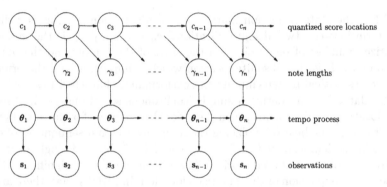

Fig. 4.7. Directed acyclic graph of the jump-Markov linear system beat model. The dependence between c_n and γ_n is deterministic, while other dependencies are stochastic.

The tempo process has an initial prior, $p(\boldsymbol{\theta}_0)$, associated with it. For the purposes of a general beat-tracking algorithm, it is assumed that the likely tempo range is 60 bpm to 200 bpm and that the prior is uniform within this range.

So far, the model for tempo evolution and proposing a set of onset times has been considered. Finally, the observation model must be specified. s_n is the nth observed onset time and therefore corresponds to the ρ_n in $\boldsymbol{\theta}_n$. Thus, $\mathbf{H}_n = \begin{bmatrix} 1 & 0 \end{bmatrix}$. The state evolution error, $\mathbf{v}_n$, and observation error, ϵ_n, are given suitable distributions—usually for mathematical convenience, these are zero-mean Gaussians with appropriate covariances [26]. The overall model can be summarized by a directed acyclic graph (DAG) as shown in Fig. 4.7. It should be noted that even spurious onsets are assigned a score location.

When working with real-world audio signals, more information than just the onset times can be extracted from the signal, and this can aid the analysis of the rhythm. The most obvious example is the amplitude of onsets while others include a measure of chordal change and other 'salience' features as postulated by Parncutt [497] and Lerdahl and Jackendoff [404]. Hainsworth [263] utilized these in his model as a separate jump-Markov linear system for amplitude and a zero-order Markov model for salience (here, the salience is only a function of the current state and has no sequential dependency). There has been little research into appropriate measures of salience for extracting accents in music; other than the papers mentioned above, Seppänen [573] and Klapuri [349] also proposed features which perform this function.

Given the above system, various estimation procedures exist. Cemgil [77] described the implementation of MCMC methods as well as particle filters to estimate the *maximum a posteriori* (MAP) estimate for the rhythm process, while Hainsworth [263] utilized particle filters to find the MAP estimate for the posterior of interest given by $p(c_{1:n}, \boldsymbol{\theta}_{1:n}, \alpha_{1:n} | s_{1:n}, a_{1:n}, \mathcal{S}_{1:n})$, where $\alpha_{1:n}$ was the underlying amplitude process observed as $a_{1:n}$, and $\mathcal{S}_{1:n}$ was the observed set of saliences. Full details can be found in either of the publications.

Other similar methods include an earlier approach of Cemgil's [79] where what he termed the 'tempogram' (which convolved a Gaussian function with the onset time vector and then used a localized tempo basis-function[7] to extract a measure of tempo strength over time) was tracked with a Kalman filter [41] to find the path of maximum tempo smoothness.

Raphael's methods [537] were based around hidden Markov models where a triple-layered dependency structure was used: quantized beat locations informed a tempo process which in turn informed an observation layer. The Markov transitions were learned between states from training data, and then the rhythmic parse evaluated in a sequential manner to decide which was the most likely tempo/beat hypothesis. This was tested on both MIDI and audio (after onset detection) and success was good on the limited number of examples, though manual correction from time to time was permitted.

Laroche [392], [393] used a maximum likelihood framework to search for the set of tempo parameters which best fit an audio data sample. The input was processed by typical energy envelope difference methods to extract a list of onset times. Inter-onset times (which are phase independent) were then used to provide likelihoods for the 2-D search space with discretized tempo and swing as the two axes. This algorithm has been included in commercially available Creative sound modules for several years. Lang and de Freitas [387] presented a very similar algorithm to that of Laroche but used a continuous signal representation and a slightly more complex estimation procedure.

---

[7]The tempo basis function was defined as a set of weighted Dirac functions $\psi(t; \tau, \omega) = \sum_{i=-\infty}^{\infty} \alpha_i \delta_{\tau + i 2^\omega}(t)$ at a delay of τ and spaced with frequency (and hence tempo) given by ω.

Hainsworth also presented a second algorithm which is essentially a reformulation of the above but using Brownian motion relations as a base [266]. It was not as successful as the above model. Others include Takeda et al. [617] and Lam and Godsill [386].

4.10.2 Continuous Signal Representations

The second approach to tracking the beat with stochastic models uses a detection function and attempts to model this directly instead of extracting onsets first. As such it must have all the elements of the above models, including a tempo process and a model for the likelihood of an onset being present at any given beat or sub-beat location; however, it must also have a model for the signal itself and what is expected at an onset and between these.

Hainsworth [263] proposed a method using particle filters whereby the tempo was modelled as a constant velocity process similar to the one described above and which proposed onsets in a generative manner at likely sub-beat locations. The signal detection function modelled was a differenced energy waveform, utilizing high-frequency information, very similar to $D_j(t)$ shown in the lower plot of Fig. 4.2.

Onset locations can clearly be seen in this signal representation, and on close examination all onsets have a very similar evolution in time which can be well modelled by a hidden Markov model (HMM; see Chapter 2 for a definition). This is performing the task of onset detection. The model used is shown in Fig. 4.8 with each state having a different output distribution (also termed likelihood). For mathematical convenience, these are Gaussians with differing means and variances but sufficiently separated so that the output distribution of state S_1 does not significantly overlap with that of S_0 or S_2, etc. This defines a generative model for the signal—by generative, it is meant that by using a random number generator and the specified distributions, a process with the same statistical properties as the original signal can be generated.

A naive scheme simply generates proposals from the prior distributions, but the Viterbi algorithm (see [654] and Chapter 2) can be used to find the best path through the HMM and also its probability, which simplifies the calculation needed once an onset is hypothesized. The model worked well on the small number of examples tried but required the expected sub-beat structure to be specified by hand for robust performance.

In comparison, Sethares et al. [577] proposed four filtered signals (time domain energy, spectral centroid, spectral dispersion, and one looking at group delay) which were then simply modelled as Gaussian noise with a higher variance at beat locations compared to between them. Looking back at Fig. 4.2, it can clearly be seen where the variance of the generative noise process used to model the signal would be higher. A model similar to those above was used and a particle filter environment chosen for the estimation procedure. The

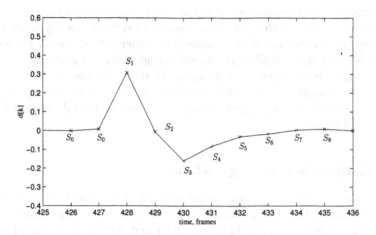

(a) Data with states superimposed.

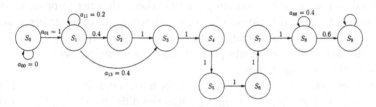

(b) Directed acyclic graph of HMM model.

Fig. 4.8. HMM for beat-tracking algorithm with Viterbi decoding included. States S_5, S_6, and S_7 are functionally equivalent to S_4, and S_8 is equivalent to S_0. The null state, S_9, has no observation associated with it, therefore making transition to it highly unattractive.

model did not explicitly include a model for sub-beats but seemed to function well on the data presented.

A somewhat different method for tracking the beat through music was presented by Klapuri et al. [348], [349]. A four-dimensional observation vector (as a function of time) was generated by applying a similar method to that of Scheirer [564] to generate resonator outputs but using different frequency bands and a different method for extracting the energy signal which also captures harmonic onsets. A measure of salience, dependent upon the normalized instantaneous energies of the comb-filter resonators, was also attached to this.

A problem with Scheirer's method was that it was prone to switch between different tempo hypotheses (usually doubling or halving), and Klapuri addressed this using an HMM to impose some smoothness to the tempo evolution. He proposed a joint density for the estimation of the period-lengths of the tatum, tactus, and measure level processes, applying a combination of sensible priors and dependencies learned from data. The phase of the tatum and tactus pulse were estimated to maximize the observed salience at beats. In estimating the phase of the super-beat (measure) structure, a key assumption made was the expectation of two simple beat patterns which occur frequently in so-called 4/4 time. While this should considerably aid performance with music in this time signature, performance in the super-beat estimation was degraded for examples with a ternary metre (e.g. 3/4). Nevertheless, the algorithm was tested on a significant database and was successful. A comparison is presented below.

4.11 Comparison of Algorithms

If the focus is restricted to beat tracking in musical audio signals, then the methods discussed above in Sections 4.5 to 4.10 have various strengths and weaknesses. This section will highlight them and then present a comparison of several methods.

Rule-based approaches have never been applied to audio and have solely been used to code sensible but simple music theoretic rules in order to model music psychology expectations. The reason that they have never been used on audio signals is possibly because they are not easily expanded to cope with erroneous data and hence would perform poorly on the inexact data produced by onset detection algorithms.

Autocorrelative and histogram methods have much in common; they are both methods of obtaining a tempo profile, the difference being that autocorrelation works with a sampled signal while histogramming works with discrete onset times. They are therefore useful for finding the tempo but are not immediately applicable to extracting the beat phase (this is a secondary task).

Adaptive oscillators are particularly suited to causal operation and have some psychoacoustic justification [390]. However, they have not been applied to audio signals. This may be because the update routines required on adaptive single filters are not easily adaptable to real data or possibly because they are not well able to cope with sub-beats. Many of the systems also required manual initiation to set the correct tempo and phase. Comb-filters as implemented by Scheirer [564] and used by Klapuri et al. [349] have been applied to audio signals.

This leaves multiple agent approaches and probabilistic, model-based methods. These two bear some significant similarities, but the latter delimits the underlying assumptions from estimation procedures whereas they are intermixed by multiple agent methods. This makes the adaption and optimization of the probabilistic models easier, though reasonable success has been reported with both approaches.

4.11.1 Tests

There has been a move in recent years towards testing algorithms with a large database of audio samples collated from all genres and usually from standard, commercially available sources. This was begun with Seppänen [572] with a database of 330 audio samples, while Klapuri [349] used 478. A comparison was also undertaken by Gouyon et al. [248] into tempo induction from audio signals using a large dataset of 3199 examples from three databases and is currently the most extensive.

The comparison below used a hand-labelled database of 222 samples of around one minute divided into six categories: rock/pop, dance, jazz, folk, classical, and choral. The tempos were limited to the range 60–200 bpm with the exception of the choral samples. Several examples exhibited significant rubato, 8 had a *rallantando* (slowing down), and 4 had a sudden tempo change. Forty-two also had varying amounts of swing added. Full details of the database can be found in [263].

Another problem is how to evaluate the performance of a beat-tracking algorithm. As of this writing, no study has yet made a serious attempt to notate the complete rhythm and idealized score locations of every onset present in the audio sample;[8] rather the assessment has been limited to 'tapping in time' to the sample and producing an output of beat times that agrees with those of trained human musicians.

Klapuri [349] gives two criteria, which are adopted here, to judge the performance of an algorithm on a particular example. The first is 'continuous length' (C-L), by which it is meant the longest continually correctly tracked segment, expressed as a percentage of the whole. Thus, a single error in the middle of a piece gives a C-L result of 50%. Another, looser criterion is simply the total percentage of the whole which is correctly tracked (defined as 'TOT' from now on). Here, both are expressed as percentages of the manually detected beats which are correctly tracked, rather than of the time stretches these represent. Using Klapuri's definitions once again, a beat is determined to be correctly tracked if the phase is within ± 15% and the tempo period is correct to within ± 10%.

Here, the trackers[9] of Scheirer [564], Klapuri [349], and Hainsworth [263] are compared and the results are shown in Table 4.2. The columns under 'Raw' are base results according to the above criteria; however, it is sometimes found that the beat tracker tracks something which is not the predefined beat but is a plausible alternative. Usually, this is half the correct tempo (in the case of fast samples) or double (for particularly slow examples). When swing is encountered, it is occasionally possible for the trackers to even track at one and a half times the tempo (i.e., tracking three to every two correct beats). Doubling or halving of tempo is psychologically plausible and hence acceptable; however the errors encountered with swing are not. The second set of columns compares results once doubling and halving of tempo are allowed. Performance on individual genres is shown graphically in Fig. 4.9 for Hainsworth's and Klapuri's algorithms.

---

[8]The closest is probably Goto and Muraoka [234].

[9]The beat trackers tested were all the original authors' own.

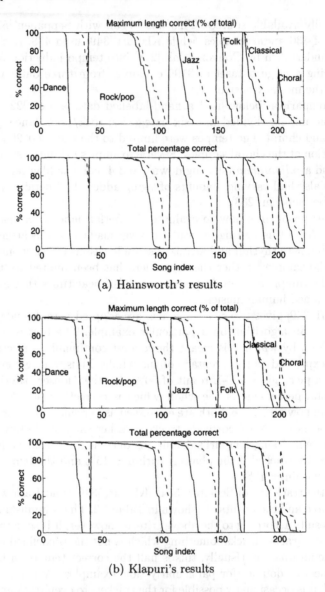

(a) Hainsworth's results

(b) Klapuri's results

Fig. 4.9. Graphical display of the results for Hainsworth's (top) and Klapuri's beat tracker. The solid line is the raw result while the dashed line is the 'allowed' result. Note that ordering is strictly by performance for each genre under any particular criteria.

Table 4.2. Comparison of results on the database. The three beat trackers use audio adata as inputs.

	Raw		Allowed	
	C-L (%)	TOT (%)	C-L (%)	TOT (%)
Hainsworth	45.1	52.3	65.5	80.4
Scheirer	23.8	38.9	29.8	48.5
Klapuri	55.9	61.4	71.2	80.9

It can be seen that Klapuri's model performs the best in terms of raw results and continuous tracking, while the performance of Hainsworth when considering total number of beats with allowed tempo mistakes is about equivalent. Klapuri's method performs better than Hainsworth's with rock/pop and dance, though it fails somewhat with jazz. Hainsworth's outperforms Klapuri's on choral music, probably because of the onset detection algorithm used by Hainsworth (described above in Section 4.4), which gives superior performance for these choral samples.

Both Klapuri's and Hainsworth's models significantly outperform Scheirer's. Klapuri [349] compared his model to Scheirer's and also Dixon's [148] modified MIDI beat-tracker. Seppänen [572] reported that his program was less successful than Scheirer's, tested on a large database that was a subset of Klapuri's. Also, on the related issue of tempo induction, the comparison by Gouyon et al. [243] showed that Klapuri's method performed the best at this task.

Finally, performance of one of the stochastic models which uses a signal representation is shown on a single example in Fig. 4.10. This shows Hainsworth's second stochastic model (described above in Section 4.10.2) with a swing example. The model is very successful at extracting onsets and is good at tempo tracking. The limitation is that the expected sub-beat structure has to be specified in advance. Thus, the model cannot be considered pan-genre.

4.12 Conclusions

This chapter has discussed a number of differing approaches to the generic task of 'beat tracking'. Under this catch-all term, there are actually a number of possible goals, from replicating human tempo preference to a full labelling of every onset as to its correct quantized score location. Recent methods have aimed to extract the correct tempo and beat phase from audio signals ('tapping in time to the music').

Current methods such as Klapuri's [349] or Hainsworth's [263], [266] are, starting to achieve a reasonable level of success over databases of significant size and complexity. However, they are less successful on certain genres such as jazz (where part of the appeal of the style is its rhythmic complexity)

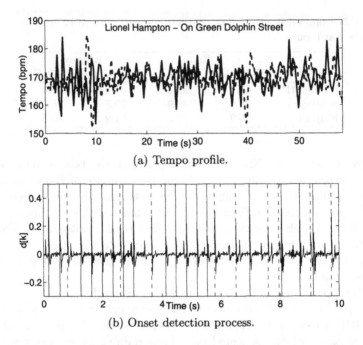

(a) Tempo profile.

(b) Onset detection process.

Fig. 4.10. Output of Hainsworth's second stochastic beat tracker (see Section 4.10.2) for a swing example. a) shows tracked tempo (dashed) and hand-labelled tempo (solid); b) shows the onset detection process for the first 10 seconds with solid vertical lines denoting detected beats and dashed vertical lines showing the detected swung quavers.

and classical music (which is prone to radical rhythmic evolution and also has fewer easily extractable beat cues). Classical music particularly seems to require pitch analysis in order to extract reliable beat cues. Thus, while the aim is obviously to have a generic beat tracker which works equally well with all genres, it is likely that in the short term, style-specific cues will have to be added. Klapuri [353] and Goto [221] both apply knowledge of typical drum patterns in popular music to their algorithms. Dixon [149] goes a step further and uses rhythmic energy patterns extracted from audio samples to aid classification of ballroom dance examples, a process which could easily be reversed to aid beat tracking.

In addition to better modelling specific styles and the rhythmic expectations therein, the second area for expansion is to look at better signal representations for extracting the cues needed to perform beat tracking. Rock and pop music with its drum-heavy style is easily processed using energy measures; classical music is much harder to process and only relatively recently have methods been applied to extract note changes where there is little transient energy. These will need to be improved.

In conclusion, the field of beat tracking or rhythmic analysis is one area of musical audio processing where some significant success has been achieved and there is much to build upon. However, there is also room for improvement and further accomplishments.

5

Unpitched Percussion Transcription

Derry FitzGerald[1] and Jouni Paulus[2]

[1] Cork Institute of Technology, Rossa Avenue, Bishopstown, Cork, Ireland
derry.fitzgerald@cit.ie
[2] Institute of Signal Processing, Tampere University of Technology,
Korkeakoulunkatu 1, 33720 Tampere, Finland
jouni.paulus@tut.fi

5.1 Introduction

Up until recently, work on automatic music transcription has concentrated mainly on the transcription of pitched instruments, i.e., melodies. However, during the past few years there has been a growing interest in the problem of transcription of percussive instruments. This chapter aims to give an overview of the methods used in this field ranging from the pioneering works of the 1980s to more recent systems.

There currently exists a large family of percussive instruments, many of which have been in existence for quite a long time [293]. The work presented in this chapter concentrates on the transcription of unpitched[3] percussive instruments. In particular, the vast majority of research to date has focused on the unpitched percussion instruments found in Western popular music, with a particular focus on the drums found in the standard rock/pop drum kit, namely snare drum, kick drum (also known as a bass drum), tom-toms, hi-hats, and cymbals. A notable exception to this is the work on tabla transcription by Gillet and Richard [211]. However, many of these methods could be utilized for transcribing other percussive sounds if suitable modifications are made.

The percussion instruments mentioned in this chapter can be divided into two main types: membranophones and idiophones. Membranophones, including drums such as snare drums, kick drums, and tom-toms, typically consist of a membrane or skin stretched across a frame. Idiophones, including instruments such as hi-hats and cymbals, are typically rigid bodies, such as a metal plate. In both cases, sound is produced by striking the membrane or plate. In a standard rock/pop drum kit, striking is usually done with a wooden drumstick, except for the kick drum which is struck using a beater made of epoxy or rubber mounted on a foot pedal. The striking of a given drum can be modelled

---

[3] The word 'unpitched' is used here to emphasize the fact that the instruments are normally not used to play melodies, even though many drums can be tuned and their sound evokes a perception of pitch.

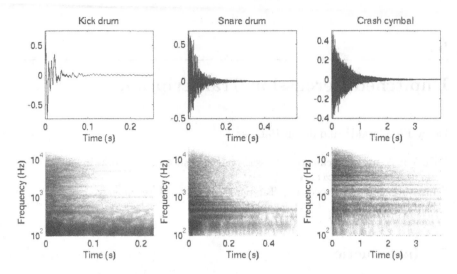

Fig. 5.1. Example waveforms. The images on the top row are the time domain waveforms of a kick drum, a snare drum, and a crash cymbal, from left to right. The lower row contains the corresponding spectrograms in the same order. The sound samples are from the RWC Musical Instrument Sound Database [230].

as an impulse function and so a broad range of frequencies will occur in the impact. As a result, all possible modes of vibration of the plate or membrane are excited simultaneously, and the narrower the frequency band associated with a given mode, the longer it sounds. The interested reader is referred to [193] for a mathematical discussion of the properties of ideal membranes and plates.

Many of the membranophones used in a standard rock/pop drum kit can be tuned by adjusting the tension of the membrane. In conjunction with the different sizes available for each drum type, this results in considerable variations in the timbre obtained within a given drum type. Nevertheless, it can be noted that the membranophones have most of their spectral energy contained in the lower regions of the frequency spectrum, typically below 1000 Hz, with the snare usually containing more high-frequency energy than other membranophones. Also, in the context of a given drum kit, the kick drum will have a lower spectral centroid than the other membranophones. It can also be noted that idiophones consisting of a metal plate will typically have their spectral energy spread out more evenly across the frequency spectrum than the membranophones, resulting in more high-frequency content.

Examples of three different drum instruments' time domain waveforms and spectrograms are shown in Fig. 5.1. A kick drum is purely a membranophone, containing a lot of low-frequency energy. A snare drum is also a membranophone, but it has a snare belt attached below the lower membrane. When the drum is hit, the lower membrane interacts with the snare belt, resulting in a distinct sound also containing high-frequency energy. This can be observed in the figures. A crash cymbal is a plate-like idiophone producing wide-band energy. It is also worth noticing that the crash cymbal sound rings about ten times longer than the example membranophone sounds.

The systems discussed in this chapter may process either signals that consist of percussive instruments only, or complex music signals where the presence of pitched instruments is allowed. In many cases, the input to percussion transcription systems will not contain recordings of live drums. Often the input will consist of either sampled real drums, drum loops taken from another recording, synthetic drums or even mixtures of the above. If a relatively small set of drum samples are used, the resulting signal's waveform will exhibit less variation between hits, making the signal easier to transcribe for many of the presented systems. The same applies to the use of sampled drum loops and synthetic drum sounds. Synthetically generated drum sounds tend to mimic the overall timbral characteristics of real drums to some extent, but their spectral characteristics tend to differ considerably from those of real drums. This may cause problems if the transcription system is trained with real drums.

Approaches to percussion transcription can be roughly divided into two categories: pattern recognition applied to sound events and separation-based systems. The former segments the signal into meaningful events and recognizes the contents of the segments with pattern recognition methods. This approach is described in Section 5.2. The latter approach tries to separate the mixture containing drum sounds in such a way that each segregated stream contains only hits from one given drum type. Onsets are then sought from the streams. Methods relying on this approach are described in Section 5.3. In addition, a distinction can be made between systems that use a supervised approach, through the use of trained classifiers or instrument templates, and systems that use an unsupervised approach, such as clustering similar segments followed by recognition of the clusters. Low-level signal analysis may not always yield a satisfying result, so some attempts have been made to utilize musicological modelling to take into account the predictability of drum patterns in music. These are described in Section 5.4. Finally, some conclusions about the presented methods are given in Section 5.5.

5.2 Pattern Recognition Approaches

In general, the percussion transcription problem can be characterized with two questions: When did something happen in the music, and what was the event that took place? The majority of the drum transcription systems developed so far operate by answering these two questions in this same order. Here, such systems are referred to as 'event-based' systems, and they tend to operate with roughly the following steps:

1. Segment the input signal into events by
   a) locating potential sound event onsets in the input signal, or
   b) generating a regular temporal grid over the signal.

2. Extract a set of features from each segment.
3. Classify the contents of each segment based on the extracted features.
4. Combine the segment time stamps with information about their content to yield the transcription.

These steps will be described in more detail in the following sections.

5.2.1 Temporal Segmentation

There are two main approaches to segmenting a signal containing percussive instruments. The first of these is based on detecting prominent sound event onsets in an input signal, which is a difficult task in itself. The detection method should be able to identify all the beginnings of meaningful sound events and still be robust against noise which might generate extraneous onsets. This has led to the development of a number of different onset detection methods which are detailed in Chapter 4 on p. 107.

An alternative to the onset detection-based segmentation is to generate a temporal grid over the whole signal and to use it for the segmentation, as suggested by Gouyon et al. [247]. The grid spacing is determined by the fastest rhythmic pulse present in the signal, so that almost all the events in the piece coincide with a grid point. A number of different names have been given for this rhythmic pulse: Gouyon et al. called it the tick [247], Schloss, the attack-point [567], and Bilmes, the tatum [37]. Here we use the term tatum.

There are several ways for estimating the tatum pulse, and they are discussed in more detail in Chapter 4. Gouyon et al. [247] and Seppänen [573] detected onsets in the input signal, calculated time intervals between all onset pairs, and determined the period of the tatum as the (approximate) greatest common divisor of the time intervals. The phase of the grid was estimated by aligning it with the located onsets. Another method of tatum estimation, described by Klapuri et al., utilized a bank of comb filter resonators and probabilistic modelling to find the tatum period and phase [349].

In practice, using a fixed equidistant grid has some problems. Even though musical metre analysis procedures have developed considerably in recent years and can often produce a reliable estimate of the tactus pulse (a.k.a the beat), the tatum estimate tends to be unreliable. In particular, overestimating the tatum period causes severe errors in the segmentation. Another drawback is that expressive playing causes deviations from the equidistant grid points (see for example [37] where Bilmes tried to construct a system capable of creating similar expressive deviations to rhythmic patterns). An advantage of the grid representation is that it is less prone to errors caused by inserting and deleting sound onsets than the onset detection-based approach.

Before the subsequent feature extraction, meaningful parts of the signal need to be segmented. The simplest approach is to take a part of the signal starting at the located onset or grid point, and ending at the next located onset or grid point [245], [209]. However, if the consecutive onsets are far apart or do not have a constant time difference, it is good to limit the minimum and maximum length of the segments. This guarantees that each of them contains enough information for extracting relevant features. For example, good initial guesses for the minimum and maximum lengths could be 50 ms and 200 ms, respectively. A window function can be used in connection to this. However, a traditional Hamming or Hanning window, for example, is not appropriate since it smooths out the informative attack part at the beginning of the segment. A half-Hanning window which starts from a unity value and decays to zero at the end of the segment is more suitable, but often windowing is omitted completely, assuming that events decay to small amplitude naturally and the signal does not contain sustained sounds at all.

5.2.2 Feature Extraction

The aim of feature extraction is to obtain numerical values describing the segments so that they can be recognized or grouped together, while reducing the amount irrelevant information in the time-domain signal. Quite a lot of research has been carried out to find good descriptors for percussive instrument classification [250], [245], [247], [287], [511]. Many of these features are used also in pitched musical instrument recognition and are discussed in more detail in Chapter 6. Here we will introduce only the features that are most commonly used in percussion transcription. Ideally, the features should be such that they are robust in the presence of other simultaneously occurring sounds, but in practice this is difficult to achieve.

Mel-frequency cepstral coefficients (MFCCs) describe the rough shape of the signal spectrum and are widely used in speech recognition [536]. Similarly, they are often encountered in percussion transcription algorithms. A detailed description can be found in Chapter 2, p. 25. Usually the coefficients are calculated in short (about 20 ms) partially overlapping frames over the analysed segment, and 5 to 15 coefficients are retained in each frame. Instead of using these directly as features, typically the mean and variance of each coefficient over the segment are used [211], [504], [209], [608]. In addition, the first- and second-order temporal differences of the coefficients, and the means and variances of these, are commonly used as features.

Another commonly used set of spectral features are bandwise energy descriptors. The energy content of the sound is calculated in a few frequency bands and their relations to the total signal energy are used as features. The number of bands and their spacing depends greatly on the desired frequency resolution, and systems have used from 6 [209] to 24 bands [285].

In addition to the rough spectrum described by the MFCCs and bandwise energy descriptors, more simple spectral shape features are also useful. These include the first four moments of the spectrum, i.e., spectral centroid, spectral spread, spectral skewness, and spectral kurtosis [125], [307], [514]. Let us denote the normalized magnitude spectrum by

$$\tilde{X}(k) = \frac{|X(k)|}{\sum_{k\in\mathcal{K}_+} |X(k)|}, \qquad (5.1)$$

where $X(k)$ denotes the discrete Fourier spectrum, k is a frequency index, and the set $\mathcal{K}_+$ contains only non-negative frequency indices. The spectral centroid is then defined as

$$C_f = \sum_{k\in\mathcal{K}_+} k\tilde{X}(k). \qquad (5.2)$$

The bandwidth of the spectrum is described by spectral spread,

$$S_f^2 = \sum_{k\in\mathcal{K}_+} (k - C_f)^2 \tilde{X}(k). \qquad (5.3)$$

The spectral skewness describes the asymmetry of the frequency distribution around the spectral centroid,

$$\gamma_1 = \frac{\sum_{k\in\mathcal{K}_+} (k - C_f)^3 \tilde{X}(k)}{S_f^3}. \qquad (5.4)$$

Finally, spectral kurtosis describes the peakiness of the frequency distribution,

$$\gamma_2 = \frac{\sum_{k\in\mathcal{K}_+} (k - C_f)^4 \tilde{X}(k)}{S_f^4}. \qquad (5.5)$$

The smaller the kurtosis, the flatter the spectrum. The quantities (5.2)–(5.5) can also be calculated using a logarithmic frequency scale, as suggested in the MPEG-7 standard [307].

In comparison with the spectral features, relatively few time-domain features have been used in percussive sound classification. Instead, temporal evolution of the sound is often modelled using differentials of spectral features extracted in short frames over the segment. Among the features that can be computed in the time domain, the two most commonly used are temporal centroid and zero crossing rate. The temporal centroid, a direct analogue to the spectral centroid, describes the temporal balancing point of the sound event energy by

$$C_t = \frac{\sum_t tE(t)}{\sum_t E(t)}, \qquad (5.6)$$

where $E(t)$ denotes the root-mean-square (RMS) level of the signal in a frame at time t, and the summation is done over a fixed-length segment starting at the onset of the sound event. The feature enables discrimination between short, transient-like sounds and longer ringing sounds. The zero crossing rate describes how frequently the signal changes its sign. It correlates with the spectral centroid and the perceived brightness of the signal. Usually, noise-like sounds tend to have a larger zero crossing rate than more clearly pitched or periodic sounds [250].

Feature set is generally selected through trial and error, though some automatic feature selection algorithms have also been evaluated by Herrera et al. [287]. It was noticed that in most cases, using a feature set that has been chosen via some feature selection method yielded better results than using all the available features. Also, a dimension reduction method such as principal component analysis can be applied to the set of extracted features prior to classification. For a more detailed description of feature selection methods and possible transformations, refer to Chapter 2 and Chapter 6.

5.2.3 Segment Recognition

The extracted features are then used to recognize the percussive sounds in each segment. There are at least two different ways to do this. The first is to try to detect the presence of a given drum, even if other drums occur at the same time, and the other is to attempt to recognize drum combinations directly. For example, if an input signal consists of snare and hi-hat sounds, the first approach will attempt to recognize the presence of both instruments independently from each other, while the latter will attempt to recognize whether 'snare', 'hi-hat', or 'snare + hi-hat' has occurred, treating sound combinations as unitary entities.

A problem that arises when recognizing drum combinations instead of individual drums is that the number of possible combinations can be very large. Given M different drum types which may all occur independently, there are 2^M possible combinations of them. That is, the number of combinations increases rapidly as a function of M, and it becomes difficult to cover them all. In practice, however, only a small subset of these combinations are found in real signals. Figure 5.2 illustrates the relative occurrence frequencies of the ten most common drum event combinations in a popular music database. These contribute 95% of the drum sound events in the analysed data. When focusing on the transcription of the drums commonly used in Western popular music, the number of possible sound types M has usually been limited to the range of two to eight. Some systems have concentrated on transcribing only the kick and snare drum occurrences [235], [250], [221], [693], [608], [683], whereas some others have extended the instrument set with hi-hats or cymbals [620], [505], or added even further classes such as tom-toms and various percussion instruments [506], [209].

Classification algorithms can be roughly divided into three different categories:

- decision tree methods,
- instance-based methods, and
- statistical modelling methods.

With the exception of the work by Herrera et al. [287], there has not been an extensive comparison of different classification methods as applied on percussive sounds. Also, the experiments done in [287] concentrated on the classi-

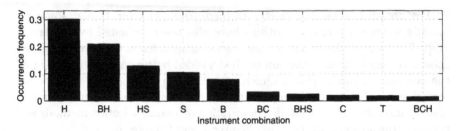

Fig. 5.2. The relative frequencies of the ten most frequently occurring drum sound combinations in the RWC Popular Music Database [229]. These combinations contribute 95% of all drum sound events present in the database. A total of five drum classes were used in the calculations, and are denoted as following: H is hi-hat, B is kick drum, S is snare drum, C is cymbal, and T is tom-tom.

fication of isolated drum sounds instead of handling entire percussive sound sequences. The results suggest that the instance-based algorithms perform best among the tested methods, and the decision tree algorithms the worst. Here we will only present briefly the methods that have been used in unpitched percussion transcription. For more detailed descriptions or other possible algorithms, refer to Chapter 2 and Chapter 6.

Decision tree classifiers operate by asking a sequence of questions about the sample to be classified. Such a question could be, for example, 'Is the spectral centroid of the signal above 500 Hz?' Each answer rules out some of the possible classification results and defines the next question. Thus, the classification process could be written out as a large conditional expression. For a more detailed description, see [533], [161]. In addition to the recognition of isolated drum sounds by Herrera et al. [287], Sandvold et al. [561] used a decision tree classifier for the transcription of continuous percussive tracks (see p. 142 below).

Instance-based methods store the given training samples (or a selected subset of them), and determine the label for the analysed sample by comparing it to the stored training data. In practical use, the principal weakness of traditional instance-based algorithms (e.g., k-nearest neighbours) is the memory required for storing the example events and the computational load needed to compute distances to all the training samples. Support vector machines (SVMs) remove these shortcomings by pre-processing the training data so as to retain only the samples that actually have an effect on the classification. The basic structure of a SVM is to locate a decision surface or a hyperplane that has the maximal margin with respect to the two classes it is trained to separate [59], [568]. The decision surface is parameterized by the sample patterns that have the smallest margin, called the support vectors.

SVMs have been used successfully in a number of classification tasks [568], and also in percussion transcription [209], [608], [210]. Originally SVMs were binary classifiers, so they have been used to detect whether or not the segment contains a percussive sound event [506] and as a detector-like recognizer for

individual drums [209], [608]. Subsequent extensions to enable multiple classes
have been developed; see [629]. Gillet and Richard evaluated the use of both
M binary SVMs and one 2^M class SVM in [209]. The results suggest that the
multiclass version performs slightly better than several binary classifiers. More
recently, the system was improved through decorrelating the used features
with principal component analysis, and adding implicit short-time context
modelling (see p. 155 for further details) in [210].

Gaussian mixture models (GMMs), a member of the statistical modelling
family, model the distribution of the feature values in each class as a sum
of Gaussian distributions. GMMs have proven to be a useful tool in pattern
recognition as they can be used to approximate arbitrary distributions. Further details on GMMs can be found in Chapter 2 or in [536]. Paulus and
Klapuri used GMMs in percussion transcription by training a separate GMM
for each of the 127 non-silent combinations of different drum instruments [506].

More commonly, GMMs are used in conjunction with hidden Markov models (HMM) to represent the feature distributions in their states. A basic HMM
consists of two parallel processes: a hidden state process which is assumed
to be a Markov chain and cannot be directly observed, and an observation
process (features). The observed features are conditioned on the hidden state
by using a GMM in each state and, based on the observations, the hidden
state sequence can be inferred. Details on HMMs can be found in Chapter 2.

HMMs have been used in percussion transcription by Gillet and Richard,
who considered target signals consisting of Indian tablas[4] [211] or Western
drum sounds [209]. In [211] the different strokes were modelled by GMMs,
and HMMs were used to model event sequences in the tabla recordings. A
recognition rate of 94% was achieved on the test database. A similar approach
was used in [209], where GMMs in conjunction with HMMs were used to
transcribe drum loops. It was also noted that a SVM classifier without any
sequence modelling performed better than a HMM-based approach [209].

A recognition approach that cannot be put directly into any of the above
categories is the template matching method proposed by Goto et al. in
what was the first attempt at transcribing polyphonic drum mixtures [241],
[236]. The system aimed to identify mixtures of kick drums, snares, tom-
toms, hi-hats, and cymbals. The templates for each drum type were obtained
from an short-time Fourier transform-based power spectrogram of each drum
type in isolation. The templates were scaled to account for amplitude differences between the templates and the mixture signal. A distance measure
was used to detect the presence of the template in the mixture. Detection of
membranophones was carried out using a logarithmic frequency resolution,
while idiophones were detected using a linear frequency resolution.

Later work by Goto identified snare and kick drum events in polyphonic
music as a means of tracking the beat in a piece of music [235]. This was done

---

[4]Tablas consist of a metallic bass drum and a wooden treble drum. Different
hand strokes on these drums produce different sounds.

by first identifying potential onsets and then creating a histogram of spectral peaks at these onset locations. The characteristic frequency of the kick drum was identified as the lowest peak of the histogram, while the characteristic frequency of the snare was identified as the largest peak in the histogram above the characteristic frequency of the kick drum. It was then adjudged that a kick drum or snare had sounded at times when an onset was detected and the onset's peak frequency coincided with the characteristic frequency of the drum in question. This method of detecting the snare was later replaced by identifying noise components widely distributed in the frequency range of 1.4 kHz to 7.5 kHz as the snare drum. Unfortunately, no evaluation of the system's performance in identifying the snare and kick drum was carried out as it focused on beat tracking, not transcription.

5.2.4 Instrument Model Adaptation

Even though isolated percussive sounds can be identified quite reliably [285], real-world recordings are not as easy to analyse. This is due to other simultaneously occurring interfering sounds, both other drums and melodic instruments, as well as the fact that drum sounds can vary between occurrences, depending on how and where they are struck. As a consequence, it is difficult to construct general acoustic models that would be applicable to any data and still discriminate reliably between different instruments.

A way to overcome this problem is to train the models with data that is as similar as possible to the target mixture signals. However, this is not possible if the exact properties of the target signals are not known in advance or they vary within the material. Model adaptation has been proposed to alleviate this problem. In this approach, the idea is to adapt general models to the mixture signal at hands, instead of using fixed models for each and every target signal. To date, only three event-based drum transcription systems have been proposed that take this approach [693], [561], [683].

The earliest percussion transcription system utilizing model adaptation was that of Zils et al. [693], which used an analysis-by-synthesis approach. Initially, simple synthetic percussion sounds $z_i(n)$ were generated from lowpass and bandpass-filtered impulses. These represented very simple approximations to kick drums and snares respectively, and were then adapted to the target signal to obtain more accurate models. The algorithm operated with the following steps:

1. Calculate correlation function between a synthetic sound event $z_i(n)$ and the polyphonic input signal $y(n)$

$$r_i(\tau) = \sum_{n=0}^{N_i-1} z_i(n)y(n+\tau), \qquad (5.7)$$

where N_i is the number of samples in the sound i, and $r_i(\tau)$ is defined for $\tau \in [0, N_y - N_i]$, where N_y is the number of samples in $y(n)$.

2. Locate occurrences of the sound $z_i(n)$ in $y(n)$ by picking peaks in $r_i(\tau)$ and by retaining only the most reliable peaks.
3. Update the sound $z_i(n)$ with

$$z_i(n) \leftarrow \frac{1}{2} \left[z_i(n) + \frac{1}{U} \sum_{j=1}^{U} y(\tau_j + n) \right], \qquad n = 0, \dots, N_i - 1, \qquad (5.8)$$

where U is the number of reliable peaks detected and τ_j contains their locations.
4. Repeat Steps 1 to 3 until convergence.

The above procedure was applied separately for the snare and kick drum. When both drums occurred simultaneously, priority was given to the kick drum, thus limiting the system to monophonic transcription of snare and kick drum in the presence of pitched instruments. The system was tested on 100 examples from various music genres, and gave a success rate of over 75% in cases where the percussion instruments were louder or as loud as the other instruments in the mixture. In cases where the percussion instruments were quieter, a 40% success rate was reported.

The system described by Yoshii et al. transcribed snare and kick drums in the presence of pitched instruments by using an analysis-by-synthesis approach in the time-frequency domain [683]. First, onset detection was carried out and excerpts $P_j(t, k)$ of the mixture power spectrogram were extracted from the positions of the onsets. These were then used to adapt snare and kick drum templates (models) which were initialized using the spectrograms of isolated examples. The following template-adaptation algorithm was repeated for both target drum classes:

1. Calculate the Euclidean distances $\Gamma_{i,j}$ between a template T_i and the spectrogram excerpts P_j with

$$\Gamma_{i,j}^2 = \sum_{t=0}^{T-1} \sum_{k=0}^{K-1} [H_{\text{LP}}(k)(T_i(t, k) - P_j(t, k))]^2, \qquad (5.9)$$

where T is the number of frames in the template and spectrogram excerpt, K is the number of frequency bins, and the response $H_{\text{LP}}(k)$ is used to attenuate high frequencies.
2. Order the spectrogram excerpts by their distances $\Gamma_{i,j}$ in ascending order.
3. Choose a fixed fraction (10%) of the excerpts with the smallest distances, and calculate a new template spectrogram with

$$T_i(t, k) \leftarrow \text{median}_{j \in S} \{ P_j(t, k) \}, \qquad (5.10)$$

where the set S contains the chosen excerpts.
4. Repeat Steps 1 to 3 until convergence.

The median was used in an attempt to suppress the effect of the presence of pitched instruments on the drum templates.

The actual segment recognition was done with a template-matching method stemming from the work of Goto and Muraoka [236] (see p. 139). The system was tested on 10 songs from the RWC database [229], and gave a recall rate[5] of 90% and a precision rate of 90% for the kick drum, and recall and precision rates of 83% and 93% for the snare drum. The use of adaption improved the results significantly over simple template matching.

In the system described by Sandvold et al. [561], the adaption took place at a higher abstraction level than in the previous systems. Here the models used were based on extracted features, rather than a time domain or time-frequency domain template. After onset detection, segmentation and feature extraction, the algorithm consisted of the following steps:

1. Classify the extracted n segments with a decision tree classifier and general instrument models (see Section 5.2.3).
2. Evaluate the reliability of the classification results, and choose $m < n$ most reliable results.
3. Create signal-specific (localized) models using the chosen m events.
4. Classify the contents of all the n segments using the localized models.

The general instrument models were constructed by extracting 115 spectral and temporal features from the training data and by choosing the most suitable subset of them with a correlation-based feature selection algorithm (see Chapter 6, p. 182). The reduced feature set consisted on average of less than 25 features for each class. When constructing the localized models in Step 3, all the 115 features were reconsidered and the best subset was again chosen with the same method. In this case, nine features per class were sufficient on the average. The ranking and selection of the most reliably classified events in Step 2 was not done fully automatically. Instead, the result from the classification with general models was corrected and the reliability ranking was done manually, thus making the overall system operate semi-automatically. The classification algorithm used with the localized models was the k-nearest neighbours algorithm with $k = 1$, i.e., a simple instance-based algorithm.

The method was evaluated with seventeen 20-s excerpts from polyphonic audio recordings with manually annotated ground truth. The results suggest that the use of localized models reduces the required number of features (the average feature set size drops from 25 to 9), and improves the recognition result, with the average accuracy improving from 72% to 92%.

5.3 Separation-Based Approaches

Having explored event-based approaches for percussion transcription, an alternative way of approaching the problem is through the use of source

---

[5]Recall rate is defined as the number of correctly transcribed events divided by the number of events in the input. Precision rate is defined as the ratio of correctly transcribed events to the total number of events at the output of the system.

separation-based techniques. These have some advantages over the event-based techniques, while at the same time having drawbacks of their own. Separation-based techniques can facilitate the analysis of mixtures of drums or percussion instruments by separating the instruments into distinct streams, which makes the task of transcribing easier. However, a drawback of many separation-based methods is that the streams have to be identified after separation. Further, these techniques often have difficulties in detecting low-intensity sounds. Despite these limitations, separation-based methods have proven to be a viable way of approaching the problem of percussion transcription.

The most commonly used source separation technique is independent component analysis (ICA) [96]. Further details on ICA can be found in Chapter 9. However, the basic ICA requires as many sensors (microphones) as there are sources for separation to occur, whereas in the drum transcription task there are usually at most two channels available; in other words, a stereo signal. Further, some drums will often be of the same amplitude in both channels, particularly snare and kick drums, and so in effect there is only a single channel available to separate these drums. As a result, most work using separation-based approaches has focused on single-channel separation.

Various methods for separating sound sources from single-channel recordings have been used, such as independent subspace analysis (ISA) [73], non-negative sparse coding (NNSC) [299], and non-negative matrix factorization (NMF) [400], and all have found uses in the area of percussion transcription. ISA has also been used for rhythmic analysis [488]. A detailed description of these methods can be found in Chapter 9. Here, discussion of these source separation techniques is limited to how they relate to the problem of percussion transcription. It is important to note that separation in the context of transcription means the separation of frequency and amplitude characteristics associated with each source in order to identify and transcribe them, as opposed to resynthesis of the separated sources, though resynthesis of the separated sources is also possible using these techniques. Methods of source resynthesis are presented by FitzGerald in [186].

All these techniques assume that the mixture spectrogram matrix $\mathbf{X}$ of size $(K \times T)$, where K is the number of frequency bins and T is the number of time frames, results from the superposition of J source spectrograms $\mathbf{Y}_j$ of the same size. Further, it is assumed that each of the spectrograms $\mathbf{Y}_j$ can be uniquely represented by the outer product of an invariant frequency basis function $\mathbf{b}_j$ of length K and a corresponding invariant amplitude basis function (or time-varying gain) $\mathbf{g}_j$ of length T which describes the gain of the frequency basis function over time. This yields

$$\mathbf{X} = \sum_{j=1}^{J} \mathbf{Y}_j = \sum_{j=1}^{J} \mathbf{b}_j \mathbf{g}_j^{\mathsf{T}}. \tag{5.11}$$

The above decomposition is typically applied to a magnitude spectrogram, though power spectrograms can also be used.[6] Where the techniques differ is in how the decomposition of the spectrogram $\mathbf{X}$ into frequency and amplitude basis functions is achieved. The basic ISA method proposed by Casey did this by performing principal component analysis (PCA) [320] on the spectrogram, keeping only a small number of decorrelated frequency lines, and then performing ICA on the retained components [73]. Further details on PCA can be found in Chapter 2, while details on ISA and ICA can be found in Chapter 9. In effect, ISA performs ICA on a low-dimensional representation of the original spectrogram. NNSC attempts to balance modelling accuracy with the sparseness of the recovered sources while enforcing non-negativity on the sources [299], whereas NMF attempts to reconstruct the data using non-negative basis functions and a Poisson or Gaussian noise model [400].

In practice, the use of invariant frequency basis functions means that no pitch changes are allowed over the course of individual spectrograms $\mathbf{Y}_j$. However, this is valid for most drum sounds, where the pitch of the drum does not change from event to event, making this type of decomposition particularly suited for analysing percussive tracks in polyphonic music.

Figure 5.3 shows the magnitude spectrogram of a drum loop containing snare and kick drum hits, while Fig. 5.4 shows the associated amplitude and frequency basis functions recovered using NMF. As can be seen from the amplitude basis functions, the amplitude envelopes of both the snare and the kick drum have been well separated, though some evidence of the kick drum is still visible in the snare basis function. The frequency basis functions can be seen to have captured the overall spectral characteristics of the sources, with the kick drum having more low-frequency energy than the snare, and the snare having its energy spread over a wider frequency range. This demonstrates the usefulness of these techniques for the purposes of drum transcription.

Nonetheless, from the point of view of percussion transcription, there are certain problems with the above-mentioned techniques, regardless of how the decomposition is achieved. These can be summarized as follows.

- Indeterminate source order: sources have to be identified after separation.
- Estimation of the optimal number of basis functions: keeping a small number of basis functions results in more recognizable features, while recovering low-energy sources requires increased numbers of basis functions.

Further, ISA recovers basis functions which may have negative elements, which does not reflect the assumption that the overall magnitude or power spectrogram results from the summation of independent spectrograms which are non-negative by definition. This can lead to errors in transcription. However, this is not a problem for NNSC and NMF, which constrain non-negativity of the sources. FitzGerald discusses these problems in greater detail in [186].

---

[6]The term 'magnitude spectrogram' refers to a representation which consists of the absolute values of the discrete Fourier transform in successive time frames. A power spectrogram is obtained as the element-wise square of this.

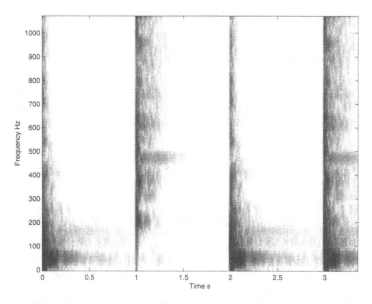

Fig. 5.3. Magnitude spectrogram of a drum loop containing snare and kick drum.

A number of techniques have been developed in an attempt to overcome the above problems, all using prior knowledge about the drums to be transcribed [187], [188], [505]. The first of these added sub-band pre-processing to ISA [187]. As noted in Section 5.1, the drums in a standard rock kit can be divided into membranophones and idiophones. Membranophones have most of their energy in the low end of the frequency range, while metal plate idiophones have most of their energy spread over the spectrum above 2 kHz. Filtering the signal using a lowpass filter with a 1-kHz cut-off frequency and a highpass filter with a 2-kHz cut-off frequency allows the membranophones and idiophones to be emphasized, respectively.

This was demonstrated in a system described in [187] that transcribed mixtures of snare, kick drum, and hi-hats. Two basis functions were found to be sufficient to recover the snare and kick drum from the lowpass-filtered signal using ISA, and similarly two basis functions recovered the hi-hats and snare from the highpass-filtered signal. To overcome the problem of source ordering, it was assumed that the kick drum had a lower spectral centroid than the snare drum, and that hi-hats occurred more frequently than the snare drum.

The system was tested on a set of 15 drum loops consisting of snare, kick drum, and hi-hats. The drums were taken from a number of different drum sample CDs and were chosen to cover the wide variations in sound within each type of drum. A range of different tempos and metres were used and the relative amplitudes between the drums varied between 0 dB and −24 dB. In total, the test set contained 133 drum events. The success rate was evaluated using $c = (t - u - i)/t$ where c is the percentage correct, t is the total number of drums, u is the number of undetected drums, and i is the number of

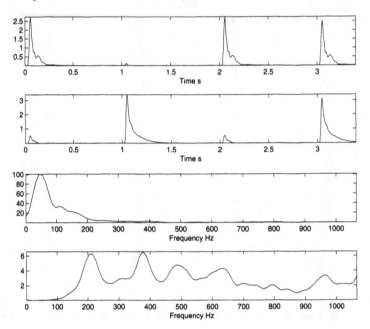

Fig. 5.4. Basis functions recovered from the spectrogram in Fig. 5.3. From top to bottom, they are the kick drum amplitude basis function, the snare drum amplitude basis function, the kick drum frequency basis function, and the snare drum frequency basis function.

incorrectly detected drums. Using this measure, an overall success rate of 90% was achieved. However, more effective means of incorporating prior knowledge were subsequently developed, and are discussed in the following subsections.

5.3.1 Prior Subspace Analysis

Prior subspace analysis was proposed as a means of tackling the problem of percussion transcription by combining prior models of drum sounds with the source separation techniques mentioned above [188]. PSA uses the same signal model as shown in (5.11) earlier. PSA then assumes that there exists known frequency subspaces or basis functions $\mathbf{b}_{\mathbf{pr}j}$ that are good initial approximations to the actual subspaces (here the terms subspace and basis function are used as synonyms). In other words, it is assumed that frequency basis functions such as shown in Fig. 5.4 are available before any analysis of the signal takes place, and that these are good approximations to the actual frequency basis functions that could be recovered from the signal. Substituting $\mathbf{b}_j$ with these prior subspaces yields

$$\mathbf{X} \approx \sum_{j=1}^{J} \mathbf{b}_{\mathbf{pr}j} \mathbf{g}_j^{\mathsf{T}}. \tag{5.12}$$

In matrix notation, this becomes

$$\mathbf{X} \approx \mathbf{B_{pr}G}, \tag{5.13}$$

where $\mathbf{B_{pr}} = [\mathbf{b_{pr1}}, \ldots, \mathbf{b_{pr}}_J]$ and $\mathbf{G} = [\mathbf{g}_1, \ldots, \mathbf{g}_J]^{\mathsf{T}}$. Estimates of the amplitude basis functions can be obtained from

$$\hat{\mathbf{G}} = \mathbf{B_{pr}}^{+}\mathbf{X}, \tag{5.14}$$

where $^{+}$ denotes the pseudoinverse. The pseudoinverse of a matrix $\mathbf{B}$ can here be computed as $\mathbf{B}^{+} = (\mathbf{B}^{\mathsf{T}}\mathbf{B})^{-1}\mathbf{B}^{\mathsf{T}}$, assuming that $\mathbf{b_{pr}}_j$ are linearly independent. The estimated amplitude basis functions $\hat{\mathbf{g}}_j$ are not independent, and may contain information related to more than one source. This is as a result of the broad-band noise-based nature of drum sounds, where the occurrence of a given drum will cause a partial match with the prior subspace for another drum.

To overcome this problem and to recover amplitude basis functions which are related to only a single source, ICA is carried out on the estimated amplitude basis functions $\hat{\mathbf{G}}$ to give

$$\mathbf{G} = \mathbf{W}\hat{\mathbf{G}}, \tag{5.15}$$

where $\mathbf{W}$ is the unmixing matrix obtained from ICA and $\mathbf{G}$ contains the independent amplitude basis functions. This results in amplitude basis functions which are generally associated with a single source, though there will still be some small traces of the other sources. Improved estimates of the frequency basis functions can then be obtained from

$$\mathbf{B} = \mathbf{X}\mathbf{G}^{+}. \tag{5.16}$$

In this case, the use of the pseudoinverse is justified in that the columns of $\mathbf{G}^{+}$ are orthogonal and do not share any information, and the pseudoinverse can be calculated as $\mathbf{G}^{+} = \mathbf{G}^{\mathsf{T}}(\mathbf{G}\mathbf{G}^{\mathsf{T}})^{-1}$. The overall procedure can be viewed as a form of model adaptation such as is described in Section 5.2.4.

Figure 5.5 shows a set of priors for snare, kick drum, and hi-hat, respectively. These priors were obtained by performing ISA on a large number of isolated samples of each drum type and retaining the first frequency basis function from each sample. The priors shown then represent the average of all the frequency basis functions obtained for a given drum type. Priors could be obtained in a similar way using some other matrix-factorization technique such as NMF (see Chapter 9 for further details). It can be seen that the priors for both kick drum and snare have most of their energy in the lower regions of the spectrum, though the snare does contain more high-frequency information, which is consistent with the properties of membranophones, while the hi-hat has its frequency content spread out over a wide range of the spectrum.

The use of prior subspaces offers several advantages for percussion instruments. First, the number of basis functions is now set to the number of

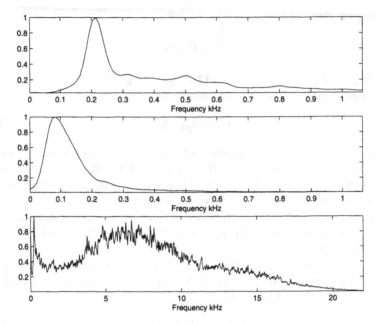

Fig. 5.5. Prior subspaces for snare, kick drum, and hi-hat.

prior subspaces used. Second, the use of prior subspaces alleviates the bias towards sounds of high energy inherent in blind decomposition methods, allowing the recovery of lower-energy sources such as hi-hats. Thus, the use of prior subspaces can be seen to go some distance towards overcoming some of the problems associated with the use of blind separation techniques, and so is more suitable for the purposes of percussion transcription.

A drum transcription system using PSA was described in [188]. Again, the system only transcribed signals containing snare, kick drum, and hi-hats without the presence of any other instruments. Prior subspaces were generated for each of the three drum types, and PSA performed on the input signals. Once good estimates of the amplitude basis functions had been recovered, onset detection was carried out on these envelopes to determine when each drum type was played. To overcome the source-ordering problem inherent in the use of ICA, it was again assumed that the kick drum had a lower spectral centroid than the snare, and that hi-hats occurred more frequently than the snare. When tested on the same material as used with sub-band ISA (see p. 145 for details), a success rate of 93% was achieved.

5.3.2 Non-Negative Matrix Factorization-Based Prior Subspace Analysis

More recently, an improved formulation of PSA has been proposed for the purposes of drum transcription [505], based on using an NMF algorithm with priorly fixed frequency basis functions $\mathbf{B_{pr}}$. The NMF algorithm estimates non-negative amplitude basis functions $\hat{\mathbf{G}}$ so that the reconstruction error of the model (5.13) is minimized.

This offers a number of advantages over the original formulation of PSA. First, the non-negative nature of NMF is more in keeping with the data being analysed, in that the spectrogram is non-negative, and so a decomposition that reflects this is likely to give more realistic results. Second, keeping $\mathbf{B_{pr}}$ fixed eliminates the permutation ambiguities inherent in the original PSA algorithm. This allows the elimination of the assumptions necessary to identify the sources after separation, and permits the algorithm to function in a wider range of circumstances.

The NMF-based algorithm estimates $\hat{\mathbf{G}}$ by first initializing all its elements to a unity value and then iteratively updating the matrix using the rule

$$\hat{\mathbf{G}} \leftarrow \hat{\mathbf{G}}. \times \left[\mathbf{B_{pr}}^{\mathsf{T}} \left(\mathbf{X}./(\mathbf{B_{pr}}\hat{\mathbf{G}}) \right) \right] ./ \left[\mathbf{B_{pr}}^{\mathsf{T}} \mathbf{1} \right], \qquad (5.17)$$

where $.\times$ and $./$ denote elementwise multiplication and division, respectively, $\mathbf{X}$ is the signal spectrogram, and $\mathbf{1}$ is an all-ones matrix of size equal to $\mathbf{X}$. When tested on the same test material as subband ISA (see p. 145), a success rate of 94% was obtained.

A transcription system using the NMF-based algorithm was presented by Paulus and Virtanen [505]. The mid-level signal representation was a magnitude spectrogram with only five frequency bands (20–180 Hz, 180–400 Hz, 400–1000 Hz, 1–10 kHz, and 10–20 kHz), with an analysis frame length of 24 ms and frame overlap of 75%. The fixed instrument priors were calculated from a large set of training samples, by factorizing each sample with NMF into an outer product of an amplitude basis function and a frequency basis function. The frequency basis functions of all samples of a given individual sound type were averaged to yield $\mathbf{b_{pr}}$. The time-varying gains were calculated by iteratively applying (5.17) while the source spectra were kept fixed. The onsets were detected from the calculated gains with a variation of the onset detection algorithm described in [349].

The performance of the transcription system was evaluated with acoustic material consisting of different drum kits recorded in different environments, in total giving four different recording sets. In each recording set, five different sequences were recorded, resulting in a total of 20 signals. The sequences were fairly simple patterns containing mostly kick drum, snare drum, and hi-hat. Two monophonic mix-downs of the signal were produced: a dry, unprocessed mix with only levels adjusted, and a wet, 'production grade' mix with compression, reverb and other effects, which attempted to resemble drum tracks on commercial recordings. Also evaluated using this material were the SVM classifier-based event recognition method [209], as implemented by Paulus, and the earlier PSA method [188] as implemented by the original author.

All three systems were trained with the same material to the extent that was possible. The spectral basis functions for the NMF system and PSA were calculated from the same set of unprocessed samples. For the NMF-based system, onset-detection thresholds were obtained by analysing a set of training signals and setting the threshold to a value minimizing the number of detection errors. For PSA, the source-labelling rules and fixed threshold values from the original publication were used. The SVM-based classifier was trained so that when analysing the dry (wet) signals, the training features were also extracted from the dry (wet) signals. This may have given the SVM method a slight advantage compared to the other systems.

The NMF-based system performed best of the methods, with the dry mix material having a hit rate of 96% compared to the 87% of the SVM method and 67% of PSA. The performance gap became smaller with the production-grade mixes, but the NMF-based method still had a hit rate of 94% compared to the 92% of SVM method and 63% of PSA [505].

5.3.3 Prior Subspace Analysis in the Presence of Pitched Instruments

A modified version of the original PSA algorithm was found to be effective in transcribing snare, kick drum, and hi-hats/ride cymbals from recordings of polyphonic music [189]. The modifications were required because the presence of a large number of pitched instruments will cause a partial match with the prior subspace used to identify a given drum, which can make detection of the drums more difficult. However, it should be noted that pitched instruments have harmonic spectra with regions of low intensity between the overtones or partials. As a result, when pitched instruments are playing there will be regions in the spectrum where little or no energy is present due to pitched instruments. It was observed that good frequency resolution reduced the interference due to the pitched instruments, and so improved the likelihood of recognition of the drums.

In the case of snare and kick drum, setting all values in the initial estimates of the amplitude basis functions $\hat{G}$ below a fixed threshold to zero was found to be sufficient to eliminate the interference. ICA was then performed on these thresholded basis functions to obtain better estimates for snare and kick drum. In the case of hi-hats/ride cymbals, the interference was often considerably worse, and simple thresholding was insufficient. However, most of the energy in pop and rock sounds is in the lower region of the spectrum, and, as already noted, the energy of the hi-hats is spread across the spectrum. Therefore, the average power spectral density (PSD) [639] over the whole duration of each excerpt was calculated, and the spectrogram then normalized by it. This is equivalent to highpass filtering the signal, but in each case the filter takes into account the characteristics of the signal being filtered. The hi-hat prior is then multiplied with the PSD-normalized spectrogram to obtain the hi-hat subspace.

The modified algorithm then consisted of carrying out ICA on the thresholded snare and kick drum subspaces, followed by onset detection to identify these drums, and then onset detection on the hi-hat subspace recovered from the PSD-normalized spectrogram. As the hi-hat subspace no longer undergoes ICA with the other drums, the algorithm loses the ability to distinguish between a snare on its own and a snare and hi-hat occurring simultaneously. Fortunately, in many cases these drums do occur simultaneously and so this results in only a small reduction in the efficiency of the algorithm. When tested on a database of 20 excerpts from pop and rock songs taken from commercial CDs, an overall success rate of 83% was achieved.

Attempts to extend the basic PSA method to include other drums such as tom-toms and cymbals met with mixed success. Extensive testing with synthetic signals revealed that this was due to the fact that when the main regions of energy of different sources overlap, as is often the case with drums such as snares and tom-toms, then the sources will not be separated correctly [186].

5.3.4 Input-Generated Priors

Attempts have been made to overcome the limitations of PSA by generating frequency basis functions from the input signal which can then be used as priors in a PSA-type framework, rather than using general priors. These basis functions should then be better able to exploit differences in the spectra of the drums present. To date, two systems have made use of this approach [190], [147].

The system described by FitzGerald in [190], which worked on signals containing drums only, models each event that occurs in the signal and generates similarity measures between each event, which are used to group the events. These groups are then used to generate frequency basis functions from the input signal which can then be used as priors.

Membranophone events were identified by onset detection in the amplitude envelopes obtained from

$$\hat{\mathbf{G}} = \mathbf{B_{pr}}^{+}\mathbf{X}, \qquad (5.18)$$

where, in this case, $\mathbf{B_{pr}}$ contains only snare and kick drum priors. This was also sufficient to recover tom-tom onsets. Onset times were then determined, and the sections of the spectrogram between each event obtained. PCA was then used to obtain a frequency basis function to represent each section. The resulting frequency basis functions were then clustered and an average frequency basis function was obtained for each of the groups.

To overcome the most commonly occurring membranophone overlap, that of snare and kick drum, the groups most likely to correspond to snare and kick drum were identified. All tom-tom events in the original spectrogram were then masked. PSA was performed on the resulting spectrogram, and the snare and kick drum events identified. Other membranophone overlaps were assumed not to occur as they are not very common, as can be seen in Fig. 5.2.

To detect idiophones, (5.18) is applied to a PSD-normalized spectrogram using only a hi-hat prior subspace to yield an idiophone amplitude envelope.

As traces of both snare and tom-toms will occur in the idiophone envelope, an amplitude envelope for snare/tom-toms is obtained by masking kick drum events in the original spectrogram, and multiplying the resulting spectrogram by a snare frequency subspace, again using (5.18). ICA is then performed to separate the snare/tom-tom amplitude envelope and the idiophone amplitude envelope. Onset detection on the resulting independent idiophone envelope then yields the idiophone events.

Grouping is then carried out on the idiophones. If two large groups occured that did not overlap in time, then both hi-hat and ride cymbal were assumed to be present; otherwise all events were allocated to the same drum. The justification for this is detailed in [190]. Unfortunately, though the algorithm distinguished between ride cymbal and hi-hats, it did not identify which was which. When tested on a database of 25 drum loops, a success rate of 90% was obtained using the same measure as sub-band ISA.

Dittmar et al. described a system which attempted to transcribe drums in the presence of pitched instruments [147]. To enable recovery of low-energy sources such as hi-hats and ride cymbals, the high-frequency content of the signal was boosted in energy. A magnitude spectrogram of the processed signal was obtained, and then differentiated in time. This suppressed some of the effects of the sustained pitched instruments present in the signal, because their amplitudes are more constant on a frame-by-frame basis than that of transient noise, and so when differentiated will have a smaller rate of change.

Onset detection was then carried out and the frame of the difference spectrogram at each onset time extracted. As the extracted frames contain many repeated drum events, PCA was used to create a low-dimensional representation of the events. J frequency components were retained and non-negative ICA [526] performed on these components to yield $\mathbf{B}$, a set of independent basis functions which characterized the percussion sources present in the signal. The amplitude envelopes associated with the sources were obtained from

$$\mathbf{G} = \mathbf{B}^\mathsf{T}\mathbf{X}, \qquad (5.19)$$

where $\mathbf{G}$ are the recovered amplitude envelopes, and $\mathbf{X}$ is the original spectrogram. A set of differentiated amplitude envelopes was then recovered from

$$\hat{\mathbf{G}} = \mathbf{B}^\mathsf{T}\mathbf{X}', \qquad (5.20)$$

where $\hat{\mathbf{G}}$ are the differentiated amplitude envelopes and $\mathbf{X}'$ is the differentiated spectrogram. Correlation between $\mathbf{G}$ and $\hat{\mathbf{G}}$ was used to eliminate recovered sources associated with harmonic sounds, as sustained harmonic sources will tend to have lower correlation than percussive sources.

The problem of determining the required number of components was approached by retaining a set number of components and then matching the recovered components to percussion instrument templates. Drum onsets were then identified by onset detection on the recovered amplitude envelopes.

5.3.5 Non-Negative Sparse Coding

The system proposed by Virtanen made use of NNSC to transcribe drums in the presence of pitched instruments [650]. The NNSC algorithm used was similar to that described by Hoyer in [299], with additional constraints to encourage temporal continuity of the sources. The optimization was carried out to obtain sparse amplitude basis functions, but sparse spectra were not assumed. A power spectrogram was used as the input to the system.

The described transcription system attempted to transcribe snare and kick drums in the presence of pitched instruments, and was tested using signals synthesized from MIDI files. The system separated a pre-defined number of basis functions, and then searched for frequency basis functions that matched well with the template spectra of either the snare or kick drum. The goodness of fit of a given source to a given template was calculated as

$$g(j, i) = \sum_{k=1}^{K} [\mathbf{b_{pri}}]_k \log \left| \frac{[\mathbf{b}_j]_k + \epsilon}{[\mathbf{b_{pri}}]_k + \epsilon} \right|, \tag{5.21}$$

where $\mathbf{b_{pri}}$ denotes the template spectrum of drum i, $\mathbf{b}_j$ is the frequency basis function for source j, k is the frequency bin index, K is the number of frequency bins, and ϵ is a small positive value to ensure that the logarithm is robust for small values.

Once sources associated with both snare and kick drum had been identified, onset detection was done on the amplitude envelopes of these sources. An overall success rate of 66% was obtained for this system. The templates were obtained by separating a number of sources in training material, and averaging the frequency basis functions of the sources whose amplitude basis functions showed temporal correlation with the kick or snare drums in a reference annotation.

As can be seen from the above, the use of source separation methods has shown great utility for the purposes of unpitched percussion transcription, overcoming some of the disadvantages of event-based systems, but at the cost of some other problems. Further, despite the success of such systems to date, there still remains much work to be done in improving the performance of separation-based percussion transcription systems.

5.4 Musicological Modelling

Many of the presented transcription systems concentrate only on low-level recognition, without any processing at a higher abstraction level. When compared to methods for automatic speech recognition, this resembles the situation where each phoneme is recognized independently of its context. It was noted quite early that by utilizing linguistic knowledge, speech recognition performance could be improved considerably [321]. It is reasonable to assume that incorporating similar ideas in music transcription can result in improved performance. Here we use the term 'musicological modelling' to refer to attempts to model the statistical dependencies of event sequences. There has been relatively little work done utilizing such models in percussion transcription. However, some methods have been proposed, and these are discussed in the following [590], [506], [504], [211], [209].

5.4.1 Enabling High-Level Processing

Usually, high-level processing consists of some sort of probabilistic modelling of the temporal relations between sound events. In practice, this requires that the low-level recognition results must also be presented as a probability or likelihood value. Some classification methods, for example Gaussian mixture models, produce probabilities as their output directly, whereas the result of the other methods has to be mapped to a probability value. There has not been much research on how to do this with different classifiers, with the exception of the work done on moderating SVM outputs into probability values [381], [524], [423], [424], [156]. The different methods have been evaluated by Rüping [556], and the results suggest that Platt's method [524] performs best.

The output of a binary SVM classifier is a real number, where the decision border between the classes resides at the value zero. Usually, the classification is done according to the sign of the output value. Given the SVM output value f_i of the ith sample, Platt proposed that the probability of the sample to be from the positive class can be calculated with a sigmoid

$$p_i = \frac{1}{1 + e^{\alpha f_i + \beta}}, \qquad (5.22)$$

where α and β are sigmoid parameters estimated by Platt's algorithm.

Another possible way of estimating likelihood values from the numerical result of a non-probabilistic classifier is to utilize a histogramming method called binning, described by Zadrozny and Elkan [686]. The output values f_i of the classifier are analysed for the training data, and a histogram of B bins is constructed. At the recognition stage, each output value of the classifier can be assigned to one of the bins. The probability of the classified sample to belong to class c is then calculated as the fraction of the training samples in the assigned bin that actually were from the class c.

5.4.2 Short-Term Context Modelling

Contextual information can be represented with a model separate from the low-level signal analysis, or it can be taken into account within the low-level recognition algorithm. In the former, the context modelling is done explicitly, whereas in the latter, it takes place more implicitly. Several methods for modelling sequential dependencies are reviewed by Dietterich [146]. Among the implicit methods, only the sliding window technique has been used in percussion transcription [210].

The sliding window method utilizes the feature values of the neighbouring segments in the recognition of a given segment. In addition to the feature vector o_i of the segment i, the features $o_{i-N_w}, \ldots, o_i, \ldots, o_{i+N_w}$ of the surrounding $2N_w$ segments are used by concatenating them into one large feature vector which is used for the classification. The window of features is slid over the whole signal to be analysed. The sliding window method was utilized by Gillet and Richard in [210] to improve the performance of their earlier method [209]. The applied window was very short, utilizing only the features of one preceding event. Combined with decorrelating the features with PCA, the gained performance improvement was on average 3 to 4 percentage units.

The simplest form of explicit musicological modelling is the use of prior probabilities for events. Depending on the selected approach, this means either the probabilities of individual drum events or different drum combinations to be present. Paulus and Klapuri applied the latter type of priors in a system which was trained to recognize 127 different drum combinations (see p. 139 above) [506]. By using prior probabilities that resembled the profile in Fig. 5.2, the error rate of the system decreased from 76% to 50%.

Even though the prior probabilities themselves are helpful in the overall task, they cannot model the sequential dependencies between consecutive events. In percussive sequences, as with natural languages, the preceding context makes some continuations more probable than the others. This sequential dependence can be modelled with N-grams.

The N-grams rely on the Markov assumption that the event w_k, occurring at the time instant k, depends only on the preceding $N-1$ events $w_{k-N+1}, \ldots, w_{k-1}$, instead of the complete history. Using $w_{1:k}$ to denote a sequence of events $w_1, w_2, \ldots, w_k$, the Markov assumption can be written as

$$p(w_k|w_{1:k-1}) = p(w_k|w_{k-N+1:k-1}). \tag{5.23}$$

To enable the use of N-grams, the concept of an 'event' itself has to be defined. For example, Paulus and Klapuri [506] estimated a grid of equidistant tatum pulses over the performance and considered the combination of drum sounds starting at grid point k as a mixture-event w_k, for which N-gram models were estimated and used.

Usually the N-gram probabilities are estimated from a large body of training material by

$$p(w_k|w_{k-N+1:k-1}) = \frac{C(w_{k-N+1:k})}{C(w_{k-N+1:k-1})}, \tag{5.24}$$

where $C(w_{k-N+1:k})$ is the number of occurrences (count) of the sequence $w_{k-N+1:k}$ in the training material [321]. The set of probabilities for all possible sequences of length N is called the N-gram model. These can be used to assess the likelihoods of different event sequences in percussion transcription, or to predict the next event.

If the events in the sequence are drawn from a dictionary of size D, there are D^N probabilities that need to be estimated for an N-gram of length N. This imposes requirements on the size of the training data set in order that the resulting N-grams do not contain too many zero-probability entries. Usually, such entries cannot be completely avoided, so methods for reducing their effect have been developed. The zero-probabilities can either be smoothed (given a non-zero value) with a discounting method like Witten–Bell discounting [673] or Good–Turing discounting [218], [90], or the required probability can be estimated from lower-order N-grams with the back-off method suggested by Katz [336] or with the deleted interpolation algorithm by Jelinek and Mercer [317]. An interested reader is referred to the cited publications for details.

The sound event N-grams in music analysis are directly analogous to the word N-grams in speech recognition. Moreover, as the words in speech are constructed from individual letters or phonemes, the mixture-events in percussive tracks may consist of multiple concurrent sounds from different instruments. The main difference between these two is that in speech recognition, the order of individual letters is important and the letters in consecutive words rarely have any direct dependence, whereas in musicological N-grams, the mixture-events consist of co-occurring sound events which alone exhibit dependencies between the same sound event in the neighbouring mixtures. This observation can be utilized to construct N-gram models for individual instruments, as suggested by Paulus and Klapuri [506].

When the set of possible instruments $\{u_1, u_2, \ldots, u_M\}$ is defined, with the restriction that each instrument can only occur once in each mixture-event, the problem of estimating mixture-event N-grams can be converted into the problem of estimating N-grams for individual instruments. In other words, at each time instant k the instrument u_i has the possibility of being present in the mixture w_k or not. By using this assumption, the probability estimate from (5.23) becomes

$$p(w_k | w_{k-N+1:k-1}) = \prod_{u_i \in w_k} p(u_{i,k} | u_{i,k-N+1:k-1})$$
$$\times \prod_{u_j \notin w_k} (1 - p(u_{j,k} | u_{j,k-N+1:k-1})), \quad (5.25)$$

where $p(u_{i,k})$ is the probability of the instrument u_i to be present at the temporal location k.

Separate N-grams for individual instruments have a clear advantage over the N-grams for mixture-events when considering the training of the model. If there are M different instruments available, then there are 2^M different mixture-events, requiring a total of 2^{MN} probabilities to be estimated for the mixture-event N-grams. When using separate N-grams for the instruments, each model has only a binary dictionary, leading to a total of $M2^N$ probabilities to be estimated. This alleviates the zero-frequency problem significantly: the sharply concentrated prior distribution of different mixture-events (see Fig. 5.2) means that some of them occur too rarely for reliable probability estimation, even in a large training set.

The main problem with individual instrument N-grams is that they lack information about simultaneously occurring events. Each N-gram 'observes' only the presence of its own instrument without any knowledge of the other co-occurring instruments. As a result, the model may give overly optimistic or pessimistic probabilities to different mixture-events. One possible way to address this problem is to use the prior probabilities of the mixture-events in connection with (5.25), as proposed in [506].

Musicological prediction can also be done using simpler modelling. For example, if two occurrences of the same event type took place with time interval t_Δ, its occurrence can be predicted again after another interval of t_Δ. A system relying on this type of modelling was proposed by Sillanpää et al. in [590].

5.4.3 Modelling of Periodic Dependencies

The fact that N-grams only use the directly preceding events to predict the next event is a minor drawback, considering their usage in music or percussive sound analysis. In particular, the percussive content of music generally exhibits repeating patterns. Even though they contain the same sequential data within the musical piece, the patterns tend to vary between pieces. As a result, temporal prediction operating on immediately preceding events may not be the most efficient way to model repeating rhythmical patterns.

Based on the above observation, Paulus and Klapuri proposed the use of periodic N-grams [506] where, instead of using the directly preceding $N - 1$ events, the idea is to take the earlier events separated by an interval L. That is, when predicting the event at temporal location k, instead of using the events at locations $k - N + 1, k - N + 2, \ldots, k - 1$. use the events at the locations $k - (N - 1)L, k - (N - 2)L, \ldots, k - L$. The N-gram model of (5.23) is then reformulated as

$$p(w_k|w_{1:k-1}) = p(w_k|w_{k-(N-1)L}w_{k-(N-2)L}\cdots w_{k-L}). \qquad (5.26)$$

It is easy to see that by assigning $L = 1$, the model reduces to the standard N-gram of (5.23). The repeating patterns tend to occur at integer multiples of the musical-measure length of the the piece. Due to this, it was suggested that the interval L should be set to correspond to the measure length λ of the piece under analysis.

The idea of the normal and the periodic N-grams are illustrated in Fig. 5.6. The horizontal arrow represents the use of a normal trigram (N-gram of length 3) to predict the event at the location of the question mark, whereas the

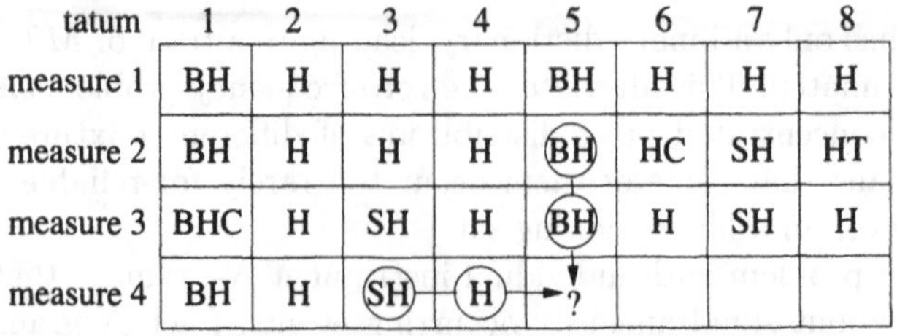

tatum	1	2	3	4	5	6	7	8
measure 1	BH	H	H	H	BH	H	H	H
measure 2	BH	H	H	H	(BH)	HC	SH	HT
measure 3	BHC	H	SH	H	(BH)	H	SH	H
measure 4	BH	H	(SH)—(H)→?					

Fig. 5.6. The idea of the normal and periodic N-grams illustrated. Time is quantized to a tatum grid, each box representing a segment between two grid points. Time flows from left to right continuing on the next row, each row being one musical measure. The letters represent the drum instruments played at the corresponding time instants (B is kick drum, S is snare drum, H is hi-hat, T is tom-tom, and C is cymbal). The horizontal arrow represents a normal trigram prediction, and the vertical arrow represents a periodic trigram prediction. The measure length here is eight tatum periods and $L = 8$ is set accordingly.

vertical arrow illustrates the use of a periodic N-gram. It should be noted that both the N-gram types can be constructed either for mixture events or individual instruments. The latter type of models can be visualized by looking at only one hit type (letter) at a time. For example, when estimating the probability of a hi-hat event occurring at the position of the question mark, the model observes only that the two preceding mixtures contain a hi-hat hit.

Use of different N-gram models was evaluated in [506]. The low-level analysis was done by creating a tatum grid over the signal, based on manually annotated information. Then a set of features was extracted at each segment between two grid points, and a combination of GMMs and an SVM was used to recognize its contents. A separate GMM was trained for each possible mixture-event, and an SVM was used as a silence detector. At each segment, the low-level recognition produced likelihoods for different mixture events. The actual transcription was done simply by choosing the mixture having the highest likelihood at each grid point. The acoustic recognition result was relatively poor, with an instrument event error rate of 76%. All the high-level processing methods were added on top of this. On adding the mixture-event prior probabilities, the error rate dropped to 50%. With mixture-event bigrams and trigrams, the periodic version performed slightly better than the standard ones. Also, the longer N-grams for individual instruments with $N = 5$ and $N = 10$ performed better than the shorter mixture-event N-grams. The best result, obtained with traditional non-periodic N-grams for individual instruments using $N = 10$, still had a 46% error rate. This demonstrates that musicological models cannot greatly improve the overall result if the low-level recognition results are poor to start with. Instead, they require sufficiently accurate information to be available from low-level recognition to be able to make proper assumptions about the musical structure and improve the overall result.

5.4.4 Modelling of Compositional Rules

Musicological models have been most commonly used in connection with supervised classification methods, such as GMMs and HMMs, but they can also be employed with unsupervised methods, such as event clustering. This was done by Paulus and Klapuri, who used a musicological model to label percussive sound events with instrument names of the regular rock/pop drum kit, even though the input signal was produced using arbitrary sounds [504]. In other words, the input signal was a percussive sequence created with nondrum sounds, like speech sounds or tapping different objects, and the output transcription was intended to map these to a normal drum kit. Because the labelling or mapping could not be determined from the acoustic properties of the sounds, it was done with a simple musicological model of drum loops. The model consisted of the probabilities of different percussive instruments to be present at different temporal locations within a musical measure. This requires that the boundaries of musical measures are located in the input signal, which can be done with the methods discussed in Chapter 4.

The system presented by Paulus and Klapuri [504] operated with the following steps. First, temporal segmentation was carried out by onset detection. Acoustic features were extracted at each segment and, based on these, the segments were clustered into K categories. The onset times were then quantized to an estimated tatum grid. At this point, each tatum grid point contained information as to which cluster the sound event at that point belonged to. Finally, suitable labels for the clusters were determined so that the generated sequence would make sense when performed with a drum kit.

The above labelling problem was formulated as finding a mapping $\mathcal{M}$ from the K clusters to the available labels $\mathcal{M}: \{0, 1, 2, \ldots, K\} \rightarrow \{\emptyset, B, S, H\}$, where the labels denote silence, kick drum, snare drum and hi-hats, respectively.[7] As the total number of different mappings was relatively small, all of them could be tested and the best one chosen. The likelihood of a certain mapping $\mathcal{M}$ was evaluated with

$$p(\mathcal{M}) = \prod_i p(q_i|n_i, \lambda), \qquad (5.27)$$

where $p(\mathcal{M})$ is the likelihood of the mapping, and $p(q|n, \lambda)$ is the probability of the label q to be present at the temporal location $n \in \{0, \ldots, \lambda - 1\}$ when the length of the musical measure is λ. The total likelihood is calculated over the whole signal containing all the events i.

The system was evaluated with acoustic signals synthesized from a commercial MIDI database comprising a wide variety of different percussive tracks [305]. The synthesis was done by using sampled speech sounds and the sounds of tapping different objects in an office environment. There were fifteen samples for each sound type, and each synthesized hit was randomly selected

---

[7]In the general case, K does not need to be equal to the number of available labels.

from this set to produce realistic acoustic variation to the synthesis result. The overall error rate of the system was 34%. Error analysis revealed that there were large differences in performance between different genres, and the genres with simpler rhythmic patterns were labelled more accurately.

It has been established that musicological modelling is useful in the context of percussive sound transcription. However, the low-level analysis has to be done sufficiently accurately before the musicological modelling can really improve the results obtained. Further, the methods required to combine low-level acoustic recognition with the high-level modelling still need development. As percussive patterns tend to be different in different time signatures, styles, and genres, specific models for each of these could be developed.

5.5 Conclusions

An overview of the current state of the art in unpitched percussion transcription has been presented. This encompassed both event-based and separation-based systems, as well as efforts to include high-level language modelling to improve system performance. As can be seen, there has been considerable effort expended on tackling the problem of percussion transcription in the past few years, and a summary of the important systems to date is presented in Table 5.1.

At present, the best performance has been obtained on systems that focus on a reduced number of drums: snare, kick drum, and hi-hats in the drums-only case, and snare and kick drum in the presence of pitched instruments. This is unsurprising in that the complexity of the problem is greatly reduced by limiting the number of target instruments. Nonetheless, these systems do deal with the most commonly occurring drums, and so represent a good starting point for further improvements.

As noted above, many of the systems do not take into account the predictability of percussion patterns within a given piece of music. However, it has been established that the use of musicological modelling does considerably improve the performance of a system using only low-level processing. In particular, it should be feasible to integrate musicological modelling to many of the separation-based models.

There has also been a trend towards adaptive systems that take into account the characteristics of the signals being analysed when attempting transcription, both in event-based and separation-based systems. This is an attempt to overcome the large variances in the sounds obtained from a given drum type such as a snare drum. For example, drums in a disco-style genre have a totally different sound to those in a heavy metal-style piece. These adaptive systems are to be encouraged, as a system that can be tailored to suit individual signals is more likely to produce a successful transcription than a system which makes use of general models.

Table 5.1. Summary of percussion transcription systems. The column *Classes* contains the number of percussion classes covered by the system. *Method* describes the overall approach, E for event-based systems, S for separation-based systems, M for systems including musicological modelling, and A for systems using adaptive modelling. An X in *Drums only* indicates that a system operates on signals containing drums only. *Mono/Poly* shows whether the system can detect two or more simultaneous sounds

Authors	Classes	Drums only	Mono/ Poly	Method	Main algorithms
Dittmar & Uhle [147]	5		p	S+A	Non-negative ICA
FitzGerald et al. [187]	3	X	p	S	Sub-band ISA
FitzGerald et al. [188]	3	X	p	S+A	PSA
FitzGerald et al. [189]	3		p	S	PSA
FitzGerald et al. [190]	7	X	p	S+A	Adaption & PSA
Gillet & Richard [209]	8	X	p	E+M	HMM, SVM
Goto & Muraoka [235]	9	X	p	E	Template matching
Goto [221]	2		p	E	Frequency histograms
Gouyon et al. [250]	2	X	m	E	Feature extraction
Herrera et al. [287]	9	X	m	E	Various
Herrera et al. [285]	33	X	m	E	Various
Paulus & Klapuri [506]	7	X	p	E+M	GMMs & N-grams
Paulus & Virtanen [505]	3	X	p	S	NMF
Sandvold et al. [561]	3		p	E+A	Localized models
Sillanpää et al. [590]	7		p	E+M	Template matching
Van Steelant et al. [608]	2		p	E	SVMs
Virtanen [650]	2		p	S	Sparse coding
Yoshii et al. [682]	2		p	E+A	Template matching
Zils et al. [693]	2		m	E+A	Template matching

A problem with the research on percussion transcription to date has been the lack of comparability of results. This is due to varying problem formulations, and to the use of disparate test sets and evaluation measures. The two different comparative evaluations mentioned above (see pp. 148 and 149) are among the few that have compared different systems on the same test data. Recently, however, a number of different transcription systems were evaluated in the framework of the Audio Drum Detection contest[8] organized in connection with the 6th International Conference on Music Information Retrieval. The contest is a clear step towards an extensive and regular comparative evaluation of different methods.

Given the noticeable progress in percussion transcription in the past few years, it is hoped that the future will see a further improvement in the performance of the systems, and, as the performance of the systems improves, in the range of percussion instruments covered. The problem is still far from

---

[8]The results are available at www.music-ir.org/mirexwiki.

solved, and it is hoped that this chapter reflects only the beginning of the study of unpitched percussion transcription.

5.6 Acknowledgements

Derry FitzGerald was supported in this work by the Irish Research Council for Science, Engineering and Technology.

6

Automatic Classification of Pitched Musical Instrument Sounds

Perfecto Herrera-Boyer,[1] Anssi Klapuri,[2] and Manuel Davy[3]

[1] Institut Universitari de l'Audiovisual, Universitat Pompeu Fabra,
  Pg. Circumval·lació 8, 08003 Barcelona, Spain
  pherrera@iua.upf.es
[2] Institute of Signal Processing, Tampere University of Technology,
  Korkeakoulunkatu 1, 33720 Tampere, Finland
  Anssi.Klapuri@tut.fi
[3] LAGIS/CNRS, BP 48, Cité Scientifique, 59651 Villeneuve d'Ascq Cedex, France
  Manuel.Davy@ec-lille.fr

6.1 Introduction

This chapter discusses the problem of automatically identifying the musical instrument played in a given sound excerpt. Most of the research until now has been carried out using isolated sounds, but there is also an increasing amount of work dealing with instrument-labelling in more complex music signals, such as monotimbral phrases, duets, or even richer polyphonies. We first describe basic concepts related to acoustics, musical instruments, and perception, insofar as they are relevant for dealing, with the present problem. Then, we present a practical approach to this problem, with a special emphasis on methodological issues. Acoustic features, or, descriptors, as will be argued, are a keystone for the problem and therefore we devote a long section to some of the most useful ones, and we discuss strategies for selecting the best features when large sets of them are available. Several techniques for automatic classification, complementing those explained in Chapter 2, are described. Once the reader has been introduced to all the necessary tools, a review of the most relevant instrument classification systems is presented, including approaches that deal with continuous musical recordings. In the closing section, we summarize the main conclusions and topics for future research.

6.1.1 Classification in Humans and Machines

Classification, in practical terms, refers to the process of assigning a class label to a given observation. This observation is typically described as a numerical vector that represents some features of the observation. In the case of music analysis systems, the features are computed from the raw audio signal by means of signal processing techniques. Class labels are by definition categorical data types representing some kind of generalization about the observable world. They are usually organized by means of structures, such as taxonomies and ontologies, that exhibit the following ideal features: (1) consistency in the principles of classification, (2) mutual exclusiveness of categories, and (3) completeness, i.e., total coverage of the universe that they attempt to describe [46].

The interest in automatic classification of musical instrument sounds is manifold:

- From the acoustics point of view, to understand what makes the sound of a given instrument 'identifiable' among other instruments;
- From the perceptual perspective, to understand what makes the sounds of two different instruments 'similar';
- From the sound librarian perspective, to automatically provide labels for retrieving desired items from sound sample databases or synthesizer libraries;
- From the musicological point of view, to locate solo sections or the appearance of a certain instrument in a musical recording.

Classification learning can proceed in a *supervised* way, where the system is provided with training data and the corresponding labels that should be associated with them, or in an *unsupervised* way, where there is no label to be associated with a given example but the system groups objects according to some similarity or homogeneity criteria. In the latter case, we are effectively doing *clustering*, a task that is slightly different from classification [674].

The supervised classification procedure can in general be described as follows (see Fig. 6.1):

1. Acoustic features are selected to describe sound samples.
2. Values of these features are computed for a labelled training database.
3. A learning algorithm that uses the selected features to learn to discriminate between instrument classes is applied and fine tuned, exploiting the training database.
4. The generalization capabilities of the learning procedure are evaluated by classifying previously unseen sound samples (cross-validation).

Research on the automatic classification of musical instruments sounds has focused, for a long time, on classifying isolated notes from different instruments. This is an approach that has a very important trade-off: we gain simplicity and tractability, as there is no need to first separate the sounds from a mixture, but we lose contextual and time-dependent cues that can be exploited as relevant features when classifying musical sounds in complex mixtures.

Human discrimination of sound sources is based on spectral and temporal cues that are extracted by the early processing in the cochlea and up to the primary auditory cortex, whereas class decisions seem to be made in

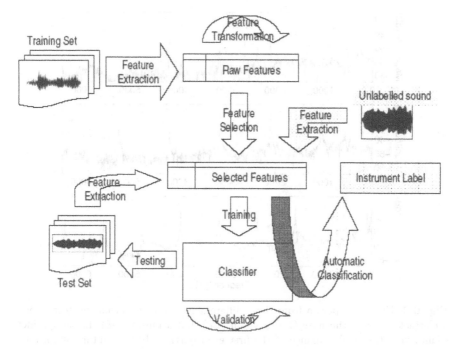

Fig. 6.1. Diagram of the different operations involved in setting up an automatic classification system for musical instrument sounds. The *training set* is described by extracting an *initial set of features* that is refined by means of feature transformation and selection algorithms. The resulting set of *selected features* is used to train and validate (i.e., fine-tune) the classifier. *Validation* can be done using a different set of sounds (not shown here) or different partitions of the training set. When fine-tuning is finished, the classifier is tested with a *test set* of sounds in order to assess its expected performance. At the right side of the diagram, using dotted elements, the automatic classification of an unlabelled (i.e., previously unseen) sound file is illustrated.

higher auditory centres [260]. In the case of pitched musical instruments, the relative strengths of the overtone partials (see Fig. 6.2) determine, to a certain extent, timbre sensations and identification. It seems that, for sustained sounds, the steady segment provides much more information than the attack, though the latter cannot be completely neglected [270]. Timbre discrimination experiments, where sounds are altered in subtle or dramatic ways and the listeners indicate whether two different versions sound the same, have provided cues concerning the relevant features for sound classification. Grey and Moorer [254] found that microvariations in amplitude and frequency are usually of little importance, and that the frequency and amplitude envelopes can be smoothed and approximated with line segments without being noticed by the listeners. Changes in temporal parameters (i.e., attack time, modulations) may have a dramatic impact on the discrimination of timbres [83], [560],

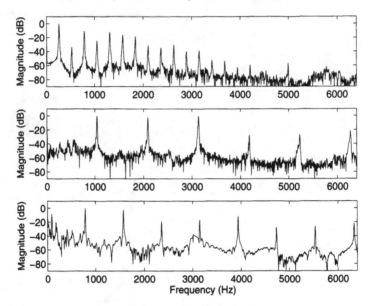

Fig. 6.2. Example spectra from three different instruments. From top to bottom: a clarinet (playing the note C4), a violin (C6) and a guitar (G5). In the clarinet sound, note the predominance of the first even partials. In the guitar sound, note the existence of plucking noise as energy unevenly distributed below 700 Hz.

though probably not in deciding the instrument name. The above facts were also supported by McAdams et al. [446], who, additionally, found that the spectral envelope shape and the spectral flux (time-variation of the spectrum; see Section 6.3.2) were the most salient physical parameters affecting timbre discrimination. Also, it has been noticed that the human sensitivity for different features depends on the sound source in question.

Very few studies have investigated the human ability to discriminate between the sounds of different musical instruments. However, some trends can be identified based on the reviews and experiments by Martin [442] and Srinivasan et al. [606]. First, humans, even those with musical training, rarely show performance rates better than 90%. The number of categories in the cited experiments varied from 9 to 39, and in the most difficult cases the recognition rate dropped to 40%. Second, confusion between certain instruments are quite usual, for example, between the French horn and the trombone. Third, the discrimination performance can be improved by musical instruction and by exposure to the acoustic material, especially to pairs of sounds from different instruments. Fourth, instrument families are easier to identify than individual instruments. Finally, contextual information (i.e., listening to instruments playing phrases, instead of isolated notes) substantially improves the identification performance [340], [57].

Table 6.1. A simplified taxonomy of pitched musical instruments that is partially based on Hornbostel and Sachs [296].

Family	Subfamily	Playing mode	Instrument
Chordophones		Bowed	Contrabass, Violoncello, Viola, Violin
		Plucked	Guitar, Harp, Ukulele, Harpsichord
		Struck	Piano
Aerophones	Woodwind	Air reed	Flute
		Double reed	Oboe, English Horn, Bassoon
		Single reed	Saxophone, Clarinet
	Brass	With valves	Tuba, French Horn, Trumpet
		W/o valves	Trombone, Bugle
Idiophones		Struck	Xylophone, Marimba, Celesta, Bell
		Plucked	Kalimba, Mbira
Membranophones		Struck	Timpani

6.1.2 Taxonomies of Musical Instruments

Taxonomies are a way to organize or classify musical instruments from the point of view of organology, which is the discipline dealing with the cultural, historical, technological, and practical issues of musical instruments [329]. When building taxonomies, we generate a tree where each category-branch can be broken down into more detailed categories. As we will see below, taxonomies are an important tool not only for conceptualizing similarities and differences between instruments, but also for helping our algorithms to discriminate between classes of sounds based on their acoustic properties. Taxonomies can be also termed *classification schemes* in certain technical domains, such as MPEG-7 [432].

One of the most extended and exhaustive taxonomies of musical instruments is that of Hornbostel and Sachs [296], which considers instruments according to how their sound is produced (see Table 6.1). Using this criterion, instruments are divided into chordophones (instruments that produce sound by acting on strings), aerophones (instruments that produce sound by acting on a wind column), idiophones (instruments that produce sound by acting on their own bodies), and membranophones (instruments that produce sound by acting on an elastic membrane). A recently added category is that of electrophones, which includes instruments that produce their sound by electrical means, such as electronic organs and synthesizers.

An additional taxonomic distinction to complement that of Hornbostel and Sachs comes from considering pitch: there are instruments that cause a clear pitch sensation (chordophones and aerophones), whereas the sounds of some other instruments do not have a definite pitch (most of the idiophones and membranophones). We denote the former ones as 'pitched', 'tuned', or 'with determinate pitch', whereas the latter ones are considered 'unpitched', 'untuned', or 'with indeterminate pitch'. Only pitched instruments are shown in

Table 6.1. Other criteria for elaborating subclasses can be the playing method, the shape of the instrument, the relationships of the exciting element to the resonating element, or the method used to put the exciting element into motion. For a detailed account on the acoustics of musical instruments, the reader is referred to Fletcher and Rossing [193] and Rossing [551].

6.1.3 Timbre Spaces

The perceived similarity or dissimilarity of the sounds of different musical instruments can be characterized with an abstract representation called 'timbre space'. Timbre spaces reflect human perception and are not necessarily optimal from the viewpoint of partitioning the space into separable classes. However, they can reveal the acoustic properties that enable computing perceptual similarities between instrument classes.

In order to derive timbre spaces, we first have to ask human listeners to generate similarity judgements by listening to pairs or triads of sounds. Similarity is recorded using a continuous or discrete scale that goes from 'very similar' to 'very dissimilar'. Similarity scores are then processed using a data dimensionality reduction technique known as multidimensional scaling (MDS). MDS is a method that represents measurements of dissimilarity between pairs of objects as distances between points in a space with a small number of dimensions (usually two or three), where each object (here, sound) corresponds to a point within the space. The MDS algorithm is explained in detail in Section 6.3.5.

When using timbre spaces to represent similarities between sounds, the obtained continuous space is delimited by axes that correspond to perceptual or acoustic factors having an important role in defining the timbre sensation. By looking at the correlations between the axes and the acoustic features that are computed from the sounds evaluated by the listeners, meaningful interpretations of the space can be derived.

Grey [253] was one of the pioneers in the elaboration of timbre spaces. He used synthetic sounds emulating twelve orchestral instruments from the string and the wind families. After requesting twenty listeners to judge the similarity between pairs of sounds, he applied MDS to the similarity judgements to derive a timbre space with three dimensions. The qualitative description of the axes he obtained when looking at their acoustic correlates were: (1) the spectral energy distribution, (2) the amount of synchronicity in the beginning transients, and (3) the temporal variations in the spectral envelope of a sound.

Further research by Wessel [665], Krumhansl [378], Krimphoff et al. [374], McAdams et al. [448], and Lakatos [385] have supported the predominance of a 'brightness' dimension (which is related to the first moment of the spectral energy distribution) as the main perceptual attribute for organizing sounds in a timbre space. Another important attribute seems to be related to the attack time, and a third dimension, receiving less clear support, could be the 'spectral flux', which is related to the temporal variation of the spectrum. These and other related features were included for timbre description in the MPEG-7 standard (see Section 6.3) [513].

Although most of the research on timbre spaces has been done in order to characterize human perception tendencies, it is also possible and useful to build them from purely acoustic measures. In this case, they reflect similarities inherent to the feature vectors that are used to characterize the sounds [105], [285], [510], [419].

6.2 Methodology

In this section, we discuss the methodological issues involved in building and evaluating classifiers which learn classes from labelled training data.

6.2.1 Databases

Databases are one of the crucial elements needed for developing a successful classification system, as they have to include enough 'representative examples' in order to grant the generalizability of the models built upon them [112]. Proper data modelling requires the careful preparation of up to three different and independent data sets (see Fig. 6.1). The first one, usually termed a *training set*, is used to build the models, whereas the second one, usually termed a *testing set*, is only used to test the model (or the system using it) and to get an estimate of its efficacy when it will be running in a real-world system. A third set, usually termed a *validation set*, is sometimes used during the design, improvement, and tweaking of a given model. In that case, the model, as it evolves and improves, is tested using the validation set, and only when the model preparation phase is finished (i.e., when the performance improvement on the training data is no longer matched by the performance improvement on the validation set) is it evaluated against the testing set, which is kept untouched until then. Of course, the three sets should be sampled from the same population of sounds.

A testing set that could be shared among research teams would help them to compare their respective improvements. Unfortunately, most of commercial audio files, MIDI files, and digitalized score files cannot be shared. In the automatic classification of musical instruments, the commercial McGill University Master Samples collection (MUMS) [487] has been frequently used, though it has not achieved the status of 'reference test set'. More recently, the University of Iowa sample collection[4] and, especially, the RWC database [230] are attracting the attention of researchers. The latter contains a wide variety of music files to be used in several music processing problems, and

---

[4]http://theremin.music.uiowa.edu/index.html

it is available to researchers under reasonable restrictions. Also, the International Music Information Retrieval Systems Evaluation Laboratory [154] is intended to provide remote access to a protected commercial music library and to a computational infrastructure that hosts algorithms for music content processing. In this way evaluation and comparison of different algorithms is possible, as the data sets and the computer systems are kept invariable. A simpler option, though, is now possible, thanks to the Creative Commons[5] initiative, which is a scheme for licensing and managing intellectual property rights. Several music providers are running their businesses partly or entirely on Creative Commons and therefore the music they distribute can be 'shared' by the scientific community.

6.2.2 Validation and Generalization

When we develop a classification system that learns from labelled examples, it is necessary to assess its generalization capabilities, i.e., its performance when classifying previously unseen instances. This process is called *validation*, and there are different procedures that can be followed: the leave-one-out method, the N-fold cross-validation, and the independent or holdout test set. The first one, recommended when we have a small number of instances available, tests a system with the same instances that are used to learn the classification model: If there are K instances in the data set, then K different models are built, each time leaving one instance out from the training and keeping it apart to test the model [379]. The results are then averaged over the K tests. In the N-fold cross-validation, we randomly divide the data set into N subsets or folds (usually 10) and then use $N - 1$ folds for training and one for testing. This is repeated N times and the final result is again averaged over the N runs (see also the alternatives and recommendations issued by Dietterich [145]). When a very large set of labelled instances is available, the recommended validation procedure consists of using a holdout set that has been kept unused up until then. Livshin and Rodet [414] have also presented interesting variations that can be used when we have several large sound collections available.

Another methodological element is the assessment of the statistical significance of the differences that have been found when comparing different sets of features or different classification algorithms. Empirical differences in performance can be obtained by chance or, alternatively, can be the consequence of true substantial differences in the goodness of the compared methods. The Student–Fisher t-test is frequently used to compare the error averages of different classification strategies [145]. It can be computed, in the case of N-fold cross-validation, as

$$ t = \frac{\bar{p}\sqrt{N}}{\sqrt{\frac{1}{N-1}\sum_{i=1}^{N}(p^{(i)} - \bar{p})^2}}, \tag{6.1} $$

---

[5]http://creativecommons.org

where N is the number of folds, $p^{(i)}$ is the average at fold i, and $\bar{p} = \frac{1}{N}\sum_{i=1}^{N} p^{(i)}$. Under the null hypothesis (i.e., the hypothesis that the average performances of two methods are not different) this statistic has a t-distribution with $n - 1$ degrees of freedom. The t-distribution is a symmetric bell-shaped distribution with the tails spread wider than those of the normal distribution. Its exact shape depends on the degrees of freedom: the greater the sample size, the more degrees of freedom and the narrower the bell shape (a large sample size makes the t-distribution resemble the normal distribution). The significance of the value t in (6.1) can be evaluated by looking it up in a table of the t-distribution and, based on this, the null hypothesis can be either rejected or accepted.

In spite of the pervasiveness of t-tests, it is sometimes recommended that one use the McNemar's test instead, as it provides a greater chance of obtaining a statistically significant result when the null hypothesis is false, and thus gives more power to our conclusions. The value for McNemar's test is computed as

$$X^2 = \frac{(|m_{01} - m_{10}| - 1)^2}{m_{01} + m_{10}}, \tag{6.2}$$

where m_{01} is the number of instances misclassified by algorithm A but not by algorithm B, and m_{10} is the number of instances misclassified by algorithm B but not by algorithm A [145]. Again, the significance of the obtained value has to be evaluated by looking at its distribution table, and this tells us whether to reject the null hypothesis or not.

Other methodological issues to be considered are the scalability of the algorithms, or their efficiency (i.e., what happens to the computation time when we double the number of features or the number of instances to be processed?). For methodological issues in computer experiments, the interested reader is referred to Santner et al. [562].

6.3 Features and Their Selection

Classification would be not possible without acoustic features that capture differences or similarities between two or more objects. In the case of automatic classification of instrument sounds, the features are extracted from the raw audio signal. As we will see, choosing good features is more crucial than the choice of the classification algorithm, and the classification itself becomes easier if the features chosen are informative enough [216].

The term 'feature' denotes a quantity or a quality describing an object. Features are also known as *attributes* or *descriptors*. Since the overall goal of classification is to distinguish between examples that belong to different classes, one of the goals of feature extraction is to reduce the variability of feature values for those examples that are associated with the same class, while increasing the variability between examples that come from different classes. Another informal requirement is to achieve a good capability of predicting the class after observing a given value of the feature (i.e., a high 'diagnosticity').

Finally, when using multiple features to predict a class label, it is desirable that the features be as little redundant as possible. Knowledge about the acoustic and perceptual qualities of the specific classes to be discriminated should guide, if possible, the choice of the features to use. Even lacking such knowledge, however, powerful feature selection and projection algorithms can be used to improve the discrimination capability.

Acoustic features are usually computed directly from the time-domain signal or, alternatively, from a representation obtained by transformations such as the discrete Fourier transform (DFT) or the wavelet transform, which have been discussed in Chapters 2 and 3, respectively. In order to grasp the micro-temporal evolution of the feature values, they are typically calculated every few milliseconds, in short analysis windows. Features representing the properties of instrument sounds are typically extracted within frames of length 20–50 ms, with a 50–75% overlap between successive frames. Macro-temporal features, such as attack time or vibrato rate, can be computed by using a longer segment of audio or by summarizing micro-temporal features over longer segments (for example with averages and variances).

For the purpose of presentation, in the following we have organized the described features under the subtitles of energy features, spectral features, temporal features, and harmonicity features (see [286], [511], [406], [694] for other feature taxonomies). As more than one hundred different features have been tested in the context of our problem, we only present the ones that constitute a basic kit to start with (see [363], [511] for more extensive accounts). An interesting set of features that has recently received some attention is the one included in MPEG-7, an ISO standard for multimedia content description (see Table 6.2) [432]. The interested reader is referred to [342] for in-depth explanations of the MPEG-7 audio descriptors and applications. These descriptors have been tested for the classification of musical instrument sounds in [71], [615], [367].

6.3.1 Energy Features

Signal energy or power can be measured at different time scales and used as an acoustic descriptor. Although these have not shown a high discriminative power when compared to other features used for musical instrument classification, they provide basic information that can be exploited to derive more complex descriptors, or to filter out potential outliers of a sound collection.

The *root mean square* (RMS) level of a signal is often used to represent the perceptual concept of *loudness*. The RMS level of a discrete time signal $x(n)$ is calculated as

$$E(n) = \sqrt{\frac{1}{N} \sum_{i=0}^{N-1} x(n+i)^2},$$

(6.3)

where n is a discrete time index and N is the size of the analysis frame.

Table 6.2. MPEG-7 descriptors, organized according to category. Each of these can be used to describe an audio segment with a summary value or with a series of sampled values. Timbral spectral descriptors are computed after extracting the relevant harmonic peaks from the spectrum of the signal. Spectral basis descriptors are a spectral representation of reduced dimensionality.

Category	Descriptors
signal parameters	fundamental frequency, harmonicity
basic	instantaneous waveform, power values
basic spectral	log-frequency power spectrum envelopes, spectral centroid, spectral spread, spectral flatness
timbral spectral	harmonic spectral centroid, harmonic spectral deviation, harmonic spectral spread, harmonic spectral variation
timbral temporal	log attack time, temporal centroid
spectral basis	spectrum basis, spectrum projection

6.3.2 Spectral Features

Different approaches can be considered in order to measure the shape of the spectral envelope or certain characteristics of the spectral fine structure. Spectral features can be computed using the DFT spectrum, or by applying a mel-scale filterbank and using the RMS levels at the mel bands. In order to do the latter, we first construct a bank of about 40 filters that are spaced uniformly on the mel frequency scale and have triangular magnitude responses (see Fig. 2.4 on p. 27). RMS levels within each mel band are then calculated by multiplying the magnitude spectrum of a given analysis frame with each triangular response, by squaring and summing the resulting DFT bin magnitudes, and by taking the square root of the result. The mel frequency scale approximates the way humans perceive pitch, having less resolution at high frequencies and a finer resolution at low frequencies. It can be computed as

$$f_{\text{mel}} = 2595 \log_{10} \left[\frac{f}{700} + 1 \right]. \tag{6.4}$$

Another psychoacoustically motivated frequency scale is the Bark scale, where usually 24 sub-bands are used.

A selection of the most useful spectral features is presented in the following. The *spectral centroid, spectral skewness, spectral kurtosis*, and *spectral spread* have been described in Chapter 5, p. 136, and are therefore not included here. In the presented equations, $X(k)$ may refer to a DFT spectrum or to a vector of RMS levels in Mel bands—depending on the pre-processing that has been adopted—and accordingly, the symbol k may index a frequency bin or a mel sub-band, and B denotes the total number of frequency bins or sub-bands used.

Spectral flatness indicates how flat (i.e. 'white-noisy') the spectrum of a sound is. A low value indicates a noisy sound, whereas a high value is indicative of tonal sounds. It is computed as the ratio of the geometric mean to the arithmetic mean of the spectrum:

$$\text{SFM}(t) = 10 \log_{10} \left[\frac{\left(\prod_{k=1}^{B} |X_t(k)| \right)^{\frac{1}{B}}}{\frac{1}{B} \sum_{k=1}^{B} |X_t(k)|} \right], \tag{6.5}$$

where t is the index of the analysis frame.

The *spectral flux*, also known as the delta spectrum magnitude, is a measure of local spectral change. It is defined as the squared ℓ^2 norm of the frame-to-frame spectral difference:

$$\text{SFX}(t) = \sum_{k=1}^{B} \left(|\tilde{X}_t(k)| - |\tilde{X}_{t-1}(k)| \right)^2, \tag{6.6}$$

where $\tilde{X}_t(k)$ and $\tilde{X}_{t-1}(k)$ are energy-normalized Fourier spectra in the current frame and in the previous frame, respectively.

Spectral irregularity measures the 'jaggedness' of the spectrum [374]. It is computed as

$$\text{SI}(t) = \frac{\sum_{k=1}^{B-1} (|X_t(k)| - |X_t(k+1)|)^2}{\sum_{k=1}^{B} |X_t(k)|^2}. \tag{6.7}$$

Spectral roll-off is defined as the frequency index R below which a certain fraction γ of the spectral energy resides:

$$\sum_{k=1}^{R} |X(k)|^2 \geq \gamma \sum_{k=1}^{B} |X(k)|^2. \tag{6.8}$$

Typically, either $\gamma = 0.95$ or $\gamma = 0.85$ is used.

The *Zero crossing rate* (ZCR) measures the number of times that the time-domain signal changes its sign. Even though it is computed in the time domain, it describes the amount of high-frequency energy in the signal (i.e., 'brightness') and correlates strongly with the spectral centroid mentioned above. ZCR has also proven to be quite discriminative for classes of percussion instruments [250]. The ZCR of a time-domain signal $x(n)$ is computed as

$$\text{ZCR}(n) = \frac{1}{2N} \sum_{i=1}^{N} |\text{sign}[x(n+i)] - \text{sign}[x(n+i-1)]|, \tag{6.9}$$

where

$$\text{sign}(x) = \begin{cases} +1 & \text{if } x > 0 \\ 0 & \text{if } x = 0 \, . \\ -1 & \text{if } x < 0 \end{cases} \tag{6.10}$$

Mel-frequency cepstral coefficients (MFCC) were introduced in Chapter 2, p. 26. To summarize the computational procedure: (1) a pre-emphasis filter is applied to flatten the spectrum; (2) audio is windowed and transformed using the DFT; (3) a mel-scale filterbank is applied in the frequency domain, as explained above; (4) the power within each sub-band is computed by squaring and summing frequency bin magnitudes within bands; (5) the dynamic range of the spectrum is compressed by taking a logarithm of the bandwise power values; and (6) cepstral coefficients are computed by applying, to the log filterbank powers, the discrete cosine transform (DCT) which decorrelates the coefficients. The dimensionality of the representation can be reduced by retaining only approximately 15 lowest-order DCT coefficients, which usually carry the relevant timbral information.

MFCCs have proven to be useful not only in speech processing tasks [536] but also in instrument recognition as shown, for example in [675], [56], [437], [176], [178]. Sometimes the delta ('velocity') and delta-delta ('acceleration') coefficients are also used, in order to capture information about their temporal evolution. The delta-MFCCs are usually computed as a least-squares approximation to the local slope, or

$$\Delta \mathsf{Cep}_i(t) = \frac{\sum_{m=-M}^{M} m \mathsf{Cep}_i(t+m)}{\sum_{m=-M}^{M} m^2}, \tag{6.11}$$

where t is the frame index, $\mathsf{Cep}_i(t)$ is the ith coefficient in frame t, and usually M is 1 or 2. The delta-delta, in turn, can be computed by substituting $\mathsf{Cep}_i(t)$ by $\Delta \mathsf{Cep}_i(t)$ in the above equation.

6.3.3 Harmonic Features

Pitched instruments usually have harmonic or nearly harmonic spectra. Therefore, it is interesting, as reported in several papers [369], [199], [688], [514], [670], [495], to use descriptors that take this property into account. Different options exist to extract the harmonic partials from a signal, such as computing the autocorrelation of the signal and generating an autoregressive (AR) model, selecting prominent peaks at integer multiples of the fundamental frequency from a long-term average spectrum, applying pattern-matching techniques to the peaks of the DFT spectrum, or estimating a sinusoid plus noise model [574] (see also Chapter 3). In this section, we denote the amplitude of the qth partial by $a(q)$, and the number of extracted partials by Q.

Inharmonicity describes the average deviation of spectral components from perfectly harmonic frequency positions. We can compute it as

$$\mathrm{IH} = \frac{2}{F_0} \times \frac{\sum_{q=1}^{Q} |f_q - qF_0| \times a^2(q)}{\sum_{q=1}^{Q} a^2(q)}, \tag{6.12}$$

where f_q is the frequency of partial q, and F_0 is the estimated fundamental frequency. An inharmonicity value of zero means that the signal is perfectly harmonic, whereas larger values indicate inharmonicity.

The *odd-to-even ratio* describes the relationship between odd and even partials. Some instruments, such as clarinets, have odd partials that are more prominent than the even ones. Other instruments, such as trumpets, have more balanced spectra. The odd-to-even ratio is given by

$$ \text{OER} = \frac{\sum_{q \text{ odd}} a^2(q)}{\sum_{q \text{ even}} a^2(q)}. \tag{6.13} $$

Tristimulus consists of three features that describe the relative weights of different harmonics [527]. These are computed as follows:

$$ \text{T1} = \frac{a^2(1)}{\sum_{q=1}^{Q} a^2(q)}, \tag{6.14} $$

$$ \text{T2} = \frac{a^2(2) + a^2(3) + a^2(4)}{\sum_{q=1}^{Q} a^2(q)}, \tag{6.15} $$

$$ \text{T3} = \frac{\sum_{q=5}^{Q} a^2(q)}{\sum_{q=1}^{Q} a^2(q)}. \tag{6.16} $$

By plotting the temporal evolution of T2 (x-axis) against T3 (y-axis), we can visualize the relative strengths of the fundamental and the high- and mid-frequency partials, and the evolution of these through time (see Fig. 6.3).

6.3.4 Temporal Features

The time dimension is usually less represented in the feature sets proposed for the automatic classification of musical sounds. The evolution of a given feature over time can be partially characterized by computing its variance, or the first- and second-order differences. Apart from these, specialized descriptors, such as the attack time, the temporal centroid, or the rate and depth of frequency modulation have proven to be useful for discriminating between instrument sounds [442], [199], [514], [615], [367].

The term *amplitude envelope* is generally used to refer to a temporally smoothed version of the signal level as a function of time. In practice, it can be calculated by lowpass filtering (with a 30-Hz cut-off frequency) the vector of RMS levels $E(n)$ of a signal. In the case of analysing isolated notes, once the envelope is computed, it is possible to segment it into attack, sustain, and release sections, as shown in Fig. 6.4 (though percussion and plucked string sounds do not have the sustain part). Specific descriptors for each of these segments can also be computed.

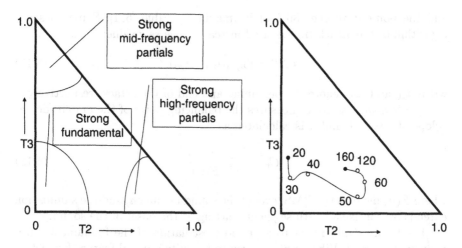

Fig. 6.3. Geometric interpretation of the tristimulus. The left figure shows the regions where, depending on the energy balance, the values of of T2 and T3 will be found. The figure on the right illustrates the temporal evolution (in milliseconds) of a clarinet note: it starts with a strong fundamental, then high frequency partials progressively dominate the sound, and finally, after 60 milliseconds, the high frequencies start to decay until the end of the sound.

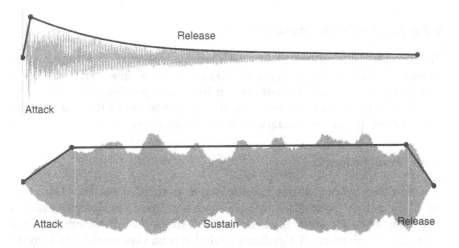

Fig. 6.4. Simplified amplitude envelopes of a guitar tone (above) and a violin tone (below). Different temporal segments of the tones are indicated in the figure.

The *attack time* is sometimes also called the 'rise time', and its definition varies slightly depending on the author. An often-used definition is the time interval between the point the audio signal reaches 20% of its maximum value and the point it reaches 80% of its maximum value [511]. Sometimes the logarithm of the attack time is used instead of the raw value:

$$\text{LAT} = \log_{10}(t_{80} - t_{20}), \tag{6.17}$$

where t_{20} and t_{80} denote the beginning and end of the attack, respectively.

The *temporal centroid* measures the balancing point of the amplitude envelope of a sound, and it is calculated as

$$C_\text{t} = \frac{\sum_n nE(n)}{\sum_n E(n)}, \tag{6.18}$$

where $E(n)$ denotes the RMS level of the sound at time n, and the summation extends over a fixed-length segment starting at the onset of the sound.

The term *vibrato* refers to a periodic oscillation of the fundamental frequency of a sound. Vibrato has proven to be quite a useful feature for instrument discrimination, whereas this does not seem to be the case with *tremolo*, which refers to a periodic oscillation in amplitude. Vibrato is characteristic for string instruments, reeds, and the human singing voice. Vibrato can be described by its rate, which is usually between 4 and 8 Hz, and its depth, which is usually less than one semitone. Techniques for estimating the rate and depth of vibrato are described in Chapter 12, Section 12.4.4.

6.3.5 Feature Transformation

Transforming features can enhance classification performance because it helps to reshape feature probability distributions or reveal interesting structures and, consequently, class boundaries can be more separable. Scale transform and feature projection are two strategies to change the features in order to get a 'better' feature set, compared to the original one.

Scale Transformation

One of the goals of features transformation is to improve their Gaussianity. Many pattern recognition techniques such as linear discriminant analysis (see Chapter 2) are optimal whenever the features have been sampled from Gaussian distributions. A significant deviation from Gaussianity can increase the classifier error rate. Osborne [490] discusses several transforms that change the features scale non-linearly and can help in making them more Gaussian:

- *Square root*, that is, substituting the original features by their square roots;
- *Logarithm*, that is, taking natural logarithm of the original features, and sometimes adding 1 before that in order to avoid having to compute logarithms of 0;
- *Inverse*, that is, substituting $1/x$ for x, where x is the original feature;

- *Arcsine-root* [410], which consists of taking the arcsine of the square root of the initial feature value; this transform spreads the feature values to the tails of the distribution (that is, preferably away from its central part). Therefore, this transform is indicated when the features are proportions such as the ratio of band energy to the total spectral energy.

Apart from these simple transforms, the Box–Cox power transform [47] is a standard tool for increasing the features Gaussianity, hence it is especially recommended when working with Gaussian mixture classifiers [510]. It can be computed as

$$h_\lambda(x) = \begin{cases} \frac{x^\lambda - 1}{\lambda} & \text{if} \quad \lambda \neq 0, \\ \log(x) & \text{if} \quad \lambda = 0, \end{cases} \tag{6.19}$$

where the value of λ is chosen in order to maximize the Gaussianity of the feature distribution.

Projection

Projections aim at reducing the dimensionality of the feature vectors by keeping a reduced set of dimensions. In other words, given a feature vector $\mathbf{x}$ composed of several features, projections only retain a subset of the initial features (or linear combinations of them), resulting in a lower-dimensional feature vector. When applied in the lower-dimension feature space, the classification algorithms are expected to have improved performance. In addition to multidimensional scaling and canonical discriminant analysis, which are presented below, other techniques are principal component analysis (see Chapter 2), independent components analysis (see Chapter 9 and [304]), projection pursuit [197], and random projection [39].

Multidimensional Scaling

The motivation behind multidimensional scaling (MDS) is to produce a space with typically 2 or 3 dimensions where projected, lower-dimensional feature vectors are arranged topographically so as to yield a visual representation of the similarity or dissimilarity of these feature vectors. Borg and Groenen [45] provide a thorough presentation of this technique.

Multidimensional scaling first requires the definition of *similarities*, or *proximities*, p_{ij} between any two objects—here between two sounds i and j $(i, j = 1, \ldots, m)$. These similarities may be obtained by numerical computations from the sounds (via the features), or by subjective evaluation by listeners. Given these proximities, MDS tries to represent the objects as points in a Euclidean space $\mathcal{X}$ with low dimension $d_\mathcal{X}$ such that a large distance between any two points in $\mathcal{X}$ represents a small proximity between the sounds, and conversely. Of course, this representation is not obtained by an exact mapping of the proximities to distances, which generally do not exist since the initial space is larger dimensional than $\mathcal{X}$. In order to implement MDS, we need to select the distance measure to be used in $\mathcal{X}$ to represent the inverse proximities; usually one selects the Euclidean distance between two objects $\mathbf{o}_i$ and $\mathbf{o}_j$ in $\mathcal{X}$:

$$d(\mathbf{o}_i, \mathbf{o}_j) = \left\{ \sum_{l=1}^{d_{\mathcal{X}}} \left[\mathbf{o}_i(l) - \mathbf{o}_j(l) \right]^2 \right\}^{1/2}. \qquad (6.20)$$

Given these distances, we need to learn a function h that approximately maps the proximities p_{ij} to the distances $d(\mathbf{o}_i, \mathbf{o}_j)$, and the location (the coordinates) of the objects $\mathbf{o}_i$, $i = 1, \dots, m$ in $\mathcal{X}$. This is done by minimizing the *stress function*

$$S(\mathsf{h}, \mathbf{o}_1, \dots, \mathbf{o}_m) = \left[\frac{\displaystyle\sum_{i=1}^{n} \sum_{j>i}^{n} \left(\mathsf{h}(p_{ij}) - d(\mathbf{o}_i, \mathbf{o}_j) \right)^2}{\displaystyle\sum_{i=1}^{n} \sum_{j>i}^{n} d(\mathbf{o}_i, \mathbf{o}_j)^2} \right]^{1/2}. \qquad (6.21)$$

The minimization is performed numerically with respect to h and to the objects $\mathbf{o}_i$, $i = 1, \dots, m$, by moving them around iteratively until the stress is minimum. The stress function gives the relative error that, on average, the distances $d(\mathbf{o}_i, \mathbf{o}_j)$ in $\mathcal{X}$ differ from the mapped proximities $\mathsf{h}(p_{ij})$. When using the Euclidean distance $d(\cdot, \cdot)$, the final coordinates of the $\mathbf{o}_i$'s in $\mathcal{X}$ do represent the principal coordinates which would be obtained when doing PCA on the dissimilarity matrix $[\mathsf{h}(p_{ij})]$. ALSCAL and PROXSCAL are two iterative algorithms designed to minimize stress when the assumed distances are not Euclidean.

Canonical Discriminant Analysis

Canonical discriminant analysis (CDA) is a dimension reduction technique related to principal component analysis and canonical correlation. In a canonical discriminant analysis, we find linear combinations of the features that provide maximal separation between the classes or groups. CDA derives a set of canonical variables that are linear combinations of the features, and that summarize between-class variation in much the same way that principal component analysis summarizes total variation; see Chapter 2. Canonical discriminants are computed iteratively: the first canonical discriminant function is the best single linear combination of attributes (that is, dimensions in the feature vectors) that discriminates between the classes. The second canonical discriminant function is the best single linear combination orthogonal to the first, and so on. Similar to principal component analysis, each canonical discriminant function is associated to an eigenvalue which indicates its relative discriminating power.

Canonical discriminants being linear combinations of the original features, they are projections, which are computed as follows. First, the *between-class scatter matrix* Σ_B and the *within-class scatter matrix* Σ_W are computed as

$$\Sigma_B = \frac{1}{J} \sum_{j=1}^{J} m_j(\mu_j - \mu)(\mu_j - \mu)^\mathsf{T}, \qquad (6.22)$$

$$\Sigma_W = \frac{1}{J} \sum_{j=1}^{J} \sum_{\mathbf{x}|y=j} (\mathbf{x} - \mu_j)(\mathbf{x} - \mu_j)^\mathsf{T}, \qquad (6.23)$$

where J is the number of classes, μ is the mean of all available m observations, μ_j is the mean of the observations in class j, $y \in \{1, 2, \ldots, J\}$ denotes the class label of $\mathbf{x}$, m_j is the number of observations in class j, and $\mathbf{x}$ is a given feature. Then, the optimal projection $\mathbf{W}$ is the projection matrix which maximizes the ratio of the determinant of the between-class scatter to the determinant of the within-class scatter of the projections (also called Fisher's criterion):

$$\mathbf{W} = \operatorname*{argmax}_{\mathbf{W}} \frac{|\mathbf{W}^\mathsf{T} \Sigma_B \mathbf{W}|}{|\mathbf{W}^\mathsf{T} \Sigma_W \mathbf{W}|} = [\mathbf{w}_1, \mathbf{w}_2, \ldots, \mathbf{w}_{d_\mathbf{x}}], \qquad (6.24)$$

where $\{\mathbf{w}_i | i = 1, 2, \ldots, d_\mathbf{x}\}$ is the set of generalized eigenvectors of Σ_B and Σ_W, corresponding to the $d_\mathbf{x}$ largest generalized eigenvalues $\{\lambda_i | i = 1, 2, \ldots, d_\mathbf{x}\}$ [295]. The matrix $\mathbf{W}$ can be found by solving the generalized eigenvalue problem $\Sigma_B \mathbf{A}^* = \lambda \Sigma_W \mathbf{A}^*$ by means of simultaneous diagonalization [202]. To avoid singularities, especially if the features are sparse, one can apply principal component analysis first to reduce the dimension of the feature space to $d_\mathbf{x} - J$, and then use CDA to reduce the dimension to $J - 1$.

6.3.6 Feature Selection

Using very large sets of features for building an automatic classification system is usually to be discouraged: First, some features can be redundant or irrelevant; second, the computational cost for using many of them might be high; and third, some features can be misleading or inconsistent regarding the task, and consequently the classification errors may increase. In any case, interpreting a model containing a large set of features can be very difficult or even impossible. In general, the informal recommendation is to use ten times fewer features than training instances[6] [313]. Selecting features can be done on a ranking basis (i.e., evaluating one feature after another) or on a best-set basis (i.e., evaluating subsets of features in a global way) [262].

We list below three different strategies in order to find a near-optimal number of features for a classification task [42]:

---

[6]Recent techniques such as support vector machines (see Chapter 2), however, are less subject to dimensionality concerns. In any case, including misleading features lowers the performance.

- *Embedding* makes the feature selection stage intertwined with the classification algorithm, as in the case with decision trees, or with discriminant analysis.
- *Filtering* decouples feature selection from the model learning process by first applying a feature selection over the original feature set, and then feeding the classification algorithm with the selected features only.
- *Wrapping* uses a features evaluation step which is intimately connected with the learning process: it uses the prediction performance of a given learning algorithm to assess the relative usefulness of subsets of features. Theoretically, this strategy should be the best one [42], but the price paid is a high computation time (this is an NP-hard problem).

In addition to selecting the features with respect to a classification algorithm, we must decide on an evaluation criterion. The *information gain* is the standard criterion used to build decision trees [467], but it can be also used to rank the importance of features, outside of the decision trees framework. In order to characterize the amount of information carried by a feature (e.g., the zero crossing rate), we study its influence on the *entropy* of the full set of features, via the information gain. Let $\mathbf{x}$ denote the vector made of several examples of a given feature (e.g., the zero crossing rate over several frames) and let X be the set of all the features, each of which being extracted over several frames. The entropy $H(\mathbf{a})$ of a set of random variables $\mathbf{a}$ with probability density function $p(\mathbf{a})$ is defined as[7]

$$H(\mathbf{a}) = -\int p(\mathbf{a}) \log p(\mathbf{a}) \, d\mathbf{a}, \qquad (6.25)$$

when $\mathbf{a}$ is a continuous random variable; the discrete case is obtained by replacing the integral by a sum over all possible values of X. We define the *information gain* of the feature $\mathbf{x}$ relative to the set of features X as

$$g(X, \mathbf{x}) = H(X) - H(X|\mathbf{x}), \qquad (6.26)$$

where $H(X|\mathbf{x})$ is the conditional entropy of X when the value of $\mathbf{x}$ is known. In other words, the information gain measures the amount of information added by considering the full feature set X with entropy $H(X)$, as opposed to the reduced set $X \backslash \mathbf{x}$ with entropy $H(X|\mathbf{x})$.

The *correlation-based feature selection* technique of Hall [268] has provided very good results in selecting features for automatic classification of sounds [288], [510]. We define the following merit heuristic of a feature subset X_s containing m features, each having several values extracted from various frames:

$$M(X_s) = \frac{mc_{fc}}{\sqrt{m + m(m-1)c_{ff}}}, \qquad (6.27)$$

---

[7]Random variables and probability density functions are introduced in Chapter 2.

where c_{fc} is the average feature–class correlation and c_{ff} is the average feature–feature intercorrelation.[8] We can interpret the numerator of (6.27) as an indicator of how a set of features is representative of the class, whereas the denominator indicates how much redundancy there is among the features. The correlation-based feature selection technique is not a ranking method, as it works with subsets of features. In practice, subsets are examined using either backward search or forward search of best first search. Backward search consists of starting from the full feature set and greedily removing one feature at a time as long as the evaluation does not degrade too much). Forward search starts from an empty set of features and greedily adds one feature at a time until no possible single feature addition results in a higher evaluation. Best first search starts with either no features or all the features and examines a given number of consecutive expansions or reductions of the existing subset in order to find local improvements over the current best subset [540], [267].

In the context of classification, given a training set of features grouped in J classes, Peeters [510] has proposed audio feature selection using *inertia ratio maximization using feature space projection* (IRMFSP), which seems to compare advantageously with other effective algorithms and has also been used by other researchers in sound classification [416], [180]. IRMFSP selects first the best features according to the value R[i] for the feature #i (where the index i refers to a given type of feature, e.g., the zero crossing rate) defined as

$$\text{R}[i] \;=\; \frac{\displaystyle\sum_{j=1}^{J} m_j \|\boldsymbol{\mu}_j[i] - \boldsymbol{\mu}[i]\|^2}{\displaystyle\sum_{n=1}^{N} \|\mathbf{x}_n[i] - \boldsymbol{\mu}[i]\|^2}, \tag{6.28}$$

where J_j is the number of training features in class j (j=1,...,J), $\boldsymbol{\mu}[i]$ is the empirical average of feature #i in the full training set, $\boldsymbol{\mu}_j[i]$ is the empirical average of the feature #i for features in the training set belonging to class j. The ratio in (6.28) is the between-class inertia to the total inertia: the larger R[i], the more discriminative is the feature. The feature selection proceeds as follows. First, the features are selected from the largest inertia to the smallest inertia. This selection process is intertwined with an orthogonalization of the feature space; that is, each new selected feature is made othogonal to the previous ones (this is the standard Gram–Schmidt orthogonalization; see e.g., [388]).

The *extractor discovery system* (EDS) [692] is an interesting tool that 'creates' new features by means of *genetic programming* [372] applied to

---

[8]The term 'correlation' was used by Hall in its general sense, without referring specifically to the classical correlation [267, p. 51]. However, in his implementation in the free software Weka [674] (www.cs.waikato.ac.nz/ml/weka), the author used the classical variance-normalized correlation, removing, however, the mean of the data before calculating the correlation.

the existing features and to a library of mathematical and signal processing functions. When used in instrument classification problems (percussion/no-percussion detection and 4-class string instruments discrimination), the EDS increased between 1 percent-units and 9 percent-units the performance compared to using the original 'simpler' features [691]. A somewhat similar system has also been presented by Mierswa and Morik [463].

Other interesting approaches to feature selection, based on genetic algorithms (GA) are by Fujinaga et al. [200], [199] and about *rough sets* [359], [360], [367], [669], [671]. The latter have been used to select features for instrument classifiers. Further details about genetic algorithms and rough sets may be found, respectively, in [217] and [508].

6.4 Classification Techniques

When designing a classifier for musical instrument sounds, researchers are usually faced with the problem of selecting a single classifier for all the instrument sound classes or, conversely, using several 'focused' classifiers. In the first case classifiers are termed *flat classifiers*, whereas the second case classifiers may be one of several different solutions ranging from *hierarchical classifiers* to *ensembles of classifiers*.

- In the *hierarchical case* [443], [176], [515], a classifier is used in each of the different decision nodes of a hierarchy that has been created by means of knowledge-based decision rules. A hierarchical classifier may decide first if the sound belongs to a given broad class (such as impulsive vs. continuous). Then, the sound is classified into a finer class that is conditioned by the first classification decision (for instance, in the case of a continuous sound, the second level of the hierarchy decides if it is from woodwinds, bowed strings or brass; in the case of an impulsive sound, the options could be plucked strings or percussion). The final level of the hierarchy, depending on the previous decisions, assigns the specific instrument class (guitar, oboe, etc.).
- Concerning *ensembles of classifiers*, different classification techniques may be used in parallel or serially, in order to deal with subsets of the original data, of the feature set, or with the whole set; see [379].

Some of the classification techniques that have been frequently used for the automatic classification of musical instrument sounds are explained in Chapter 2. This is the case with support vector machines (SVMs), and Gaussian mixture models (GMMs), which are therefore not included here. In this section, we discuss instance-based classifiers, discriminant analysis, decision trees, and artificial neural networks.

6.4.1 Instance-Based Learning

Instance-based learning or *lazy learning* refers to the idea of avoiding the computation of any abstract model for a class of objects. Instead, techniques under this category assume that instances that are closer to the one we try to classify (i.e., they are 'similar' in the feature space) will provide the required class label. Instance-based learning is usually the most workable option when the number of available classes is extremely large, or cannot be totally anticipated, as it is the case, for instance, in the domain of classification of sound effects [62].

The *k-nearest neighbours* (*k*-NN) algorithm is one of the most popular algorithms for instance-based learning. It first stores the feature vectors of all the training examples and then, for classifying a new instance, it finds a set of *k* nearest training examples in the feature space, and assigns the new example to the class that has the largest number of examples in the set (see Fig. 6.5). Traditionally, the Euclidean distance measure (6.20) is used to determine similarity, and the number of neighbours is determined empirically; see the description of *k*-NN in Chapter 2. Although the *k*-NN technique is easy to implement, it has some drawbacks [467]:

- It requires having all the training instances in memory in order to yield a decision for classifying a new instance;
- It may require a significant computational load each time a new query is processed;
- It is highly sensitive to irrelevant features that can dominate the distance metrics;
- It does not provide a generalization mechanism (because it is only based on local information), although several techniques (K-means clustering, linear vector quantization) can be used to compute 'prototypes' from several nearest neighbours.

The *k*-NN algorithm has probably been the most used in classification studies of musical instrument sounds, achieving very high rates of performance under most of the tested conditions; see [200], [442], [11], [174], [416], [325]. Given its simplicity, it is usual to include it as a kind of 'reference' when comparing different classification algorithms.

6.4.2 Discriminant Analysis

Discriminant analysis (DA) includes several variants of the generic idea of deriving a discrimination function (i.e., one that separates two classes of objects) that is a weighted combination of a subset of the features used to characterize a series of observations. As we have seen in previous sections, this idea can be also used for selecting and projecting features by minimizing the ratio of within-class scatter to the between-class scatter. A thorough formal

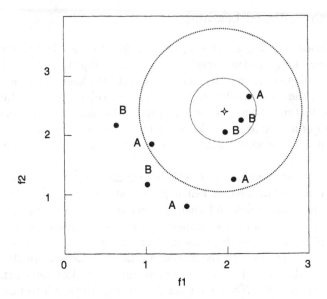

Fig. 6.5. An illustration of k-NN classification. The point marked with a star would be classified as belonging to category B when $k = 3$ (as 2 out of its 3 neighbours are from class B; but in case of using $k = 5$ classification would be A because there are 3 nearest neighbours belonging to this category and only 2 belonging to B.

and practical treatment of this techniques can be found in the monograph by McLachlan [453].

Linear discriminant analysis (LDA) is a minimum-distance classification method that uses the empirical mean and covariance matrix of the training set classes. The data are projected so as to eliminate inter-feature correlations and to standardize the variance of each variable, after which Euclidean distances are computed in the standardized space; see Chapter 2 for a presentation. LDA assumes that the classes have a common covariance matrix C_i but sometimes this cannot be true. In that case the *quadratic discriminant analysis* (QDA) can be applied, where the decision boundary between each pair of classes is described by a quadratic equation. These and other variations such as logistic or regularized discriminant analysis have been been used in several studies on classification of musical instruments [443], [11], [12], [510], [416], [552].

6.4.3 Decision Trees

Decision trees are pervasively used for different machine learning and classification tasks. One of the main reasons for their popularity may lie in the fact that they produce a simple classification procedure which can be interpreted and understood as a series of 'if-then' actions. Decision trees are constructed top-down, beginning with the feature that seems to be the most informative, that is, the one that maximally reduces entropy according to the information gain measure (6.26). For this feature, several branches are created: one for each of its possible (discrete) values. In the case of non-discrete valued features, a procedure for discretization of the value range must be defined (see, for instance, [183]). The training data are assigned to a *descendant node*, at the bottom of the branch that corresponds their value. This process is repeated recursively starting from each descendant node. An in-depth treatment of decision trees can be found in Mitchell [467] and Duda et al. [161]. Quinlan's *ID3* and *C4.5* [533] are among the most popular algorithms for building decision trees.

Decision trees used for the classification of instrument sounds have usually yielded worse results than other classification methods [288], [510]. Otherwise they have provided hints on the nature of the features and values that discriminate among pitched instrument classes [319], [668]. Recent enhancements to basic decision trees such as AdaBoost [196] or Random Forests [51] may provide results that are competitive with other cutting-edge classification algorithms.

6.4.4 Artificial Neural Networks

An *artificial neural network* (ANN) is an information-processing structure that is composed of a large number of highly interconnected elements—called neurons or units—working in unison to solve a specific problem. Neurons are grouped into layers (which can be *input*, *hidden*, and *output*) to be interconnected through different patterns (see Fig. 6.6). Each neuron has an *activation function*. The activation of each neuron depends on the activity coming from the other neurons that connect to it, and on the activation function. The full neural network learns a complex, non-linear function between *input* (here, features extracted from a given sound) and *output* vectors (here, an instrument class) by changing the interconnections between the neurons: a training set containing example input and related output vectors is presented to the ANN, which learns the non-linear function. This is a regularization problem, as described on p. 57. The interested reader is referred to the specific monograph of Bishop [40] for an in-depth presentation of the subject.

One of the most popular neural network models is the *multilayer perceptron* (MLP). Usually, the action of each neuron is modelled as a linear function of its inputs, and the activation is modelled by a sigmoid function [279]. Structure-wise, MLP is a spatially iterative neural network with several layers of hidden neuron units between the input and output neuron layers. The most commonly used learning scheme for the MLP is the back-propagation algorithm [555]. The weight updating for the hidden layers adopts the mechanism of back-propagated corrective signal from the output layer. It was demonstrated that perceptrons with one hidden layer are adequate as universal approximators of any non-linear function [666], which means that they can theoretically learn any non-linear relation between their inputs and outputs. This makes MLPs

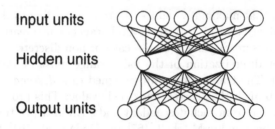

Fig. 6.6. A diagram of an artificial neural network with 8 input units, a hidden layer with 3 units, and an output layer with 8 units.

a good choice when the function to be learned is not known in advance, or it is suspected to be non-linear.

ANNs do have some disadvantages, though: the computation time for the learning phase may be very long, adjustment of user-defined parameters can be tedious and prohibitively time consuming, and data over-fitting can degrade their generalization capabilities. Interest in ANNs appears to have declined since the arrival of SVMs. This can be explained because of several advantages of the latter over ANNs: SVMs require fewer parameters to be tuned and less time to be trained/used, and are able to provide similar or greater accuracy than ANNs, with minimal over-fitting. However, when either ANNs or SVMs are learned, the classification decision is very fast when compared to other popular methods such as k-NN, and they can also learn to disregard irrelevant attributes.

Kostek [360] and Park [495] are probably the most exhaustive studies on automatic instrument classification using neural networks.

6.5 Classification of Isolated Sounds

In this section, we review research aimed at developing systems for the automatic classification of isolated sounds of pitched musical instruments. As we will see, for the more complex tasks of recognizing instruments in solo and duet phrases or in polyphonic music, the fact that each research team has been using a different test database makes it unfair and unreliable to make direct comparisons between the reported classification accuracies. The goal of this section is, hence, to provide enough information so that the reader can understand the work done and to evaluate the main outcomes of the research himself. We first examine flat systems, which provide instrument labels in a single classification step. Then we will move to systems that, instead of directly deciding the instrumental class, proceed from broader categories to the more specific ones, following a hierarchical classification scheme.

6.5.1 Flat Classification Systems

An important distinction to be made first is that between systems that have been developed and tested using several different databases—meaning that several different 'instances' of each instrument (for example, several pianos), different recording conditions, and players are considered—and those using only one database (mostly MUMS; see Section 6.2.1). The systems in the latter group are only learning to identify a given instrument instance, for example a particular violin, and therefore their generalization capabilities are suspect. Martin [442] and Eronen [174] observed this fact indirectly, and a conclusive demonstration was provided by Livshin and Rodet in [414] (but also see [285]), where a classifier trained and tested using the same large collection (IRCAM's Studio On-Line) achieved excellent performance rates (96% for instrument classes), but when tested using a different collection of sounds, performance dropped down to 26%.

A summary of classification systems for isolated instrument sounds is presented in Table 6.3. It should be noted that only systems dealing with multiple instances of each instrument are included in the table. Also, definitions of the acoustic features may vary slightly from the formulas given in Section 6.3, depending on the authors. The data shown is not repeated below, but instead, some general remarks are made in the following.

Although most of the reviewed systems based their features on a fast Fourier transform (FFT) front-end, some studies have reported results on wavelets [371], correlograms (see Chapter 8, p. 245) [442], Hough transforms [552], linear prediction [174], [682], [87], and constant Q transforms [56], [57], [325]. An advantage of FFT-based features over wavelet-based features has been reported in [361], and an advantage of linear prediction over cepstral analysis has been reported in [375]. It is clear that more comparative research is needed to make strong conclusions regarding the time-frequency analysis front-ends.

Fundamental frequency (F0) estimation is an important step as this feature by itself, and harmonic descriptors based on it, may provide substantial discriminative information. The dependence of timbre features on F0 makes it difficult to compare the feature values of two instruments playing different notes. Kitahara et al. [345] developed a musical instrument classification system using an F0-dependent multivariate normal distribution, where the mean of the distribution of each feature was represented as a cubic polynomial function of the fundamental frequency. This F0-dependent mean function was used to represent the pitch dependency of each feature, while F0-normalized covariances after substracting the pitch contribution were used to represent the non-pitch dependence of the features. Musical instrument sounds were first analysed by the F0-dependent multivariate distribution, and then identified by using a discriminant function based on the Bayes rule. A slight advantage of considering the F0-dependence was observed both at the instrument-level

190   Perfecto Herrera-Boyer, Anssi Klapuri, and Manuel Davy

Table 6.3. Summary of selected research on automatic classification of isolated instrument sounds. Only studies dealing with several different instrument instances and recording conditions have been included. The column NC shows the number of classes in each system. In the last two columns, the performance of hierarchical systems has been indicated with '(H:)'.

Author, year [ref]	Total instances[a]	NC	Acoustic features[b]	Classification algorithm	Instrument performance (%)	Family performance (%)
Eronen, 2001 [174]	5286 (MUMS, Iowa, SOL, RolandXP30, own recordings)	29	MFCC (attack-steady), F0, ATT, onset features, SC, Crest Factor, AM	k-NN	35 (H: 30)	77 (H: 75)
Livshin et al., 2003 [413]	4381 (SOL, Iowa, MUMS, Prosonus, Vitous)	16	SC, ATT, temporal decrease, TRI, HD, SKW, KUR, SV, SS, MFCC, noisiness	LDA & k-NN	47-69	62-92
Peeters, 2003 [511]	4163 (SOL, Iowa, MUMS, Microsoft MI, Prosonus, Vitous)	23	same as above	LDA & GMM (hierarchical)	54 (H: 64)	81 (H: 85)
Eronen, 2003 [175]	5895 (MUMS, Iowa, SOL, Martin, own recordings)	7	MFCC, delta-MFCC + ICA	HMM	68	n/a
Kitahara et al., 2003 [345]	6247 (RWC)	19	SC, OER, F0 relative energy, KUR, SKW, FM, amplitude envelope slope, onset energy	Bayes (k-NN after PCA & LDA)	80	91
Kostek et al., 2004 [362]	n/a (CMIS, MUMS)	12	Wavelet-based energy bands, MPEG-7 features	MLP	71	n/a
Szczuko et al., 2004 [615]	2517 (CMIS, MUMS)	16	MPEG-7 features, OER, F0	MLP (2-stage hierarchical MLP)	86 (H: 89)	n/a
Park et al., 2005 [496]	829 (several commercial instrument-sample CDs)	12	SS, SC, harmonic slope, LPC noise, harmonic expansion/contraction, spectral jitter and shimmer, spectral flux, TC, ZCR	MLP with elliptical/radial basis functions	71	88
Chétry et al., 2005 [88]	4415 (Iowa, RWC, voice)	11	Line spectrum frequencies	K-means derived codebook	95	n/a

[a]In the Total instances column, the abbreviated sound collections are McGill University Master Samples (MUMS), University of Iowa samples ('Iowa'), IRCAM's Studio On-Line (SOL), Gdansk's Catalogue of Musical Instrument Sound (CMIS), and Real World Computing (RWC). Other names refer to commercial sample CD-ROMs.

[b]In the Acoustic features column, SC denotes spectral centroid, SS is spectral spread, SV is spectral variation, SKW is skewness, KUR is kurtosis, SFM is spectral flatness, ZCR is zero crossing rate, TRI is tristimulus, OER is odd-to-even ratio, HD is harmonic deviation, ATT is attack time, TC is temporal centroid, FM denotes vibrato, and AM is amplitude modulation.

and at the family-level classification. A different approach to this problem can also be found in [688].

Evidence abounds on the positive effect of combining different types of features. This has been demonstrated for wavelet-based energy bands plus MPEG-7 audio features [362], for spectral plus temporal features [670], and for MFCC plus other spectral and temporal features [412], [510], [597].

Feature-vector dimensionality reduction by projection is another recommendable option, as the work of Peeters et al. [515], [511], Livshin et al. [413], [416], Kitahara et al. [345], and Essid et al. [177] illustrates. In all these cases, sets with hundreds of features were reduced by means of discriminant analysis (see Section 6.3.5 above) to a fraction of the original dimensionality without decreasing, or even effectively increasing, the discrimination rate of the system. Another interesting projection approach has been proposed by Kaminskyj and Czaszejko [325], who applied PCA to a vector consisting of the constant-Q transform frequency bins in the vicinity of the first 20 harmonic overtones. The PCA-transformed feature vectors in successive analysis windows formed a multidimensional abstract trajectory space, where sounds of different classes showed different trajectories.

Regarding the classification techniques, k-NN is the most-often used and one of the best performers in the this kind of problem. Being very easy to implement, it is a good point of comparison for other classifiers. LDA is sometimes suggested as another 'default' method to be included in comparative studies but unless prior feature selection and projection is properly done, it cannot compete with SVMs or GMMs. We expect to see an increasing number of studies using SVMs, as they have theoretical and practical advantages that are well understood and clearly demonstrated in other areas of data classification (see Chapter 2).

Artificial neural networks have been used almost exclusively by Kostek et al. [357], [358], [368], [360], [366], [363], [364], who also reported on performing instrument classification using *rough sets*. The latter, and other rule-based techniques such as decision trees, have not proven to be competitive with other classification techniques [597], [515], although some authors have found other reasons for adopting them [361], [671]. The ANNs used by Kostek's team were standard MLPs with one hidden layer. Explorations of parallel networks can be found in [615], and networks with radial and elliptical basis functions have been studied by Park et al. [495], [496]. Also, Park proposed a technique called *nearest centroid error clustering*, which he used together with the standard back-propagation learning algorithm for training the network. With this new technique, the system achieved 71% classification accuracy, 25%-units more than without the clustering.

A surprising observation is that hidden Markov models (HMMs) have not been very usual in automatic instrument classification systems, given their prevalence in speech recognition and speaker recognition (see Chapter 2). Casey [71] provided an interesting classification framework, in connection to MPEG-7 spectral bases, but the most advanced HMM-based system for classifying pitched instrument sounds seems to be that of Eronen [175]. He proposed to use independent component Analysis for transforming the original feature vectors (which improved the performance by 6%) and to use a discriminative training algorithm for the HMM, which also slightly improved the performance.

To conclude, we have observed that the difficulty of classification is directly related to the number of classes [442], [12]. We should then expect very good performance for systems dealing with less than 10 classes. On the other hand, working with more than 20 classes becomes an extremely difficult problem, not only in the case of instrument sounds but also in any other automatic learning domain. Hierarchical classifiers have been proposed as a means of maintaining the generalization ability of a classifier when the number of classes is increased [442]. These are discussed in the next section.

6.5.2 Classification Based on Hierarchical Decisions

One of the influential ideas of Martin [441], [442], [443] was to use a hierarchical procedure consisting of (1) initial discrimination between *pizzicato* (plucked) and sustained sounds, (2) discrimination between different instrument families (e.g., strings, woodwind, and brass), and (3) depending on the previous decisions, final classification into specific instrument categories. Other hierarchical systems have been developed since then by Eronen et al. [176], [174], Agostini et al. [11], Szczuko et al. [615] and finally, Peeters et al. [515], [510], which is probably a fair representative of the current state of the art in instrument classification.

Table 6.3 summarizes some facts about the systems mentioned in this section. As can be seen, the results using a hierarchical instead of a flat classification scheme have not been conclusive: some authors report moderate improvements, whereas others report moderate deterioration in error rates. On the other hand, what is consistent is a trend of increasing difficulty when the categorization of an instrument goes from the most abstract to the most specific: the pizzicato versus sustained decision is very easy, the family classification problem is a bit more difficult, and finally, the specific assignment of instrument labels still leaves some room for improvement.

Although the use of hierarchies is conceptually appealing, it is worth noting that errors at each level are 'carried over' in a multiplicative way. Therefore, if there is an error at the topmost level, it will be propagated to the lower levels of the hierarchy. One rule of thumb for trying a hierarchical system would be to look at the between-class confusion matrices and search for a large number of confusions between instruments from different families. In this case, provided a very efficient family discriminator, these confusions could be reduced by means of a series of hierarchical decisions.

Instead of embedding some taxonomic knowledge into the classifier (i.e. hardwiring the family and subfamily taxonomy), Agostini [10] and Kitahara et al. [346] have approached the automatic building of instrument taxonomies by means of clustering techniques, although the reported results were worse than those using a predefined taxonomy. A more elaborate proposal was presented by Essid et al. [179], but the authors did not use their induced taxonomy for classification purposes.

A related and promising thread for future explorations is an approach called 'mixtures of experts' [593], where a sound is analysed by a number of specialized classifiers, and the results are then combined by a voting mechanism which attempts to neutralize individual errors made by one of the classifiers. For example, if a sound comes from a bowed violin, but the most generic classifier erroneously classifies it as 'not-sustained', while the bowed-string and violin experts assign it to the correct classes, then the sound would be correctly labelled provided a conflict resolution mechanism is properly implemented. Fanelli et al. [181] presented a modular neuro-fuzzy network for classifying instrument families, where different submodules operated on distinct subsets of input features. Similarly, Kostek et al. [371] presented a system of parallel ANNs, where each network generated a decision based on a single energy band and then all decisions were combined to produce a coherent output. This idea of combining specially trained subsystems has not been pursued so far in any other of the studies we have reviewed here, but it deserves more attention (see also Essid et al. [177] in the next section).

6.6 Classification of Sounds from Music Files

One of the research trends in the classification of musical sounds is moving toward more *ecological* scenarios, that is, working with real music files, addressing practical applications, and utilizing musicological information in the process. Three different classification contexts, in the order of increasing complexity, are reviewed here: solo phrases, duets, and complex musical mixtures.

6.6.1 Classification of Instruments in Solo Phrases

Detecting the instrument played in a solo phrase should in principle be feasible by applying the same algorithms and models that are used when working with isolated notes. Even though we know that articulation and expressive phrasing elements may convey cues for instrument identification [340], the existing research has not exploited these yet. Research has also disregarded, up until now, the problem of how to locate the solo sections in an entire musical piece, and all the works reviewed here process solo phrases that have been manually extracted.

Table 6.4 summarizes some studies dealing with this problem. The target material, features, and other details of the systems are not repeated here; instead we highlight other aspects that also deserve attention.

One important decision in solo instrument identification is how long it takes to decide on the instrument label. Using longer time segments and calculating averages or choosing the most frequent class decision should provide

Table 6.4. Summary of research on instrument classification from solo phrases. The column *NC* shows the number of classes in each system. See the footnotes below Table 6.3 for interpreting the abbreviated feature names and sound collections.

Author, year [Ref]	Acoustic material	NC	Acoustic features	Classifi-cation algo-rithm	Instru-ment Perfor-mance (%)
Martin, 1999 [442]	1500 excerpts (MUMS, CDs, own recordings)	14	ATT, SC, SI, FM, AM, ZCR, onset features	Hierarchi-cal k-NN	39
Marques & Moreno, 1999 [437]	1638 (train) + 171 (test) seconds from 100 music tracks on CDs	8	MFCC	SVM	70
Brown et al., 2001 [57]	14 commercial CDs	4	Cepstral coefficients, autocorrelation, SC, constant-Q band differences	GMM	80
Ventura-Miravet et al., 2003 [640]	1800 (train) + 900 (test) seconds per class from CDs (7 different players per class)	6	Perceptual linear prediction, Linear-prediction derived cepstral coefficients	HMM	94
Livshin et al., 2003 [416]	108 solos (commercial CDs)	7	SC, KUR, SKW, SS, slope, TRI, MFCC	LDA + k-NN	88 (85 real-time)
Vincent & Rodet, 2004 [647]	Train: sounds from RWC; test: 2 × 5 seconds excerpts from 10 solo recordings per class	5	non-linear ISA	GMM	90
Essid et al., 2004 [177]	900 (train) + 90 (test) seconds per class from MUMS, Iowa, SOL, and commercial music CDs	5	MFCC	GMM	67
Essid et al., 2004 [178]	900 (train) + 150 (test) seconds per class from same as above (+instruments added)	10	MFCC, delta-MFCC, SC, SS, SKW, KUR, bandwise SFM	SVM	77
Krishna & Sreenivas, 2004 [375]	Train: Iowa; test: RWC	3	Line Spectrum Frequencies, MFCC, LPC	GMM	74
Chétry et al., 2005 [87]	300 (train) + 300 (test) seconds per class from CDs	6	Line Spectrum Frequencies plus their velocity and acceleration	SVM	78

better results, even leading to better discrimination rates than those in classifying isolated sounds. The optimal reported durations vary between two and ten seconds, depending on the authors [437], [640], [178]. As the time dimension has a more prominent role here than in identifying isolated sounds, HMMs may have some advantage over 'static' modelling techniques such as those used in the previous section. The HMM-based system reported by Ventura-Miravet et al. [640] provided impressive evidence on that.

An original proposal that goes back to the idea of 'mixture of experts' is using a 'one against one' classification strategy, where a series of classifiers are devised that only make binary decisions between an instrument A and instrument B, and the decisions are then combined by voting [177]. In addition to improving the classification accuracy under certain circumstances, the use of pair-wise classifiers makes it possible to find some characteristic features that are very useful in a given pair-wise discrimination but remain useless in others.

Exploring new features and increasing their number is another trend observed in this research context. The log-energy within octave sub-bands and the logarithm of the energy ratio of adjacent sub-bands have been used as a way to characterize the spectral energy distribution of a sound [177]. Line spectral frequencies have also been incorporated as a way to model the resonances and peaks of the power spectrum with higher precision and robustness than linear prediction coefficients (LPC) [375], [87].

Independent subspace analysis (ISA) has been successfully used for percussive sound classification (see Chapter 5) but it has only been tested for classifying pitched sounds by Eronen [175] and by Vincent and Rodet [647]. The latter represented the short-time spectrum of musical excerpts as a non-linear weighted sum of typical spectra plus noise, which were learned using files from a database containing isolated notes and solo recordings. These templates were then used to determine, with a very high accuracy, the instrument played in the solos of commercial recordings (even when they were artificially distorted with reverberation or noise). The authors showed that their model has some theoretical advantages over methods based on GMMs or on linear ISA and that it worked successfully even for the classification of instruments in duets.

A surprising observation is that any of the reviewed systems do not segment solo phrases into notes. The addition of a reliable onset detector would allow the inclusion of envelope-related temporal features and the reduction of computational load by doing the actual classification only once per note onset. Onset detection is, however, a hard problem in the case of music signals that do not contain percussion instruments [95].

6.6.2 Classification of Instruments in Duets

Classification of instruments in musical duets requires more complex techniques than those for isolated samples or solo phrases. The added difficulty lies in achieving some kind of sound source separation, or robust extraction of the features of a given source in the interference of the other source. As we will see, the few existing studies that have been summarized in Table 6.5 are still more or less exploratory in nature: in addition to the sound separation issue, they use few duet types and a small music collection, so their reported high performance estimations should be taken with care.

Two different approaches have been examined here. In the first one, primitive source separation (or *source reduction*, as Livshin and Rodet [416] termed it) is performed by estimating the F0 of one or both instruments and then substracting the partials associated with the first sound in order to estimate

Table 6.5. Summary of research on automatic identification of instruments from duets.

Author, year [Ref]	Acoustic material	Number of pairs	Acoustic features	Separation	Classifier	Performance (%)
Livshin & Rodet, 2004 [416]	Training: isolated sounds and sounds from solo phrases. Test: 18 music pieces	8 (7 different instruments)	SC, ATT, temporal decrease, TRIS, HD, SKW, KUR, SV, SS, MFCC, noisiness	Source reduction (spectral subtraction)	LDA projection + k-NN	40–100, depending on the pairs
Kostek et al., 2004 [362]	Training: 400 sounds, plus 400 sounds for cross-validation of the timbre model. Test: artificial pairs	8 (4 different instruments)	Partials' amplitudes	Spectral subtraction	MLP	n/a (only average activation of the MLP output layer reported)
Eggink & Brown, 2004 [167]	Training: isolated sounds mixed in pairs and artificially mixed solo phrases. Test: mixed pairs of sounds	10 (5 different instruments)	Energy within 120 sub-bands	Missing feature	GMM	74

the spectrum of the second one. Using this approach, harmonic cancellations and erroneous enhancement of partials should be expected, depending on the degree of overlap between the spectra of the two sounds.

Livshin and Rodet [416] approached the classification of instruments in duets by using their real-time solo recognition system (see Table 6.4). In order to detect instruments in duets, the system first estimated the two F0s using an algorithm by Yeh and Röbel [681], and also computed their corresponding harmonic partials. The fundamental frequencies were quantized to the nearest musical note, and contiguous frames with the same value were chunked together. Each chunk was then used twice in a phase-vocoder filtering process of source reduction: in the first pass, all the harmonics of the estimated fundamental were kept, whereas in the second pass, the sustained note's harmonics were filtered out and the 'residual' partials were kept. Overlapping harmonics of the two notes were not filtered out. Finally, the partials of the fundamental (intended to correspond to one instrument) and the residual partials (intended to correspond to the other instrument) were sent to a classifier in order to generate their corresponding labels.

Kostek et al. [365], [362] proposed the decomposition of duet sounds based on the modified frequency envelope distribution (FED) analysis, which was originally described in [370]. The FED algorithm decomposes a signal into a linear expansions of sinusoids with time-varying amplitudes and phases. The first step of the duet analysis method is the estimation of the F0 of the lower-pitched instrument. The input signal is divided into short overlapping blocks, and F0 is estimated for each block separately to deliver the F0 contour. Using the F0 information and the FED algorithm, the time-varying amplitudes and phases of the first ten harmonics of the sound are estimated and cancelled from the signal, in order to obtain a residual where the harmonics of the second sound are analysed. The estimated spectra of the two sounds are then fed to a neural classifier to recognize the two instruments.

Another approach to analysing duet signals is based on a so-called *missing feature theory* that was developed for speech processing and speaker identification. The main idea consists of using only the spectro-temporal regions which are dominated by the target sound, and ignoring those that are dominated by background noise or interfering tones. This approach is motivated by a model of auditory perception, proposed by Cooke et al. [99], which postulates a similar process in listeners. In polyphonic music, partials of one instrument often overlap with those of another one. Consequently, the observed amplitudes of these partials no longer correspond to those of any individual instrument. Within the missing feature approach, these corrupted or unreliable features can be excluded from the recognition process. The remaining information is therefore incomplete, but the hope is that it is still sufficient to enable robust instrument classification (additionally, it is possible to partially reconstruct the missing values by exploiting known correlations between the missing and the reliable values). Eggink and Brown [168], [167] used sub-band energies as features, although other features could be utilized, too.

6.6.3 Classification of Instruments in Complex Mixtures

If classifying instruments in duets is still in its infancy, we should say that identifying them in more complex polyphonic mixtures is a kind of newborn subject. Recognizing all the instruments that are played at a given time in a complex musical mixture is a tough problem. The pioneering proposal by Kashino and Murase [331], which is closely related to sound source separation (see Chapters 9 and 10), required that a representative waveform of each note of each instrument to be identified was stored beforehand in order to perform a template adaptation and matching process. Once the onsets and F0s were estimated (in their original system this was done manually), the most prominent instrument tone was determined by comparing the mixture with phase-adjusted example waveforms. In an iterative processing cycle, the energy of the corresponding waveform was then subtracted to find the next most prominent instrument tone in the residual. The system also successfully exploited high-level musical knowledge related to note transitions and voice leading.

A closely related system was presented by Kinoshita et al. [343]. It included improvements over the spectral template matching method of Kashino and Murase [330]. F0 estimation was performed prior to the instrument classification process to determine where partials of concurrent sounds overlap. Features used for instrument identification were heuristically evaluated according to their reliability and diagnosticity with relation to the stored templates of each instrument class. The overlapping features (e.g. power in bands) were adapted according to the known templates, and those reliable and very characteristic features of an instrument were selected for performing the classification. Both the system of Kinoshita et al. and that of Kashino and Murase were evaluated using combinations of three instruments, and it is not clear if they could be used with more dense mixtures.

Approaches other than source separation are feasible by considering some simplifications to the problem. One of them consists of focusing on a given instrument or instrument type only. This is the case, for instance, when we concentrate on percussion instruments, which include very idiosyncratic sounds (i.e., 'noises') that stand apart from the remaining 'harmonic' sounds generated by pitched instruments (see Chapter 5). Also, detecting a singing voice above the rest of the instrumentation is an example of this approach [630]. Heittola applied a similar detection-without-separation strategy to detect the presence of certain instruments (bowed strings, electric guitar, piano, saxophone, vocals) in polyphonic music, but found that this straightforward approach did not produce very reliable results [281].

Deciding if a musical piece involves certain instruments in a 'predominant' role, even though they cannot be accurately located in time, is another way to address the identification of instruments in polyphonic music. An instrument can be considered as predominant if it is present for a significant part of a piece and it is louder than the others most of the time. For instance, in popular music the singer's voice is usually louder than most of the other instruments. Following this path, the systems proposed by Eggink and Brown [170], and by Zhang [688] assume that F0 estimation provides a reliable cue for identifying a predominant instrument. The most dominant F0 is determined by a frequency-domain harmonic pattern matching process (see Chapter 8) and the spectrum of the predominant instrument is then estimated and used to compute discriminative features. Finally, a classifier that has been trained using isolated sounds and solo phrases makes the decision on the instrument class. It is important to note that in both systems it is required to decide on only one instrument at each given time. In Zhang's system, an 'instrument profile' was extracted for the whole piece by computing the proportion of notes that were played by each instrument (hence, onset detection was performed). In the case of Eggink and Brown, their system only attempted to identify which instrument was, among 5 classes, the main one played in accompanied sonatas and concertos (but not where it was played).

Identifying combinations of instruments directly (for example *piano + violin*), instead of separating them, has been addressed by Essid et al. [180]. The authors presented a multi-instrument recognition scheme capable of processing jazz piano quartets based on prior knowledge of the musical context. Here, the number of possible combinations can be reduced by building super-classes consisting of unions of classes having similar acoustic features. The classification can be performed hierarchically in the sense that a given test segment can be first classified among the top-level classes, and then more precisely (when needed) at a lower level. For example, if a test segment involves double bass, drums, and trumpet, then it is first identified as belonging to the combined category 'Bass-Drums-MonoPitchedWind' and subsequently as 'Bass-Drums-Trumpet'. A substantial amount of data from each of the effective sources (i.e., individual instruments but also their usual combinations) is required to train the classifier.[9]

6.7 Conclusions

In this chapter we have provided a review of the theoretical, methodological, and practical issues involved in the automatic classification of pitched musical instruments. Assigning instrument labels to analysis frames, sounds, or musical segments requires, first, a solid knowledge of the acoustic features of the instruments, and of the ways we can exploit signal processing techniques to convert them into numerical features. Additionally, a systematic methodology comprising the collection of data sets for training and testing, the selection and transformation of features, and the comparison of the results obtained using different approaches, defines the right path for obtaining a robust classification system.

In the five years from our first review of the field of automatic classification of musical sounds [283] to the present moment, the number of studies exceeded twice the number of those published in 1990s. Some general tendencies can be noted in these recent works: use of larger and more varied databases, interest for unpitched percussion sounds, improvements in the methodological aspects, and an increasing concern for practical applications and for dealing with truly musical fragments.

The performance of the systems dealing with a large amount of isolated sounds and pitched instrument classes achieves correct decisions nearly 70% of the time, whereas a simpler decision on instrument family rises a bit beyond 80% accuracy. This leaves some room for improvement that could be achieved, among other options, by carefully looking at the discriminative acoustic and perceptual features that each class of sounds may have and then devising feature extractors that capture them properly.

On the other hand, systems dealing with the classification of instruments in musical excerpts are the actual hot-spot of the field even though the achievements are still quite modest. For the classification of solo phrases, the achievable performance can be a bit better than that for isolated sounds, but when duets or more complex combinations are considered, the performance drops substantially. In those cases, systems become more complex as they rely on multiple-F0 estimation and on incomplete or noisy estimation of spectral and temporal information. The current approaches try to avoid 'hard' source separation and exploit contextual or musical knowledge. This makes the problem more manageable, even though the provided solutions have still been limited in terms of sound combinations or musical styles.

---

[9]See also Chapter 5, p. 137, where the idea of recognizing combinations of sounds directly is discussed from the viewpoint of percussion transcription.

In this review we have identified several open issues that could provide interesting returns when properly addressed: (1) The need of a reference test collection, containing enough variability in instruments, recording conditions and performers to be considered as an unbiased sample of the real population of instrument sounds, and granting that any proposed system can be fairly compared to other alternative proposals. Fortunately, RWC is currently a serious candidate that should gain wider acceptance among research groups. (2) The need to develop better features and instrument-specific features. (3) The need to investigate possibilities to embed some general knowledge about the task into the classification system, such as the usual frequency ranges of the instruments, voice leading rules etc., but also very specific knowledge, for example by crafting ensembles of specialized classifiers. (4) The need to evaluate the robustness of a system under reverberant, noisy, or other distortion conditions. (5) Incorporating instruments outside the typical orchestral ones: singing voice and some electrophones, for instance, the electric guitar, would deserve specific studies by themselves, given their broad timbral registers. (6) The need to the develop systems dealing with realistic polyphonic music signals.

The automatic classification of pitched sounds of musical instrument has progressed a lot in the past five years. Even though real time [199], [415] and commercial systems for instrument sound classification have been devised,[10] we expect to find more of them soon, in connection with applied problems posed by personal digital music players. This means that now is time to exploit the knowledge we have gained working with isolated sounds, in order to address the identification of instruments played in polyphonic music. This is, without doubt, the challenge for the forthcoming years.

Acknowledgements

The writing of this chapter was partially supported by the EU project SIMAC (Semantic Interaction with Music Audio Contents) EU-FP6-IST-507142. The first author wishes to acknowledge the input, feedback, and help received during the preparation of the manuscript from Eduard Aylon, Emilia Gómez, Fabien Gouyon, Enric Guaus, Eulalia Montalvo, Bee Suan Ong, and Sebastian Streich.

---

[10]http://www.musclefish.com, http://www.soundfisher.com, http://cuidadosp.ircam.fr, http://www.audioclas.net,

Part III

Multiple Fundamental Frequency Analysis

7

Multiple Fundamental Frequency Estimation Based on Generative Models

Manuel Davy

LAGIS/CNRS, BP 48, Cité Scientifique, 59651 Villeneuve d'Ascq Cedex, France
Manuel.Davy@ec-lille.fr

Western tonal music is highly structured, both along the time axis and along the frequency axis. The time structure is described in other chapters of this book (see Chapter 4), and it may be exploited to build efficient beat trackers, for example. The frequency structure is also quite strong in tonal music. It has been shown since Helmholtz (and probably before) that an individual note is composed of one fundamental and several overtone partials [451], [193]. Though acoustic waveforms may vary from one musical instrument to another, and even from one performance to another with the same instrument, they can be modelled accurately using a unique mathematical model, with different parameters.

In addition to a mathematical model that describes the waveform generation, the frequency structure of music can be used to derive priors over the model parameter values. Here, we understand *frequency structure* in terms of fundamental and partials structure, pitch/F0 structure, etc. For example, assume the instrument playing is a piano; then the note frequencies cannot be just any frequencies; they have to match the piano key frequencies. Also, the piano overtone partial frequencies are slightly inharmonic (that is, they are not integer multiples of the fundamental partial frequency), and their frequencies are described by a specific model [451], [193] which can be used to build parameter priors. More generally, the structure of tonal music may be exploited to build a *Bayesian model*, that is, a mathematical model embedded into a probabilistic framework that leads to the simplest model that explains a given waveform (see Chapter 2 for an introduction).

The Bayesian setting is quite natural for this problem as it enables the use of many heuristics within a rigorous framework. Moreover, acoustic waveform models generally have many parameters, which cannot be accurately estimated without regularizing assumptions, such as parameter priors. Bayesian models for multiple F0 estimation have received, however, relatively little attention. A possible cause is that such models are complex, and this makes their use difficult—though achievable—when confronted with real data. Sometimes they are also computationally heavy. However, such models enable much more than multiple F0 tracking. They do model the acoustic waveform: the parameters which are estimated from real musical records may be used for multiple F0 estimation, but also for monaural source separation, sound compression, pitch correction, etc.

In this chapter, we present several approaches to multiple F0 estimation that rely on a *generative model* of the acoustic waveform. More precisely, we present a *noisy sum-of-sines model*, which has been studied by many authors in various contexts. This model and some of its variants are presented in Section 7.1. Section 7.2 introduces a Bayesian off-line processing method which requires notewise processing (that is, processing is performed on a complete waveform section which does not include note changes). In Section 7.3, we present the on-line processing model of Cemgil et al. [78], Dubois and Davy [158], and Vincent and Plumbley [646]. Section 7.4 is devoted to other on-line multiple F0 tracking algorithms that rely on incomplete or indirect acoustic waveform modelling: for example, the approach of Thornburg et al. [625] and Sterian et al. [609] models the time evolution of time-frequency energy peaks. Dubois and Davy [159] model the signal spectrogram (which comes down to estimating on-line the acoustic waveform up to the initial phase parameter, though). Section 7.5 presents some conclusions.

7.1 Noisy Sum-of-Sines Models

In this section, we first present some simple models which were developed for single F0 acoustic signals. The earliest noisy sum-of-sines acoustic waveform models were developed for speech synthesis; see e.g. [449]. These models did not assume, however, frequency relations between the fundamental partial and overtone partials. Laroche et al. introduce a *harmonic plus noise model* [394] which assumes such relations. These models were soon used for music processing; see [575] for a review of early methods.

7.1.1 Single F0 Stationary Models

The frequency structure of tonal music acoustic waveforms has been observed for many years. As can be seen in Fig. 7.1, these waveforms are almost periodic and their Fourier transforms reduce to (approximately) sums of sine waves whose frequencies are multiples of a given frequency. For a perfectly periodic (infinite length) signal x, with discrete time $n = 1, 2, \ldots$,

$$x(n) = \sum_{m=1}^{M} \alpha^s \sin(2\pi m k_1 n) + \alpha^c \cos(2\pi m k_1 n). \qquad (7.1)$$

In the following, we name the sine + cosine component with frequency k_1 the *fundamental* and components with harmonic number $m = 2, \ldots, M$ the *overtone partials.*[1] The model in (7.1) is quite simple, but it is rather theoretical. Real signals always include components which cannot be modelled as individual sines or cosines: for example, a flute player breathing can be heard in recorded signals, and this is highly non-periodic [219]. As such components are quite different from one occurence to another, they can be jointly modelled in terms of their statistical distribution as a *noise component* ϵ, yielding the model

$$x(n) = \sum_{m=1}^{M} \alpha^s \sin(2\pi m k_1 n) + \alpha^c \cos(2\pi m k_1 n) + \epsilon(n). \qquad (7.2)$$

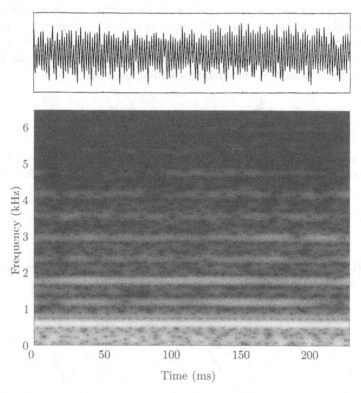

Fig. 7.1. Flute acoustic waveform (top) together with its spectrogram (bottom).

Two noise statistical models have received some attention. The simplest assumes $\epsilon(n)$ to be a white noise with Gaussian distribution (see for example [394], [616], [124]). This is the less informative assumption, as in this case,

---

[1]As pointed out in Chapter 1, the frequency of the fundamental partial, denoted here by k_1, is different from the fundamental frequency F0, which is the inverse of the acoustic waveform period.

$\epsilon(n)$ is a purely random sequence with a flat power spectrum. Another popular model takes the form of an autoregressive process [576], [575], [292], [609], [122] (see Chapter 2 for a presentation of autoregressive models), which also corresponds to random sequences, but with non-flat power spectrum. A review of possible choices for ϵ may be found in [310], [308].

In addition to non-harmonic components, acoustic waveforms produced by real musical instruments also have another important characteristic: they are not strictly periodic. This is explained by two phenomena: *inharmonicity* (or *partial de-tuning*) and *partials amplitude nonstationarity*. Inharmonicity appears whenever the frequency of the partial with harmonic number m is not exactly mk_1. For the example in Fig. 7.2, where several periods of the acoustic waveform produced by a piano, a flute, and a clarinet are superimposed, the non-periodicity appears clearly. This may be caused by amplitudes decay (for all three examples), but also by inharmonicity (piano example). Note that in Fig. 7.2, two of the three periods plotted are contiguous, whereas one is taken further apart. A more general model enabling inharmonicity is

$$x(n) = \sum_{m=1}^{M} \alpha^s \sin(2\pi k_m n) + \alpha^c \cos(2\pi k_m n) + \epsilon(n), \qquad (7.3)$$

where the partial frequencies k_m $(m = 2, \ldots, M)$ are related to k_1 by a more elaborate relation than $k_m = mk_1$. For example, Fletcher and Rossing [193, p. 363] propose the following piano inharmonicity model:

$$k_m = mk_1 \sqrt{\frac{1 + m^2 B}{1 + B}}, \qquad (7.4)$$

with $B \in [10^{-5}, 10^{-2}]$. For $B = 0.0004$, this shifts the 17th partial at the frequency position of the 18th partial.

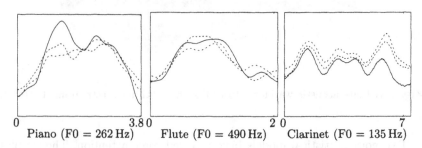

0   3.8 0   2 0   7
   Piano (F0 = 262 Hz)   Flute (F0 = 490 Hz)   Clarinet (F0 = 135 Hz)

Fig. 7.2. Three superimposed periods of the acoustic waveforms played by a piano, a flute, and a clarinet. The waveforms are not strictly periodic and the three periods represented (in solid, dashed, and dashed-dotted lines) are not exactly superimposed. The time scale is in milliseconds.

Aside from partial frequency models, it may be useful to model the partial-to-partial amplitude profile (referred to as the *spectrum envelope* in the following). The power spectrum of a note is formed by the instrument body response which modulates the partial frequency peaks. Of course, this modulation depends on the instrument and it may be characterized by a *spectrum envelope* (see the saxophone family example [193, p. 497]) which may be modelled. Spectrum envelope models need to be quite flexible, though, because some instruments have special behavior. For example, clarinets have almost zero amplitude for every other low frequency partials (partials with even harmonic numbers m); see Fig. 6.2 p. 166. Godsill and Davy [213] propose a statistical model where the amplitudes are assumed to be approximately constant below some cut-off frequency k_{cutoff}, and decay exponentially for higher frequencies. The parameters defining the model (k_{cutoff} and the exponential decay rate) are to be estimated from the processed acoustic waveform. Cemgil [78] uses an exponential decay where the amplitude α_m^s of partial m equals $\left[\alpha_1^s\right]^m$ (and similarly with α_m^c). Alternative models may be proposed, based on the spectral smoothness principle; see Klapuri [351].

7.1.2 Single F0 Non-Stationary Models

The models presented above permit quite good modelling of very short signal portions, insofar as the amplitudes and frequencies do not vary too much over time. As explained in Chapter 4, however, the amplitudes do vary quickly enough so that the above models cannot be used to process musical segments longer than 20 to 50 ms. A more general non-stationary model is

$$x(n) = \sum_{m=1}^{M} \alpha_m^s(n) \sin\left(2\pi k_m(n)n\right) + \alpha_m^c(n) \cos\left(2\pi k_m(n)n\right) + \epsilon(n), \quad (7.5)$$

where the amplitudes $\alpha_m^s(n)$ and $\alpha_m^c(n)$, the frequencies $k_m(n)$ for $m = 1,\ldots,M$, and the noise statistics now depend on the time. This model is quite flexible, but it is no more a sine-plus-noise model: to understand this, assume $k_m(n)$ is an independent sequence of random frequencies with $M = 1$; then the wave generated is not a sine wave at all! This shows that the model in (7.5) should be constrained so as to be suited to tonal music. This can be done in many ways, but the simplest is certainly to assume a smooth time evolution of the amplitudes and frequencies, and assume a statistically (almost) stationary noise.

Many amplitude and frequency evolution models may be found. The amplitude evolution model should mimic the way notes appear and disappear in music (onset and decay), whereas the frequency models should adapt to stationary cases (for instruments such as the piano where the performer has quite limited influence on the note frequencies evolution) or vibrato (e.g., for violins). Many such models may be found in the music synthesis literature, and they are generally quite instrument specific; see e.g. [543], [628]. These models may be used for musical signal analysis, in particular when the analysis conditions are well controlled. Here, we present more general models which can adapt to different instruments and different kinds of music.

Amplitude Evolution Models

As pointed out above, amplitude evolution models need to allow quick enough variations in order to fit note onset and decay. However, amplitudes should not vary too quickly: to make this point clearer, assume the acoustic waveform is a sine wave and let $M = 1$ and $k_1 = 0$. Then, the only way for the model to fit the data is for the amplitude itself to be a sine wave. This illustrates that, given an acoustic waveform, there is not a unique frequency/amplitude parameter set for the model in (7.5); rather, there are many. It is thus important to prevent the time-varying amplitudes from fitting the sine waves. This can be easily done by selecting amplitude evolution models that do not permit oscillations with frequencies over some 10 Hz, for example.

A relevant model is that of *damped amplitudes*, where the amplitude evolves according to a decreasing exponential; that is (where we drop the partial index m for notation clarity),

$$\alpha^s(n) = \tilde{\alpha}^s \exp(-\lambda n) \quad \text{and} \quad \alpha^c(n) = \tilde{\alpha}^c \exp(-\lambda n), \qquad (7.6)$$

where the damping factor λ tunes the amplitude decay rate, and $\tilde{\alpha}^s$, $\tilde{\alpha}^c$ are fixed initial amplitudes. When substituted into (7.5), this yields a *damped sinusoids model*, as used by Hilands and Thomopoulos [292] for multiple sinusoids frequency estimation or by Cemgil et al. [78] for music transcription; see Section 7.3 below. Note that such an amplitude time evolution may be coupled with a spectrum envelope as described in Section 7.1.1.

Another amplitude evolution model consists of assuming a random walk or an autoregressive process, as proposed in [106] for chirp signals. This reduces the amplitude parameters to the set of AR coefficients, which may be small. These coefficients should be chosen, however, so as to ensure a smooth amplitude.[2]

The last model presented here writes the time-varying amplitude as a sum of weighted smooth, time-localized functions with time-domain shape $\phi(n)$

$$\alpha^s(n) = \sum_{I=0}^{I} \tilde{\alpha}_i^s \phi[n - i\Delta_n] \quad \text{and} \quad \alpha^c(n) = \sum_{I=0}^{I} \tilde{\alpha}_i^c \phi[n - i\Delta_n], \qquad (7.7)$$

where $\tilde{\alpha}_i^s$, $\tilde{\alpha}_i^c$ are the amplitudes (also called *weights*) associated to each time-localized function $\phi[n - i\Delta_n]$. The step Δ_n sets the spacing between two such successive functions. The shape ϕ is typically chosen so as to obtain smooth amplitude profile; in general it is one of the standard 'sliding windows':

---

[2]This can be obtained by choosing the AR coefficients whose corresponding characteristic polynomial has zeros with relatively small amplitude [339].

Gaussian, Hamming, Hanning, etc. Here also, the amplitude evolution from one frame to another may follow a random walk [159] or may be unrelated a priori [616]. An important remark is that the resulting time-varying sum-of-sines model is closely related to a *Gabor representation*, where a signal is decomposed into windowed sine/cosine waves whose time-frequency locations are determined by a regular lattice [184]. Here, the lattice is regular along the time axis and irregular along the frequency axis. Gabor-style amplitude models were used by Davy and Godsill [122] (irregular lattice) and Wolfe et al. [676] (regular lattice).

Frequency Evolution Models

A simple frequency evolution model may assume constant frequencies. This is actually quite a realistic assumption insofar as that the instrument player has no means of changing note frequencies during their emission. This is the case for many instruments, including pianos, flutes, oboes, etc. It can also be a sought-after effect to keep it constant in other instruments such as violins, cellos, or trumpets.

Aside this simple model, a possibility is to assume random walk frequency evolution, that is

$$k_m(n) = k_m(n-1) + v_m(n) \quad \text{for } n = 2, \dots, T, \tag{7.8}$$

where $v_m(n)$ is a Gaussian white random noise with some fixed variance. In order to be valid, though, the noises $v_m(n)$, $m = 1, \dots, M$ have to be correlated so that the frequencies of related partials follow similar evolutions.

In the case of abrupt frequency changes, the models presented here are out of their validity domain. However, a standard assumption is that no note changes occur within the acoustic waveform segments processed.

7.1.3 Multiple F0 Non-Stationary Models

It is straightforward to extend from single to multiple F0 models by simply adding J single F0 models. Equation (7.5) becomes

$$x(n) = \sum_{j=1}^{J} \sum_{m=1}^{M_j} \alpha_{j,m}^{s}(n) \sin\left(2\pi k_{j,m}(n)n\right) + \alpha_{j,m}^{c}(n) \cos\left(2\pi k_{j,m}(n)n\right) + \epsilon(n),$$

$$\tag{7.9}$$

where the amplitudes and frequencies are modelled as in the single F0 models discussed above. In specific contexts, it may also be useful to model links between the frequencies and amplitudes of different notes. For example, the sound of an electromechanical organ may be modulated by rotating loudspeakers; the strings of an electric guitar may be jointly tightened by a moving bridge. Such links may be either deterministic or probabilistic.

The mathematical models described above are characterized by many parameters to be estimated. Depending on the specific model selected, the number of parameters for a 100 ms excerpt with 3 notes may vary from about 50 to more than 3000. In order to make the estimation feasible and non-trivial, it is necessary to define a measure of the fit between the data and the model and to favour simple models. This can be easily done with a Bayesian model. The following two sections define such probabilistic models first in the off-line case, and then in the on-line case.

7.2 Off-line Approaches

The off-line approaches are characterized by their *global* viewpoint on the parameter estimation problem. In the Bayesian estimation settings, we need a likelihood term (which describes the model-to-data fit) and priors, because the quantity of interest (the posterior) is proportional to the product of the former two terms; see Chapter 2. An important assumption is made in this section: the processed acoustic waveforms have been segmented so that they do not include transitions between the notes. This can be achieved by applying some onset detection technique; see Chapter 4.

7.2.1 Likelihood

The likelihood term is completely described by the mathematical model selected, and by the probability density function of the noise $\epsilon(n)$. As an illustrative example, we assume here that this noise is zero-mean white Gaussian with variance σ^2, and we assume the model of (7.9) with the amplitude evolution model of (7.7) and constant frequencies. In order to simplify the notation, we write the waveform mathematical model in vector form. Let $\mathbf{x} = [x(1) \ldots x(T)]^\mathsf{T}$ and $\boldsymbol{\epsilon} = [\epsilon(1) \ldots \epsilon(T)]^\mathsf{T}$. Moreover, we define $\boldsymbol{\alpha}$ to be the vector of sine/cosine amplitudes. Finally, let $\mathbf{D}$ be a matrix with T lines, whose rows contain the sine/cosine Gabor atoms for each note partial frequency, in the same order as in $\boldsymbol{\alpha}$ (see [122], [124] and (2.68) in Chapter 2 for explicit definitions of $\boldsymbol{\alpha}$ and $\mathbf{D}$). With these notations, the acoustic waveform model can be written as

$$\mathbf{x} = \mathbf{D}\boldsymbol{\alpha} + \boldsymbol{\epsilon}. \tag{7.10}$$

All of the models presented above may be written in this vector form, where the basis function matrix $\mathbf{D}$ and the amplitude vector $\boldsymbol{\alpha}$ may have different structures. In any case, the dimension of $\mathbf{D}$ depends on J, on $\mathbf{M} = [M_1, \ldots, M_J]$, and on I, and its rows depend on the frequencies $\mathbf{k}$ (where $\mathbf{k}$ is the vector that contains all the frequency parameters $k_{j,m}(n)$, $m = 1, \ldots, M_j$ and $j = 1, \ldots, J$).

It is now possible to write the likelihood of this model. Of course, it depends on the probabilistic assumptions made about the noise $\boldsymbol{\epsilon}$. Let us define $\tilde{\mathbf{x}}$ and

$\tilde{\mathbf{D}}$ such that $\tilde{\mathbf{x}} = \mathbf{x}$ and $\tilde{\mathbf{D}} = \mathbf{D}$ whenever ϵ is a zero-mean white Gaussian noise of variance σ^2 and $\tilde{\mathbf{x}} = \mathbf{Ax}$, $\tilde{\mathbf{D}} = \mathbf{AD}$ if the noise is autoregressive of order p and coefficients $\boldsymbol{a} = [a_1, \ldots, a_p]$, with Gaussian noise of variance σ^2. The matrix $\mathbf{A}$ is defined as

$$
\mathbf{A} = \begin{bmatrix}
1 & 0 & \cdots & \cdots & \cdots & \cdots & 0 \\
-a_1 & 1 & 0 & \cdots & \cdots & \cdots & 0 \\
\vdots & \ddots & \ddots & \ddots & & & \vdots \\
-a_p & \cdots & -a_1 & 1 & 0 & \cdots & 0 \\
0 & \ddots & & \ddots & \ddots & \ddots & \vdots \\
\vdots & & \ddots & & \ddots & \ddots & 0 \\
0 & \cdots & 0 & -a_p & \cdots & -a_1 & 1
\end{bmatrix}. \tag{7.11}
$$

Using these notations, the likelihood is written

$$
\mathsf{p}(\mathbf{x}|J, \mathbf{M}, \boldsymbol{\alpha}, \mathbf{k}, \sigma^2, (\boldsymbol{a})) = (2\pi\sigma^2)^{-T/2} \exp\left(-\frac{1}{2\sigma^2}\|\tilde{\mathbf{x}} - \tilde{\mathbf{D}}\boldsymbol{\alpha}\|^2\right), \tag{7.12}
$$

where the possible AR coefficients $\boldsymbol{a}$ are only included in the likelihood when ϵ is autoregressive. In the following, the optional parameter $\boldsymbol{a}$ is omitted for the sake of simplicity.

7.2.2 Prior Distributions Selection

The next step towards a Bayesian model is that of *prior distributions selection*. Priors are essential in this problem, because they make the estimation of the numerous parameters feasible: without priors, many possible solutions would be likely, including trivial, meaningless, and over-complex ones. In particular, priors are useful to avoid having the model feature too many notes. There are many possible priors for J, $\mathbf{M}$, $\boldsymbol{\alpha}$, $\mathbf{k}$, σ^2 (and possibly $\boldsymbol{a}$). However, computational load is reduced when selecting a Gaussian amplitude prior $\mathsf{p}(\boldsymbol{\alpha})$ and an inverse gamma prior distribution for the variance parameter σ^2. Thus, the overall prior should be written as

$$
\mathsf{p}(J, \mathbf{M}, \boldsymbol{\alpha}, \mathbf{k}, \sigma^2, \psi) = \mathsf{p}(\boldsymbol{\alpha}|J, \mathbf{M}, \mathbf{k}, \sigma^2|\psi)\,\mathsf{p}(J, \mathbf{M}, \mathbf{k}|\psi)\,\mathsf{p}(\sigma^2|\psi), \tag{7.13}
$$

where ψ is called the *hyperparameter vector*; it contains the prior parameters, called *hyperparameters* ([122], [124] and Chapter 2). Similar to (2.74) and (2.75) in Chapter 2, we select here

$$
\mathsf{p}(\boldsymbol{\alpha}|J, \mathbf{M}, \mathbf{k}, \sigma^2|\psi) = \mathcal{N}(\boldsymbol{\alpha}; \mathbf{0}, \sigma^2 \boldsymbol{\Sigma}_\alpha), \tag{7.14}
$$

$$
\mathsf{p}(\sigma^2|\psi) = \mathcal{IG}\left(\sigma^2; \frac{\nu_0}{2}, \frac{\nu_1}{2}\right), \tag{7.15}
$$

which enables us to write the posterior as a product of three terms

$$
\mathsf{p}(J, \mathbf{M}, \boldsymbol{\alpha}, \mathbf{k}, \sigma^2|\mathbf{x}, \psi) \propto \\
\mathsf{p}(\sigma^2|J, \mathbf{M}, \mathbf{k}, \mathbf{x}, \psi)\,\mathsf{p}(\boldsymbol{\alpha}|J, \mathbf{M}, \mathbf{k}, \sigma^2, \mathbf{x}, \psi)\,\mathsf{p}(J, \mathbf{M}, \mathbf{k}|\mathbf{x}, \psi). \tag{7.16}
$$

The three terms of the posterior distribution are

$$\mathsf{p}(\sigma^2|J,\mathbf{M},\mathbf{k},\mathbf{x}) = \mathcal{IG}\left(\sigma^2; \frac{T+\nu_1}{2}, \frac{\nu_0 + \widetilde{\mathbf{x}}^\mathsf{T}\mathbf{P}\widetilde{\mathbf{x}}}{2}\right), \tag{7.17}$$

$$\mathsf{p}(\boldsymbol{\alpha}|J,\mathbf{M},\mathbf{k},\sigma^2,\mathbf{x}) = \mathcal{N}(\boldsymbol{\alpha}; \mathbf{S}\widetilde{\mathbf{D}}^\mathsf{T}\widetilde{\mathbf{x}}, \sigma^2\mathbf{S}), \tag{7.18}$$

$$\mathsf{p}(J,\mathbf{M},\mathbf{k}|\mathbf{x},\boldsymbol{\psi}) \propto \frac{[\nu_0 + \widetilde{\mathbf{x}}^\mathsf{T}\mathbf{P}\widetilde{\mathbf{x}}]^{-\frac{T+\nu_1}{2}}}{\det(\mathbf{S}^{-1}\boldsymbol{\Sigma}_{\boldsymbol{\alpha}})^{1/2}}\mathsf{p}(J,\mathbf{M},\mathbf{k}|\boldsymbol{\psi}), \tag{7.19}$$

where the matrices $\mathbf{S}$ and $\mathbf{P}$ are defined as in Chapter 2, page 47.

Equations (7.17), (7.18), and (7.19) define the posterior, up to the prior $\mathsf{p}(J,\mathbf{M},\mathbf{k}|\boldsymbol{\psi})$ which still needs to be selected. The parameters ν_0, ν_1 tune the prior over the noise variance σ^2 and they should be set so as to favour small values of σ^2. This ensures that the residual $\widetilde{\mathbf{x}} - \widetilde{\mathbf{D}}\boldsymbol{\alpha}$ is as small as possible. This can be achieved by selecting $\nu_0 \ll 1$ and $\nu_1 \ll 1$ (typically, smaller than 10^{-2}).

The covariance matrix $\boldsymbol{\Sigma}_{\boldsymbol{\alpha}}$ controls the amplitude coefficients. Standard choices are

- The *g-prior*, which consists of setting $\boldsymbol{\Sigma}_{\boldsymbol{\alpha}} = \gamma^2(\widetilde{\mathbf{D}}^\mathsf{T}\widetilde{\mathbf{D}})^{-1}$. This prior has two advantages: First, the amplitudes of sinusoids with very close frequencies are constrained to small values. This shares the energy between those two neighbour components. Second, $20\log_{10}(\gamma)$ can be interpreted as the expected signal-to-noise ratio (in dB). Third, the numerical computations are even easier because the matrix $\mathbf{S}$ becomes really simple. The hyperparameter γ should be considered unknown, and might be estimated. Actually, its value may be quite critical for correct estimation of the number of notes/partials. A standard choice consists of selecting a prior $\mathsf{p}(\gamma^2)$ as an inverse gamma distribution [21], [656], [122].

- A block diagonal matrix, made of 2×2 blocks. This choice enables the implementation of a fast estimation algorithm [127], [124]. Each block $\boldsymbol{\Sigma}_{\boldsymbol{\alpha}}[n]$ located at lines/rows $(2n, 2n+1)$ can be set as, e.g., $\boldsymbol{\Sigma}_{\boldsymbol{\alpha}}[n] = \gamma_n^2(\widetilde{\mathbf{D}}_n^\mathsf{T}\widetilde{\mathbf{D}}_n)^{-1}$, where $\widetilde{\mathbf{D}}_n$ denotes the two rows of $\widetilde{\mathbf{D}}$ located at indexes $2n$ and $2n+1$. The hyperparameter γ_n^2 may be set to the same value for all blocks, which assumes the same prior amplitude for all partials and all notes. It may also have the same value for all blocks that are related to the same note and have different values from one note to another. Finally, γ_n^2 may also be used to model the amplitude decay from one partial to another, according to the given spectrum envelope; see Section 7.1.1.

- A diagonal matrix, which may be proportional to the identity, i.e., $\boldsymbol{\Sigma}_{\boldsymbol{\alpha}} = \gamma^2\mathbf{I}$, where γ^2 is an unknown hyperparameter with inverse gamma prior, or $\boldsymbol{\Sigma}_{\boldsymbol{\alpha}}$ is a diagonal matrix where each diagonal term γ_n^2 is set for each partial, according to some spectrum envelope.

When additional prior information is available (for example, the instruments playing are known), it is possible to design an even more specific prior over the amplitudes by setting the γ_n^2's to some preset values, multiplied by a common scale parameter to be estimated.

The remaining parameters joint prior $\mathsf{p}(J, \mathbf{M}, \mathbf{k}|\boldsymbol{\psi})$ should also be selected. This prior may be decomposed into individual terms as

$$\mathsf{p}(J, \mathbf{M}, \mathbf{k}|\boldsymbol{\psi}) \; = \; \mathsf{p}(\mathbf{k}|J, \mathbf{M}, \boldsymbol{\psi})\, \mathsf{p}(J, \mathbf{M}|\boldsymbol{\psi}). \tag{7.20}$$

Similar to the amplitude prior, the frequencies prior may have various shapes, depending on the level of prior information available. A very general prior is

$$p(\mathbf{k}|J, \mathbf{M}, \boldsymbol{\psi}) \; = \; \prod_{j=1}^{J}\left[\mathsf{p}(k_{j,1}|M_j, \boldsymbol{\psi}) \prod_{m=1}^{M_j} \mathsf{p}(k_{j,m}|k_{j,1}, \boldsymbol{\psi})\right]. \tag{7.21}$$

In (7.21), the fundamental partial frequency prior $\mathsf{p}(k_{j,1}|M_j, \boldsymbol{\psi})$ may be instrument specific. For instruments with keys (pianos, clarinets, etc.) this prior may have the shape depicted in Fig. 7.3. In the general case, this kind of prior may also be used to perform *frequency quantization* (see Chapter 12); in this case, considering the $A4$ frequency as an additional unknown parameter to be estimated around 440 Hz makes the quantization more robust. Another

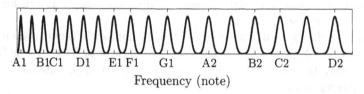

A1 B1C1 D1   E1 F1   G1   A2   B2 C2   D2

Frequency (note)

Fig. 7.3. Prior distribution of the fundamental frequencies $\mathsf{p}(k_{j,1}|M_j, \boldsymbol{\psi})$ in a 'key instruments' model. The spread of the Gaussian function increases with frequency, i.e., with the note label in {A1,A1#,B1, ...}. This is aimed at limiting areas where the prior is zero, in order to make the numerical estimation easier.

efficient fundamental partial frequency prior is the uniform distribution over the interval $[0, k_s/2M_j]$ $(j = 1, \ldots, J)$. The upper limit $k_s/2M_j$ ensures that no overtone partial has a frequency higher than the Nyquist frequency $k_s/2$, as this would cause aliasing.

The overtone partial frequency prior terms $\mathsf{p}(k_{j,m}|k_{j,1}, \boldsymbol{\psi})$ need to take possible inharmonicity into account.

A good, instrument unspecific, partial frequency prior is given by a truncated Gaussian distribution

$$\mathsf{p}(k_{j,m}|k_{j,1}, \boldsymbol{\psi}) \; = \; \mathcal{N}\big(k_{j,m}; mk_{j,1}, (mk_{j,1})^2\sigma_{j,m}^2\big)\mathbb{I}_{\left[-\frac{k_{j,1}}{2}, \frac{k_{j,1}}{2}\right]}, \tag{7.22}$$

where $\sigma^2_{j,m}$ is a small variance parameter (which can be chosen equal for all partials and all notes). In (7.22), the indicator function $\mathbb{I}$ (see p. 28) restricts $k_{j,m}$ to the range $[-\frac{k_{j,1}}{2}, \frac{k_{j,1}}{2}]$ to avoid overtone partial frequencies switching.

The last prior term to be defined, $\mathsf{p}(J, \mathbf{M}|\boldsymbol{\psi})$, is also the most critical. This term should be strong enough to avoid models with too many notes and too many partials. However, it should not overly penalize models with many notes and partials, because some important partial/note components may be missed.

Overall, there are two standard choices for $\mathsf{p}(J, \mathbf{M}|\boldsymbol{\psi})$ (others may be designed, though). The first consists of using the hierarchical structure $\mathsf{p}(J, \mathbf{M}|\boldsymbol{\psi}) = \mathsf{p}(J|\boldsymbol{\psi}) \prod_{j=1}^{J} \mathsf{p}(M_j|\boldsymbol{\psi})$, where each term $\mathsf{p}(M_j|\boldsymbol{\psi})$ is a Poisson distribution with parameter Λ_j

$$\mathsf{p}(M_j|\boldsymbol{\psi}) = \mathcal{P}(M_j; \Lambda_j) = e^{-\Lambda_j} \frac{\Lambda_j^{M_j}}{M_j!}, \qquad (7.23)$$

and with $\mathsf{p}(J|\boldsymbol{\psi}) = \mathcal{P}(J; \Lambda)$. The hyperparameters Λ_j $(j = 1, \ldots, J)$ and Λ play a crucial role in the estimation. Indeed, a model with too many allowed partials may tend to underestimate the fundamental partial frequency by one or several octaves, that is, by setting it to, e.g., half its true value. This adds fake partials in between the existing ones. It is thus important to let the hyperparameters Λ_j $(j = 1, \ldots, J)$ and Λ adapt to the analysed acoustic waveform. A robust approach consists of defining gamma priors for these hyperparameters; see [124].

The second approach consists of defining a prior for the model total number of partials $R = \sum_{j=1}^{J} M_j$. Here again, a Poisson prior $\mathsf{p}(J, \mathbf{M}|\boldsymbol{\psi}) = \mathsf{p}(R|\boldsymbol{\psi}) = \mathcal{P}(R; \Lambda)$ may be used, where the hyperparameter Λ is considered unknown with a gamma prior.

The probabilistic model being completely defined, the important issue is now to design an efficient estimation algorithm.

7.2.3 Estimation Issues

Parameter estimation from a Bayesian model may be achieved in many ways. The main two approaches are minimum mean squared error (MMSE) and maximum a posteriori (MAP) estimators (see Chapter 2, page 41). Here, the structure of the posterior distribution $\mathsf{p}(J, \mathbf{M}, \boldsymbol{\alpha}, \mathbf{k}, \sigma^2|\mathbf{x}, \boldsymbol{\psi})$ is especially complex, with discrete parameters (J and M), positive parameters ($\mathbf{k}$ and σ^2), and real-valued vector parameters ($\boldsymbol{\alpha}$ and the optional parameter $\boldsymbol{a}$).

A possible solution consists of estimating J and $\mathbf{M}$ by marginal MAP, that is, by selecting the number of notes $\widehat{J}_{\mathrm{MAP}}$ and number of partials $\widehat{\mathbf{M}}_{\mathrm{MAP}}$ for which $\mathsf{p}(J, \mathbf{M}|\mathbf{x}, \boldsymbol{\psi})$ is maximum—this is typically computed by Monte Carlo techniques; see Chapter 2. Then, the other parameters may be estimated by marginal MMSE, conditional to $\widehat{J}_{\mathrm{MAP}}$ and $\widehat{\mathbf{M}}_{\mathrm{MAP}}$, that is,

$$[\widehat{\alpha}, \widehat{k}, \widehat{\sigma}^2]_{\text{MMSE}} = \mathbb{E}_{p(\alpha, k, \sigma^2 | x, \widehat{J}_{\text{MAP}}, \widehat{M}_{\text{MAP}}, \psi)} [\alpha, k, \sigma^2].$$ (7.24)

This kind of estimator requires, however, the computation of marginal distribution, of maxima, and of integrals such as the expectation in (7.24). This requires dedicated algorithms, as outlined in the next section.

7.2.4 Computational Issues

The computation of estimates from the posterior $p(J, M, \alpha, k, \sigma^2 | x, \psi)$ requires the 'exploration' of this multidimensional probability distribution. Several works address similar problems. Andrieu and Doucet [21] propose a Markov chain Monte Carlo (MCMC) algorithm for a simple noisy sum-of-sines Bayesian model. Walmsley et al. [657], [658], [656], derive an MCMC algorithm for single F0 Bayesian models. Finally, Davy and Godsill [122], [126], [124] propose various MCMC approaches for multiple F0 models, with a fast implementation [127], [124].

As explained in Chapter 2, the aim of MCMC algorithms is to produce a chain of samples $\widetilde{J}^{(i)}, \widetilde{M}^{(i)}, \widetilde{\alpha}^{(i)}, \widetilde{k}^{(i)}, \widetilde{\sigma}^{2(i)}, \widetilde{\psi}^{(i)}$, for $i = 1, 2, 3, \ldots$. These samples are used to estimate the various quantities described in Section 7.2.3 above. The derivation of such algorithms being quite lengthy, the interested reader may refer to the publications cited for details. We present below (Algorithm 7.1) the general structure of an MCMC algorithm dedicated to a multiple F0 noisy sum-of-sines model, with a two-level hierarchical structure, from the individual partial level up to the multiple-note level.

Algorithm 7.1: Overall MCMC Algorithm for Multiple F0 Models.

---

1. Initialization.
   - Step 7.1.1 Initialize the parameters J, M, α, k, σ^2, and ψ.
   - Sample $\widetilde{\psi}^{(1)}$ from its prior distribution.
   - Sample $\widetilde{J}^{(1)}$ according to some initial distribution $q_{\text{init}}(J)$.
   - For $j = 1, \ldots, \widetilde{J}^{(1)}$ sample $\widetilde{M}_j^{(1)}$ according to its Poisson prior distribution.
   - Sample $\widetilde{k}^{(1)}$ according to $q_{\text{init}}(k|x)$ where $q_{\text{init}}(k|x)$ is the probability distribution proportional to the Fourier spectrum of x (see [21] for a similar implementation).
   - Sample the noise variance parameter $\widetilde{\sigma}^{2(1)}$ according to its posterior distribution $p(\sigma^2 | \widetilde{J}^{(1)}, \widetilde{M}^{(1)}, \widetilde{k}^{(1)}, (\widetilde{\alpha}^{(1)}), x, \widetilde{\psi}^{(1)})$ given in (7.17).
   - Sample the amplitudes $\widetilde{\alpha}^{(1)}$ according to their posterior distribution $p(\alpha | \widetilde{J}^{(1)}, \widetilde{M}^{(1)}, \widetilde{k}^{(1)}, \widetilde{\sigma}^{2(1)}, (\widetilde{\alpha}^{(1)}), x, \widetilde{\psi}^{(1)})$ given in (7.18).
2. For $i = 1, 2, \ldots, N$, do
   - Step 7.1.2 Sample the note parameters $\widetilde{J}^{(i)}$, $\widetilde{M}^{(i)}$, $\widetilde{k}^{(i)}$.
   - With probability μ_J, try to add a new note using a *note birth move*, which consists of generating a set of note parameters (number of partials, frequencies, amplitudes), and testing it using a Metropolis–Hastings reversible jump.

   – Otherwise, with probability ν_J, try to remove a note using a *note death move*,
     which consists of selecting one of the existing notes at iteration $i-1$ and testing
     its possible removal using a Metropolis–Hastings reversible jump.
   – Otherwise, with probability $1 - \mu_J - \nu_J$, try the *note update move* as follows:
     · Set $\tilde{J}^{(i)} \leftarrow \tilde{J}^{(i-1)}$.
     · For $j = 1, \ldots, \tilde{J}^{(i)}$, update the parameters of note $\#j$ by possibly changing
       the number of partials or their frequencies and amplitudes, yielding $\widetilde{M}_j^{(i)}$,
       $\tilde{\mathbf{k}}_j^{(i)}$ and $\tilde{\boldsymbol{\alpha}}_j^{(i)}$.
   • Step 7.1.3 Sample σ^2 from $p(\sigma^2 | \tilde{J}^{(i)}, \widetilde{\mathbf{M}}^{(i)}, \tilde{\mathbf{k}}^{(i)}, \mathbf{x})$; see (7.17).
   • Step 7.1.4 Sample $\tilde{\psi}^{(i)}$ from its posterior distribution (either directly, or us-
     ing a Metropolis–Hastings test).
   – Set $i \leftarrow i + 1$.

Algorithm 7.1 deserves several comments. First, the hierarchical levels appear clearly. The note death/birth/update moves correspond to the highest
level (that of notes). The middle level corresponds to adding/removing partials
inside a given note, or changing its fundamental partial frequency. The lowest
level is that of individual partials, whose frequencies and amplitudes may be
updated. Second, the Metropolis–Hastings moves as well as the initialization
require proposal distributions $q(\cdot)$. In order for the distribution of the samples
$\tilde{J}^{(i)}$, $\widetilde{\mathbf{M}}^{(i)}$, $\tilde{\boldsymbol{\alpha}}^{(i)}$, $\tilde{\mathbf{k}}^{(i)}$, $\tilde{\sigma}^{2(i)}$, $\tilde{\psi}^{(i)}$ to converge quickly to the posterior distribution, these proposals may be built on heuristics. For example, the spectrum of
$\mathbf{x}$ is used in order to build the frequencies proposal distribution. Other similar
heuristics may be included in the same way; see [124]. Overall, such MCMC
algorithms may be seen as a rigorous way to use various heuristics for multiple
F0 estimation.

7.2.5 Performance

The above class of Bayesian models being based on a generative model, they
can be used for many tasks, including multiple F0 estimation and signal compression. The performance estimation thus depends on the task assigned to
the algorithm. The computation for the full inference problem reported in
Davy et al. [124] is about 1.35 seconds per MCMC iteration for 0.5-second
excerpts, where 800 iterations are necessary on average.

In terms of F0 estimation, Davy et al. [124] report 100% accuracy for
single F0 (that is, for $J = 1$), about 85% for $J = 2$, 75% accuracy for $J = 3$,
and 71% accuracy for $J = 4$. The test samples were random mixes of single
F0 acoustic waveforms from Western classical music instruments. In terms
of residual energy and reconstruction accuracy, all experiments showed very
good reconstruction,[3] in spite of some octave errors.

---

[3] It is much more difficult to evaluate the reconstruction accuracy with a psychoacoustically relevant measure. The residual total energy may yield such a performance
estimator, though.

7.2.6 Extensions

The above models may be extended in many ways. For example, it is possible to use binaural (stereophonic) records in order to make the estimation more robust [120]. This is implemented by defining two models, one for the left channel and one for the right channel. These models are connected through their parameters prior. This can be extended to multi-track recordings, as in standard source separation approaches; see Chapter 9.

Another possible extension comes from a sequential view of music processing. The off-line approach presented here concerns music waveform segments which come from a longer excerpt, though they could be applied framewise. It is possible to use the parameter estimation results of previous segments to design the prior for the current waveform segment, in terms of instruments playing, A4 frequency tuning, etc.

7.3 On-Line Approaches

The off-line approaches described above have the major advantage of being quite accurate, because they have the full acoustic waveform available for processing. The drawback is that the computational requirements may be high. On-line approaches, on the other hand, only use the current acoustic waveform sample $x(n)$—or the current frame—at each processing step n. They also use information from the past, thanks to sequential priors (also termed *transition probability distributions*) over the model parameters. Finally, it should be noted that sequential models are quite attractive for music transcription, because they do not require a separate onset/offset detection mechanism, and they provide some flexibility to formulate completely on-line inference schemas, such as those presented in this section.

In this section, we first introduce the approach of Cemgil et al. [78], and then we describe two approaches based on an explicit sliding window-based model, proposed by Dubois and Davy [158], [160] and Vincent and Plumbley [646]. Note that Irizarry [308] also proposes such a model, but it is restricted to single F0.

7.3.1 Approach of Cemgil et al.

The noisy sum-of-sines models presented above are written in a non-sequential form. Sequential approaches, however, require the noisy sum-of-sines model to be written in a sequential form.

Sequential Noisy Sum-of-Sines Model

Cemgil et al. [78] propose to write a sine/cosine wave with frequency k and unity amplitude as follows:

$$\boldsymbol{\theta}(n) = \mathbf{B}(k)\boldsymbol{\theta}(n-1), \qquad n = 1,\ldots,T, \tag{7.25}$$

where $\boldsymbol{\theta}(n)$ is a two-dimensional vector and $\mathbf{B}(k)$ is a rotation matrix written as

$$\mathbf{B}(k) = \begin{bmatrix} \cos(2\pi k) & -\sin(2\pi k) \\ \sin(2\pi k) & \cos(2\pi k) \end{bmatrix}, \tag{7.26}$$

which rotates the vector $\boldsymbol{\theta}(n-1)$ by $2\pi k$ radians counter-clockwise. The first dimension of $\boldsymbol{\theta}$ carries the cosine component, whereas the second carries the sine component.

It is now possible to write many sum-of-sines models in a sequential form, including time-varying amplitude/frequency cases. Cemgil et al. [78] focus on a damped amplitude, constant frequency model as

$$\boldsymbol{\theta}(n) = \lambda(n)\mathbf{B}(k)\boldsymbol{\theta}(n-1), \tag{7.27}$$

$$x(n) = [1\ 0]\ \boldsymbol{\theta}(n), \tag{7.28}$$

where λ is the damping factor that tunes the amplitude decay rate, with $0 \le \lambda(n) \le 1$. The vector $[1\ 0]$ is used to project the vector $\boldsymbol{\theta}(n)$ onto one axis; see Fig. 7.4. The initial phase is tuned by the initial vector $\boldsymbol{\theta}(0)$ at time 0. In order to model the note partials, and their frequency/amplitude rela-

Rotating vector $\boldsymbol{\theta}(n)$ Projection: damped sine $x(n)$

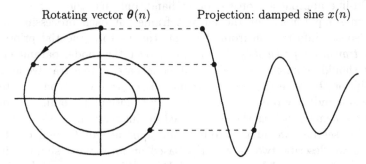

Fig. 7.4. The model based on a rotation matrix $\mathbf{B}(k)$ rotates the vector $\boldsymbol{\theta}(n)$ by $2\pi k$ radians counter-clockwise, and applies a damping factor. When projected onto one axis (here, the sine axis), this produces a damped sine wave with frequency k (adapted from [78]).

tions, the vector $\boldsymbol{\theta}$ may be extended to M sines/cosines, that is, $\boldsymbol{\theta}(n)^{\mathsf{T}} = [\boldsymbol{\theta}_1(n)^{\mathsf{T}}, \boldsymbol{\theta}_2(n)^{\mathsf{T}} \ldots \boldsymbol{\theta}_M(n)^{\mathsf{T}}]$ (of dimension $2M$), and the sequential model becomes

$$\boldsymbol{\theta}(n) = \mathbf{A}(k)\boldsymbol{\theta}(n-1), \tag{7.29}$$

$$x(n) = \mathbf{C}\boldsymbol{\theta}(n), \tag{7.30}$$

where $\mathbf{C} = [1\ 0\ 1\ 0\ \ldots\ 1\ 0]$. The $2M \times 2M$ matrix $\mathbf{A}(k)$ is defined as

$$\mathbf{A}(k) = \begin{bmatrix} \lambda_1(n)\mathbf{B}(k_1) & 0 & \cdots & 0 \\ 0 & \lambda_2(n)\mathbf{B}(k_2) & & \vdots \\ \vdots & & \ddots & 0 \\ 0 & \cdots & 0 & \lambda_M(n)\mathbf{B}(k_M) \end{bmatrix}. \tag{7.31}$$

Cemgil et al. assume harmonic frequency relationships of the partials and exponentially decaying spectrum envelope, from one partial to another. In other words, $k_m = mk_1$ and $\lambda_m(n) = \lambda^m(n)$ for $m = 1, \ldots, M$, where M is assumed to be known.

From this sequential setting, Cemgil et al. derive a probabilistic model. First, it is assumed that the note frequencies belong to a frequency grid with K nodes. The note fundamental partial frequencies are assumed to be one of the K predefined frequency positions among, for example, note frequencies of the tempered scale over several octaves. Second, each of the grid frequencies is assumed to be in one of the two states 'mute' and 'sound' at each time instant. In the following, we denote by $e_k(n)$ the state of the frequency grid point with index k at time n, and we have either $e_k(n) = $ 'sound' or $e_k(n) = $ 'mute'.

Coming back to the model in (7.29)–(7.31), the 'sound' case corresponds to a damping parameter, denoted λ^{sound}, and the 'mute' case corresponds to another damping factor λ^{mute}, with $\lambda^{\text{sound}} > \lambda^{\text{mute}}$. From these damping factors, Cemgil et al. define two matrices $\mathbf{A}(k, e_k(n))$ for $e_k(n) = $ 'sound' and $e_k(n) = $ 'mute' by replacing $\lambda(n)$ in (7.31) with either λ^{mute} or λ^{sound}. The probabilistic model is written for each of the K note frequencies in the grid. For $n = 1, \ldots, T$ and for a frequency k in the grid,

$$e_k(n) \sim \mathsf{p}(e_k(n)|e_k(n-1)). \tag{7.32}$$

Moreover, if no onset occurs, that is, if $e_k(n) = e_k(n-1)$ or if $e_k(n) = $ 'mute' whereas $e_k(n-1) = $ 'sound', then the damped sum-of-sines model with damping factor λ^{sound} or λ^{mute} is active:

$$\boldsymbol{\theta}_k(n) \sim \mathcal{N}\big(\boldsymbol{\theta}_k(n); \mathbf{A}(k, e_k(n-1))\boldsymbol{\theta}_k(n-1), \boldsymbol{\Sigma}_{\text{no onset}}\big), \tag{7.33}$$

or, if an onset occurs, that is, if $e_k(n) = $ 'sound' whereas $e_k(n-1) = $ 'mute', then the amplitude of the sine wave at this frequency is re-initialized by sampling a new initial vector $\boldsymbol{\theta}_k(n)$ as follows:

$$\boldsymbol{\theta}_k(n) \sim \mathcal{N}\big(\boldsymbol{\theta}_k(n); \mathbf{0}_{2M}, \boldsymbol{\Sigma}_{\text{onset}}\big). \tag{7.34}$$

The acoustic waveform is modelled overall as the sum over all the frequencies in the grid as

$$x(n) = \sum_k \mathbf{C}\boldsymbol{\theta}_k(n) + \epsilon(n), \tag{7.35}$$

where $\epsilon(n)$ is a Gaussian white noise with variance σ^2. Equation (7.35) yields the model likelihood at each time n (also termed the *observation probability distribution*), and the sequential priors given in (7.32)–(7.33) enable writing of the posterior in a sequential form. The multiple F0 model being completely defined, we next describe the estimation algorithm.

Parameter Estimation and Algorithmic Issues

Cemgil et al. propose to implement MAP estimation, that is, estimate the sequence $\mathbf{e}_k = [e_k(1), \ldots, e_k(T)]$ for each frequency k in the grid, where $\mathbf{e}_k$ is called a *piano roll*, as explained in Chapter 1. Note that the piano roll contains all the necessary information for multiple F0 estimation. Let us denote by $\widehat{\mathbf{e}_K}_{\text{MAP}}$ the estimated piano-roll sequence. It is computed as

$$\widehat{\mathbf{e}_K}_{\text{MAP}} = \arg\max_{\mathbf{e}_K} \mathsf{p}(\mathbf{e}_K, \boldsymbol{\psi}|\mathbf{x}), \qquad (7.36)$$

where the hyperparameter vector $\boldsymbol{\psi}$ includes σ^2, $\boldsymbol{\Sigma}_{\text{onset}}$, $\boldsymbol{\Sigma}_{\text{no onset}}$, λ^{sound} and λ^{mute}. The posterior $\mathsf{p}(\mathbf{e}_K|\mathbf{x})$ is obtained by integrating out the parameters $\boldsymbol{\theta}_k(n)$ for all k and all $n = 1, \ldots, T$—denoted by the shorthand $\boldsymbol{\theta}_K(1:T)$—as follows:

$$\mathsf{p}(\mathbf{e}_K, \boldsymbol{\psi}|\mathbf{x}) = \int_{\boldsymbol{\theta}} \mathsf{p}(\mathbf{e}_K, \boldsymbol{\theta}_K(1:T), \boldsymbol{\psi}|\mathbf{x}) \, d\boldsymbol{\theta}_K(1:T), \qquad (7.37)$$

where the full parameter posterior is given by Bayes's rule

$$\mathsf{p}(\mathbf{e}_K, \boldsymbol{\theta}_K(1:T), \boldsymbol{\psi}|\mathbf{x}) \propto \mathsf{p}(\mathbf{x}|\mathbf{e}_K, \boldsymbol{\theta}_K(1:T))\mathsf{p}(\mathbf{e}_K, \boldsymbol{\theta}_K(1:T)|\boldsymbol{\psi})\mathsf{p}(\boldsymbol{\psi}). \tag{7.38}$$

The terms $\mathsf{p}(\mathbf{x}|\mathbf{e}_K, \boldsymbol{\theta}_K(1:T))$ and $\mathsf{p}(\mathbf{e}_K, \boldsymbol{\theta}_K(1:T)|\boldsymbol{\psi})$ are computed sequentially as follows:

$$\mathsf{p}(\mathbf{x}|\mathbf{e}_K, \boldsymbol{\theta}_K(1:T)) = \prod_{n=1}^{T} \mathcal{N}\left(x(n); \sum_k \mathbf{C}\boldsymbol{\theta}_k(n), \sigma^2\right), \qquad (7.39)$$

$$\mathsf{p}(\mathbf{e}_K, \boldsymbol{\theta}_K(1:T)|\boldsymbol{\psi}) = \prod_k \left[\mathsf{p}(e_k(1), \boldsymbol{\theta}_K(1)) \prod_{n=2}^{T} \mathsf{p}(e_k(n)|e_k(n-1), \boldsymbol{\psi}) \right.$$
$$\left. \times \prod_{n=2}^{T} \mathsf{p}(\boldsymbol{\theta}_k(n)|\boldsymbol{\theta}_k(n-1), \boldsymbol{\psi}) \right].$$
$$\tag{7.40}$$

In (7.40) above, the transition pdf $\mathsf{p}(\boldsymbol{\theta}_k(n)|\boldsymbol{\theta}_k(n-1), \boldsymbol{\psi})$ may correspond to either the 'onset' given in (7.34) or the 'no onset' pdf, (7.33). The initial pdf of the states and state parameters is given by $\mathsf{p}(e_k(1), \boldsymbol{\theta}_K(1))$ for all k in the frequency grid.

The maximization in (7.36) is made complicated by the 'nuisance' parameters $\boldsymbol{\theta}_K(1:T)$ and $\boldsymbol{\psi}$, which need to be integrated out. Actually, the piano roll contains all the information for multiple F0 estimation, though the hyperparameters $\boldsymbol{\psi}$ may contain information useful for, e.g., instruments classification.

Of course, the integration of the nuisance parameters, as well as the maximization in (7.36), cannot be performed analytically. Cemgil et al. propose an algorithm based on message passing, expectation-maximization, and Kalman filtering on a sliding window which runs along the whole acoustic waveform. This algorithm jointly estimates the piano roll, the parameters, and the hyperparameters. In the special case of single F0 estimation, a more efficient estimation procedure can be implemented. Details about these algorithms may be found in [78].

7.3.2 Approaches Based on a Sliding Window

The approaches based on noisy sum-of-sines models and sliding windows make the sliding window $w_n(i)$, located at time n, appear explicitly in the model. The model in (7.9) becomes, inside the frame located at time n, denoted $s_n^w(i)$ where i is the local time index

$$s_n^w(i) = \sum_{j=1}^{J(n)} \sum_{m=1}^{M_j} \alpha_{j,m}^s(n) w_n(i) \sin\left(2\pi k_{j,m}(n)i\right) \\ + \alpha_{j,m}^c(n) w_n(i) \cos\left(2\pi k_{j,m}(n)i\right) + \epsilon_n(i), \tag{7.41}$$

where the statistics of the noise $\epsilon_n(i)$ are defined framewise. The model in (7.41) is local in the sense that its parameters have a local interpretation. In order to fully define the on-line model from a Bayesian viewpoint, the sequential prior has to be designed.

Approach of Dubois and Davy

Dubois and Davy [158], [160] propose to use Gaussian random walks for both the frequencies and the amplitude,

$$\mathbf{k}(n) = \mathbf{k}(n-1) + \mathbf{v_k}(n), \tag{7.42}$$

$$\boldsymbol{\alpha}(n) = \boldsymbol{\alpha}(n-1) + \mathbf{v_\alpha}(n), \tag{7.43}$$

where $\mathbf{k}(n)$ (respectively $\boldsymbol{\alpha}(n)$) denotes the vector of frequencies (respectively amplitudes) for all the partials and notes. The variances of the transition noises $\mathbf{v_k}(n)$ and $\mathbf{v_\alpha}(n)$ are also assumed to follow a random walk with log-Gaussian distribution (that is, the logarithm of these variances follows a Gaussian random walk). The number of partials M_j of note j is set to a fixed (large) number, and the amplitudes of highest partials is forced to zero whenever they are above the frequency $k_s/2$. The number of notes is assumed to be unknown and time varying. A sequential prior is defined as follows:

$$J(n) = J(n-1) + \begin{cases} +1 & \text{with probability } 1/10 \\ 0 & \text{with probability } 8/10 \, , \\ -1 & \text{with probability } 1/10 \end{cases} \tag{7.44}$$

which allows the number of notes to increase, decrease, or remain constant.[4] Equations (7.41) and (7.42)–(7.44) define a jump Markov system [19].

For the sake of simplicity, the frequency evolution model in (7.42) does not assume links between the frequencies of a given note's partials. Such links may be built in several ways:

- Perfect harmonicity is assumed, and the frequency vector only includes the fundamental partial frequency. The overtone partial frequencies are computed deterministically from the fundamental partial frequency.
- Inharmonicity is allowed, and each partial frequency evolves according to its own scheme. At the creation of a new note (that is, whenever $J(n) = J(n-1)+1$), the new partial frequencies are initialized according to some probabilistic technique which includes heuristics from deterministic F0 estimation methods.
- An inharmonicity model is assumed, for example the piano model, or the model proposed in [213]. The parameters of this model are assumed to follow a random walk, for example. The frequency vector only includes the fundamental partial frequencies.

The probabilistic model being defined, parameter estimation can be performed from the posterior distribution at time n denoted by

$$p\big(\mathbf{k}(1:n), \boldsymbol{\alpha}(1:n), J(1:n), \boldsymbol{\psi}(1:n)|x(1:n)\big),$$

where the notation $\mathbf{a}(1:n)$ is used to denote $[\mathbf{a}(1),\ldots,\mathbf{a}(n)]$ for any vector/scalar $\mathbf{a}$. The hyperparameter $\boldsymbol{\psi}$ contains all the model hyperparameters, such as the time-varying variances of $\mathbf{v_k}(n)$ and $\mathbf{v_\alpha}(n)$, the inharmonicity model parameters, etc.

An interesting feature of this model is that, conditional on $J(1:n)$, $\mathbf{k}(1:n)$, and $\boldsymbol{\psi}(1:n)$, the model is linear and Gaussian: this is because the model in (7.41) defines a linear and Gaussian likelihood (The noise $\epsilon_n(i)$ is assumed to be white and Gaussian). From an algorithmic viewpoint, this permits an efficient estimation strategy based on parallel Kalman filters, each running with different values of $J(1:n)$ and $\boldsymbol{\psi}(1:n)$. More precisely, it is possible to implement a *Rao-Blackwellized particle filter*[5] in order to estimate the frequencies, amplitudes, and number of notes from the posterior distribution at time n. This algorithm is summarized below, and details may be found in [158]. The overall principle of particle filtering is to run N particles side-by-side at each time, compute their weights (which are related to their likelihood) and use these to weight them in the parameter estimation. Particles with low weights correspond to low-probability parameter successive values, and thus they are deleted. Particles with high weights have high probability parameter

---

[4]The probabilities in (7.44) are indicative, and may be adjusted at will, in particular when $J(n)$ reaches its minimum/maximum value.

[5]The reader interested in particle filtering may refer to Chapter 2 for a short introduction, and to [152] for a full survey.

successive values, and thus they are duplicated in order to better explore the parameter space around the most probable values.

Algorithm 7.2: Harmonic Tracking Rao–Blackwellized Particle Filter

1. Initialization
   - Step 7.2.1 Initialize the particles at time $n = 1$.
   - For particles $j = 1, \ldots, N$, sample the initial number of notes $\tilde{J}^{(j)}(1)$, the initial frequency vector $\tilde{\mathbf{k}}^{(j)}(1)$, and the partial amplitudes $\tilde{\alpha}^{(j)}(1)$, using some probability density possibly derived from deterministic estimation algorithm applied to the initial frame s_1^w. Sample the initial hyperparameter $\tilde{\psi}^{(j)}(1)$.
2. Iterations, for $n = 1, 2, \ldots$
   - Step 7.2.2 The particles are updated.
   - For particles $j = 1, \ldots, N$, sample the number of notes using the proposal distribution q_J as follows: $\tilde{J}^{(j)}(n) \sim \mathsf{q}_J\big(\tilde{J}^{(j)}(n)|\tilde{J}^{(j)}(n-1), \tilde{\psi}^{(j)}(n-1), x(1:n)\big)$.
   - For $j = 1, \ldots, N$, sample the frequencies using the proposal distribution q_k as follows: $\tilde{\mathbf{k}}^{(j)}(n) \sim \mathsf{q}_k\big(\tilde{\mathbf{k}}^{(j)}(n)|\tilde{\mathbf{k}}^{(j)}(n-1), \tilde{J}^{(j)}(n), \tilde{\psi}^{(j)}(n-1), x(1:n)\big)$.
   - For particles $j = 1, \ldots, N$, update the amplitude vector $\tilde{\alpha}^{(j)}(n)$ using a Kalman filter applied to the amplitude evolution model in (7.43) and the likelihood defined in (7.41), where the other parameters are set to $\tilde{J}^{(j)}(n)$, $\tilde{\mathbf{k}}^{(j)}(n)$ and $\tilde{\psi}^{(j)}(n-1)$.
   - For $j = 1, \ldots, N$, update the hyperparameter vector using the proposal distribution q_ψ as follows: $\tilde{\psi}^{(j)}(n) \sim \mathsf{q}_\psi\big(\tilde{\psi}^{(j)}(n)|\tilde{\psi}^{(j)}(n-1), \tilde{\alpha}^{(j)}(n), \tilde{\mathbf{k}}^{(j)}(n), \tilde{J}^{(j)}(n)\big)$.
   - For particles $j = 1, \ldots, N$, compute the importance weight $\tilde{\omega}^{(j)}(n)$ according to the rules of importance sampling (see (2.84) in Chapter 2).
   - Step 7.2.3 The particles are resampled.
   - Duplicate particles with large weights and suppress particles with small weights.
   - Step 7.2.4 The parameters are estimated.
   - Use the particles in order to compute estimates of the number of notes, the frequencies, and the related amplitudes.

In Algorithm 7.2, the particles are updated using proposal distributions, which are designed so as to favour likely parameter values. For example, the frequencies proposal distribution q_k may be build on the current frame frequency spectrum, or from a randomized deterministic F0 estimation procedure; see [158]. The performance of the algorithm may also be improved by adding, at each time n, a particle update step based on one iteration of the MCMC algorithm presented in Section 7.2.4.

Approach of Vincent and Plumbley

The approach proposed by Vincent and Plumbley [646] uses a model equivalent to that of (7.41), but written in a different form: a cosine and an initial phase component are used instead of a sine and a cosine, namely

$$s_n^{\mathrm{w}}(i) = \sum_{j=1}^{J(n)} \sum_{m=1}^{M_j(n)} \alpha_{j,m}(n)\mathrm{w}_n(i)\cos\left(2\pi k_{j,m}(n)i + \phi_{j,m}(n)\right) + \epsilon_n(i), \quad (7.45)$$

where harmonicity is assumed in each frame, i.e., the partial frequencies are $k_{j,m}(n) = mk_{j,1}(n)$. Similar to the approach by Cemgil et al., it is assumed that the fundamental frequency belongs to a fixed grid; here, the MIDI semitone scale. The noise probabilistic model is defined in the frequency domain: roughly, it is Gaussian where the variance in a given auditory frequency band is proportional to the loudness of $s_n^{\mathrm{w}}(i)$ in that frequency band; see [646].

The parameter estimation procedure is made of two steps, using the Bayesian setting. First, parameter priors are defined in each frame, independently of neighbouring frames. Unknown parameters are estimated framewise using MAP. Then in the second step, the parameters in different frames are linked together using sequential priors, and they are re-estimated.

The framewise local priors are defined as follows. Assume $k_{j,1}^{\mathrm{grid}}(n)$ is the frequency in the grid corresponding to the fundamental partial of note j, in the current frame. The true fundamental partial frequency is assumed to be close to $k_{j,1}^{\mathrm{grid}}(n)$, namely,

$$p\left(k_{j,1}(n)|k_{j,1}^{\mathrm{grid}}(n)\right) = \mathcal{LN}\left(k_{j,1}(n); k_{j,1}^{\mathrm{grid}}(n), \sigma_{\mathrm{freq}}^2\right), \quad (7.46)$$

where $\mathcal{LN}(\cdot;\cdot,\cdot)$ is the log-Gaussian distribution.[6] In each frame, the number of partials is set so that $M_j(n)k_{j,1}^{\mathrm{grid}}(n)$ is below $k_{\mathrm{s}}/2$. The initial phase prior is uniform; that is, $p\left(\phi_{j,m}(n)\right) = \mathcal{U}_{[0,2\pi]}\left(\phi_{j,m}(n)\right)$ and the amplitudes are assumed to follow a given spectrum envelope denoted $\lambda_{j,m}(n)$ for $m = 1, \ldots, M_j(n)$. Each partial amplitude is assumed to follow a log-Gaussian distribution

$$p\left(\alpha_{j,m}(n)|\lambda_{j,m}(n)\right) = \mathcal{LN}\left(\alpha_{j,m}(n); \gamma_j^2(n)\lambda_{j,m}(n), \sigma_{\mathrm{ampli}}^2\right), \quad (7.47)$$

where $\gamma_j^2(n)$ is a scale factor with log-Gaussian prior distribution. This completes the framewise Bayesian model. Such a model is defined for each frequency in the grid, though only a small number of these frequencies are 'active' in a given frame. Again, similar to Cemgil et al. [78], each grid frequency is associated to a state $e_k(n)$ which may be either 'sound' or 'mute' according to given prior probabilities. The number of notes $J(n)$ is simply the number of grid frequencies in the state 'sound'. The framewise Bayesian estimation procedure consists of finding the state vector $\widehat{e}_{K\,\mathrm{MAP}}(n)$ such that

$$\widehat{e}_{K\,\mathrm{MAP}}(n) = \arg\max_{e_K(n)} p(e_K(n)|s_n^{\mathrm{w}}), \quad (7.48)$$

where $s_n^{\mathrm{w}} = [s_n^{\mathrm{w}}(1), s_n^{\mathrm{w}}(2), \ldots]$ and $e_K(n)$ is defined as in Section 7.3.1. In practice, the MAP estimation in (7.48) is based on the joint posterior probability of the vector $e_K(n)$ together with the actual frequencies $k_{j,1}$ related to

---

[6] '$\mathcal{LN}(a; \mu, \Sigma)$ is a log-Gaussian distribution' means that $\log(a)$ is Gaussian with mean $\log(\mu)$ and covariance matrix Σ.

grid frequencies in the 'sound' state. The other parameters (amplitudes and γ_j^2 factor) have been integrated out using the Laplace approximation method; see [646]. The actual estimation procedure is based on a greedy, local search for the states $e_k(n)$, and a gradient-style optimization for the frequencies $k_{j,1}$.

The note parameters being estimated independently in each frame, they may now be connected across the frames. Vincent and Plumbley define a sequential prior $p(e_k(n)|e_k(n-1))$, and the Viterbi algorithm is used to estimate the sequence $\widehat{e}_{K\,\text{MAP}}(1:T)$ from the individual framewise estimates $\widehat{e}_{K\,\text{MAP}}(n)$. In other words, a grid frequency can be in the state 'sound' in $\widehat{e}_{K\,\text{MAP}}(1:T)$ at time n only if it is already in that state in the local estimate $\widehat{e}_{K\,\text{MAP}}(n)$. The final step consists of re-estimating the actual frequencies and amplitudes from the posterior $p(k_{j,1}(1:T), \alpha_{j,m}(1:T), \gamma_j(1:T)|x(1:T), \widehat{e}_{K\,\text{MAP}}(1:T))$—that is, given the optimal sequence $\widehat{e}_{K\,\text{MAP}}(1:T)$—where the priors in (7.46) and (7.47) and the prior over $\gamma_j(n)$ have been redefined as log-Gaussian random walks.

7.4 Other On-Line Bayesian Approaches

Aside from direct modelling of the acoustic waveform, as presented in the previous sections, indirect models have been studied. In this section, we review approaches where the signal spectrogram is modelled, and where *harmonic trajectories* are searched.

7.4.1 Methods Designed for Music and General Audio

We present four approaches proposed for general audio processing. All of them follow *harmonic trajectories* in the acoustic waveform spectrogram.

Method of Yeh and Röbel

The method of Yeh and Röbel [681] applies framewise a model similar to that of (7.9). This model assumes parallel evolution of the partial amplitudes, together with the spectral smoothness principle of Klapuri [351] and the inharmonicity model of Davy and Godsill [122]. The parameter estimation algorithm is computationally simpler than MCMC, though it is also based on the generation of candidate notes. Each candidate is evaluated using a score function, and the best candidates are kept in the final list. This method is applied framewise.

Spectrogram Modelling Method of Dubois and Davy

Dubois and Davy [159], [158] extend their method presented in the previous section to the case where the model is written in the spectrogram domain. More precisely, the model in (7.41) is changed into

226   Manuel Davy

$$\left|\mathrm{DFT}_{s_n^{\mathrm{w}}(\cdot)}(l)\right|^2 = \left|\sum_{j=1}^{J(n)}\sum_{m=1}^{M_j}\alpha_{j,m}^{\mathrm{s}}(n)\mathrm{DFT}_{\mathrm{w}_n(\cdot)\sin(2\pi k_{j,m}(n)\cdot)}(l)\right|^2 + \epsilon_n(l),$$

(7.49)

where the noise $\epsilon_n(l)$ is assumed to be zero-mean white Gaussian. The cosine term, which was previously used to represent the initial phase, would
be redundant as we consider the power spectrum, and it has been omitted.
The sequential priors are defined as in (7.42)–(7.44). This model is no longer
linear in the amplitudes, and a particle filter close to that in Algorithm 7.2 is
devised; see [158].

Method of Thornburg et al.

Thornburg et al. [625] propose a sequential estimation procedure which uses
the main energy peaks in the spectrogram as input information. It should
be noted that this procedure is aimed at melody extraction, and thus it is
restricted to single F0. The fundamental partial frequency is assumed to belong to the MIDI semitone scale, with some unknown, possibly time-varying,
frequency shift from the A4 440 Hz tuning. The amplitudes are also assumed
to belong to a grid with exponential spacing, and the method is designed so
as to associate a state $e(n)$ to each frame s_n^{w} at a given time, where $e(n)$
takes its values in the set {'transient frame onset', 'transient frame continuation', 'pitched frame onset', 'pitched frame continuation', 'silent frame'}.
The transition probability $p(e(n)|e(n-1))$ is assumed to be given, and only
allows some transitions between the five possible state values. Conditional
on $e(n)$, various transition probabilities are defined for the frequencies, the
amplitudes, and the MIDI scale frequency shift parameter. The unknown parameters are estimated by MAP Bayesian estimation, using an approximate
forward-backward Viterbi algorithm. The output also includes the sequence
of state $e(1:T)$ which may be used for multiple F0 estimation, but also music
segmentation and rhythm quantization.

Method of Sterian et al.

An earlier work is that of Sterian et al. [609]. A Kalman filter was used to
extract sinusoidal partials from the signal. Then, these partials were grouped
into their sources by implementing the grouping principles of Bregman [49] in
terms of individual likelihood functions aimed at evaluating, e.g, the harmonic
concordance. The implementation was based on multiple hypothesis tracking.

7.4.2 Methods for Multiple-Speakers Pitch Tracking

In the speech processing literature, several multiple F0 tracking algorithms
have been proposed. A special feature of speech is that the sounds are not

always *voiced*; that is, portions of a speech signal may not be well modelled as a noisy sum of sines. Algorithms for multiple-speakers speech tracking take this feature into account. Moreover, voiced speech is usually quite harmonic, and thus speech partial frequency models do not incorporate inharmonicity. Apart from Tabrikian et al. [616] which is for single pitch, Bach and Jordan [25] and Wu et al. [677] (see Chapter 8) propose multipitch algorithms for speech processing based on the acoustic waveform spectrogram, or correlogram. The inference is based on variants of the Viterbi algorithm.

Though developed for a problem different from multiple F0 estimation in musical signals, methods developed for speech yield interesting frameworks that may be adapted to music.

7.5 Conclusions

In this chapter, several methods for multiple F0 estimation have been presented. Most of them rely on a generative model of the acoustic waveform, and some model the signal spectrogram. Probabilistic models are defined in order to make estimation of the generative model parameters feasible.

These methods are quite powerful in the sense that they capture a large fraction of the information from the acoustic waveform, and this information may be used for tasks other than multiple F0 estimation. Moreover, Bayesian approaches clearly distinguish model construction from inference (that is, parameter estimation algorithms). The drawback is that they can be computationally intensive. However, their computational cost may be reduced by plugging as many heuristics as possible into the algorithms, for example to build the proposal distributions in Monte Carlo algorithms, as this makes the convergence faster. Designing a method of this kind can also be viewed as a principled way to built algorithms with known theoretical properties (convergence speed, estimation error, etc.). Heuristics can be also used to design the unknown parameter priors. Finally, generative models may be specifically designed for various transcription tasks. Models for melody transcription in complex polyphonic music can surely be based on spectrogram modelling (see Section 7.4) and are simpler than precise generative models aimed at estimating the subtle expressive controls of a guitar player.

Work in this vein is in its infancy. More research has to be done to use as much prior information as possible, and to define more elaborate models, which could also be used for, e.g., percussion transcription. This will probably result in quite complex algorithms, but complexity is the main issue in multiple F0 estimation, and it may not be avoided.

8

Auditory Model-Based Methods for Multiple Fundamental Frequency Estimation

Anssi Klapuri

Institute of Signal Processing, Tampere University of Technology,
Korkeakoulunkatu 1, 33720 Tampere, Finland
Anssi.Klapuri@tut.fi

8.1 Introduction

This chapter describes fundamental frequency (F0) estimation methods that make use of computational models of human auditory perception and especially pitch perception. At the present time, the most reliable music transcription system available is the ears and the brain of a trained musician. Compared with any artificial audio processing tool, the analytical ability of human hearing is very good for complex mixture signals: in natural acoustic environments, we are able to perceive the characteristics of several simultaneously occurring sounds, including their pitches [49]. It is therefore quite natural to pursue automatic music transcription and multiple F0 estimation by investigating what happens in the human listener. Here the term multiple F0 estimation means estimating the F0s of several concurrent sounds.

Fundamental frequency is the measurable physical counterpart of *pitch*. In Chapter 1, pitch was defined as the perceptual attribute of sounds which allows them to be ordered on a frequency-related scale extending from low to high. More exactly, the pitch of a sound was said to be the frequency of a sine wave that is matched to a target sound by human listeners. The importance of pitch for hearing in general is indicated by the fact that the auditory system tries to assign a pitch frequency to almost all kinds of acoustic signals. Not only sinusoids and periodic signals have a pitch, but even noise signals of various kinds can be consistently matched with a sinusoid of a certain frequency. For a steeply lowpass- or highpass- filtered noise signal, for example, a weak pitch is heard around the spectral edge. Amplitude modulating a random noise signal causes a pitch perception corresponding to the modulation frequency. Also, the sounds of bells and vibrating membranes have a pitch, although their waveform is not clearly periodic and their spectrum does not have a regular structure. A complete review of this 'zoo of pitch effects' can be found in [275], [474], [297]. The auditory system seems to be strongly inclined towards using a single frequency value to summarize certain aspects of sound events. Computational models of pitch perception attempt to replicate this phenomenon.

Practical multiple F0 estimation methods have a slightly different purpose than pitch perception models. The set of acoustic signals of interest is narrower since the physical concept of F0 is defined only for periodic and nearly periodic sounds. Also, the evaluation criteria are different: multiple F0 estimation methods are judged based on their reliability in the given task, F0 estimation in a mixture signal, whereas an auditory model should faithfully reproduce the mechanisms and the behaviour of the auditory system.

For musical sounds, the F0 and the perceived pitch are practically equivalent. However, there are ambiguous situations such as the octave ambiguity, where it is not clear if the F0 of a sound is x Hz or half or twice that value. From the music transcription point of view, it would be desirable to solve these ambiguities so that the estimated F0 would correspond to the perceived pitch. This is one of the reasons why auditory model-based methods have been employed. Other reasons include the aim of achieving robustness for diverse kinds of musical sounds (these are discussed in Section 8.2) and obtaining a good time/F0 resolution by using a time-frequency decomposition similar to that in human hearing. The advantages and disadvantages of auditory model-based methods are summarized later in this chapter. In general, perceptually motivated methods have been quite successful in audio content analysis.

The primary focus of this chapter is on practical multiple F0 estimation and not so much on auditory modelling. More comprehensive introductions to pitch perception models can be found in [297], [522], [132]. Also, the emphasis is laid on *multiple* F0 estimation methods: some perceptually motivated methods are omitted that are purported to be useful for single F0 estimation in noisy speech signals. The aim of this chapter is twofold: to give a compact description of pitch perception models so that the reader will be able to develop auditorily motivated analysis methods of his own and, secondly, to describe already-existing multiple F0 estimators that are based on and motivated by these models.

This chapter is organized as follows. Section 8.2 discusses the basic acoustic characteristics of pitched musical sounds and how these can be used to compute the F0 of the sounds. Section 8.3 describes computational models of pitch perception. Section 8.4 introduces music transcription systems which use an auditorily model as a 'front end'. That is, the systems apply a perceptually-motivated data representation but the emphasis is laid on the inference that follows the auditory modelling stage, instead of proposing changes to the auditory model itself. Section 8.5 describes multiple F0 estimation methods which extend or modify pitch perception models in order to make them better applicable to F0 estimation in polyphonic music signals. In the end, two algorithms are described which can be directly used for this purpose. Finally, Section 8.6 summarizes the main conclusions.

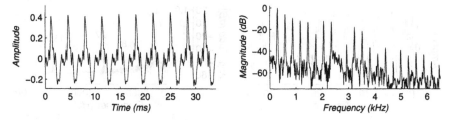

Fig. 8.1. A harmonic sound in the time and frequency domains. The example represents a violin sound with fundamental frequency 290 Hz and fundamental period 3.4 ms.

8.2 Musical Sounds and F0 Estimation

This section discusses the acoustic characteristics of pitched musical sounds and F0 estimation when the sounds are presented in isolation. This provides the background for describing pitch perception models and multiple F0 estimation methods in the subsequent sections.

8.2.1 Pitched Musical Sounds

Musical sounds usually consist of several frequency components. The relative amplitudes of the overtone partials and their time evolution determines the timbre of the sound. Here we are primarily interested in the *frequencies* of the partials since F0 estimation methods try to normalize away the timbre information. From this point of view, pitched musical sounds can be divided into two main classes: sounds that are harmonic and sounds that are not. The methods to be described in this chapter are concerned with both of these.

Most Western musical instruments produce harmonic sounds.[1] These sounds have a spectral structure where the dominant frequency components, called *harmonics*, are approximately regularly spaced. Figure 8.1 illustrates a harmonic sound in the time and frequency domains. The F0 of the sound is the inverse of its time-domain period and the frequency spacing between the overtone partials corresponds approximately to the F0. Usually the overtone components are not perceived separately but only the pitch and the timbre of the entire sound are heard.

For an ideal harmonic sound, the frequencies of the overtone partials are integer multiples of the F0. However, it should be noted that the spectra of harmonic sounds are not always perfectly harmonic; the higher-order overtones of plucked and struck string instruments deviate slightly from their ideal

---

[1]More exactly, all instruments in the chordophone and aerophone families (see Table 6.1 on p. 167).

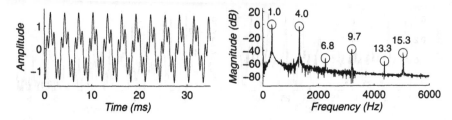

Fig. 8.2. A vibraphone sound (F0 330 Hz) illustrated in the time and frequency domains. In the right panel, the frequencies of the most dominant spectral components are shown in relation to the F0.

harmonic positions. For these classes of instruments, the partial frequencies obey the formula

$$f_j = jF\sqrt{1 + B(j^2 - 1)}, \tag{8.1}$$

where F is the fundamental frequency, $j = 1, 2, \ldots$ is the partial index, and B is an inharmonicity factor [193, p. 363]. Typical values of B are of the order 10^{-4} or 10^{-3} for the middle pitch range of the piano, for example. This makes the higher-order partials gradually shift upwards in frequency, but the structure of the spectrum is in general very similar to that in Fig. 8.1, and the sounds can be classified as harmonic. The inharmonicity is due to the stiffness of real strings, which contributes a restoring force along with the string tension [193], [315].

Figure 8.2 shows an example of a sound which does not belong to the class of harmonic sounds although it is nearly periodic in the time domain and has a clear pitch. In Western music, mallet percussion instruments are a case in point: these instruments produce pitched sounds which are not harmonic. The most common instruments in this family are the marimba, the vibraphone, the xylophone, and the glockenspiel. The sound production mechanism in all of these is a vibrating bar. A bar of uniform thickness with free ends has vibration modes whose frequencies are not in integral ratios. However, by making the bar thinner at the middle of its length, the overtones can be tuned. The first overtone of the marimba and the vibraphone is typically tuned to be four times the F0 and that of the xylophone to be three times the F0.

8.2.2 Basic Principles of F0 Estimation

There are a large number of different methods for monophonic F0 estimation [289]. Comparative evaluations of these can be found e.g. in [535], [290], [134]. The aim of this section is not to make an exhaustive coverage of these, but merely to point out the main acoustic features that different algorithms are built upon: time-domain periodicity and frequency-domain periodicity, and to provide a few representative examples of each approach.

The majority of F0 estimation methods are based on measuring the periodicity of an acoustic signal in the time domain (see e.g. [618], [135]). This makes sense, since all the pitched musical sounds described above are periodic or almost periodic in the time domain. As reported in [134], quite accurate single F0 estimation can be achieved simply by an appropriate normalization of the short-time autocorrelation function (ACF), defined as

$$r(\tau) = \frac{1}{N} \sum_{n=0}^{N-1} x(n)x(n + \tau).$$ (8.2)

The F0 of the signal $x(n)$ can be computed as the inverse of the lag τ that corresponds to the maximum of $r(\tau)$ within a predefined range. To avoid detecting an integer multiple of the period, short lags have to be favoured over longer ones.

An implicit way of measuring time-domain periodicity is to match a *harmonic pattern* to the signal in the frequency domain. According to the Fourier theorem, a periodic signal with period τ can be represented with a series of sinusoidal components at the frequencies j/τ, where j is a positive integer. This can be observed for the musical sounds in Figs. 8.1 and 8.2. Algorithms that are based on frequency-domain harmonic pattern matching have been proposed in [153], [54], [428], for example.

Another class of F0 estimators measure the periodicity of the Fourier spectrum of a sound [384], [380]. These methods are based on the observation that a harmonic sound has an approximately periodic magnitude spectrum, the period of which is the F0. In its simplest form, the autocorrelation function $\rho(m)$ over an N-length magnitude spectrum is calculated as

$$\rho(m) = \frac{2}{N} \sum_{k=0}^{N/2-m-1} |X(k)||X(k + m)|.$$ (8.3)

In the above formula, any two frequency components with a certain spectral interval m support the corresponding F0. The spectrum can be arbitrarily shifted without affecting the output value. An advantage of this is that the calculations are somewhat more robust against the imperfect harmonicity of plucked and struck string instruments since the intervals between the overtone partials do not vary as much as their absolute frequencies deviate from the harmonic positions. However, in its pure form this approach has more drawbacks than advantages. In particular, estimating low F0s is not reliable since the F0 resolution of the method is linear whereas the time-domain ACF leads to $1/F$ resolution.

An interesting difference between the F0 estimators in (8.2) and (8.3) is that measuring the periodicity of the time-domain signal is prone to errors in F0 halving because the signal is periodic at twice the fundamental period too, whereas measuring the periodicity of the magnitude spectrum is prone to errors in F0 doubling because the spectrum is periodic at twice the F0 rate, too. The two approaches can be combined using an auditory model, as will be described in Section 8.3.2.

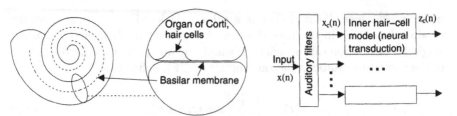

Fig. 8.3. An illustration of the cochlea (*left*) and its cross-section (*middle*). The right panel shows a rough computational model of the cochlea.

8.3 Pitch Perception Models

This section describes computational models of pitch perception and discusses the advantages that an auditory model-based method may have in multiple F0 estimation.

The human auditory system can be divided into two main parts: peripheral hearing and the auditory cortex in the brain. Both of these play an important part in pitch perception. The peripheral part consists of the outer ear, the middle ear, and the inner ear. The first two of these essentially contribute to directional hearing and impedance matching of sound. From the pitch analysis point of view, the interesting part starts from the inner ear, where there is an organ called the *cochlea.*

The cochlea is a sophisticated organ where pressure variations are transformed into properly coded neural impulses in the auditory nerve. Physiologically, the cochlea is a long, coiled, tubular structure which is filled with liquid and tapers towards its end (see Fig. 8.3). The cochlea is divided into two main sections by the *basilar membrane* that runs its entire length. When the mechanical vibrations of the eardrum are transmitted via the middle ear to the inner ear, hydraulic pressure waves are caused in the cochlea and the basilar membrane starts to vibrate. The waves propagate along the basilar membrane so that high frequencies peak in amplitude (resonate) near the beginning and low frequencies get their largest amplitude at the far end.

On the basilar membrane, there is the *organ of Corti* which contains two types of *hair cells.* Outer hair cells are active elements which contribute to the resolution of the cochlear frequency analysis, making different places along the basilar membrane more sharply tuned to their characteristic frequencies than they would be by the acoustic properties of the membrane alone. Inner hair cells register the movement of the basilar membrane. They respond to mechanical displacement by generating nerve impulses into the auditory nerve fibres that are attached to them and lead to the brain [680].

Computational models of the cochlea comprise two main parts which can be summarized as follows (see Fig. 8.3):

1. An acoustic input signal is passed though a bank of bandpass filters, called *auditory filters*, which model the frequency selectivity of the inner ear.

Typically about 100 filters are used with centre frequencies uniformly distributed on a nearly logarithmic frequency scale (details in Section 8.3.1). The outputs of individual filters simulate the mechanical movement of the basilar membrane at different points along its length.

2. The signal at each band, or *auditory channel*, is processed to model the transform characteristics of the inner hair cells which produce neural impulses in the auditory nerve. In signal processing terms, this involves three main characteristics: compression and level adaptation, half-wave rectification, and lowpass filtering (details in Section 8.3.2).

In the following, the acoustic input signal is denoted by $x(n)$ and the impulse response of an auditory filter by $g_c(n)$, where c is the channel index. The output of the auditory filter at channel c is denoted by $x_c(n)$ and functions as an input to the second step. The output of the inner hair cell model is denoted by $z_c(n)$ and represents the probability of observing a neural impulse at channel c.

The processing mechanisms in the brain can be studied only indirectly and are therefore not as accurately known. Typically the relative merits of different models are judged according to their ability to predict the perception of human listeners for various acoustic stimuli in psychoacoustic tests. Different theories and models of the central auditory processing will be summarized in Section 8.3.3, but in all of them, the following two processing steps can be distinguished:

3. Periodicity analysis of some form takes place for the signals $z_c(n)$ within the auditory channels. Phase differences between channels become meaningless.
4. Information is integrated across channels.

In the above processing chain, the auditory nerve signal $z_c(n)$ represents a nice 'interface' between the Steps 2 and 3 and thus between the peripheral and central processes. The signal in the auditory nerve has been directly measured in cats and in some other mammals and this is why the stages 1 and 2 are quite well known. Computational models of the peripheral hearing can approximate the auditory-nerve signal quite accurately, which is a great advantage since an important part of the processing already takes place at these stages. However, central processes and especially Step 3 are (arguably) even more crucial in pitch perception. The above four steps are now described in more detail.

8.3.1 Cochlear Filterbank

Frequency analysis is an essential part of the cochlear processing. Frequency components of a complex sound can be perceived separately and are coded independently in the auditory nerve (in distinct nerve fibres) provided that their frequency separation is sufficiently large [473]. This frequency analysis

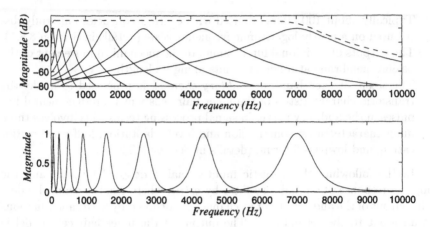

Fig. 8.4. Frequency responses of a few auditory filters shown on the logarithmic (*top*) and on the linear magnitude scale (*bottom*). The dashed line in the upper panel shows the summary response of the filterbank when 70 auditory filters are distributed between 60 Hz and 7 kHz.

can be modelled with a bank of linear bandpass filters: Figure 8.4 shows an example of such a filterbank.

The bandwidths and the shape of the power response of the auditory filters have been studied using the *masking* phenomenon [192], [499]. Masking refers to a situation where an audible sound becomes inaudible in the presence of another, louder sound. In particular, if the distance between two spectral components is less than a so-called *critical bandwidth*, one easily masks the other. The situation can be thought of as if the components would go to the same auditory filter, or to the same channel in the auditory nerve. If the frequency separation is larger, the components are coded independently and are both audible.

The bandwidths of the auditory filters can be conveniently expressed using the equivalent rectangular bandwidth (ERB) concept. The ERB of a filter is defined as the bandwidth of a perfectly rectangular filter which has a unity magnitude response in its passband and an integral over the squared magnitude response which is the same as for the specified filter. The ERB bandwidths b_c of the auditory filters have been found to obey

$$b_c = 0.108 f_c + 24.7 \text{ Hz,} \tag{8.4}$$

where f_c is the centre frequency of the filter at channel c [473].

The centre frequencies of the auditory filters are typically assumed to be uniformly distributed on a critical-band scale. This frequency-related scale is derived by integrating the inverse of (8.4), which yields

$$\xi(f) = 21.4 \log_{10}(0.00437 f + 1). \tag{8.5}$$

In the above expression, f denotes frequency in Hertz and $\xi(f)$ gives the critical-band scale. When f varies between 0 Hz and 20 kHz, $\xi(f)$ varies between 0 and 42. Intuitively, this means that approximately 42 critical bands (or auditory filters) would fit within the range of hearing if the passbands of the filters were non-overlapping and rectangular in shape. Conversion from the critical-band scale back to Hertz units is given by

$$f(\xi) = 229 \times (10^{\xi/21.4} - 1). \tag{8.6}$$

For example, let us distribute 70 filters uniformly on the critical-band scale between 100 Hz and 10 kHz. Using (8.5), we find that the corresponding frequency boundaries on the critical-band scale are 3.36 and 35.3, respectively, and that the distance between each two centre frequencies has to be $(35.3 - 3.36)/69 = 0.463$ on this scale. The centre frequencies on the critical-band scale can then be converted to Hertz units using (8.6).

When a lot of auditory filters are uniformly distributed on the scale $\xi(f)$, power responses of the filters sum approximately to a flat response, as indicated by the dashed line in Fig. 8.4. Typically about 100 filters are used to obtain a good sampling of centre frequencies along the cochlea and a sufficiently flat summary response. Note that in this case, the passbands of the filters overlap considerably. In F0 estimation, only the filters up to about 5 to 8 kHz need to be used, as the most significant harmonic components are below this.

The time-domain impulse responses of the auditory filters have been studied using a so-called reverse correlation method. In the study by de Boer and de Jongh [128], the ear of a cat was stimulated with white noise and the resulting action potentials of individual auditory nerve fibres were recorded simultaneously. Using the input signal and the recorded train of neural impulses, the impulse response of the corresponding auditory filter was derived. The impulse response relates the input signal to the *firing probability* of the nerve fibre under study, that is, to the probability of an inner hair cell generating an impulse to the fibre.

A so-called gammatone filter provides an excellent fit to the experimentally found impulse responses. The filter is defined by its impulse response as [502]

$$g_c(t) = at^{n-1}e^{-2\pi bt} \times \cos(2\pi f_c t + \theta), \tag{8.7}$$

where the normalization factor $a = (2\pi b)^n/\Gamma(n)$ ensures a unity response at the centre frequency, $\Gamma(n)$ is the gamma function, and the parameter value $n = 4$ leads to a shape of the power response that matches best with real auditory filters. The parameter $b = 1.019b_c$ is used to control the bandwidth of the filter.

Figure 8.5 illustrates the impulse responses of two gammatone filters with centre frequencies 100 Hz and 1.0 kHz, and with bandwidths obtained from (8.4). The impulse response consists of a sinusoidal tone at the centre

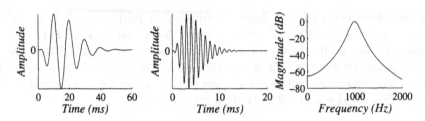

Fig. 8.5. Impulse responses of two gammatone filters with centre frequencies 100 Hz (*left*) and 1.0 kHz (*middle*). The frequency response of the latter filter is shown on the right.

frequency of the filter, f_c, windowed with a function that is precisely the gamma distribution from statistics. Frequency responses of several gammatone filters are shown in Fig. 8.4.

The gammatone filters can be implemented efficiently using a cascade of four second-order IIR filters. A detailed description of the design of the filterbank and the corresponding source code can be found in the technical report by Slaney [591].

8.3.2 Mechanical-to-Neural Transduction

Inner hair cells (IHC) are the elements which convert the mechanical motion of the basilar membrane into firing activity in the auditory nerve. Each IHC rests at a certain point along the basilar membrane and thus follows its movement at this position. Correspondingly, in the computational models the output of each auditory filter is processed by an IHC model.

The IHCs produce neural impulses, or 'spikes', which are binary events. However, since there is a large population of the cells, it is conventional to model the firing *probability* as a function of the basilar membrane movement. Thus the input to an IHC model comes from the output of an auditory filter, $x_c(n)$, and the output of the IHC model represents the time-varying firing probability denoted by $z_c(n)$.

Several computational models of the IHCs have been proposed. An extensive comparison of eight different models was presented by Hewitt and Meddis in [291]. In the evaluation, the model of Meddis [456] outperformed the others by showing only minor discrepancies with the empirical data and by being also one of the most efficient computationally. An implementation of this model is available in the AIM [501] and HUTear [273] auditory toolboxes, for example.

A problem with the realistic IHC models is that they depend critically on the absolute level of their input signal. The dynamic range of the model of Meddis [456], for example, is only 25 dB and the firing rate saturates at the 60 dB level. This limitation of individual IHCs is real, and it seems that the auditory system uses a population of IHCs with different dynamic ranges to achieve the good intensity discrimination performance over a dynamic range of about 120 dB [523, pp. 137–142]. This has not been included in the computational models of the individual IHCs [291].

For the above-described reason and for the sake of simplicity, many practical systems have replaced a realistic IHC model by a cascade of (i) compression, (ii) half-wave rectification, and (iii) lowpass filtering [171], [327], [677], [354]. As mentioned in the beginning of this section, these are the main characteristics of the IHCs. An advantage of doing this is that the behaviour of the overall system becomes easier to analyse and the signal-level dependency is removed. As a disadvantage, the longer-term level adaptation properties of more realistic IHC models are lost. This is also the approach followed here: instead of going into the details of realistic IHC models, we analyse the basic characteristics of the IHC in order to understand their function in pitch perception and practical F0 estimation.

(i) The compression step has taken slightly different forms in different implementations, but a common theme in all of these has been to scale the sub-band signals $x_c(n)$ inversely proportional to their variance. Ellis scaled the variances of the sub-band signals to unity [171]. Klapuri generalized this approach by scaling the sub-band signals by a factor $\sigma_c^{\nu-1}$, where σ_c is the standard deviation of $x_c(n)$ and $0 \leq \nu \leq 1$ is a compression coefficient [354]. Tolonen and Karjalainen omitted compression at sub-bands but pre-whitened the spectrum of an input signal using inverse warped-linear-prediction filtering, which leads to a very similar result [627].

(ii) Half-wave rectification (HWR) is the clearly non-linear processing step in the mechanical-to-neural transduction. It is defined as

$$\mathrm{HWR}(x) = \begin{cases} x, & x \geq 0, \\ 0, & x < 0. \end{cases} \tag{8.8}$$

As simple as it seems, rectification within the sub-bands plays an important part in pitch perception and in practical F0 estimation. In particular, it allows a synthesis of the time and the frequency-domain periodicity analysis methods introduced in (8.2) and (8.3), respectively.

Figure 8.6 illustrates the HWR operation for a narrow-band signal which consists of five overtones of a harmonic sound. Most importantly, the rectification generates spectral components which correspond to the frequency *intervals* between the input partials. The spectral components generated below 1 kHz represent the amplitude envelope of the input signal, as shown in the lowest panels. A signal that consists of more than one frequency component exhibits periodic fluctuations, *beating*, in its time-domain amplitude envelope. That is, the partials alternately amplify and cancel each other out, depending on their phase. The rate of beating caused by each pair of frequency components depends on their frequency difference and, for a harmonic sound, the frequency interval corresponding to the F0 dominates.

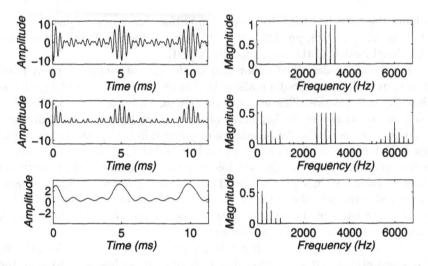

Fig. 8.6. Upper panels show a signal consisting of the overtone partials 13–17 of a sound with F0 200 Hz (fundamental period 5 ms) in the time and frequency domains. Middle panels illustrate the signal after half-wave rectification. Lower panels show the result of lowpass filtering the rectified signal with a 1 kHz cut-off.

The complex Fourier spectrum $Y(k)$ of a rectified signal $y(n) = \text{HWR}(x(n))$ can be approximated by

$$\hat{Y}(k) = \frac{\sigma_x}{\sqrt{8\pi}}\delta(k) + \frac{1}{2}X(k) + \frac{1}{\sigma_x\sqrt{8\pi}}\sum_{j=-N/2+k}^{N/2-k} X(j)X(k-j), \qquad (8.9)$$

where $\delta(k)$ is the unit impulse function, and $X(k)$ and σ_x are the complex Fourier spectrum and the standard deviation of $x(n)$, respectively [353, p. 38], [117]. The approximation assumes that $x(n)$ is a zero-mean Gaussian random process but it is sufficiently accurate for signals such as that in Fig. 8.6, too. On the right-hand side of (8.9), the first term is a dc-component, the second term represents the spectrum of the input signal, and the last term, the convolution of the spectrum $X(k)$ with itself, represents the beating components of the amplitude-envelope spectrum. In addition, the last term generates a harmonic distortion spectrum centred on twice the centre frequency of the input narrow-band signal $x(n)$ in Fig. 8.6. Periodicity analysis of the resulting signal in the time domain (see the next subsection) leads to a combined use of the time and frequency domain periodicity because the rectified signal consists of both the input partials and partials that correspond to their difference frequencies.

Another important property of the HWR is that a series of partials with approximately *uniform amplitudes* cause strong beating. This is because the magnitude of beating caused by each two frequency components is determined by the smaller of the two amplitudes. In the spectrum of a harmonic sound, each pair of neighbouring harmonics contributes to the beating at the fundamental-frequency rate, but the 'minimum amplitude' property filters out individual higher-amplitude partials. This phenomenon is well known in hearing: if the amplitude of one of the overtones of a harmonic sound rises clearly above the others, it is perceptually segregated and stands out as an independent sound [49]. In computational multiple F0 estimation, this is a desirable characteristic since it makes the F0 computations more immune to the partials of other, co-occurring sounds. Especially when processing the higher overtones of a sound, this partly prevents stealing the energy of the partials of other sounds.

(iii) Lowpass filtering the rectified signal can be used to balance the weight between the amplitude envelope versus the input narrow-band signal. Most systems have used a fixed low-order lowpass filter with a cut-off frequency around 1 kHz at all channels. The sub-band signal after compression, rectification, and lowpass filtering is denoted by $z_c(n)$.

8.3.3 Periodicity Analysis at Sub-Bands and Cross-Band Integration

The auditory nerve signal, modelled by $z_c(n)$, $c = 1, \ldots, C$, is further processed in the brain. Although the central processing mechanisms are not accurately known, it has been convincingly shown that periodicity analysis of some kind takes place within each auditory channel and the results are then combined across channels to yield a pitch perception [457], [67]. This amount of knowledge is already very useful and almost carries us to a situation where only parameter optimization is left in order to process pitch in a way similar to that of the human brain.

The first pitch model of the above-described type was proposed by Licklider [409]. He proposed to computed short-time autocorrelation functions $r_c(\tau)$ within the auditory channels c and to derive pitch from the resulting two-dimensional $(c \times \tau)$ representation. This became known as the 'duplex theory' of pitch perception because it involved both frequency analysis (by the cochlear filterbank) and autocorrelation analysis. Further development with this class of models was made by Lyon [422], Weintraub [663], and Slaney and Lyon [594].

Meddis and Hewitt implemented Licklider's model using a gammatone filterbank and a realistic IHC model and carried out extensive simulations to investigate if the pitch estimate of the model agreed with human listeners for various audio signals [457]. The authors computed ACFs within the auditory channels as

$$r_c(n, \tau) = \sum_{i=0}^{n} z_c(n - i) z_c(n - i - \tau) w(i), \qquad (8.10)$$

where $z_c(n)$ is the output of the IHC model in channel c and at time n, $r_c(n, \tau)$ is the ACF, and an exponentially decaying window function $w(i) = (1/\Omega) e^{-i/\Omega}$ was applied to give more emphasis to the most recent samples [457], [459].[2] It should be noted that the data structure at this stage was three dimensional ($c \times \tau \times n$). Across-channel information integration was then done simply by summing across channels, resulting in a summary ACF

$$s(n, \tau) = \sum_c r_c(n, \tau). \tag{8.11}$$

Pitch at time n was estimated by searching the highest peak in $s(n, \tau)$ within a predefined lag range [457, p. 2884].

Meddis and Hewitt demonstrated that the model was able to predict the perceived pitch for a large set of test stimuli used previously in psychoacoustic tests [457]. Moreover, Meddis and O'Mard later noted that the implementation is a special case of a more general model consisting of four stages: (i) cochlear bandpass filtering, (ii) half-wave rectification and lowpass filtering, (iii) within-channel periodicity extraction, and (iv) across-channel aggregation of periodicity estimates [459]. This became known as the *unitary model* of pitch perception because the single model was capable of simulating a wide range of pitch perception phenomena. Different variants of the unitary model have been used since then in a number of signal analysis systems [171], [133], [627], [677].

Cariani and Delgutte carried out a direct experiment to find out the characteristics in the auditory nerve signals that correlate with the perceived pitch [67]. Instead of using a simulated cochlea, the authors studied the signal in the auditory nerve of a cat in response to complex acoustic waveforms. They found that the *time intervals* between neural spikes are particularly important in encoding pitch. The authors computed histograms of time intervals between both successive and non-successive impulses in individual auditory nerve fibres, and summed the histograms of 507 fibres to form a pooled histogram. What the authors noticed was that, for a diverse set of audio signals, the perceived pitch correlated strongly with the most frequent interspike interval in the pooled histogram at any given time [67]. This suggests that the pitch of these signals could result from central auditory processing mechanisms that analyse interspike interval patterns. Computational models of the cochlea do not produce discrete neural spikes but rather real-valued signals $z_c(n)$, which represent the probability of a neural firing (in different nerve fibres). However, Cariani and Delgutte noted that the interspike interval codes are closely related to autocorrelation operations [67, p. 1712]. For a real-valued signal, ACF can replace the interval histogram.

Despite the above strong evidence, it seems that the ACF is not *precisely* the mechanism used for periodicity estimation in the central auditory system, but some experimental and neurophysiological findings contradict the ACF (see e.g. [322] and the brief summary in [131, p. 1262]). Meddis and Hewitt, for example, used the ACF but wanted to 'remain neutral about the exact

---

[2]In practice, the windowing and summing can be implemented very efficiently using a leaky integrator.

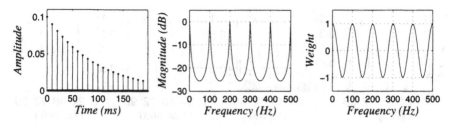

Fig. 8.7. The impulse response (*left*) and frequency response (*middle*) of a comb filter with the feedback delay of 10 ms and feedback gain 0.9. For comparison, the right panels shows the power response of the ACF for 10 ms lag.

mechanism by which temporal information is extracted from the activity of the auditory nerve fibres' [457, p. 2879].

A number of alternative mechanisms to the ACF have been proposed [500], [68], [66], [132]. Although none of these really surpass the modelling power of the ACF for a large class of signals, *comb filter*-like solutions have been proposed by several authors and are therefore discussed in the following. The output of a comb filter for an input signal $z_c(n)$ is given by

$$y_c(n, \tau) = (1 - \alpha)z_c(n) + \alpha y_c(n - \tau, \tau), \qquad (8.12)$$

where τ is the feedback delay and $0 < \alpha \le 1$ is the feedback gain.

Figure 8.7 shows the impulse response and the frequency response of a comb filter with a feedback delay $\tau = 10$ ms. For comparison, the power response of the ACF for the corresponding lag τ is shown in the rightmost panel.[3] As can be seen, the comb filter is more sharply tuned to the harmonic frequencies of the period candidate and no negative weights are applied between these.

Periodicity analysis with comb filters can be accomplished by invoking a bank of such filters with different feedback delays τ and by computing locally time-averaged powers at the outputs of the filters. Figure 8.8 illustrates the output powers of a bank of comb filter for a couple of test signals. In the case of a periodic signal, all comb filters that are in rational-number relations to the period of the sound show response to it, as seen in panel (b).

A bank of comb filters has been proposed for auditory processing e.g. by Cariani [66, Eq. (1)], who used the filterbank to separate concurrent vowels with different F0s. Cariani also proposed a non-linear mechanism which consisted of an array of delay lines, each associated with its characteristic delay and a non-linear feedback mechanism instead of the linear one in (8.12). Periodic sounds were reported to be captured by the corresponding delay loop and thus became segregated from the mixture signal. The *strobed temporal*

---

[3]As a non-linear operation, the ACF does not have a frequency response. However, since the ACF of a time-domain signal is the inverse Fourier transform of its power spectrum, the power response of the ACF can be depicted for a single period value.

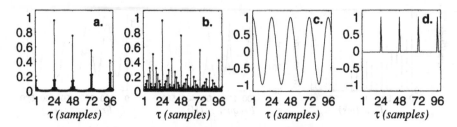

Fig. 8.8. Normalized output powers of a bank of comb filters for (a) a sinusoidal with 24-sample period and (b) an impulse train with the same period. The feedback delays of the filters are shown on the x-axis and all the feedback gains were 0.9. The panels (c) and (d) show the ACFs of the same signals, respectively.

integration (STI) mechanism of Patterson [502], [500, p.186] is closely related to comb filters too, although the relation is less direct and full details of the method are beyond the scope of this chapter.

In all the above-described models, the across-band integration has received a rather small role. For example, Meddis and Hewitt [457] and Cariani and Delgutte [67] suggest simply summing the autocorrelation functions or periodicity histograms across channels (see (8.11)). More complex ways of integrating the information across channels have been proposed, though. These will be discussed in more detail in Section 8.5.1, in connection with the estimation of multiple pitches. In particular, a technique called *channel selection* will be discussed which attempts to identify the spectro-temporal regions that represent the target sound and to reject the channels which contain noise or interference. Here it suffices to note that the across-channel information integration takes place in the central auditory system and may thus employ almost any complex technique. One curious consequence of this is that the pitch of a sound can be perceived even when two overtone partials of the sound are fed to the different ears of a listener [298].

8.4 Using an Auditory Model as a Front End

This section discusses music transcription systems which use an auditory model as a front end. That is, the systems apply a perceptually motivated data representation but the emphasis is laid on higher-level processing instead of proposing changes to the auditory model itself. Section 8.5 will discuss systems which do the latter and, as will be seen, often some practical modifications are needed in order to make the models more robust in polyphonic music signals. However, putting transcription systems under these two sections primarily serves the purpose of presentation instead of representing two clear categories.

The intermediate data representations employed between an input signal and the transcription result are of great importance. An appropriate representation facilitates the design of algorithms that use it and often improves

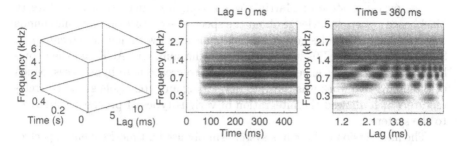

Fig. 8.9. Illustration of the log-lag correlogram of Ellis [171]. Input signal in this case was a trumpet sound with F0 260 Hz (fundamental period 3.8 ms). The left panel illustrates the three-dimensional correlogram volume. The middle panel shows the zero-lag face of the correlogram which is closely related to the power spectrogram. The right panel shows one time slice of the volume, from which the summary ACF can be obtained by summing over frequency.

the analysis result in practice. The idea of using the same data representation as the human auditory system is therefore very appealing. The aim of this section is to investigate the advantages and disadvantages of doing this and to introduce a few auditory-model implementations that have been employed. For this purpose, three different music transcription systems are briefly introduced. A discussion of other mid-level data representations in acoustic signal analysis can be found in [173] and in Chapter 3.

8.4.1 Martin's Transcription System

Martin proposed a system for transcribing piano performances of four-voice Bach chorales [440], [439]. As a front end of his system, Martin used the log-lag correlogram model of Ellis [171] which is closely related to the unitary model of Meddis and Hewitt described above. A bank of 40 gammatone filters was applied, the output of each filter was half-wave rectified and lowpass filtered, and then subjected to autocorrelation analysis. Specific to Ellis's model is that the within-channel ACFs are computed only for a set of logarithmically distributed lag values, 48 lags per an octave. This makes it computationally feasible to estimate the ACFs continuously over time and not just in discrete frames. For each lag τ and channel c, the signal $\hat{r}_c(n, \tau) = z_c(n)z_c(n - \tau)$ is computed and then lowpass filtered in the time dimension, analogous to (8.10). Summary ACFs are obtained by normalizing each ACF by the value at lag zero and by summing across channels. Figure 8.9 illustrates Ellis's model.

Martin utilized the good time resolution of Ellis's model by tracking summary ACF peaks through time and by combining temporally continuous peaks into musical notes. Simple pruning mechanisms were introduced to eliminate spurious subharmonic peaks in the summary ACF.

The overall system of Martin's was a complex inference architecture (a blackboard) where knowledge about the spectral structure of harmonic sounds was combined with rules governing tonal music and with heuristic techniques. Support for different F0s was sought for in the summary ACF and then combined with the power envelope information to create note hypotheses. Much of the innovative work was put into developing an extendable software architecture which allowed the integration of various types of processing modules to the system.

The first version of Martin's system simply used a time-frequency spectrogram as its input [440], but later the author switched to using the auditory model [439]. Interestingly, Martin mentions a specific reason for switching to an auditorily motivated data representation: he suspected that the log-lag correlogram would facilitate the detection of notes in an octave relationship without introducing explicit instrument models. Although some evidence for this was presented, no extensive simulations were carried out to support this conclusion. Also, Martin reported that the correlogram representation indicated chord roots very clearly and that the analysis did not require resolving individual higher-order harmonic partials in the spectrum [439, p. 10]. Although Martin's transcription system was never formally evaluated, it was among the first systems to be able to process signals with more than two simultaneous sounds and thus had a strong influence on subsequent research.

8.4.2 Auditory Scene Analysis Approach of Godsmark and Brown

Godsmark and Brown proposed a system for modelling the auditory scene analysis (ASA) function in humans, that is, our ability to perceive and recognize individual sound sources in mixture signals [215]. The authors used music signals as their test material. ASA is usually viewed as a two-stage process where a mixture signal is first *decomposed* into time-frequency components of some kind, and these are then *grouped* to their respective sound sources. In humans, the grouping stage has been found to depend on various acoustic properties of the components, such as their harmonic frequency relationships, common onset times, or synchronous frequency modulation [49].

Godsmark and Brown used the auditory model of Cooke [100] for the decomposition stage. This auditory model also uses a bank of gammatone filters at its first stage. Notable in Cooke's model is that rectification and lowpass filtering are not applied at the filterbank outputs but only the compression and level adaptation properties of the IHCs are modelled, amounting to an auditorily motivated bandwise gain control. Thus the overall model can actually be viewed as a sophisticated way of extracting sinusoidal components from an input signal, instead of being a complete and realistic model of the auditory periphery. The frequency of the most prominent sinusoidal component at the output of each auditory filter is tracked through time using median-smoothed instantaneous-frequency estimation [100, p. 36] and, in addition, the instantaneous amplitudes of the components are calculated. Since the passbands of the gammatone filters overlap, usually several adjacent filters show response to the same frequency component. This redundancy is removed by combining the outputs of adjacent channels so as to form 'synchrony strands' which represent the time-frequency behaviour of dominant spectral components in the input signal.

The main focus in the work of Godsmark and Brown was on developing a computational architecture which would facilitate the integration of different spectral organization (grouping) principles [215]. The synchrony strands were used as the elementary units that were grouped to sound sources. The authors reported that these were particularly suitable for modelling the ASA because the temporal continuity of the strands is made explicit and they are sufficiently few in number to perform the grouping for every strand.[4] Godsmark and Brown computed various acoustic features for each strand and then performed grouping according to onset and offset synchrony, time-frequency proximity, harmonicity, and common frequency movement.

Godsmark and Brown evaluated their model by investigating its ability to segregate polyphonic music into its constituent melodic lines. This included both multiple F0 estimation and organization of the resulting notes into melodic lines according to the applied musical instruments. The latter task was carried out by computing pitch and timbre proximities between successive sounds. Although transcription accuracy as such was not the main goal, promising results were obtained for musical excerpts with polyphonies ranging from one to about four simultaneous sounds.

8.4.3 Marolt's Transcriber for Piano Music

Marolt proposed a system for the automatic transcription of piano music [434]. His system was composed of two main parts: a partial tracking module and a note recognition module. Input to the partial tracking part was provided by a model of the peripheral hearing where an input signal was passed through a bank of 200 gammatone filters and the output of each filter was processed by Meddis's IHC model [456]. Adaptive oscillators were then used to track partials at the outputs of the IHC models, one oscillator per channel. The oscillators employed were similar to those proposed by Large and Kolen in [391], locking their period and phase to the incoming signal. In order to track harmonically related partials, the oscillators were interconnected to *oscillator nets*, one per each candidate musical note.

Time-delay neural networks (NNs) were trained to recognize musical notes at the output of the partial tracking module. Each NN was specifically trained to recognize a certain piano note in its input. The input to the NNs consisted of the outputs of all the oscillator networks in a few recent time frames and of the amplitude envelopes at the outputs of the auditory filterbank. Supervised learning with a large amount of piano music was used to train the NNs.

---

[4]Cooke designed his model exactly for this purpose: to support the grouping activities in ASA [100, p. 14].

Good transcription results were reported for a test set of three real and three synthesized piano performances. Concerning the use of the auditory model, Marolt reported that the compression and level adaptation properties of Meddis's IHC model were important to the system as they reduced the dynamic range of the signal and thus enabled the system to track small-amplitude partials.

8.4.4 Summary of Using an Auditory Front End

Specific advantages of using a perceptually motivated data representation were reported in the above systems. Martin observed that the log-lag correlogram is a good indicator of chord roots and that the analysis with the model does not require resolving individual higher-order harmonics, allowing a better time resolution. Some evidence for detecting two notes in an octave relationship was presented. Godsmark and Brown reported that the model of Cooke was particularly suitable for computational ASA since it produced temporally continuous sinusoidal components which were relatively few in number. Marolt reported that the dynamic compression and level adaptation properties of Meddis's IHC model facilitated the use of small-amplitude partials in the analysis. Finally, an important feature of auditory models that is not explicitly mentioned by any of the above authors is that the compression properties of the IHC models remove timbral information efficiently and thus make the models more robust for different musical instruments.

The disadvantages of employing an auditory model were not specifically reported. However, compared to the use of the Fourier spectrum, for example, it is fair to say that the computational load of an auditory model is significantly higher and that the output of the model is not as straightforward to interpret and understand.

8.5 Computational Multiple F0 Estimation Methods

The pitch perception models described in Section 8.3 are not sufficient as such for accurate multiple F0 estimation in real-world music signals. The purpose of this section is to describe different approaches to extending the models so that they become applicable in the present task.

The most obvious shortcoming of the pitch perception models is that they typically account for a single pitch only. Several pitches in a mixture signal cannot be detected simply by picking several local maxima in the summary ACF, for example. The models have been tested using very diverse kinds of acoustic signals but usually not with sound mixtures. Another shortcoming, related to the first one, is that the models are not robust in polyphonic signals. Even the global maximum of the summary ACF does not necessarily correspond to any of the actual pitches in a mixture signal; certain pitch relationships can confuse the model. In a typical situation, the constituent notes of a musical chord match the overtones of a non-existing chord root and the highest peak in the summary ACF indicates the chord root instead of one of the component sounds.[5] Further, the pitch models do not address robustness against additive noise: drum sounds often accompany the pitched sounds in music. Finally, the computational complexity of the models is rather high since they involve periodicity analysis at a large number of sub-bands.

On the other hand, there are several issues that are quite efficiently dealt with using a pitch model. These were summarized in Section 8.4.4 above.

In the following, a number of different methods are described that aim at overcoming the above-mentioned shortcomings. Some of these were designed for two-speaker speech signals but are included here in order to cover the substantial amount of work done in the analysis of multiple-speaker speech signals. This is followed by a more detailed description of two multiple F0 estimation methods for music signals. It should be noted that the main interest in this section is not to model hearing but to address the practical task of multiple F0 estimation.

8.5.1 Multiple F0 Estimation in Speech Signals

Multiple F0 estimation is closely related to sound separation. An algorithm that is able to estimate the F0 of a sound in the presence of other sounds is, in effect, also assigning the respective spectral components to their sound sources [49, p. 240]. Separation of speech from interfering speech for the purpose of its automatic recognition is an important area of sound separation. Here we look at methods that have utilized pitch information to carry out this task. A couple of state-of-the-art methods are described, with the aim of discussing the basic mechanisms that have been used to extend an auditory model to process multiple pitches.

Multiple F0 estimation in speech signals is in many ways a more constrained task than in music: the F0 range is limited to about three octaves and the described methods attempt to estimate only two simultaneous F0 tracks. However, the described basic mechanisms are not restricted to speech signals, and many of them can be generalized to the case of more than two simultaneous sounds.

Channel Selection

Meddis and Hewitt extended their pitch model (see p. 241) to simulate the human ability to identify two concurrent vowels with different F0s [458]. The proposed method included a template-matching process to recognize the vowels too, but here only the F0 estimation part is summarized. It consists of the following steps:

---

[5]Examples of such chords are the major triad and the interval of a perfect fifth.

1. The pitch model of Meddis and Hewitt is applied [457]. This involves a bank of gammatone filters, Meddis's IHC simulation, within-channel ACF computation, and across-channel summing. The highest peak in the summary ACF within a predefined lag range is used to estimate the F0 of the more dominant sound.
2. Individual channel ACFs that show a peak at the period of the first detected F0 are removed. If more than 80% of the channels get removed, only one F0 is judged to be present and the algorithm terminates.
3. The ACFs of the remaining channels are combined into a new summary ACF from which the F0 of the other vowel is derived.

The authors did not give statistics on the F0 estimation accuracy, but reported clear improvements in vowel recognition as the F0 difference of the two sounds was increased from zero to one semitone or beyond.

Time-Domain Cancellation

The above channel selection scheme can be seen as an instance of a more general iterative approach where F0 estimation is followed by the cancellation of the detected sound from the mixture, and the estimation is then repeated for the residual signal. This generalization was pointed out by de Cheveigné, who further proposed that the cancellation can take place in the time domain [129], [130]. When the period τ_0 of one sound in the mixture has been found, the sound can be removed by applying a cancellation filter with the impulse response

$$h_{\tau_0}(n) = \delta(n) - \delta(n - \tau_0), \qquad (8.13)$$

where $\delta(n)$ is the unit impulse function. Convolving an input signal $x(\tau)$ with $h_{\tau_0}(n)$ yields $h_{\tau_0}(n) \otimes x(n) = x(n) - x(n - \tau_0)$ and, if the detected sound is perfectly periodic, the above filter completely removes it from the mixture. As a side-effect, however, the filter also removes the partials of other sounds that coincide with those of the sound being cancelled. Also, a more sophisticated filter is needed to cancel a sound whose period is not precisely a multiple of the sampling interval [382].

An advantage of the time-domain cancellation is that it is not bound to the resolution of the cochlear filterbank and, in principle, it works even when all the channels are dominated by a single period. The filtering can be done directly for the input signal or within the channels of an auditory model. These two are equivalent unless the within-channel filtering is done after the non-linear IHC simulation stage.

De Cheveigné used the cancellation principle for the actual F0 estimation, too. He proposed to calculate a squared difference function (SDF) which is defined for an input signal $x(n)$ as

$$\text{SDF}(n, \tau) = \sum_{i=0}^{N-1} (x(n-i) - x(n-i-\tau))^2, \qquad (8.14)$$

where N is the analysis frame size [131].[6] By expanding the square, it can be seen that $\text{SDF}(n, \tau) = E(n) + E(n - \tau) - 2r(n, \tau)$, where $E(n)$ denotes the signal power at time n and $r(n, \tau)$ is the ACF. Thus the SDF and the ACF are functionally equivalent, and period estimation can be carried out by searching for minima in the SDF instead of maxima in the ACF. De Cheveigné also proposed a joint cancellation model, where two cancellation filters with periods τ_A and τ_B were applied in a cascade so as to cancel two periodic sounds. By computing the power of the resulting signal as a function of the two periods, the F0s were found by locating the minumum of the two-dimensional function [129], [133].

De Cheveigné evaluated both the iterative and the joint F0 estimation method for mixtures of two-voiced speech segments [129]. The iterative algorithm was reported to produce estimates which were correct within 3% accuracy in 86% of the frames and the exhaustive joint estimator produced correct estimates in 90% of the frames. Computational complexity is a drawback of the joint estimator.

Channel and Peak Selection

Wu, Wang, and Brown proposed an algorithm for tracking the F0s of two simultaneous speakers, taking particular interest in noise robustness [677]. Their method employed a computational model of the peripheral auditory system, after which the channels significantly corrupted by noise were excluded. From the remaining channels, ACF peaks were selected so that peaks judged to give misleading information were rejected. This led to an intermediate data representation which consisted of only the lag values and channel labels of the selected ACF peaks (discarding peak amplitudes). The information was then processed using statistical models.

In more detail, the channel and peak selection process was the following. First, a gammatone filterbank was applied and the resulting channels were classified as 'low-frequency' or 'high-frequency' channels depending on whether their centre frequency was below or above 800 Hz. Normalized ACFs were then computed for the low-channel signals directly and for the amplitude envelopes of the high-channel signals. Low channels were selected (i.e., included in further computations) if the highest peak of the normalized ACF exceeded a given threshold value. High-frequency channels were selected if the shapes of the normalized ACFs computed in 16 ms and in 32 ms frames were sufficiently similar. Peak selection, in turn, consisted of two main rules. First, an acceptable peak (peak not due to noise) was required to show a submultiple peak at twice its lag value. At high-frequency channels, envelope beating at the F0 rate was assumed and, therefore, subharmonics of any peak higher than a threshold value were removed. Full details can be found in [677].

---

[6]The SDF is closely related to the average magnitude difference function (AMDF) that has been used to estimate the F0 of speech [549]. The AMDF is obtained by summing absolute values instead of their squares in (8.14).

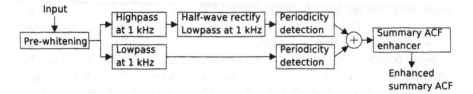

Fig. 8.10. Block diagram of the pitch analysis method proposed by Karjalainen and Tolonen [627]. ©2005 IEEE, reproduced here by permission.

The remaining channels and peaks were subjected to statistical modelling. Using clean speech as training material, the difference $\delta_c = \tau_c - \tau_0$ between a true (annotated) fundamental period τ_0 and the period of the closest selected peak τ_c at channel c was studied. The statistical distribution of δ_c was used to determine the likelihood of the observed peaks at channel c given a fundamental period candidate τ. Different observation likelihood functions were defined for the cases of zero, one, and two F0s (two F0s were jointly estimated). Finally, a hidden Markov model was employed to model the dynamic aspects of the F0 contours. This included both the continuity of the F0 tracks and jump probabilities between the state spaces of zero, one, or two F0s.

In evaluations, Wu et al. used ten voiced utterances to generate mixtures of two voices. These were mixed with realistic noise signals, including harmonic interference and interfering speech signals. Five utterances were used for training and five for testing. Good results were reported for this database and an implementation of the method is publicly available [677].

8.5.2 Multiple F0 Estimator of Karjalainen and Tolonen

Karjalainen and Tolonen proposed a computationally efficient version of the unitary pitch model (see p. 241) and extended it to the multiple F0 estimation of musical sounds. [327], [627] Figure 8.10 shows the block diagram of their method. The most obvious difference from the original auditory model is that the method divides an input signal into two channels only, below and above 1 kHz, and then analyses the periodicity of the low-channel signal and of the envelope of the high-channel signal. Despite the drastic reduction in computation load compared to the unitary pitch model, many important characteristics of the model were preserved.

The method included several features to address practical robustness issues. Robustness against timbral variation (different musical instruments for example) was achieved by *pre-whitening* the input signal using inverse warped-linear-prediction filtering [272]. In essence, this flattens the spectral energy distribution but does not affect the spectral fine structure.

Periodicity analysis in the method of Karjalainen and Tolonen was carried out using a *generalized ACF*, originally proposed by Indefrey et al. in [306]. According to the Wiener–Khintchine theorem, the ACF of a time-domain signal $\mathbf{x}$ is the inverse Fourier transform of its power spectrum [276, p. 334]. The generalized ACF, then, is defined as

$$\hat{r}(\tau) = \mathrm{IDFT}(|\mathrm{DFT}(\mathbf{x})|^{\alpha}), \qquad (8.15)$$

where DFT and IDFT denote the discrete Fourier transform and its inverse, and α is a free parameter which determines the frequency domain compression.[7] The standard ACF is obtained by substituting $\alpha = 2$. Definition of the *cepstrum* of $\mathbf{x}$ is analogous to ACF and is obtained by replacing the second power with the logarithm function. The difference between the ACF and cepstrum-based F0 estimators is quantitative: raising the magnitude spectrum to the second power emphasizes spectral peaks in relation to noise but, on the other hand, further aggravates spectral peculiarities of the target sound. Applying the logarithm function causes the opposite for both. And indeed, ACF-based F0 estimators have been reported to be relatively noise immune but sensitive to formant structures in speech, and vice versa for cepstrum-based methods [535]. As a trade-off, Karjalainen and Tolonen suggested using the value $\alpha = 0.67$.

Extension to multiple F0 estimation was achieved by cancelling subharmonics in the summary ACF (SACF) by clipping the SACF to positive values, time-scaling it to twice its length, and by subtracting the result from the original clipped SACF. This cancellation operation was repeated for time-scaling factors up to about five. From the resulting *enhanced SACF*, all F0s were picked without iterative estimation and cancellation. In more detail, the enhancing procedure was as follows:

Algorithm 8.1: Enhancing Procedure of Karjalainen and Tolonen

1. The enhanced SACF $\tilde{s}(\tau)$ is initialized to be equal to the SACF $s(\tau)$. The scaling factor m is initialized to value 2.
2. The original SACF is time-scaled to m times its length and the result is denoted by $s_m(\tau)$. Using linear interpolation,

$$s_m(\tau) = s(d) + \frac{\tau - md}{m}\left(s(d+1) - s(d)\right), \qquad (8.16)$$

where $d = \lfloor \tau/m \rfloor$ and $\lfloor \cdot \rfloor$ denotes rounding towards negative infinity.
3. The enhanced SACF is updated as

$$\tilde{s}(\tau) \leftarrow \max(0, \tilde{s}(\tau) - \max(0, s_m(\tau))). \qquad (8.17)$$

4. Increment m by 1. If m is smaller than 6, return to Step 2.

The above enhancing procedure is surprisingly efficient in removing spurious peaks from the SACF and in revealing more than one F0 in it. Also,

---

[7]In practice, the analysis frame $\mathbf{x}$ has to be zero-padded to twice its length before the first transform.

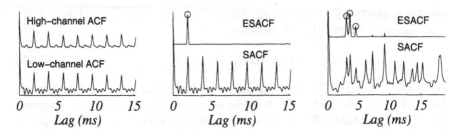

Fig. 8.11. Left: The ACFs at the low and the high channel for a violin sound (F0 523 Hz). Middle: SACF and enhanced SACF for the same sound. Right: SACF and enhanced SACF for a major triad chord played by the trumpet (F0s 220 Hz, 277 Hz, and 330 Hz). The circles indicate the correct fundamental periods.

it partly solves the 'chord root' problem mentioned in the beginning of Section 8.5 since the enhancing procedure scales the true F0 peaks to the position of the chord root and, if a note does not truly appear at the root, the spurious peak becomes cancelled. The only place where care has to be taken is in setting values of the original SACF to zero in the lag range $[0, f_s/1000$ Hz$]$ before the enhancing (here f_s denotes the sampling rate). This ensures that the values on the $\tau = 0$ hill do not spread and wipe away important information. Zeroing the mentioned lags causes no harm for the analysis since the algorithm cannot detect F0s above 1 kHz.

Figure 8.11 illustrates the enhancing procedure for an isolated sound and for a musical chord. As mentioned by Martin [439], the SACF indicates the non-existing F0 of the chord root in the latter case. After enhancing, however, the true F0s are revealed.

Overall, the method of Karjalainen and Tolonen is quite accurate and it has been described in sufficient detail to be exactly implementable based on [627] and on the Matlab toolbox for frequency-warped signal processing by Härmä et al. [272]. A drawback of the method as stated by the authors is that it is 'not capable of simulating the spectral pitch' [627, p. 713], i.e., the pitch of a sound whose first few harmonics are above 1 kHz. In practice, the method is most accurate for F0s below about 600 Hz. Later, Karjalainen and Tolonen also proposed an iterative approach to multiple F0 estimation using the described simplified auditory model [328].

8.5.3 Multiple F0 Estimator of Klapuri

Klapuri's multiple F0 estimator for music signals was originally described in [353, Ch. 4] and later improved and simpified in [354]. The method consists of a model of the peripheral auditory system followed by a periodicity analysis mechanism where F0s are iteratively estimated and cancelled (Fig. 8.12).

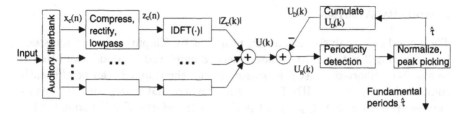

Fig. 8.12. Block diagram of the multiple F0 estimator of Klapuri [354]. ©2005 IEEE, reproduced here by permission.

Model of the Peripheral Auditory System

In the peripheral hearing model, an input signal was first passed through a bank of gammatone filters with centre frequencies uniformly distributed on the critical-band scale (see (8.5)) between 60 Hz and 5.2 kHz. A total of 72 filters were employed using the implementation of Slaney [591].

Hair cell transduction was modelled by compressing, half-wave rectifying, and lowpass filtering the sub-band signals. The compression was implemented by simulating the full-wave νth law compression (FWC), which is defined as

$$\text{FWC}(x) = \begin{cases} x^\nu, & x \geq 0, \\ -(-x)^\nu, & x < 0. \end{cases} \qquad (8.18)$$

For a narrow-band signal, such as the output of an auditory filter, the effect of the FWC within the passband of the filter can be accurately modelled by simply scaling the signal with a factor

$$\gamma_c = a(\sigma_c)^{\nu-1}, \qquad (8.19)$$

where σ_c is the standard deviation of the signal at channel c and the scalar a depends on ν but is common to all channels and can thus be omitted [353, p. 37]. In addition to the scaling mentioned, FWC generates small-amplitude distortion components at odd multiples of the channel centre frequency. These were avoided by using the model (8.19) instead of (8.18) directly.

The FWC provides a single parameter ν which determines the degree of spectral whitening applied on an input signal. The scaling factors γ_c normalize the variances of the sub-band signals towards unity when $0 \leq \nu \leq 1$. Here, the value $\nu = 0.33$ was applied.

The compressed sub-band signals were half-wave rectified by constraining negative values to zero. As shown in Fig. 8.6, this generates spectral components near zero frequency and on twice the channel centre frequency. The rectified signal at each channel was steeply lowpass filtered with a cut-off frequency 1.5 times the channel centre frequency in order to attenuate the distortion spectrum at twice the centre frequency but to pass the sub-band signal along with its amplitude envelope spectrum. The rectified and lowpass filtered signals $z_c(n)$ were then subjected to periodicity analysis.

Periodicity Analysis

The periodicity analysis mechanism proposed by Klapuri is best understood by comparing it with the ACF-based method employed by Meddis and Hewitt (see p. 241). Short-time ACF estimates within the channels can be efficiently computed as $r_{c,n}(\tau) = \text{IDFT}(|Z_{c,n}(k)|^2)$, where IDFT denotes the inverse Fourier transform and $Z_{c,n}(k)$ is the Fourier transform of $z_c(n)$ computed in a time frame that is centred at time n and zero-padded to twice its length before the transform. The within-band ACFs are then summed to obtain the summary ACF, $s_n(\tau) = \sum_c r_{c,n}(\tau)$.

Because the IDFT and the summing are linear operations, their order can be reversed and we can write $s_n(\tau) = \text{IDFT}(S_n(k))$, where $S_n(k) = \sum_c |Z_{c,n}(k)|^2$. The spectra of real-valued (audio) signals are conjugate symmetric and the IDFT can therefore be written out as

$$s_n(\tau) = \text{IDFT}(S_n(k)) = \frac{2}{K} \sum_{k=0}^{K/2-1} \cos\left(\frac{2\pi\tau k}{K}\right) S_n(k), \qquad (8.20)$$

where K is the length of the transform frame after the zero-padding.

Klapuri made three modifications to (8.20). First, as seen in Fig. 8.12, magnitude spectra were summed across channels instead of power spectra. Analogous to the generalized ACF in (8.15), it was observed that raising the magnitude spectra to the second power accentuates timbral peculiarities that cannot be completely removed by bandwise compression in polyphonic signals. Therefore, within-band magnitude spectra were summed to obtain a summary magnitude spectrum (SMS),

$$U(k) = \sum_c |Z_c(k)|, \qquad (8.21)$$

where the time index n has been omitted to simplify the notation in the following. The SMS functioned as an intermediate data representation and all the subsequent processing took place using it only.

Figure 8.13 illustrates the bandwise magnitude spectra $|Z_c(k)|$ for a saxophone sound. As can be seen, the within-channel rectification maps the contribution of higher-order partials to the position of the F0 and its few multiples in the spectrum. Most importantly, the degree to which an individual overtone partial j is mapped to the position of the fundamental increases as a function of j. This is because the auditory filters become wider at higher frequencies and the partials thus have larger-magnitude neighbours with which to generate the difference frequencies (beating) in the envelope spectrum. Klapuri's method was largely based on this observation, as will be explained below.

The second modification concerned the function $\cos(\cdot)$ in (8.20), which can be seen as a harmonic template that picks overtone partials of the frequency K/τ in the spectrum (see the rightmost panel of Fig. 8.7 on p. 243). The function was replaced by a response that is more sharply tuned to the frequencies

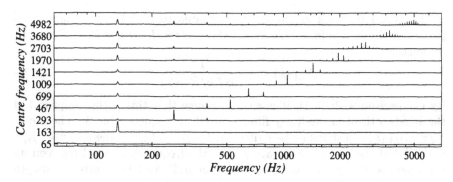

Fig. 8.13. The spectra $|Z_c(k)|$ at a few channels for a tenor saxophone sound (F0 131 Hz).

of the harmonic overtones of a F0 candidate and employs no negative weights between the partials. In practice, the frequency response resembled that of a comb filter shown in Fig. 8.7. This modification alleviates the interference of other, co-occurring sounds. Moreover, instead of pointwise multiplying the complete spectrum $U(k)$ with a comb filter response and then summing, it was found sufficient to sum up spectral components near the positions of the peaks of the comb-filter response (see (8.22) below). This led to a very efficient implementation computationally and is closely related to the *harmonic selection* methods reviewed by de Cheveigné in [129], and to the harmonic transform of Walmsley et al. [657].

The relative strength, or *salience*, $\lambda(\tau)$ of a fundamental period candidate τ was calculated in Klapuri's system as

$$\lambda(\tau) = \frac{f_s}{\tau} \sum_{j=1}^{\tau/2} \left(\max_{k \in \kappa_{j,\tau}} [H_{\mathrm{LP}}(k)U(k)] \right), \tag{8.22}$$

where f_s denotes the sampling rate and the factors f_s/τ and $H_{\mathrm{LP}}(k)$ are related to the third modification to be explained later. The set $\kappa_{j,\tau}$ defines a narrow range of frequency bins in the vicinity of the jth overtone partial of the F0 candidate f_s/τ. More exactly, $\kappa_{j,\tau} = [k_{j,\tau}^{(0)}, k_{j,\tau}^{(1)}]$, where

$$k_{j,\tau}^{(0)} = \lfloor jK/(\tau + \Delta\tau/2) \rfloor + 1, \tag{8.23}$$

$$k_{j,\tau}^{(1)} = \max(\lfloor jK/(\tau - \Delta\tau/2) \rfloor, k_{j,\tau}^{(0)}). \tag{8.24}$$

In the above formulas, K is the transform length and the scalar $\Delta\tau = 1$ denotes spacing between successive period candidates τ. A uniform sampling of lag values was used, analogous to the ACF. Equations (8.23)–(8.24) define the sets $\kappa_{j,\tau}$ so that, for a fixed partial index j, all the spectral components belong to the range of at least one period candidate τ, and the ranges of adjacent period candidates cannot overlap by more than one frequency bin.

The third modification in (8.22) compared to (8.20) is that individual partials in the sum in (8.22) are weighted by $f_s/\tau \times H_{\mathrm{LP}}(k)$, where the lowpass response is

$$H_{\mathrm{LP}}(k) = \frac{1}{0.108 f_s k/K + 24.7}. \qquad (8.25)$$

By comparison with (8.4), it is easy to notice that this is the reciprocal of the bandwidth of an auditory filter centred at frequency bin k. The factor $f_s/\tau \times H_{\mathrm{LP}}(k)$ can therefore be written as $F(\tau)/b_c(jF(\tau))$, where $F(\tau) = f_s/\tau$ is the F0 of the period candidate τ (i.e., the frequency interval between its overtones) and $b_c(jF(\tau))$ is the width of an auditory filter centred at its jth overtone. The ratio of these two was interpreted as the *resolvability* of the partial j [353, p. 45]. The lower-order overtones of a harmonic sound are resolved into separate auditory channels, whereas the higher-order overtones go to the same auditory channel with their neighbours and their frequencies cannot be perceived separately (resolved). Actually the lowpass filter $H_{\mathrm{LP}}(k)$ would belong to the within-band IHC modelling stage but, since the filter is the same for all channels, it is equivalent to apply it after the channels have been combined. The higher the centre frequency of an auditory channel, the more the filter attenuates the spectrum at the passband of the auditory filter and thus gives it a smaller weight in relation to the envelope spectrum, which is around zero frequency and not much affected. This corresponds to the fact that, at higher auditory channels, the neural firing activity more and more follows the amplitude envelope of the sub-band signal and not its fine structure—this is directly related to the concept of resolvability. Discrete categorization into 'low' and 'high' channels is not needed.

The degree of resolvability as modelled above (and thus the weight of a partial in the sum in (8.22)) is approximately inversely proportional to the harmonic index j when τ is fixed. As a consequence, the sum in (8.22) can be limited to $j \approx 20$ since weights beyond this are relatively small.

Taken together, the computation of the salience function $\lambda(\tau)$ can be seen as a process where partials are picked from harmonic positions of the spectrum $U(k)$, their magnitudes are weighted by the estimated resolvability $f_s/\tau \times H_{\mathrm{LP}}(k)$, and then summed. What makes all the difference is that the within-channel rectification maps the contribution of higher-order partials to the position of the fundamental and its few multiples in the spectra $Z_c(k)$, and the degree to which an individual overtone partial j is mapped to the position of the fundamental increases as a function of j, as explained above. As a consequence, the whole harmonic series of a sound contributes to its salience, despite the weighting with resolvability.

The above-described benefit of bandwise rectification cannot be overemphasized. Assigning the higher-order partials to their respective sound sources in polyphonic music signals is a nightmare. The rectification operation accomplishes this 'automatically' by mapping the support from higher-order harmonics to the position of F0 and its few multiples in $U(k)$. Figure 8.14

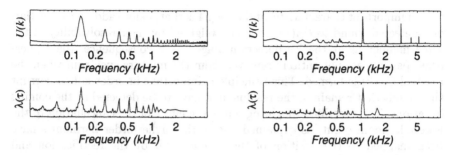

Fig. 8.14. The upper panels show the summary magnitude spectrum $U(k)$ for a saxophone sound with F0 131 Hz (*left*) and a violin sound with F0 1050 Hz (*right*). The lower panels show the corresponding salience functions $\lambda(\tau)$.

illustrates the calculation of $\lambda(\tau)$ for the saxophone sound shown in Fig. 8.13, and for a violin sound with the F0 1050 Hz.

Iterative Estimation and Cancellation

The global maximum of the function $\lambda(\tau)$ was found to be a robust indicator of one of the correct F0s in polyphonic signals. As with most F0 estimators, however, the next-highest salience was often assigned to half or twice that of the first detected F0. Similarly to de Cheveigné (see p. 250), Klapuri employed an iterative technique where F0 estimation was followed by the cancellation of the detected sound from the mixture and the estimation was then repeated for the residual signal. Algorithm 8.2 summarizes the applied technique [354].

Algorithm 8.2: Multiple F0 Estimator of Klapuri

1. A *residual SMS* $U_R(k)$ is initialized to be equal to $U(k)$. A summary spectrum of all detected sounds, $U_D(k)$, is initialized to zero.
2. A fundamental period $\hat{\tau}$ is estimated using $U_R(k)$ and (8.22).
3. Harmonic selection is carried out for the found period $\hat{\tau}$ according to (8.22)–(8.24). However, instead of summing up the magnitude values, the precise frequency and amplitude of each partial is estimated and used to calculate its magnitude spectrum at the few surrounding frequency bins.
4. The magnitude spectrum of the jth partial is weighted by $f_s/\tau \times H_{LP}(k_j)$ and added to the corresponding position of $U_D(k)$ which represents the cumulative spectrum of all the detected sounds.
5. The residual SMS is recalculated as

$$U_R(k) \leftarrow \max(0, U(k) - dU_D(k)), \qquad (8.26)$$

where $d = 0.5$ controls the amount of the subtraction and is a free parameter of the algorithm.
6. Return to Step 2.

An important characteristic of the Step 4 is that, before adding the partials of a detected sound to $U_D(k)$, they are weighted by their resolvability in the same manner as at the F0 detection stage. As a consequence, the higher-order partials are not entirely removed from the mixture spectrum when the residual $U_R(k)$ is formed. This principle is important in order not to corrupt the sounds that remain in the residual and have to be detected at the coming iterations. The described weighting limits the effect of the cancellation to the lowest harmonics but, as explained above, the higher-order harmonics have been mapped to the position of the fundamental by the rectification and are thus effectively cancelled, too.

8.5.4 Results

Simulation experiments were carried out to evaluate the performance of the method of Tolonen and Karjalainen [627] and that of Klapuri [354]. Implementations of the method of Wu et al. [677] and Marolt [434] are publicly available too, but these would have required a specific experimental setup since the former was designed to process continuous two-speaker speech signals and the latter to transcribe piano music only.

The acoustic material consisted of samples from the McGill University Master Samples collection [487], the University of Iowa website,[8] IRCAM Studio Online,[9] and of independent recordings for the acoustic guitar. There were altogether 32 different musical instruments, comprising brass and reed instruments, strings, flutes, the piano, the guitar, and mallet percussion instruments. The total number of samples (individual notes) was 2842.

Semi-random sound mixtures were generated by first allotting an instrument and then a random note from its playing range. This was repeated to get the desired number of simultaneous sounds, which were then mixed with equal mean-square levels. One thousand test cases were generated for mixtures of one, two, four, and six sounds.

One analysis frame immediately after the onset[10] of the sounds was fed to the multiple F0 method. The number of F0s to extract, i.e., the polyphony, was given along with the mixture signal. A correct F0 estimate was defined to deviate less than 3% from the nominal F0 of the sound, making it round to a correct note on the Western musical scale. Two different error rates were computed. *Multiple F0* estimation error rate was defined as the percentage of all F0s that were not correctly detected in the input signals. In *predominant F0* estimation, only one F0 in the mixture was being estimated and it was defined to be correct if it matched the correct F0 of any of the component sounds.

---

[8]University of Iowa samples: theremin.music.uiowa.edu/MIS.html

[9]IRCAM Studio Online: soleil.ircam.fr

[10]The onset of the sounds was defined to be at the point where the waveform reached one third of its maximum value during the first 200 ms of its playing.

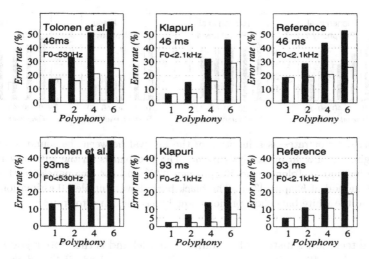

Fig. 8.15. F0 estimation error rates as a function of the number of concurrent sounds (polyphony) for the method of Tolonen and Karjalainen [627], the method of Klapuri [354], and the reference method [351]. The black bars and the white bars show the multiple F0 and the predominant F0 estimation error rates, respectively. The upper panels show the results for a 46 ms analysis frame and the lower panels for a 93 ms frame.

The left-hand panels of Fig. 8.15 show the error rates for the method of Tolonen and Karjalainen in 46 ms and 93 ms analysis frames. The F0 range in these experiments was limited to the three octaves between 65 Hz and 520 Hz, because the accuracy of the method was found to degrade rapidly above 600 Hz (see Section 8.5.2). The black bars show the multiple F0 estimation error rates and the white bars show the predominant F0 estimation error rates. The global maximum of the enhanced SACF was used for the latter purpose. The method performed robustly in polyphonic mixtures, and especially the predominant F0 estimation error rates remained reasonably low even in short time frames and in rich polyphonies. Taking into account the computational efficiency (faster than real-time) and conceptual simplicity of the method, the results are very good.

The middle panels of Fig. 8.15 show the error rates for the method of Klapuri [354]. The first detected F0 was used for the predominant F0 estimation. In these experiments, the pitch range was limited to five octaves between 65 Hz and 2.1 kHz. The method performs robustly in all cases and is very accurate, especially in the 93 ms analysis frame. Computational complexity is a drawback of this method. The calculations are clearly slower than real-time on a 2-GHz desktop computer, the most intensive part being the cochlear filterbank and the within-band DFT calculations.

The right-hand panels of Fig. 8.15 show the error rates for a state-of-the-art reference method proposed by Klapuri in [351]. This method is based on

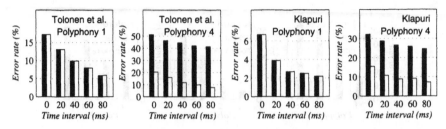

Fig. 8.16. Error rates as a function of the interval between the sound onset and the beginning of a 46 ms analysis frame. The two panels on the left show results for the method of Tolonen and Karjalainen [627] and the two panels on the right for the method of Klapuri [354]. The black bars show multiple F0 estimation error rates and the white bars show predominant F0 estimation error rates.

spectral techniques instead of an auditory model and is therefore a good point of comparison. The test cases given to Klapuri's method [354] and the reference method were identical. It was observed that the reference method requires quite a long analysis frame to resolve and process the overtones of low-pitched sounds, and mallet percussion instruments could not be reliably analysed. In addition to the differences in handling the higher-order overtones, a factor involved is that the frequency resolution of the Fourier spectrum is linear, whereas time-domain periodicity analysis within the auditory channels leads to $1/f$ frequency resolution, which enables more accurate analysis at the lower end of the logarithmic scales applied in music. The reference method is conceptually (technically) the most complex among the three.

An important factor in the above results is that the analysis frames were positioned immediately after the onsets of the sounds. Figure 8.16 shows the error rates of the two methods as a function of the time interval between the sound onset and the beginning of the analysis frame. As can be seen, the error rates improve clearly as the interval increases, and especially the predominant F0 estimation error rates shrink to about a third of the initial values after 80 ms of the onset. This is because the noisy beginning transients of many sounds die off rapidly and F0 estimation becomes easier thereafter. In music signals, however, notes are often short and such an offset cannot be applied. In Fig. 8.15, maximally realistic simulations were of interest and thus a zero offset was applied. Figure 8.16 shows results only for the 46 ms analysis frame, but the general trend is similar (although less pronounced) for the longer frame.

8.5.5 Summary of the Multiple F0 Estimation Methods

The beginning of this section listed several issues where the pitch perception models fall short of being practically applicable multiple F0 estimators. This section summarizes and discusses the various technical solutions that were proposed as improvements.

Two main approaches can be distinguished among the techniques used to extend a single-pitch model to the estimation of multiple pitches: the iterative estimation-and-cancellation approach and the joint estimation approach. Most methods fall into the former category: F0 estimation is done using the summary ACF, for example, and the F0 found is then cancelled before deciding the next one. Meddis and Hewitt performed the cancellation by removing the auditory channels associated with the first detected pitch [458]. De Cheveigné employed within-channel cancellation filtering in the time domain [130]. Klapuri subtracted the partials of a detected sound in the frequency domain and removed only the lower-order partials entirely [354].

Joint estimation methods were proposed by de Cheveigné [129], Karjalainen and Tolonen [327], and Wu et al. [677]. Among these, the method of de Cheveigné was not actually based on an auditory model, but the method applied two cancellation filters in a cascade and searched for such cancellation-filter periods that the output power was minimized. Karjalainen and Tolonen enhanced the summary ACF so that all F0s could be directly extracted from the result. In the method of Wu et al., the distribution of the peaks in the sub-band ACFs was statistically modelled in the cases of zero, one, or two pitches.

The limited robustness of the pitch perception models in polyphonic signals is another important problem addressed by the multiple F0 estimation methods. The chord-root detection problem was mentioned as an example of this. The SACF enhancing technique of Karjalainen and Tolonen [627] is rather efficient in this respect, as illustrated in Fig. 8.11. Klapuri addressed the problem by applying the lowpass response in (8.25), which suppresses the support of higher-order partials to the chord root unless the series of partials has sufficiently uniform amplitudes so as to generate strong beating at the fundamental rate. This is usually not the case if the partials are due to several different sounds (component F0s of a chord). Also, the use of harmonic selection in the frequency domain alleviated the interference of other sounds since the spectrum between the partials was not used in salience calculations. Iterative estimation and cancellation methods that estimate the first F0 directly from the summary ACF suffer from its robustness limitations [458], [130].

Robustness for different sound sources (different musical instruments) is a very important aspect in F0 estimation. Here the pitch perception models are readily very efficient. Meddis's hair-cell model compresses the sub-band signals and results in spectral whitening, that is, removal of timbral information to some extent [457], [456]. Ellis [171] and Klapuri [354] carried out this function by scaling the sub-band signals inversely proportional to their variance. Karjalainen and Tolonen pre-processed the input signals by inverse warped-linear-prediction filtering. This had the advantage that a multi-channel filterbank was not needed [327], [272]. An advantage of all these is that they flatten the spectral energy distribution without raising the noise floor in relation to spectral peaks. The latter happens for example in cepstrum pitch detection, where the logarithm function is applied bin-by-bin to the magnitude spectrum [535]. The system of Wu et al. [677] is interesting, since in this method, whitening would not have any effect at all because only the lag-values of within-channel ACF peaks are retained and not their amplitudes.

Noise robustness was not discussed in depth in this chapter. In music, percussive instruments and recording imperfections cause noise-like interference for the F0 estimation. Particular emphasis on this issue was laid by Wu et al., who performed channel and peak selection so as to avoid the spectro-temporal regions that were severely corrupted by noise. The authors remarked that they essentially treated multiple F0 tracking and noise robustness as a single problem [677, p. 240]. Karjalainen and Tolonen selected the generalized ACF power so as to make a compromise between noise robustness and spectral flattening [327].

Computational complexity of the pitch models was significantly reduced only in the method of Karjalainen and Tolonen [327]. In the other methods, the most time-consuming operation is typically the peripheral filterbank and the periodicity analysis within channels, usually leading to computation times which are 10 to 100 times slower than that of the method of Karjalainen and Tolonen. Ellis computed the within-channel ACFs only for a set of logarithmically distributed lag values, which allowed the use of a very good time resolution without causing a prohibitative computational load [171]. In the iterative methods, the peripheral analysis usually has to be computed only once [458], [354].

8.6 Conclusions

Pitch perception models and practical F0 estimators address slightly different tasks and are judged according to different criteria. The former should faithfully represent the mechanisms of the human auditory system, whereas the latter are expected to perform accurate multiple F0 estimation by any means available. The main focus of this chapter was on the practical side. However, the two aspects have significantly influenced and benefited each other and this is one reason to study auditory modelling.

Many characteristics of human pitch perception can be traced to the peripheral stages of hearing, as discussed in Section 8.3. In this sense, auditory models have a lot to say about the intermediate data representations used in acoustic signal analysis. A particularly important principle in an auditorily motivated analysis is that the higher-order overtones of a sound are processed collectively within each auditory channel; estimation and separation of individual higher-order partials is not attempted. The cochlear filterbank is 'fair' for different F0 values in this respect since the first few harmonic partials of all F0s are resolved into separate auditory channels, whereas the harmonics above about 10 go to the same channel along with their neighbours and generate amplitude envelope beating at the fundamental rate. This is an efficient mechanism for dealing with the higher-order overtones in complex polyphonic data. The other advantages and disadvantages of a perceptually motivated data representation were discussed in Section 8.4.4.

Compared with the peripheral stages of hearing, at least an equally important part of pitch perception takes place in the brain (Steps 3 and 4 in the overview on p. 235). These stages are not yet well understood and thus there is a larger variance in the proposed practical techiques as well. The biggest defect of the existing pitch perception models from the music transcription viewpoint is that they have been designed to process isolated sounds instead of polyphonic signals. Different techniques for transforming a pitch model into a multiple F0 estimator were described in Section 8.5. Both iterative methods and joint estimation methods were discussed, and different ways of cancelling a detected F0 from the mixture signal were described. Periodicity analysis techniques were presented that applied the ACF [458], [130], the generalized ACF [627], statistical modelling of the ACF peaks [677], adaptive oscillators [434], or simulation of comb filters in the frequency domain [354]. For now, none of the described methods can claim to be the 'right' or the optimal one, but they provide a wealth of technical solutions and approaches to build upon.

Unsupervised Learning Methods for Source Separation in Monaural Music Signals

Tuomas Virtanen

Institute of Signal Processing, Tampere University of Technology,
Korkeakoulunkatu 1, 33720 Tampere, Finland
Tuomas.Virtanen@tut.fi

9.1 Introduction

Computational analysis of polyphonic musical audio is a challenging problem. When several instruments are played simultaneously, their acoustic signals mix, and estimation of an individual instrument is disturbed by the other co-occurring sounds. The analysis task would become much easier if there was a way to separate the signals of different instruments from each other. Techniques that implement this are said to perform *sound source separation*. The separation would not be needed if a multi-track studio recording was available where the signal of each instrument is on its own channel. Also, recordings done with microphone arrays would allow more efficient separation based on the spatial location of each source. However, multi-channel recordings are usually not available; rather, music is distributed in stereo format. This chapter discusses sound source separation in *monaural* music signals, a term which refers to a one-channel signal obtained by recording with a single microphone or by mixing down several channels.

There are many signal processing tasks where sound source separation could be utilized, but the performance of the existing algorithms is still quite limited compared to the human auditory system, for example. Human listeners are able to perceive individual sources in complex mixtures with ease, and several separation algorithms have been proposed that are based on modelling the source segregation ability in humans (see Chapter 10 in this volume).

Recently, the separation problem has been addressed from a completely different point of view. The term *unsupervised learning* is used here to characterize algorithms which try to separate and learn the structure of sources in mixed data based on information-theoretical principles, such as statistical independence between sources, instead of sophisticated modelling of the source characteristics or human auditory perception. Algorithms discussed in this chapter are *independent component analysis* (ICA), *sparse coding*, and *non-negative matrix factorization* (NMF), which have been recently used in source separation tasks in several application areas. When used for monaural audio source separation, these algorithms usually factor the spectrogram or other short-time representation of the input signal into elementary components, which are then clustered into sound sources and further analysed to obtain musically important information. Although the motivation of unsupervised learning algorithms is not in the human auditory perception, there are similarities between them. For example, all the unsupervised learning methods discussed here are based on reducing redundancy in data, and it has been found that redundancy reduction takes place in the auditory pathway, too [85].

The focus of this chapter is on unsupervised learning algorithms which have proven to produce applicable separation results in the case of music signals. There are some other machine learning algorithms which aim at separating speech signals based on pattern recognition techniques, for example [554].

All the algorithms mentioned above (ICA, sparse coding, and NMF) can be formulated using a linear signal model which is explained in Section 9.2. Different data representations are discussed in Section 9.2.2. The estimation criteria and algorithms are discussed in Sections 9.3, 9.4, and 9.5. Methods for obtaining and utilizing prior information are presented in Section 9.6. Once the spectrogram is factored into components, these can be clustered into sound sources or further analysed to obtain musical information. The post-processing methods are discussed in Section 9.7. Systems extended from the linear model are discussed in Section 9.8.

9.2 Signal Model

When several sound sources are present simultaneously, the acoustic waveforms of the individual sources add linearly. Sound source separation is defined as the task of recovering each source signal from the acoustic mixture. A complication is that there is no unique definition for a sound source. One possibility is to consider each vibrating physical entity, for example each musical instrument, as a sound source. Another option is to define this according to what humans tend to perceive as a single source. For example, if a violin, section plays in unison, the violins are perceived as a single source, and usually there is no need to separate the signals played by each violin. In Chapter 10, these two alternatives are referred to as physical source and perceptual source, respectively (see p. 302). Here we do not specifically commit ourselves to either of these. The type of the separated sources is determined by the properties of the algorithm used, and this can be partly affected by the designer according to the application at hand. In music transcription, for example, all the equal-pitched notes of an instrument can be considered as a single source.

Many unsupervised learning algorithms, for example standard ICA, require that the number of sensors be larger or equal to the number of sources. In multi-channel sound separation, this means that there should be at least as many microphones as there are sources. However, automatic transcription of music usually aims at finding the notes in monaural (or stereo) signals, for which basic ICA methods cannot be used directly. By using a suitable signal representation, the methods become applicable with one-channel data.

The most common representation of monaural signals is based on short-time signal processing, in which the input signal is divided into (possibly over-lapping) frames. Frame sizes between 20 and 100 ms are typical in systems designed to separate musical signals. Some systems operate directly on time-domain signals and some others take a frequency transform, for example the discrete Fourier transform (DFT) of each frame. The theory and general discussion of time-frequency representations is presented in Chapter 2.

9.2.1 Basis Functions and Gains

The representation of the input signal within each frame $t = 1 \ldots T$ is denoted by an observation vector $\mathbf{x}_t$. The methods presented in this chapter model $\mathbf{x}_t$ as a weighted sum of basis functions $\mathbf{b}_n$, $n = 1 \ldots N$, so that the signal model can be written as

$$\mathbf{x}_t \approx \sum_{n=1}^{N} g_{n,t} \mathbf{b}_n, \qquad t = 1, \ldots, T, \tag{9.1}$$

where $N \ll T$ is the number of basis functions, and $g_{n,t}$ is the amount of contribution, or gain, of the nth basis function in the tth frame. Some methods estimate both the basis functions and the time-varying gains from a mixed input signal, whereas others use pre-trained basis functions or some prior information about the gains.

The term *component* refers to one basis function together with its time-varying gain. Each sound source is modelled as a sum of one or more components, so that the model for source m in frame t is written as

$$\mathbf{y}_{m,t} = \sum_{n \in S_m} g_{n,t} \mathbf{b}_n, \tag{9.2}$$

where S_m is the set of components within source m. The sets are disjoint, i.e., each component belongs to only one source.

In (9.1) approximation is used, since the model is not necessarily noise-free. The model can also be written with a residual term $\mathbf{r}_t$ as

$$\mathbf{x}_t = \sum_{n=1}^{N} g_{n,t} \mathbf{b}_n + \mathbf{r}_t, \qquad t = 1, \ldots, T. \tag{9.3}$$

By assuming some probability distribution for the residual and a prior distri-bution for other parameters, a probabilistic framework for the estimation of $\mathbf{b}_n$ and $g_{n,t}$ can be formulated (see e.g. Section 9.4). Here (9.1) without the residual term is preferred for its simplicity. For T frames, the model (9.1) can be written in matrix form as

$$\mathbf{X} \approx \mathbf{BG}, \qquad (9.4)$$

where $\mathbf{X} = [\mathbf{x}_1, \mathbf{x}_2, \ldots, \mathbf{x}_T]$ is the *observation matrix*, $\mathbf{B} = [\mathbf{b}_1, \mathbf{b}_2, \ldots, \mathbf{b}_N]$ is the *mixing matrix*, and $[\mathbf{G}]_{n,t} = g_{n,t}$ is the *gain matrix*. The notation $[\mathbf{G}]_{n,t}$ is used to denote the $(n, t)^{\text{th}}$ entry of matrix $\mathbf{G}$. The term mixing matrix is typically used in ICA, and here we follow this convention.

The estimation algorithms can be used with several data representations. Often the absolute values of the DFT are used; this is referred to as the *magnitude spectrum* in the following. In this case, $\mathbf{x}_t$ is the magnitude spectrum within frame t, and each component n has a fixed magnitude spectrum $\mathbf{b}_n$ with a time-varying gain $g_{n,t}$. The observation matrix consisting of framewise magnitude spectra is here called a *magnitude spectrogram*. Other representations are discussed in Section 9.2.2.

The model (9.1) is flexible in the sense that it is suitable for representing both harmonic and percussive sounds. It has been successfully used in the transcription of drum patterns [188], [505] (see Chapter 5), in the pitch estimation of speech signals [579], and in the analysis of polyphonic music signals [73], [600], [403], [650], [634], [648], [43], [5].

Figure 9.1 shows an example signal which consists of a diatonic scale and a C major chord played by an acoustic guitar. The signal was separated into components using the NMF algorithm described in [600], and the resulting components are depicted in Fig. 9.2. Each component corresponds roughly to one fundamental frequency: the basis functions are approximately harmonic and the time-varying gains follow the amplitude envelopes of the notes. The separation is not perfect because of estimation inaccuracies. For example, in some cases the gain of a decaying note drops to zero when a new note begins.

Factorization of the spectrogram into components with a fixed spectrum and a time-varying gain has been adopted as a part of the MPEG-7 pattern recognition framework [72], where the basis functions and the gains are used as features for classification. Kim et al. [341] compared these to mel-frequency cepstral coefficients which are commonly used features in the classification of audio signals. In this study, mel-frequency cepstral coefficients performed better in the recognition of sound effects and speech than features based on ICA or NMF. However, final conclusions about the applicability of these methods to sound source recognition have yet to be made. The spectral basis decomposition specified in MPEG-7 models the summation of components on a decibel scale, which makes it unlikely that the separated components correspond to physical sound objects.

9.2.2 Data Representation

The model (9.1) presented in the previous section can be used with time-domain or frequency-domain observations and basis functions. Time-domain

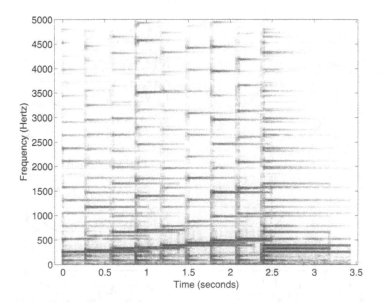

Fig. 9.1. Spectrogram of an example signal which consist of a diatonic scale from C5 to C6, followed by a C major chord (simultaneous notes C5, E4, and G5), played by an acoustic guitar. The notes are not damped, meaning that consecutive notes overlap.

observation vector $\mathbf{x}_t$ is the signal within frame t directly, whereas a frequency-domain observation vector is obtained by applying a chosen transformation to this. The representation of the signal and the basis functions have to be the same. ICA and sparse coding allow the use of any short-time signal representation, whereas for NMF, only a frequency-domain representation is appropriate. Naturally, the representation has a significant effect on performance. The advantages and disadvantages of different representations are considered in this section. For a more extensive discussion, see Casey [70] or Smaragdis [598].

Time-Domain Representation

Time-domain representations are straightforward to compute, and all the information is preserved when an input signal is segmented into frames and windowed. However, time-domain basis functions are problematic in the sense that a single basis function alone cannot represent a meaningful sound source: the phase of the signal within each frame varies depending on the frame position. In the case of a short-duration percussive source, for example, a separate basis function is needed for every possible position of the sound event within the

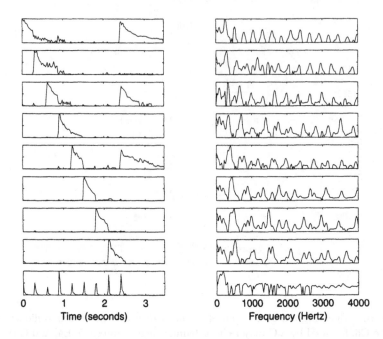

Fig. 9.2. Components estimated from the example signal in Fig. 9.1. Basis functions are plotted on the right and the corresponding time-varying gains on the left. Each component except the bottom one corresponds to an individual pitch value and the gains follow roughly the amplitude envelope of each note. The bottom component models the attack transients of the notes. The components were estimated using the NMF algorithm [400], [600] and the divergence objective (explained in Section 9.5).

frame. A shift-invariant model which is later discussed in Section 9.8 is one possible method of overcoming this limitation [43].

The time-domain signals of real-world sound sources are generally not identical at different occurrences since the phases behave very irregularly. For example, the overtones of a pitched musical instrument are not necessarily phase-locked, so that the time-domain waveform varies over time. Therefore, one has to use multiple components to represent even a single note of a pitched instrument. In the case of percussive sound sources, this phenomenon is even clearer: the time-domain waveforms vary a lot at different occurrences.

The larger the number of the components, the more uncertain is their estimation and further analysis, and the more observations are needed. If the sound event represented by a component occurs only once in the input signal, separating it from co-occurring sources is difficult since there is no information about the component elsewhere in the signal. Also, clustering the components into sources becomes more difficult when there are many of them for each source.

Separation algorithms which operate on time-domain signals have been proposed for example by Dubnov [157], Jang and Lee [314], and Blumensath and Davies [43]. Abdallah and Plumbley [3], [2] found that the independent components analysed from time-domain music and speech signals were similar to a wavelet or short-time DFT basis. They trained the basis functions using several days of radio output from BBC Radio 3 and 4 stations.

Frequency-Domain Representation

When using a frequency transform such as the DFT, the phases of the complex-valued transform can be discarded by considering only the magnitude or power spectrum. Even though some information is lost, this also eliminates the phase-related problems of time-domain representations. Unlike time-domain basis functions, many real-world sounds can be rather well approximated with a fixed magnitude spectrum and a time-varying gain, as seen in Figs. 9.1 and 9.2, for example. Sustained instruments in particular tend to have a stationary spectrum after the attack transient.

In most systems aimed at the separation of sound sources, DFT and a fixed window size is applied, but the estimation algorithms allow the use of any time-frequency representation. For example, a logarithmic spacing of frequency bins has been used [58], which is perceptually and musically more plausible than a constant spectral resolution.

The linear summation of time-domain signals does not imply the linear summation of their magnitude or power spectra, since phases of the source signals affect the result. When two signals sum in the time domain, their complex-valued DFTs sum linearly, $X(k) = Y_1(k) + Y_2(k)$, but this equality does not apply for the magnitude or power spectra. However, provided that the phases of $Y_1(k)$ and $Y_2(k)$ are uniformly distributed and independent of each other, we can write

$$E\{|X(k)|^2\} = |Y_1(k)|^2 + |Y_2(k)|^2, \qquad (9.5)$$

where $E\{\cdot\}$ denotes expectation. This means that in the expectation sense, we can approximate time-domain summation in the power spectral domain, a result which holds for more than two sources as well. Even though magnitude spectrogram representation has been widely used and it often produces good results, it does not have similar theoretical justification. Since the summation is not exact, use of phaseless basis functions causes an additional source of error. Also, a phase generation method has to be implemented if the sources are to be synthesized separately. These are discussed in Section 9.7.3.

The human auditory system has a large dynamic range: the difference between the threshold of hearing and the threshold of pain is approximately 100 dB [550]. Unsupervised learning algorithms tend to be more sensitive to high-energy observations. If sources are estimated from the power spectrum, some methods fail to separate low-energy sources even though they would be perceptually and musically meaningful. This problem has been noticed, e.g., by FitzGerald in the case of percussive source separation [186, pp. 93–100]. To overcome the problem, he used an algorithm which processed separately high-frequency bands which contain low-energy sources, such as hi-hats and cymbals [187]. Vincent and Rodet [648] addressed the same problem. They proposed a model in which the noise was additive in the log-spectral domain. The numerical range of a logarithmic spectrum is compressed, which increases the sensitivity to low-energy sources. Additive noise in the log-spectral domain corresponds to multiplicative noise in power spectral domain, which was also assumed in the system proposed by Abdallah and Plumbley [5]. Virtanen proposed the use of perceptually motivated weights [651]. He used a weighted cost function in which the observations were weighted so that the quantitative significance of the signal within each critical band was equal to its contribution to the total loudness.

9.3 Independent Component Analysis

ICA has been successfully used in several 'blind' source separation tasks, where very little or no prior information is available about the source signals. One of its original target applications was multi-channel sound source separation, but it has also had several other uses. ICA attempts to separate sources by identifying latent signals that are maximally independent. In practice, this usually leads to the separation of meaningful sound sources.

Mathematically, statistical independence is defined in terms of probability densities: random variables x and y are said to be independent if their joint probability distribution function[1] $\mathsf{p}(x, y)$ is a product of the marginal distribution functions, $\mathsf{p}(x, y) = \mathsf{p}(x)\mathsf{p}(y)$.

The dependence between two variables can be measured in several ways. Mutual information is a measure of the information that given random variables have on some other random variables [304]. The dependence is also closely related to the Gaussianity of the distribution of the variables. According to the central limit theorem, the distribution of the sum of independent variables is more Gaussian than their original distributions, under certain conditions. Therefore, some ICA algorithms aim at separating output variables whose distributions are as far from Gaussian as possible.

The signal model in ICA is linear: K observed variables $x_1, \ldots, x_K$ are modelled as linear combinations of N source variables $g_1, \ldots, g_N$. In a vector-matrix form, this can be written as

$$\mathbf{x} = \mathbf{B}\mathbf{g}, \qquad (9.6)$$

where $\mathbf{x} = \begin{bmatrix} x_1, \ldots x_K \end{bmatrix}^\mathsf{T}$ is an observation vector, $[\mathbf{B}]_{k,n} = b_{k,n}$ is a mixing matrix, and $\mathbf{g} = \begin{bmatrix} g_1, \ldots, g_N \end{bmatrix}^\mathsf{T}$ is a source vector. Both $\mathbf{B}$ and $\mathbf{g}$ are unknown.

---

[1]The concept of probability distribution function is described in Chapter 2.

The standard ICA requires that the number of observed variables K (the number of sensors) be equal to the number of sources N. In practice, the number of sensors can also be larger than the number of sources, because the variables are typically decorrelated using principal component analysis (PCA; see Chapter 2), and if the desired number of sources is less than the number of variables, only the principal components corresponding to the largest eigenvalues are selected.

As another pre-processing step, the observed variables are usually centred by subtracting the mean and their variance is normalized to the unity. The centred and whitened data observation vector $\mathbf{x}$ is obtained from the original observation vector $\tilde{\mathbf{x}}$ by

$$\mathbf{x} = \mathbf{V}(\tilde{\mathbf{x}} - \boldsymbol{\mu}), \tag{9.7}$$

where $\boldsymbol{\mu}$ is the empirical mean of the observation vector, and $\mathbf{V}$ is a whitening matrix, which is often obtained from the eigenvalue decomposition of the *empirical covariance matrix* of the observations [304]. The empirical mean and covariance matrix are explained in Chapter 2.

To simplify the notation, it is assumed that the data $\mathbf{x}$ in (9.6) is already centred and decorrelated, so that $K = N$. The core ICA algorithm carries out the estimation of an unmixing matrix $\mathbf{W} \approx \mathbf{B}^{-1}$, assuming that $\mathbf{B}$ is invertible. Independent components are obtained by multiplying the whitened observations by the estimate of the unmixing matrix, to result in the source vector estimate $\hat{\mathbf{g}}$:

$$\hat{\mathbf{g}} = \mathbf{W}\mathbf{x}. \tag{9.8}$$

The matrix $\mathbf{W}$ is estimated so that the output variables, i.e., the elements of $\hat{\mathbf{g}}$, become maximally independent. There are several criteria and algorithms for achieving this. The criteria, such as non-Gaussianity and mutual information, are usually measured using high-order cumulants such as kurtosis, or expectations of other non-quadratic functions [304]. ICA can be also viewed as an extension of PCA. The basic PCA decorrelates variables so that they are independent up to second-order statistics. It can be shown that if the variables are uncorrelated after taking a suitable non-linear function, the higher-order statistics of the original variables are independent, too. Thus, ICA can be viewed as a non-linear decorrelation method.

Compared with the previously presented linear model (9.3), the standard ICA model (9.6) is exact, i.e., it does not contain the residual term. Some special techniques can be used in the case of the noisy signal model (9.3), but often noise is just considered as an additional source variable. Because of the dimension reduction with PCA, $\mathbf{Bg}$ gives an exact model for the PCA-transformed observations but not necessarily for the original ones.

There are several ICA algorithms, and some implementations are freely available, such as FastICA [302], [182] and JADE [65]. Computationally quite efficient separation algorithms can be implemented based on FastICA, for example.

9.3.1 Independent Subspace Analysis

The idea of *independent subspace analysis* (ISA) was originally proposed by Hyvärinen and Hoyer [303]. It combines the multidimensional ICA with invariant feature extraction, which are shortly explained later in this section. After the work of Casey and Westner [73], the term ISA has been commonly used to denote techniques which apply ICA to factor the spectrogram of a monaural audio signal to separate sound sources. ISA provides a theoretical framework for the whole separation procedure described in this chapter, including spectrogram representation, decomposition by ICA, and clustering. Some authors use the term ISA also to refer to methods where some other algorithm than ICA is used for the factorization [648].

The general ISA procedure consists of the following steps:

1. Calculate the magnitude spectrogram $\tilde{\mathbf{X}}$ (or some other representation) of the input signal.
2. Apply PCA[2] on the matrix $\tilde{\mathbf{X}}$ of size $(K \times T)$ to estimate the number of components N and to obtain whitening and dewhitening matrices $\mathbf{V}$ and $\mathbf{V}^+$, respectively. A centred, decorrelated, and dimensionally reduced observation matrix $\mathbf{X}$ of size $(N \times T)$ is obtained as $\mathbf{X} = \mathbf{V}(\tilde{\mathbf{X}} - \mu\mathbf{1}^\mathsf{T})$, where $\mathbf{1}$ is a all-ones vector of length T.
3. Apply ICA to estimate an unmixing matrix $\mathbf{W}$. $\mathbf{B}$ and $\mathbf{G}$ are obtained as $\mathbf{B} = \mathbf{W}^{-1}$ and $\mathbf{G} = \mathbf{WX}$.
4. Invert the decorrelation operation in Step 2 in order to get the mixing matrix $\tilde{\mathbf{B}} = \mathbf{V}^+\mathbf{B}$ and source matrix $\tilde{\mathbf{G}} = \mathbf{G} + \mathbf{WV}\mu\mathbf{1}^\mathsf{T}$ for the original observations $\tilde{\mathbf{X}}$.
5. Cluster the projected components to sources (see Section 9.7.1).

The above steps are explained in more detail below. Depending on the application, not all of them may be necessary. For example, prior information can be used to set the number of components in Step 2.

The basic ICA is not directly suitable for the separation of one-channel signals, since the number of sensors has to be larger than or equal to the number of sources. Short-time signal processing can be used in an attempt to overcome this limitation. Taking a frequency transform such as DFT, each frequency bin can be considered as a sensor which produces an observation in each frame. With the standard linear ICA model (9.6), the signal is modelled as a sum of components, each of which has a static spectrum (or some other basis function) and a time-varying gain.

The spectrogram factorization has its motivation in invariant feature extraction, which is a technique proposed by Kohonen [356]. The short-time spectrum can be viewed as a set of features calculated from the input signal. As discussed in Section 9.2.2, it is often desirable to have shift-invariant basis

---

[2]Singular value decomposition can also be used to estimate the number of components [73].

functions, such as the magnitude or power spectrum [356], [303]. Multidimensional ICA (explained below) is used to separate phase-invariant features into *invariant feature subspaces*, where each source is modelled as the sum of one or more components [303].

Multidimensional ICA [64] is based on the same linear generative model (9.6) as ICA, but the components are not assumed to be mutually independent. Instead, it is assumed that the components can be divided into disjoint sets, so that the components within each set may be dependent on each other, while dependencies between sets are not allowed. One approach to estimating multidimensional independent components is to first apply standard ICA to estimate the components, and then group them into sets by measuring dependencies between them.[3]

ICA algorithms aim at maximizing the independence of the elements of the source vector $\hat{\mathbf{g}} = \mathbf{W}\mathbf{x}$. In ISA, the elements correspond to the time-varying gains of each component. However, the objective can also be the independence of the spectra of components, since the roles of the mixing matrix and gain matrix can be swapped by $\mathbf{X} = \mathbf{B}\mathbf{G} \Leftrightarrow \mathbf{X}^\mathsf{T} = \mathbf{G}^\mathsf{T}\mathbf{B}^\mathsf{T}$. The independence of both the time-varying gains and basis functions can be obtained by using the spatiotemporal ICA algorithm [612]. There are no exhaustive studies regarding different independence criteria in monaural audio source separation. Smaragdis argued that in the separation of complex sources, the criterion of independent time-varying gains is better, because of the absence of consistent spectral characteristics [598]. FitzGerald reported that the spatiotemporal ICA did not produce significantly better results than normal ICA, which assumes the independence of gains or spectra [186].

The number of frequency channels is usually larger than the number of components to be estimated with ICA. PCA or singular value decomposition (SVD) of the spectrogram can be used to estimate the number of components automatically. SVD decomposes the spectrogram into a sum of components with a fixed spectrum and time-varying gain, so that the spectra and gains of different components are orthogonal, whereas PCA results in the orthogonality of either the spectra or the gains. The components with the largest singular values are chosen so that the sum of their singular values is larger than or equal to a pre-defined threshold $0 < \theta \leq 1$ [73].

ISA has been used for general audio separation by Casey and Westner [73], for the analysis of musical trills by Brown and Smaragdis [58], and for percussion transcription by FitzGerald et al. [187], to mention some examples.

9.3.2 Non-Negativity Restrictions

When magnitude or power spectrograms are used, the basis functions are magnitude or power spectra which are non-negative by definition. Therefore,

---

[3]ICA aims at maximizing the independence of the output variables, but it cannot guarantee their complete independence, as this depends also on the input signal.

it can be advantageous to restrict the basis functions to be entry-wise non-negative. Also, it may be useful not to allow negative gains, but to constrain the components to be purely additive. Standard ICA is problematic in the sense that it does not enable these constraints. In practice, ICA algorithms also produce negative values for the basis functions and gains, and often there is no physical interpretation for such components.

ICA with non-negativity restrictions has been studied for example by Plumbley and Oja [526], and the topic is currently under active research. Existing non-negative ICA algorithms can enforce non-negativity for the gain matrix but not for the mixing matrix. They also assume that the probability distribution of the source variables g_n is non-zero all the way down to zero, i.e., the probability $g_n < \delta$ is non-zero for any $\delta > 0$. The algorithms are based on a noise-free mixing model and in our experiments with audio spectrograms, they tended to be rather sensitive to noise.

It has turned out that the non-negativity restrictions alone are sufficient for the separation of the sources, without the explicit assumption of statistical independence. NMF algorithms are discussed in Section 9.5.

9.4 Sparse Coding

Sparse coding represents a mixture signal in terms of a small number of active elements chosen out of a larger set [486]. This is an efficient approach for learning structures and separating sources from mixed data. General discussion of sparse adaptive representations suitable for the analysis of musical signals is given in Chapter 3. In the linear signal model (9.4), the sparseness restriction is usually applied on the gains $\mathbf{G}$, which means that the probability of an element of $\mathbf{G}$ being zero is high. As a result, only a few components are active at a time and each component is active only in a small number of frames. In musical signals, a component can represent, e.g., all the equal-pitched notes of an instrument. It is likely that only a small number of pitches are played simultaneously, so that the physical system behind the observations generates sparse components.

In this section, a probabilistic framework is presented, where the source and mixing matrices are estimated by maximizing their posterior distributions. The framework is similar with the one presented by Olshausen and Field [486]. Several assumptions of, e.g., the noise distribution and prior distribution of the gains are used. Obviously, different results are obtained by using different distributions, but the basic idea is the same. The method presented here is also closely related to the algorithms proposed by Abdallah and Plumbley [4] and Virtanen [650], which were used in the analysis of music signals.

The posterior distribution of $\mathbf{B}$ and $\mathbf{G}$ given an observed spectrogram $\mathbf{X}$ is denoted by $p(\mathbf{B}, \mathbf{G}|\mathbf{X})$. The maximization of this can be formulated as [339, p. 351]

$$\max_{\mathbf{B},\mathbf{G}} p(\mathbf{B},\mathbf{G}|\mathbf{X}) \propto \max_{\mathbf{B},\mathbf{G}} p(\mathbf{X}|\mathbf{B},\mathbf{G})p(\mathbf{B},\mathbf{G}), \tag{9.9}$$

where $p(\mathbf{X}|\mathbf{B},\mathbf{G})$ is the probability of observing $\mathbf{X}$ given $\mathbf{B}$ and $\mathbf{G}$, and $p(\mathbf{B},\mathbf{G})$ is the joint prior distribution of $\mathbf{B}$ and $\mathbf{G}$. The concepts of probability distribution function, conditional probability distribution function, and maximum a posteriori estimation are described in Chapter 2.

For mathematical tractability, it is typically assumed that the noise (the residual term in (9.3)) is i.i.d.; independent from the model $\mathbf{BG}$, and normally distributed with variance σ^2 and zero mean. The likelihood of $\mathbf{B}$ and $\mathbf{G}$ (see Section 2.2.5 for the eplanation of likelihood functions) can be written as

$$p(\mathbf{X}|\mathbf{B},\mathbf{G}) = \prod_{t,k} \frac{1}{\sigma\sqrt{2\pi}} \exp\left(-\frac{([\mathbf{X}]_{k,t} - [\mathbf{BG}]_{k,t})^2}{2\sigma^2}\right). \tag{9.10}$$

It is further assumed here that $\mathbf{B}$ has a uniform prior, so that $p(\mathbf{B},\mathbf{G}) \propto p(\mathbf{G})$. Each time-varying gain $[\mathbf{G}]_{n,t}$ is assumed to have a sparse probability distribution function of the exponential form

$$p([\mathbf{G}]_{n,t}) = \frac{1}{Z} \exp\left(-f([\mathbf{G}]_{n,t})\right). \tag{9.11}$$

A normalization factor Z has to be used so that the density function sums to unity. The function f is used to control the shape of the distribution and is chosen so that the distribution is uni-modal and peaked at zero with heavy tails. Some examples are given later.

For simplicity, all the entries of $\mathbf{G}$ are assumed to be independent from each other, so that the probability distribution function of $\mathbf{G}$ can be written as a product of the marginal densities:

$$p(\mathbf{G}) = \prod_{n,t} \frac{1}{Z} \exp\left(-f([\mathbf{G}]_{n,t})\right). \tag{9.12}$$

It is obvious that in practice the gains are not independent of each other, but this approximation is done to simplify the calculations. From the above definitions we get

$$\max_{\mathbf{B},\mathbf{G}} p(\mathbf{B},\mathbf{G}|\mathbf{X}) \propto \max_{\mathbf{B},\mathbf{G}} \prod_{t,k} \frac{1}{\sigma\sqrt{2\pi}} \exp\left(-\frac{([\mathbf{X}]_{k,t} - [\mathbf{BG}]_{k,t})^2}{2\sigma^2}\right)$$
$$\times \prod_{n,t} \frac{1}{Z} \exp\left(-f([\mathbf{G}]_{n,t})\right). \tag{9.13}$$

By taking a logarithm, the products become summations, and the exp-operators and scaling terms can be discarded. This can be done since logarithm is order preserving and therefore does not affect the maximization. The sign is changed to obtain a minimization problem

$$\min_{\mathbf{B},\mathbf{G}} \sum_{t,k} \frac{([\mathbf{X}]_{k,t} - [\mathbf{BG}]_{k,t})^2}{2\sigma^2} + \sum_{n,t} f([\mathbf{G}]_{n,t}), \qquad (9.14)$$

which can be written as

$$\min_{\mathbf{B},\mathbf{G}} \frac{1}{2\sigma^2} ||\mathbf{X} - \mathbf{BG}||_F^2 + \sum_{n,t} f([\mathbf{G}]_{n,t}), \qquad (9.15)$$

where the Frobenius norm of a matrix is defined as

$$||\mathbf{Y}||_F = \sqrt{\sum_{i,j} [\mathbf{Y}]_{i,j}^2}. \qquad (9.16)$$

In (9.15), the function f is used to penalize 'active' (non-zero) entries of **G**. For example, Olshausen and Field [486] suggested the functions $f(x) = \log(1 + x^2)$, $f(x) = |x|$, and $f(x) = x^2$. In audio source separation, Benaroya et al. [32] and Virtanen [650] have used $f(x) = |x|$. The prior distribution used by Abdallah and Plumbley [2], [4] corresponds to the function

$$f(x) = \begin{cases} |x|, & |x| \geq \mu, \\ \mu(1 - \alpha) + \alpha|x|, & |x| < \mu, \end{cases} \qquad (9.17)$$

where the parameters μ and α control the relative mass of the central peak in the prior, and the term $\mu(1-\alpha)$ is used to make the function continuous at $x = \pm\mu$. All these functions give a smaller cost and a higher prior probability for gains near zero. The cost function $f(x) = |x|$ and the corresponding Laplacian prior $\mathsf{p}(x) = \frac{1}{2}\exp(-|x|)$ are illustrated in Fig. 9.3. Systematic large-scale evaluations of different sparse priors in audio signals have not been carried out. Naturally, the distributions depend on source signals, and also on the data representation.

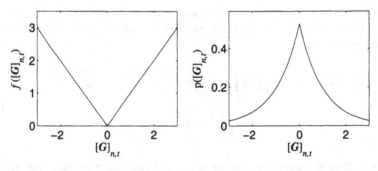

Fig. 9.3. The cost function $f(x) = |x|$ (left) and the corresponding Laplacian prior distribution $\mathsf{p}(x) = \frac{1}{2}\exp(-|x|)$ (right). Values of **G** near zero are given a smaller cost and a higher probability.

From (9.15) and the above definitions of f, it can be seen that a sparse representation is obtained by minimizing a cost function which is the weighted sum of the reconstruction error term $||\mathbf{X} - \mathbf{BG}||_F^2$ and the term which incurs a penalty on non-zero elements of $\mathbf{G}$. The variance σ^2 is used to balance between these two. This objective 9.15 can be viewed as a penalized likelihood, discussed in the Tools section (see Sections 2.2.9 and 2.3.3).

Typically, f increases monotonically as a function of the absolute value of its argument. The presented objective requires that the scale of either the basis functions or the gains is somehow fixed. Otherwise, the second term in (9.15) could be minimized without affecting the first term by setting $\mathbf{B} \leftarrow \mathbf{B}\theta$ and $\mathbf{G} \leftarrow \mathbf{G}/\theta$, where the scalar $\theta \rightarrow \infty$. The scale of the basis functions can be fixed for example with an additional constraint $||\mathbf{b}_n|| = 1$, as done by Hoyer [299], or the variance of the gains can be fixed.

The minimization problem (9.15) is usually solved using iterative algorithms. If both $\mathbf{B}$ and $\mathbf{G}$ are unknown, the cost function may have several local minima, and in practice reaching the global optimum in a limited time cannot be guaranteed. Standard optimization techniques based on steepest descent, covariant gradient, quasi-Newton, and active-set methods can be used. Different algorithms and objectives are discussed for example by Kreutz-Delgado et al. [373].

If $\mathbf{B}$ is fixed, more efficient optimization algorithms can be used. This can be the case for example when $\mathbf{B}$ is learned in advance from training material where sounds are presented in isolation. These methods are discussed in Section 9.6.

No methods have been proposed for estimating the number of sparse components in a monaural audio signal. Therefore, N has to be set either manually, using some prior information, or to a value which is clearly larger than the expected number of sources. It is also possible to try different numbers of components and to determine a suitable value of N from the outcome of the trials.

As discussed in the previous section, non-negativity restrictions can be used for frequency-domain basis functions. With a sparse prior and non-negativity restrictions, one has to use the projected steepest descent algorithms which are discussed, e.g., by Bertsekas in [35, pp. 203–224]. Hoyer [299], [300] proposed a non-negative sparse coding algorithm by combining NMF and sparse coding. His algorithm used a multiplicative rule to update $\mathbf{B}$, and projected steepest descent to update $\mathbf{G}$. Projected steepest descent alone is computationally inefficient compared to multiplicative update rules, for example.

In musical signal analysis, sparse coding has been used for example by Abdallah and Plumbley [4], [5] to produce an approximate piano-roll transcription of synthesized harpsichord music and by Virtanen [650] to transcribe drums in polyphonic music signals synthesized from MIDI. Also, Blumensath and Davies used a sparse prior for the gains, even though their system was based on a different signal model [43]. The framework also enables the use of further assumptions. Virtanen used a cost function which included a term that favoured the temporal continuity of gains by making large gain changes between adjacent frames unlikely [650].

9.5 Non-Negative Matrix Factorization

As discussed in Section 9.3.2 (see p. 277), it is reasonable to restrict frequency-domain basis functions and their gains to non-negative values. In the signal model $\mathbf{X} \approx \mathbf{BG}$, the element-wise non-negativity of $\mathbf{B}$ and $\mathbf{G}$ alone is a sufficient condition for the separation of sources in many cases, without an explicit assumption of the independence of the sources.

Paatero and Tatter proposed an NMF algorithm in which the weighted energy of the residual matrix $\mathbf{X} - \mathbf{BG}$ was minimized by using a least-squares algorithm where $\mathbf{B}$ and $\mathbf{G}$ were alternatingly updated under non-negativity restrictions [492]. More recently, Lee and Seung [399, 400] proposed NMF algorithms which have been used in several machine learning tasks since the algorithms are easy to implement and modify.

Lee and Seung proposed two cost functions and estimation algorithms to obtain $\mathbf{X} \approx \mathbf{BG}$ [400]. The cost functions are the square of the Euclidean distance d_{euc} and divergence d_{div}, which are defined as

$$d_{\text{euc}}(\mathbf{B}, \mathbf{G}) = ||\mathbf{X} - \mathbf{BG}||_F^2 \qquad (9.18)$$

and

$$d_{\text{div}}(\mathbf{B}, \mathbf{G}) = \sum_{k,t} D([\mathbf{X}]_{k,t}, [\mathbf{BG}]_{k,t}), \qquad (9.19)$$

where the function D is defined as

$$D(p, q) = p \log \frac{p}{q} - p + q. \qquad (9.20)$$

Both cost functions are lower-bounded by zero, which is obtained only when $\mathbf{X} = \mathbf{BG}$. It can be seen that the Euclidean distance is equal to the first term in (9.15). Minimization of the Euclidean distance leads to a maximum likelihood estimator for $\mathbf{B}$ and $\mathbf{G}$ in the presence of Gaussian noise. Similarly, minimization of the divergence (9.19) leads to a maximum likelihood estimator, when the observations are generated by a Poisson process with mean value $[\mathbf{BG}]_{k,t}$ [399]. When $\sum_{k,t}[\mathbf{X}]_{k,t} = \sum_{k,t}[\mathbf{BG}]_{k,t} = 1$, the divergence (9.19) is equal to the Kullback–Leibler divergence, which is widely used as a distance measure between probability distributions [400].

The estimation algorithms of Lee and Seung minimize the chosen cost function by initializing the entries of $\mathbf{B}$ and $\mathbf{G}$ with random positive values, and then by updating them iteratively using multiplicative rules. Each update decreases the value of the cost function until the algorithm converges, i.e., reaches a local minimum. Usually, $\mathbf{B}$ and $\mathbf{G}$ are updated alternately.

The update rules for the Euclidean distance are given as

$$\mathbf{B} \leftarrow \mathbf{B}.\times(\mathbf{XG}^\mathsf{T})./(\mathbf{BGG}^\mathsf{T}) \tag{9.21}$$

and

$$\mathbf{G} \leftarrow \mathbf{G}.\times(\mathbf{B}^\mathsf{T}\mathbf{X})./(\mathbf{B}^\mathsf{T}\mathbf{BG}), \tag{9.22}$$

where $.\times$ and $./$ denote the element-wise multiplication and division, respectively. The update rules for the divergence are given as

$$\mathbf{B} \leftarrow \mathbf{B}.\times\frac{(\mathbf{X}./\mathbf{BG})\mathbf{G}^\mathsf{T}}{\mathbf{1G}^\mathsf{T}} \tag{9.23}$$

and

$$\mathbf{G} \leftarrow \mathbf{G}.\times\frac{\mathbf{B}^\mathsf{T}(\mathbf{X}./\mathbf{BG})}{\mathbf{B}^\mathsf{T}\mathbf{1}}, \tag{9.24}$$

where $\mathbf{1}$ is an all-ones K-by-T matrix, and $\frac{\mathbf{X}}{\mathbf{Y}}$ denotes the element-wise division of matrices $\mathbf{X}$ and $\mathbf{Y}$.

To summarize, the algorithm for NMF is as follows:

Algorithm 9.1: Non-Negative Matrix Factorization

1. Initialize each entry of $\mathbf{B}$ and $\mathbf{G}$ with the absolute values of Gaussian noise.
2. Update $\mathbf{G}$ using either (9.22) or (9.24) depending on the chosen cost function.
3. Update $\mathbf{B}$ using either (9.21) or (9.23) depending on the chosen cost function.
4. Repeat Steps (2)–(3) until the values converge.

Methods for the estimation of the number of components have not been proposed, but all the methods suggested in Section 9.4 are applicable in NMF, too. The multiplicative update rules have proven to be more efficient than for example the projected steepest-descent algorithms [400], [299], [5].

NMF can be used only for a non-negative observation matrix and therefore it is not suitable for the separation of time-domain signals. However, when used with the magnitude or power spectrogram, the basic NMF can be used to separate components without prior information other than the element-wise non-negativity. In particular, factorization of the magnitude spectrogram using the divergence often produces relatively good results. The divergence cost of an individual observation $[\mathbf{X}]_{k,t}$ is linear as a function of the scale of the input, since $D(\alpha p, \alpha q) = \alpha D(p, q)$ for any positive scalar α, whereas for the Euclidean cost the dependence is quadratic. Therefore, the divergence is more sensitive to small-energy observations.

NMF does not explicitly aim at components which are statistically independent from each other. However, it has been proved that under certain conditions, the non-negativity restrictions are theoretically sufficient for separating statistically independent sources [525]. It has not been investigated whether musical signals fulfill these conditions, and whether NMF implement a suitable estimation algorithm. Currently, there is no comprehensive theoretical explanation of why NMF works so well in sound source separation. If a mixture spectrogram is a sum of sources which have a static spectrum with a time-varying gain, and each of them is active in at least one frame and frequency line in which the other components are inactive, the objective function of NMF is minimized by a decomposition in which the sources are separated perfectly. However, real-world music signals rarely fulfill these conditions. When two or more more sources are present simultaneously at all times, the algorithm is likely to represent them with a single component.

In the analysis of music signals, the basic NMF has been used by Smaragdis and Brown [600], and extended versions of the algorithm have been proposed for example by Virtanen [650] and Smaragdis [599]. The problem of the large dynamic range of musical signals has been addressed e.g. by Abdallah and Plumbley [5]. By assuming multiplicative gamma-distributed noise in the power spectral domain, they derived the cost function

$$D(p,q) = \frac{p}{q} - 1 + \log \frac{q}{p}, \qquad (9.25)$$

to be used instead of (9.20). Compared to the Euclidean distance (9.18) and divergence (9.20), this distance measure is more sensitive to low-energy observations. In our simulations, however, it did not produce results as good as the Euclidean distance or the divergence did.

9.6 Prior Information about Sources

Manual transcription of music requires a lot of prior knowledge and training. The described separation algorithms used some general assumptions about the sources in the core algorithms, such as independence or non-negativity, but also other prior information on the sources is often available. For example in the analysis of pitched musical instruments, it is known in advance that the spectra of instruments are approximately harmonic. Unfortunately, it is difficult to implement harmonicity restrictions in the models discussed earlier.

Prior knowledge can also be source-specific. The most common approach to incorporate prior information about sources in the analysis is to train source-specific basis functions in advance. Several approaches have been proposed. The estimation is usually done in two stages, which are

1. Learn source-specific basis functions from training material, such as monotimbral and monophonic music. Also the characteristics of time-varying gains can be stored, for example by modelling their distribution.
2. Represent a polyphonic signal as a weighted sum of the basis functions of all the instruments. Estimate the gains and keep the basis functions fixed.

It is not yet known whether automatic music transcription is possible without any source-specific prior knowledge, but obviously this has the potential to make the task much easier.

Several methods have been proposed for training the basis functions in advance. The most straightforward choice is to also separate the training signal using some of the described methods. For example, Jang and Lee [314] used ISA to train basis functions for two sources separately. Benaroya et al. [32] suggested the use of non-negative sparse coding, but they also tested using the spectra of random frames of the training signal as the basis functions or grouping similar frames to obtain the basis functions. They reported that non-negative sparse coding and the grouping algorithm produced the best results [32]. Gautama and Van Halle compared three different self-organizing methods in the training of basis functions [204].

The training can be done in a more supervised manner by using a separate set of training samples for each basis function. For example in the drum transcription systems proposed by FitzGerald et al. [188] and Paulus and Virtanen [505], the basis function for each drum instrument was calculated from isolated samples of each drum. It is also possible to generate the basis functions manually, for example so that each of them corresponds to a single pitch. Lepain used frequency-domain harmonic combs as the basis functions, and parameterized the rough shape of the spectrum using a slope parameter [403]. Sha and Saul trained the basis function for each discrete fundamental frequency using a speech database with annotated pitch [579].

In practice, it is difficult to train basis functions for all the possible sources beforehand. An alternative is to use trained or generated basis functions which are then adapted to the observed data. For example, Abdallah and Plumbley initialized their non-negative sparse coding algorithm with basis functions that consisted of harmonic spectra with a quarter-tone pitch spacing [5]. After the initialization, the algorithm was allowed to adapt these.

Once the basis functions have been trained, the observed input signal is represented using them. Sparse coding and non-negative matrix factorization techniques are feasible also in this task. Usually the reconstruction error between the input signal and the model is minimized while using a small number of active basis functions (sparseness constraint). For example, Benaroya et al. proposed an algorithm which minimizes the energy of the reconstruction error while restricting the gains to be non-negative and sparse [32].

If the sparseness criterion is not used, a matrix $\mathbf{G}$ reaching the global minimum of the reconstruction error can be usually found rather easily. If the gains are allowed to have negative values and the estimation criterion is the energy of the residual, the standard least-squares solution

$$\hat{\mathbf{G}} = (\mathbf{B}^{\mathsf{T}}\mathbf{B})^{-1}\mathbf{B}^{\mathsf{T}}\mathbf{X} \tag{9.26}$$

produces the optimal gains (assuming that the previously trained basis functions are linearly independent) [339, pp. 220–226]. If the gains are restricted to non-negative values, the least-squares solution is obtained using the non-negative least-squares algorithm [397, p. 161]. When the basis functions, observations, and gains are restricted to non-negative values, the global minimum of the divergence (9.19) between the observations and the model can be computed by applying the multiplicative update (9.24) iteratively [563], [505]. Lepain minimized the sum of the absolute value of the error between the observations and the model by using linear programming and the Simplex algorithm [403].

The estimation of the gains can also be done in a framework which increases the probability of basis functions being non-zero in consecutive frames. For example, Vincent and Rodet used hidden Markov models (HMMs) to model the durations of the notes [648].

It is also possible to train prior distributions for the gains. Jang and Lee used standard ICA techniques to train time-domain basis functions for each source separately, and modelled the probability distribution function of the component gains with a generalized Gaussian distribution which is a family of density functions of the form $\mathsf{p}(x) \propto \exp(-|x|^q)$ [314]. For an observed mixture signal, the gains were estimated by maximizing their posterior probability.

9.7 Further Processing of the Components

The main motivation for separating an input signal into components is that each component usually represents a musically meaningful entity, such as a percussive instrument or all the equal-pitched notes of an instrument. Separation alone does not solve the transcription problem, but has the potential to make it much easier. For example, estimation of the fundamental frequency of an isolated sound is easier than multiple fundamental frequency estimation in a mixture signal.

9.7.1 Associating Components with Sources

If the basis functions are estimated from a mixture signal, we do not know which component is produced by which source. Since each source is modelled as a sum of one or more components, we need to associate the components to sources. There are roughly two ways to do this. In the unsupervised classification framework, component clusters are formed based on some similarity measure, and these are interpreted as sources. Alternately, if prior information about the sources is available, the components can be classified to sources based on their distance to source models. Naturally, if pre-trained basis functions are used for each source, the source of each basis function is known and classification is not needed.

Pairwise dependence between the components can be used as a similarity measure for clustering. Even in the case of ICA, which aims at maximizing the independence of the components, some dependencies may remain because it is possible that the input signal contains fewer independent components than are to be separated.

Casey and Westner used the symmetric Kullback–Leibler divergence between the probability distribution functions of basis functions as a distance measure, resulting in an independent component cross-entropy matrix (an 'ixegram') [73]. Dubnov proposed a distance measure derived from the higher-order statistics of the basis functions or the gains [157]. Casey and Westner [73] and Dubnov [157] also suggested clustering algorithms for grouping the components into sources. These try to minimize the inter-cluster dependence and maximize the intra-cluster dependence.

For predefined sound sources, the association can be done using pattern recognition methods. Uhle et al. extracted acoustic features from each component to classify them either to a drum track or to a harmonic track [634]. The features in their system included, for example, the percussiveness of the time-varying gain, and the noise-likeness and dissonance of the spectrum. Another system for separating drums from polyphonic music was proposed by Helén and Virtanen. They trained a support vector machine (SVM) using the components extracted from a set of drum tracks and polyphonic music signals without drums. Different acoustic features were evaluated, including the above-mentioned ones, mel-frequency cepstral coefficients, and others [282].

9.7.2 Extraction of Musical Information

The separated components are usually analysed to obtain musically important information, such as the onset and offset times and fundamental frequency of each component (assuming that they represent individual notes of a pitched instrument). Naturally, the analysis can be done by synthesizing the components and by using analysis techniques discussed elsewhere in this book. However, the synthesis stage is usually not needed, but analysis using the basis functions and gains directly is likely to be more reliable, since the synthesis stage may cause some artifacts.

The onset and offset times of each component n are measured from the time-varying gains $g_{n,t}$, $t = 1 \ldots T$. Ideally, a component is active when its gain is non-zero. In practice, however, the gain may contain interference from other sources and the activity detection has to be done with a more robust method.

Paulus and Virtanen [505] proposed an onset detection procedure that was derived from the psychoacoustically motivated method of Klapuri [347]. The gains of a component were compressed, differentiated, and lowpass filtered. In the resulting 'accent curve', all local maxima above a fixed threshold were considered as sound onsets. For percussive sources or other instruments with a strong attack transient, the detection can be done simply by locating local maxima in the gain functions, as done by FitzGerald et al. [188].

The detection of sound offsets is a more difficult problem, since the amplitude envelope of a note can be exponentially decaying. Methods to be used in the presented framework have not been proposed.

There are several different possibilities for the estimation of the fundamental frequency of a pitched component. For example, prominent peaks can be located from the spectrum and the two-way mismatch procedure of Maher and Beauchamp [428] can be used, or the fundamental period can be estimated from the autocorrelation function which is obtained by inverse Fourier transforming the power spectrum. In our experiments, the enhanced autocorrelation function proposed by Tolonen and Karjalainen [627] was found to produce good results (see p. 253 in this volume). In practice, a component may represent more than one pitch. This happens especially when the pitches are always present simultaneously, as is the case in a chord, for example. No methods have been proposed to detect this situation. Whether or not a component is pitched can be estimated, e.g., from features based on the component [634], [282].

Some systems use fixed basis functions which correspond to certain fundamental frequency values [403], [579]. In this case, the fundamental frequency of each basis function is of course known.

9.7.3 Synthesis

Synthesis of the separated components is needed at least when one wants to listen to them, which is a convenient way to roughly evaluate the quality of the separation. Synthesis from time-domain basis functions is straightforward: the signal of component n in frame t is generated by multiplying the basis function $\mathbf{b}_n$ by the corresponding gain $g_{n,t}$, and adjacent frames are combined using the overlap-add method where frames are multiplied by a suitable window function, delayed, and summed.

Synthesis from frequency-domain basis functions is not as trivial. The synthesis procedure includes calculation of the magnitude spectrum of a component in each frame, estimation of the phases to obtain the complex spectrum, and an inverse discrete Fourier transform (IDFT) to obtain the time-domain signal. Adjacent frames are then combined using overlap-add. When magnitude spectra are used as the basis functions, framewise spectra are obtained as the product of the basis function with its gain. If power spectra are used, a square root has to be taken, and if the frequency resolution is not linear, additional processing has to be done to enable synthesis using the IDFT.

A few alternative methods have been proposed for the phase generation. Using the phases of the original mixture spectrogram produces good synthesis quality when the components do not overlap significantly in time and frequency [651]. However, applying the original phases and the IDFT may produce signals which have unrealistic large values at frame boundaries, resulting in perceptually unpleasant discontinuities when the frames are combined using overlap-add. The phase generation method proposed by Griffin and Lim [259] has also been used in synthesis (see for example Casey [70]). The method finds phases so that the error between the separated magnitude spectrogram and the magnitude spectrogram of the resynthesized time-domain signal is minimized in the least-squares sense. The method can produce good synthesis quality especially for slowly varying sources with deterministic phase behaviour. The least-squares criterion, however, gives less importance to low-energy partials and often leads to a degraded high-frequency content. The phase generation problem has been recently addressed by Achan et al., who proposed a phase generation method based on a pre-trained autoregressive model [9].

9.8 Time-Varying Components

As mentioned above, the linear model (9.1) is efficient in the analysis of music signals since many musically meaningful entities can be rather well approximated with a fixed spectrum and a time-varying gain. However, representation of sources with strongly time-varying spectrum requires several components, and each fundamental frequency value produced by a pitched instrument has to be represented with a different component. Instead of using multiple components per source, more complex models can be constructed which allow either a time-varying spectrum or a time-varying fundamental frequency for each component. These are discussed in the following two subsections.

9.8.1 Time-Varying Spectra

Time-varying spectra of components can be obtained by replacing each basis function $\mathbf{b}_n$ by a sequence of basis functions $\mathbf{b}_{n,\tau}$, where $\tau = 0 \ldots L - 1$ is the frame index. If a frequency-domain representation is used, this means that a static short-time spectrum of a component is replaced by a spectrogram of length L frames.

The signal model for one component can be formulated as a convolution between its spectrogram and time-varying gain. The model for a mixture spectrum of N components is given by

$$\mathbf{x}_t \approx \sum_{n=1}^{N} \sum_{\tau=0}^{L-1} \mathbf{b}_{n,\tau} g_{n,t-\tau}. \tag{9.27}$$

The model can be interpreted so that each component n consists of repetitions of an event which has a spectrogram $\mathbf{b}_{n,\tau}$, $\tau = 0...L-1$. Each non-zero value of the time-varying gain $g_{n,t}$ denotes an onset of the event and the value of the gain gives the scaling factor of each repetition. A simple two-note example is illustrated in Fig. 9.4.

The parameters of the convolutive model (9.27) can be estimated using methods extended from NMF and sparse coding. In these, the reconstruction error between the model and the observations is minimized, while restricting the parameters to be entry-wise non-negative. Also favouring sparse gains is clearly reasonable, since real-world sound events set on in a small number of

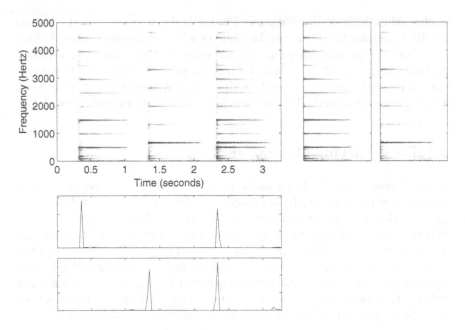

Fig. 9.4. An example of the convolutive model (9.27) which allows time-varying components. The mixture spectrogram (upper left panel) contains the notes C#6 and F#6 of the acoustic guitar, first played separately and then together. The upper right panels illustrate the learned note spectrograms and the lower panel shows their time-varying gains. In the gains, an impulse corresponds to the onset of a note. The components were estimated using a modified version of the algorithm proposed by Smaragdis in [599]. In the case of more complex signals, it is difficult to obtain such clear impulses.

frames only. Virtanen [651] proposed an algorithm which is based on non-negative sparse coding, whereas that of Smaragdis [599] aims at minimizing the divergence between the observation and the model while constraining non-negativity.

Arbitrarily long durations L may not be used if the basis functions are estimated from a mixture signal. When $NL \geq T$, the input spectrogram can be represented perfectly as a sum of concatenated event spectrograms (without separation). Meaningful sources are likely to be separated only when $NL \ll T$. In other words, estimation of several components with large L requires long input signals.

In addition, the method proposed by Blumensath and Davies [43] can be formulated using (9.27). Their objective was to find sparse and shift-invariant decompositions of a signal in the time domain. Their model allows an event to begin at any time with one sample accuracy which makes the number of free parameters in the model large. To reduce the dimensionality of the problem, Blumensath and Davies proposed an algorithm which carried out the optimization in a subspace of the parameters. They also included a sparse prior for the gains.

9.8.2 Time-Varying Fundamental Frequencies

In some cases, it is desirable to use a model which can represent different pitch values of an instrument with a single component. For example, in the case where a note with a certain pitch is present only during a short time, separating it from co-occurring sources is difficult. However, if other notes of the source with adjacent pitch values can be utilized, the estimation becomes more reliable.

Varying fundamental frequencies are difficult to model using time-domain basis functions or frequency-domain basis functions with linear frequency resolution. This is because changing the fundamental frequency of a basis function is a non-linear operation which is difficult to implement in practice: if the fundamental frequency is multiplied by a factor γ, the frequencies of the harmonic components are also multiplied by γ; this can be viewed as a stretching of the spectrum. For an arbitrary value of γ, the stretching is difficult to perform on a discrete linear frequency resolution, at least using a simple operator which could be used in the unsupervised learning framework. The same holds as well for time-domain basis functions.

A logarithmic spacing of frequency bins makes it easier to represent varying fundamental frequencies. A logarithmic scale consists of discrete frequencies $f_{ref}\beta^{k-1}$, where $k = 1 \ldots K$ is the discrete frequency index, $\beta > 1$ is the ratio between adjacent frequency bins, and f_{ref} is a reference frequency in Hertz which can be selected arbitrarily. For example, $\beta = \sqrt[12]{2}$ produces a frequency scale where the spacing between the frequencies is one semitone.

On the logarithmic scale, the spacing of the partials of a harmonic sound is independent of its fundamental frequency. For fundamental frequency f_0, the overtone frequencies of a perfectly harmonic sound are mf_0, where $m > 0$ is an integer. On the logarithmic scale, the corresponding frequency indices are $k = \log_\beta(m) + \log_\beta(f_0/f_{ref})$, and thus the fundamental frequency affects only the offset $\log_\beta(f_0/f_{ref})$, not the intervals between the harmonics.

Given the spectrum $X(k)$ of a harmonic sound with fundamental frequency f_0, a fundamental frequency multiplication γf_0 can be implemented simply as a translation $\hat{X}(k) = X(k - \delta)$, where δ is given by $\delta = \log_\beta \gamma$. Compared with the stretching of the spectrum, this is usually easier to implement.

The estimation of harmonic spectra and their translations can be done adaptively by fitting a model onto the observations.[4] However, this is difficult for an unknown number of sounds and fundamental frequencies, since the reconstruction error as a function of translation δ has several local minima

---

[4]This approach is related to the fundamental frequency estimation method of Brown, who calculated the cross-correlation between an input spectrum and a single harmonic template on the logarithmic frequency scale [54].

at harmonic intervals, which makes the optimization procedure likely to become stuck in a local minimum far from the global optimum. A more feasible parameterization allows each component to have several active fundamental frequencies in each frame, the amount of which is to be estimated. This means that each time-varying gain $g_{n,t}$ is replaced by gains $g_{n,t,z}$, where $z = 0, \ldots, Z$ is a frequency-shift index and Z is the maximum allowed shift. The gain $g_{n,t,z}$ describes the amount of the nth component in frame t at a fundamental frequency which is obtained by translating the fundamental frequency of basis function $\mathbf{b}_n$ by z indices.

The size of the shift z depends on the frequency resolution. For example, if 48 frequency lines within each octave are used ($\beta = \sqrt[48]{2}$), $z = 4$ corresponds to a shift of one semitone. For simplicity, the model is formulated to allow shifts only to higher frequencies, but it can be formulated to allow both negative and positive shifts, too.

A vector $\mathbf{g}_{n,t} = \left[g_{n,t,0}, \ldots, g_{n,t,Z} \right]^{\mathsf{T}}$ is used to denote the gains of component n in frame t. The model can be formulated as

$$\mathbf{x}_t \approx \sum_{n=1}^{N} \mathbf{b}_n * \mathbf{g}_{n,t}, \qquad t = 1 \ldots T, \tag{9.28}$$

where $*$ denotes a convolution operator, defined between vectors as

$$\mathbf{y} = \mathbf{b}_n * \mathbf{g}_{n,t} \Leftrightarrow y_k = \sum_{z=0}^{Z} b_{n,k-z} g_{n,t,z}, \qquad k = 1 \ldots K. \tag{9.29}$$

Figure 9.5 shows the basis function and gains estimated from the example signal in Fig. 9.1. In general, the parameters can be estimated by fitting the model to observations with certain restrictions, such as non-negativity or sparseness. Algorithms for this purpose can be derived by extending those used in NMF and sparse coding. Here we present an extension of NMF, where the parameters are estimated by minimizing the divergence (9.19) between the observations $\mathbf{X}$ and the model (9.28), while restricting the gains and basis functions to be non-negative.

The elements of $\mathbf{g}_{n,t}$ and $\mathbf{b}_n$ are initialized with random values and then updated iteratively until the values converge. To simplify the notation, let us denote the model with current parameter estimates by $\mathbf{v}_t = \sum_{n=1}^{N} \mathbf{b}_n * \mathbf{g}_{n,t}$, $t = 1 \ldots T$. The update rule for the gains is given as

$$\mathbf{g}_{n,t} \leftarrow \mathbf{g}_{n,t} \cdot \times \frac{\mathbf{b}_n \star \left(\frac{\mathbf{x}_t}{\mathbf{v}_t} \right)}{\mathbf{b}_n \star \mathbf{1}}, \tag{9.30}$$

where $\mathbf{1}$ is a K-length vector of ones and $\star$ denotes the correlation of vectors, defined for real-valued vectors $\mathbf{b}_n$ and $\mathbf{y}$ as $\mathbf{g} = \mathbf{b}_n \star \mathbf{y} \Leftrightarrow g_z = \sum_{k=1}^{K} b_{n,k} y_{k+z}$, $z = 0, \ldots, Z$. The update rule for the basis functions is given as

$$\mathbf{b}_n \leftarrow \mathbf{b}_n . \times \frac{\sum_{t=1}^{T} (\mathbf{g}_{n,t} \star \frac{\mathbf{x}_t}{\mathbf{v}_t})}{\sum_{t=1}^{T} \mathbf{g}_{n,t} \star \mathbf{1}}. \tag{9.31}$$

The overall optimization algorithm for non-negative matrix deconvolution is as follows:

Algorithm 9.2: Non-Negative Matrix Deconvolution

1. Initialize each $\mathbf{g}_{n,t}$ and $\mathbf{b}_n$ with the absolute values of Gaussian noise.
2. Calculate $\mathbf{v}_t = \sum_{n=1}^{N} \mathbf{b}_n * \mathbf{g}_{n,t}$ for each $t = 1 \ldots T$.
3. Update each $\mathbf{g}_{n,t}$ using (9.30).
4. Calculate $\mathbf{v}_t$ as in Step 2.
5. Update each $\mathbf{b}_n$ using (9.31). Repeat Steps (2)–(5) until the values converge.

The algorithm produces good results if the number of sources is small, but for multiple sources and more complex signals, it is difficult to get as good

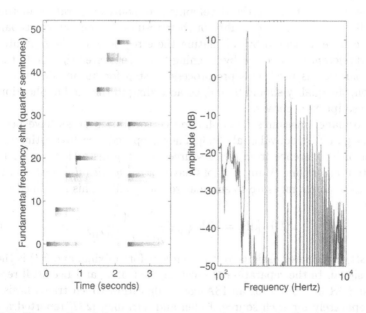

Fig. 9.5. Illustration of the time-varying gains (left) and the basis function (right) of a component that was estimated from the example signal in Fig. 9.1 containing a diatonic scale and a C major chord. On the left, the intensity of the image represents the value of the gain at each fundamental frequency shift and frame index. Here the fundamental frequencies of the notes can be seen more clearly than from the spectrogram of Fig. 9.1. The parameters were estimated using the algorithm proposed in this section.

results as those illustrated in Fig. 9.5. The model allows all the fundamental frequencies within the range $z = 0 \ldots Z$ to be active simultaneously, thus, it is not restrictive enough. For example, the algorithm may model a non-harmonic drum spectrum by using a harmonic basis function shifted to multiple adjacent fundamental frequencies. Ideally, this could be solved by restricting the gains to be sparse, but the sparseness criterion complicates the optimization.

In principle, it is possible to combine time-varying spectra and time-varying fundamental frequencies into the same model, but this further increases the number of free parameters so that it can be difficult to obtain good separation results.

When shifting the harmonic structure of the spectrum, the formant structure becomes shifted, too. Therefore, representing time-varying pitch by translating the basis function is appropriate only for nearby pitch values. It is unlikely that the whole fundamental frequency range of an instrument could be modelled by shifting a single basis function.

9.9 Evaluation of the Separation Quality

A necessary condition for the development of source separation methods is the ability to measure the quality of their results. In general, the separation quality can be measured by calculating the error between the separated signals and reference sources, or by listening to the separated signals. In the case that separation is used a pre-processing step for automatic music transcription, the quality should be judged according to the final application, i.e., the transcription accuracy.

Performance measures for audio source separation tasks have been discussed, e.g., by Gribonval et al. [258]. They proposed measures estimating the amount of interference from other sources and the distortion caused by the separation algorithm. Many authors have used the *signal-to-distortion ratio* (SDR) as a simple measure to summarize the quality. This is defined in decibels as

$$\text{SDR [dB]} = 10 \log_{10} \frac{\sum_t s(t)^2}{\sum_t [\hat{s}(t) - s(t)]^2}, \qquad (9.32)$$

where $s(t)$ is a reference signal of the source before mixing, and $\hat{s}(t)$ is the separated signal. In the separation of music signals, Jang and Lee [314] reported average SDR of 9.6 dB for an ISA-based algorithm which trains basis functions separately for each source. Helén and Virtanen [282] reported average SDR of 6.4 dB for NMF in the separation of drums and polyphonic harmonic track, and a clearly lower performance (SDR below 0 dB) for ISA.

In practice, quantitative evaluation of the separation quality requires that reference signals, i.e., the original signals $s(t)$ before mixing, be available. In the case of real-world music signals, it is difficult to obtain the tracks of each individual source instrument and, therefore, synthesized material is often used.

Generating test signals for this purpose is not a trivial task. For example, material generated using a software synthesizer may produce misleading results for algorithms which learn structures from the data, since many synthesizers produces notes which are identical at each repetition. In the case that source separation is a part of a music transcription system, quality evaluation requires that audio signals with an accurate reference notation are available (see Chapter 11, p. 355). Large-scale comparisons of different separation algorithms for music transcription have not been made.

9.10 Summary and Discussion

The algorithms presented in this chapter show that rather simple principles can be used to learn and separate sources from music signals in an unsupervised manner. Individual musical sounds can usually be modelled quite well using a fixed spectrum with time-varying gain, which enables the use of ICA, sparse coding, and NMF algorithms for their separation. Actually, all the algorithms based on the linear model (9.4) can be viewed as performing matrix factorization; the factorization criteria are just different.

The simplicity of the additive model makes it relatively easy to extend and modify it, along with the presented algorithms. However, a challenge with the presented methods is that it is difficult to incorporate some types of restrictions for the sources. For example, it is difficult to restrict the sources to be harmonic if they are learned from the mixture signal.

Compared to other approaches towards monaural sound source separation, the unsupervised methods discussed in this chapter enable a relatively good separation quality—although it should be noted that the performance in general is still very limited. A strength of the presented methods is their scalability: the methods can be used for arbitrarily complex material. In the case of simple monophonic signals, they can be used to separate individual notes, and in complex polyphonic material, the algorithms can extract larger repeating entities, such as chords. Some of the algorithms, for example NMF using the magnitude spectrogram representation, are quite easy to implement. The computational complexity of the presented methods may restrict their applicability if the number of components is large or the target signal is long.

Large-scale evaluations of the described algorithms on real-world polyphonic music recordings have not been presented. Most published results use a small set of test material and the results are not comparable with each other. Although conclusive evaluation data are not available, a preliminary experience from our simulations has been that NMF (or sparse coding with non-negativity restrictions) often produces better results than ISA. It was also noticed that prior information about sources can improve the separation quality significantly. Incorporating higher-level models into the optimization algorithms is a big challenge, but will presumably lead to better results. Contrary to the general view held by most researchers less than 10 years ago, unsupervised learning has proven to be applicable for the analysis of real-world music signals, and the area is still developing rapidly.

Part IV

Entire Systems, Acoustic and Musicological Modelling

10

Auditory Scene Analysis in Music Signals

Kunio Kashino

NTT Communication Science Laboratories,
Nippon Telegraph and Telephone Corporation
3–1, Morinosato-Wakamiya, Atsugi, 243-0198, Japan
kunio@eye.brl.ntt.co.jp

10.1 Introduction

This chapter discusses work done in the area of music scene analysis (MSA). Generally, scene analysis is viewed as the transformation of information from a sensory input (physical entity) into concepts (psychological or perceptual entities). Therefore, MSA is defined as a process that converts an audio signal into musical concepts such as notes, chords, beats, and rhythms. Related tasks include music transcription, pitch tracking, and beat tracking; however, this chapter focuses on the auditory scene analysis (ASA) related aspect of this process and does not explore the issues of pitch and beat tracking. An important idea related to this is the distinction between physical and perceptual sounds, as explained in Section 10.1.3 below.

10.1.1 Scene Analysis

We are exposed to various physical stimuli in our daily life. Our ears and eyes receive acoustic and optical stimuli, respectively. These stimuli originate in specific events or states. For example, when a ball hits a wall, vibrations in both the ball and the wall resulting from the impact travel through the air, and the air vibration arrives at our ears. We then understand that something like a ball has hit a hard surface such as a wall.

Understanding physical stimuli is an everyday experience. However, it poses an important question. An event such as a ball hitting a wall and causing physical phenomena such as air vibration is a natural process. However, how can we determine the events from the received physical phenomena? This is an inversion of the natural process, and therefore the solution to this problem is non-trivial.

Generally, a task that consists of recognizing an event or status from physical stimuli is called scene analysis, and scene analysis problems were first investigated in the visual domain.

In the 1920s, psychology researchers noticed the problem involved in determining how humans organize perceptual entities from visual stimuli. Their series of studies was referred to as 'Gestalt psychology'. In the 1960s and 1970s, pioneering work was undertaken on computer vision. For example, the problem of describing a simple scene comprising a set of building blocks with a simpler set of building blocks such as triangular and rectangular prisms was considered. Various operators for feature extraction were also proposed. In the 1970s and 1980s, many researchers addressed the problem of building generic scene analysis systems. In particular, a processing architecture comprising multiple processing modules and a common space for exchanging information among the modules was applied to scene analysis tasks. Such an architecture is called the blackboard architecture.

In the 1980s, Marr proposed a computational approach. He defined the vision problem in relation to the information processing needed to describe the visual world, and distinguished three levels of the task: computational theory, representation and algorithm, and hardware implementation [438]. Along with this computational approach, many researchers reported physical or mathematical formulations. Examples include work on 'shape-from-X' problems that address shape recovery by using various clues, denoted by X, such as motion, shading, and multiple images.

There is still a large amount of research being reported that relates to visual scene analysis. The amount of work on visual scene analysis from a generic perspective has decreased in recent years, and most of this work is dedicated to specific targets such as information retrieval, robot navigation, motion analysis, and encoding.

The history of auditory scene analysis is rather short compared with that of visual scene analysis, although the 'cocktail party problem'[1] was raised as early as the 1950s [86], [466]. In auditory psychophysics, various phenomena have been found that show that our auditory system has complex, adaptive, and active functions [662], [475], [270], [49], and auditory versions of Gestalt principles have been accumulated with respect to the way in which humans organize complex auditory stimuli into 'auditory streams' that correspond to distinct perceptual entities. However, from an engineering point of view, there are still many important problems to overcome in the auditory scene analysis field.

10.1.2 Music as an Auditory Scene

Music is a good domain for considering the auditory scene analysis problem not only from a cognitive perspective [36] but also from an engineering viewpoint. We use the term 'music scene analysis' to refer to auditory scene analysis for music.

---

[1]The cocktail party problem refers to the task of following the discussion of one's neighbours in a situation where lots of other sound sources are present, too.

The nature of music provides the first reason for considering it as an auditory scene. The overlapping of tones is a fundamental element of music. With the exception of a solo performance by a single instrument or voice, music usually comprises multiple simultaneous sounds played by single or multiple musical instruments. As humans we can appreciate such sound mixtures, implying that we recognize what is happening to a certain extent. Only a trained person is capable of transcribing music, but ordinary people can recognize a vocal line or the principal accompaniment when they listen to pop music. When we hear music played by a flute and a piano, it is easy to distinguish the two instruments.

The second reason is the usefulness of the prospective applications. Currently, computers are not as good as humans at recognizing multiple simultaneous sounds. However, if computers are developed with this capability, various useful systems will be realized including automatic music transcription systems and automatic music indexing systems for unlabelled music archives.

A research topic closely related to music scene analysis is automatic music transcription [610] and pitch tracking. As already introduced in Chapter 1, an automatic music transcription system was reported as long ago as the mid 1970s [521]. From then until the mid 1980s, several systems were built that mainly targeted the transcription of monophonic melodies such as singing, or simple polyphonic music such as guitar duets [483], [542]. The main methodology employed in such work involved signal processing techniques such as the fast Fourier transform (FFT). This period can be considered as the pioneering era of music transcription. Although the systems targeted rather simple compositions, various problems were identified such as frequency and temporal fluctuations.

From the mid 1980s to the mid 1990s, the main target moved from monophonic music to rather complicated polyphonic music, such as piano compositions. With such signals, even determining the number of simultaneous notes is a hard task let alone extracting fundamental frequencies for each note. To overcome this problem, researchers pointed out that knowledge is required for such transcription [470], [81], [80]. The main methodology consisted of integrating symbolic knowledge and signal processing. For example, Katayose et al. built a rule-based automatic music transcription system for multiple simultaneous-note performances by a single instrument [335]. The system comprises a control module, a processing module, and a music analysis module. The control module is an inference engine performing rule-based reasoning and invoking the processing module that extracts fundamental frequencies and beat times. The music analysis module analyses musical characteristics such as melody, rhythm, chord transitions, and keys, and then its results are fed back to the control and processing modules. This type of approach to some degree parallels the methodology of visual scene analysis in use at that time as mentioned above. This period can be viewed as the system-oriented era of music transcription.

At that time, in the artificial intelligence field, it was recognized that there was a bottleneck as regards knowledge acquisition for rule-based inference engines. This means that it was difficult to prepare all the knowledge required for these systems, and the systems tended to fail to work properly when they faced a situation not dealt with by the installed knowledge.

Since the mid 1990s, a lot of research has targeted polyphonic music played by *multiple* musical instruments in, for example, orchestras or commercial pop music [611], [350], [353]. Music played on multiple instruments is even harder to transcribe than music played on a single instrument, mainly because there are fewer constraints on the instruments or because we have less prior knowledge. That is, we must estimate the most plausible transcription under highly ambiguous and uncertain situations. Researchers noticed that it is hard to deal with such problems solely by the rule-based approach, and this requirement has naturally inspired researchers to apply probabilistic modelling. We call this period the model-oriented era of music transcription.

As is widely recognized, music usually has simultaneous and temporally hierarchical structures. For example, multiple simultaneous notes form a chord, and a sequence of notes across multiple bars form a phrase. Such structures are naturally incorporated in probabilistic models.

This is the point that distinguishes music transcription from music scene analysis. The object of music transcription is to create a score from a musical audio signal. On the other hand, the object of music scene analysis is to recover hierarchical structures and describe the auditory entities encoded in the structures from a musical audio signal. The recognition of structures is not a prerequisite for creating a score. However, a score can be produced once a complete music structure has been obtained. From this viewpoint, music transcription can be a specific instance of music scene analysis.

The present author used the term music scene analysis in this sense, and proposed a music scene analysis system based on a probabilistic model in the mid 1990s [332]. We defined the problem as the estimation of the posterior probability distribution given an input audio signal and a set of prior knowledge encoded in internal models. Recently, other probabilistic approaches toward music analysis tasks have been emerging. For example, Goto highlighted a sub-symbolic[2] aspect in music scene analysis and specifically termed it music scene description [228]. As discussed in Chapter 11, his descriptions of, for example, predominant fundamental frequencies and beat times correspond to a primal level of a music scene representation structure, which is very important to address.

10.1.3 Perceptual Sounds and Their Structure

To specify the problem of music scene analysis a little more precisely, here we introduce the terms 'physical sound' and 'perceptual sound'. A physical

---

[2]A sub-symbolic description here means description that involves continuous quantity.

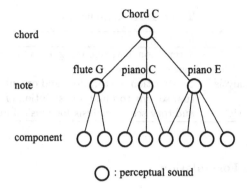

Fig. 10.1. An example snapshot of perceptual sounds in music.

sound means the vibration of the medium itself. It is a physical entity. On the other hand, there is another aspect to sound, namely *perceptual* sound, which is a psychological entity that corresponds to what human ear perceives as 'one' sound.

Suppose we are listening to ensemble music through a monaural speaker. In this case, there is one physical sound source. However, we will hear multiple perceptual sounds produced by multiple musical instruments. That is, the number of perceptual sound sources is greater than one.

In this sense, the concept of a perceptual sound is similar to that of an 'auditory stream', as used by Bregman [49]. However, a perceptual sound is not necessarily a stream but an entity in an intrinsically hierarchical structure. Acoustic energy must be organized in a structure if we are to understand the scene. In reality, what humans hear as one sound depends on time, place, occasion, and even attention. For example, when we listen to music, we hear multiple levels of perceptual sounds; sometimes we hear notes or a melody, and at other times we hear chords. As another example, when we are waiting for someone on a busy street, we may sometimes hear all the street noises as one sound, but sometimes we may hear car noises or people's footsteps as one sound. If we are specifically interested, we can focus on an individual car noise, and furthermore we may hear the specific car's engine noise, wind noise, or road noise individually.

Figure 10.1 shows an example snapshot of perceptual sounds in a music performance. In Fig. 10.1, a (perceptual) component corresponds to a (physical) frequency component, which is a continuous time-frequency region on a sound spectrogram. Figure 10.1 only shows a snapshot, but in fact there is also a time structure.

Table 10.1 lists the meanings of the terms used in this chapter.

Table 10.1. Terminology

Term	Meaning
perceptual sound	description corresponding to a sound that humans perceive as one
auditory scene analysis	construction, restoration, and organization of perceptual sounds to build a description of the world
music scene analysis	auditory scene analysis for musical audio signals

10.1.4 Problem Formulation

For a music scene analysis task, we should use multiple constraints or pieces of information stored in advance. For example, in Western tonal music, a sequence of chords does not appear at random but exhibits certain statistical characteristics, and multiple frequency components whose frequencies have a harmonic relationship tend to arise from 'one sound'.

Thus, the problem can be formulated as an a posteriori estimation. Let H be a set of random variables corresponding to internal states or perceptual sounds to be modelled, and the observation be x. Then, the task is generally written as

$$\hat{H} = \arg\max_H P(H|x) = \arg\max_H P(x|H)P(H). \qquad (10.1)$$

This is a Bayesian estimation of posteriors. Since it is very hard to calculate this in a general form, we must impose a structure on H. That is, when some elements of H can be considered to be independent, the calculations can take advantage of that fact. This point will be discussed later in this chapter.

10.2 Strategy for Music Scene Analysis

This section discusses the clues that may be used for music scene analysis. First, we review work on sound source separation. We then look at cues or information sources that have been utilized to associate time-frequency components to a same or to a different perceptual sound.

10.2.1 Sound Source Separation

From an engineering point of view, topics related to auditory scene analysis or music scene analysis include sound source separation. Since the aim of sound source separation is to separate the source signals, it is clearly a different task from auditory or music scene analysis where the goal is to obtain perceptual description of the content. However, one of the major difficulties in scene analysis is that we have to deal with mixtures of sounds. Therefore, if we can separate signals corresponding to the physical sound sources, it may assist scene analysis. To clarify the position of the music scene analysis problem, here we look at sound source separation, specifically from the viewpoint of applicability to music signals.

There are two kinds of problems considered in sound source separation: one assumes multiple input channels, and the other assumes one input channel (monaural). In the former case, it is usually further assumed that no prior knowledge is available about the source characteristics or mixing conditions. As a result, this is called blind source separation [63].

A powerful methodology for dealing with the blind source separation problem is independent component analysis (ICA), as described in Chapter 9. The basic idea is as follows. First, it is assumed that we observe linear combinations of source waveforms:

$$\mathbf{y}(n) = \mathbf{A}\mathbf{x}(n), \tag{10.2}$$

where the vector $\mathbf{y}(n)$ consists of the observed signals at time n, $\mathbf{A}$ is a time-invariant mixing matrix, and the vector $\mathbf{x}(n)$ represents the source signals at time n. Then, the problem is to estimate $\mathbf{A}^{-1}$. ICA is a method for estimating $\mathbf{A}^{-1}$ so that the source signals $\mathbf{x}(n)$ become as independent as possible from each other. Standard ICA methods require that the number of observed signals be larger or equal to the number of sources. In recent years, many researchers have been interested in sound source separation by ICA under more realistic conditions such as convolutive mixtures or cases where there are fewer microphones (i.e., observed signals $\mathbf{y}(n)$) than sound sources [28], [401].

For sound source separation with music signals, it is desirable for the separation to work even for one-channel or two-channel (monaural or stereo) signals [554], [73]. However, these cases require further research, as explored in Chapter 9.

Research on sound source separation for monaural inputs has been conducted since the 1970s. The most widely used approach is the selective resynthesis of frequency components. Specifically, selection by employing the harmonic frequency relationships as proposed by Parsons [498] has been widely used. This approach is based on prior knowledge of the sound sources, namely that sounds often consist of a fundamental frequency component and its overtones whose frequencies are integer multiples of the fundamental frequency. In reality, sounds such as voiced speech or pitched musical sounds consist of approximately harmonic components and, in such cases, the harmonic selection and resynthesis achieves reasonably good separation [482].

The selective resynthesis approach was mostly based on harmonic relations until around 1990. However, in the last two decades, rapid progress has been made on psychophysical research designed to clarify the mechanisms whereby humans recognize sounds. This has encouraged research efforts even from an engineering viewpoint. Specifically, a lot of work has tried to incorporate various clues in addition to harmonic relations for component selection, i.e., for associating components to a same source. The clues considered in such work have included: (1) common onset and offset, (2) common amplitude and frequency modulation, (3) estimated spatial locations, and (4) the

Table 10.2. Clues for the integration of simultaneous frequency components

Feature of frequency components	Promotes
harmonicity	fusion
harmonic mistuning	segregation
onset synchrony	fusion
onset asynchrony	segregation
offset synchrony	fusion
offset asynchrony	segregation
common amplitude modulation	fusion
common frequency modulation	fusion
spatial location	fusion/segregation
timbre similarity/dissimilarity	fusion/segregation

temporal and spectral proximity of frequency components. A major impetus behind the idea of using such auditory characteristics as clues for component selection or grouping was the accumulation of psychological findings, such as the ones introduced by Bregman [49]. Based on the title of his book, *Auditory Scene Analysis*, a series of engineering studies in this field was referred to as 'computational auditory scene analysis' [485].[3]

Here, we briefly introduce the clues that humans use (see also [475], [49], [108], [98] for details). It is known that certain types of frequency component behaviour can promote the perceptual fusion or segregation of those components, as listed in Table 10.2.

Many authors took these clues into account in their attempts to achieve toward sound source separation. Cooke [97] and Brown [52] incorporated psychophysical and physiological knowledge into their models. An input signal is analysed by a gammatone filterbank that simulates the frequency selectivity of the inner ear (see Chapter 8 for more details on auditory models). Then, the outputs of the filters are processed by Meddis's hair cell model to estimate the neural activity in the auditory nerve. Autocorrelation on the model output serves as a time-frequency energy representation. The system then extracts frequency components based on the cross-channel correlations in the energy representation. Onset and offset synchrony and the harmonic relations of the frequency components are used for grouping these components. Each of the resulting groups is considered to be a single sound, such as a speech, and resynthesized as a separate signal. They also applied their model to music signals [53].

Mellinger [460] also introduced a physiologically and psychologically motivated model of sound source separation. He extracted and utilized common

---

[3]This terminology is slightly misleading. Originally, scene analysis meant creating a descriptive representation from signals. However, a lot of work on computational auditory scene analysis appears to address signal separation rather than scene description.

frequency modulations of frequency components and tested the separation performance with music signals.

Nakatani [481], [480] reported a sound stream segregation system based on harmonic relations and spatial locations. Their system employs a multi-agent architecture and comprises three processing modules, namely a generator, a tracer, and a monitor. The generator generates a tracer when harmonically related and spatially localized frequency components are detected. The tracer tracks the harmonic components as long as they continue. The monitor controls the activities of the tracers to maintain global consistency. Then, the source-separated signals are presented to an automatic speech recognition system. They reported a significant improvement in the recognition accuracy compared with using the mixture signal directly.

In addition, work has also been reported based on mathematical formulations of the above-mentioned grouping rules [6], [636].

When considering sound source separation for music signals, it is important to note that frequency components in music tend to overlap other components. This is not limited to cases where the notes are exactly one octave apart in frequency. For example, Fig. 10.2 is the beginning part of a piece entitled 'Drei Duos' composed by Beethoven. The integral multiple of the fundamental frequency of the notes marked by × is the same as the fundamental frequency of a note existing at the same time in another part. Since music is often composed with notes at harmonic intervals, almost complete overlapping of frequency components frequently occurs even in a two-part composition as in this example. This means that the performance of the frequency component selection approach will be limited.

As described above, sound source separation for music signals will be more difficult than for other audio signals because the desired number of input channels is one or two, and the frequency components often overlap completely [309], [653]. This makes it hard to decompose the music transcription or the scene analysis problem into sound source separation and recognition of the separated signal. Therefore, many researchers have addressed the recognition or analysis of a sound mixture as it is, without a prior source separation process.

10.2.2 Bottom-Up Clues for Music Scene Analysis

As mentioned above, researchers have tried to incorporate the clues for human auditory scene analysis to sound source separation. It is also important to consider such psychological findings in music scene analysis tasks, because a perceptual sound is a psychological entity rather than a physical entity.

The bottom-up process of music scene analysis can be considered to be a clustering of frequency components. The idea of clustering based on such cues as harmonicity and synchrony has been employed by many authors for the bottom-up processing of auditory scene analysis methods [53], [460], [171].

To formulate the clustering, it is important to explore quantitative distance measures. There are at least two approaches to this problem: one is

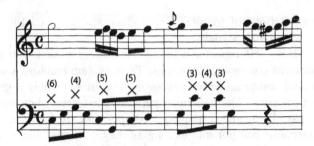

Fig. 10.2. Overlapping components in a composition: × indicates the notes whose fundamental frequencies are an integral fraction of those of the notes in the other part. The number shows the ratio of these fundamental frequencies.

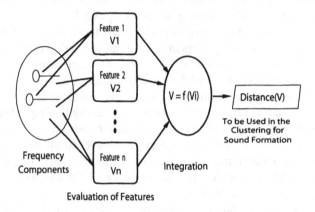

Fig. 10.3. The evaluation-integration model of sound formation.

to determine the measure directly from a psychometric function obtained for simplified stimuli, and the other is to learn the measure from many realistic sound samples. As an example, here we review an attempt using the former approach. Kashino et al. proposed the evaluation-integration model for the clustering for note formation [334]. Among the clues listed in Table 10.2, it focused on the harmonic mistuning and onset asynchrony.

As shown in Fig. 10.3, the model comprises two sequential steps. The first step involves independent evaluations of multiple features, and the second step integrates the results of these. Here, the evaluation means determining a distance measure between the clusters and the frequency components.

They determined the model parameters in three kinds of psychoacoustic experiments using stimuli comprising two frequency components. The experiments involved measuring: (1) the probability of segregation caused by harmonic mistuning, (2) the probability of segregation caused by onset asynchrony, and (3) the probability of segregation caused by both harmonic mistuning and onset asynchrony.

Two simultaneous frequency components are heard as one sound when they are harmonically related, but they tend to be heard as separate sounds when they are mistuned [476], [277]. In the first experiment, three subjects were presented with stimuli with various degree of mistuning in a random order and asked to choose whether each stimulus was one sound or two sounds. Then, a linear model that measures segregation probability was determined by the least squares fitting of the experimental results. The probability of segregation c_h was given by

$$c_h(u) = \begin{cases} -\dfrac{1}{p_-}\,u, & p_- < u < 0\,, \\ \dfrac{1}{p_+}\,u, & 0 \le u < p_+\,, \\ 1 & \text{otherwise.} \end{cases} \tag{10.3}$$

Here p_+ and p_- are parameters and u denotes the degree of mistuning in percent. That is,

$$u = \left(\frac{f_2}{2f_1} - 1\right) \times 100, \tag{10.4}$$

where f_1 and f_2 $(f_1 < f_2)$ are the frequencies of the components.

An experimental summary is shown in Fig. 10.4. The horizontal axis n in Fig. 10.4 is given by

$$n = \frac{|\log f_2 - \log(2f_1)|}{\log 1.005}. \tag{10.5}$$

No significant difference was found for p_+ and p_- and they are therefore not distinguished in Fig. 10.4.

Two harmonic frequency components are heard as one sound when they start simultaneously, but they tend to be heard as separate sounds when they are asynchronous. In the second experiment, the segregation probability was measured as a function of onset time/gradient difference, and a linear model was obtained by least squares fitting. The probability of segregation c_o was given by the following equation.

$$c_o(S) = \begin{cases} \dfrac{1}{S_p}\,S, & S < S_p\,, \\ 1 & \text{otherwise.} \end{cases} \tag{10.6}$$

$$S \ge 0.$$

Here, S is the area of the region surrounded by the amplitude envelopes of two frequency components projected onto the time-amplitude plane, as shown in Fig. 10.5. In this experiment, the amplitudes of frequency components after the onset part were chosen to be the same. The parameter S_p is given by

$$S_p = \frac{a}{f_1} + \frac{b}{g_1} + c, \tag{10.7}$$

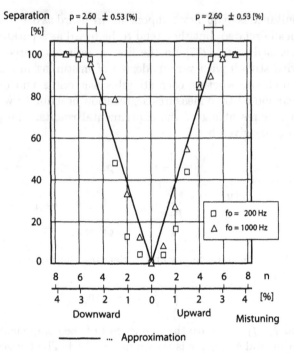

Fig. 10.4. Evaluation of harmonic mistuning.

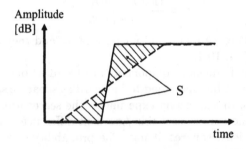

Fig. 10.5. An onset asynchrony model.

where f_1 and g_1 are the frequency and the onset gradient of the earlier frequency component, respectively. The parameters a, b, and c were obtained by regression analysis, leading to the values $a = 250$, $b = 1.11$, and $c = 0.317$. An example of the relation between the model and experimental results is shown in Fig. 10.6.

The two kinds of feature evaluations are then integrated by

$$m(u, S) = 1 - (1 - c_h(u))(1 - c_o(S)). \tag{10.8}$$

where $m(u, S)$ represents the probability of segregation. Equation (10.8) was obtained using Dempster's rule of combination [580].

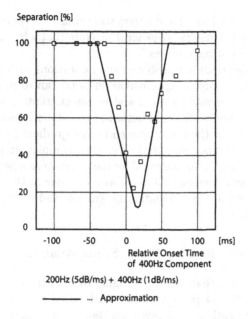

Fig. 10.6. An evaluation example of onset asynchrony.

In the third experiment, correlations were calculated between m given by (10.8) and the experimental results when there was both harmonic mistuning and onset asynchrony. The correlation values when the fundamental frequencies were 200 Hz (i.e., sounds comprising around 200 Hz and around 400 Hz components) and 1000 Hz (i.e., sounds comprising around 1000 Hz and around 2000 Hz components) were 0.84 and 0.75, respectively.

Equation (10.8) is the model when there are only two frequency components. Kashino et al. introduced an approximation, and used m in (10.8) as a distance measure for the clustering of frequency components even for more realistic conditions [334]. The clustering algorithm, performed for each processing window, was as follows.

1. Find the frequency component F_1 that has the lowest frequency and let it be the cluster centre C_1.
2. Scan frequency components from a lower frequency to a higher frequency, and let the next unscanned component be F_i. Then, calculate m_{ij}, which is the distance between F_i and the existing cluster centre C_j. If m_{ij} is greater than m_θ for all j, then the F_i is chosen as a new cluster centre C_{j+1}.
3. Repeat Step 2 until all frequency components have been scanned.
4. For each cluster centre, let all frequency components belong to the cluster if the distance between the components and the cluster centre is less than m_θ.

Here, m_θ is a threshold for sound segregation chosen between 0 and 1. If it is close to 0, then many clusters are generated. Note that a frequency component can be shared by multiple clusters.

It is obvious that there are problems and limitations with the above model. First, the function is only an approximation of human responses in a specific and simplified experimental setting, and is not guaranteed to be sufficiently accurate in other, more realistic, situations. Second, with real music signals, it is difficult to determine the onset time and onset gradient for each component because frequency components tend to exhibit complex amplitude patterns. Third, a sound without a fundamental frequency, such as a percussive sound or a missing-fundamental sound, will not form a cluster in the above algorithm. Fourth, other cues such as sequential integration and timbre memories [333] are not incorporated.

10.2.3 Top-Down Clues for Music Scene Analysis

Bregman pointed out that hearing a signal in a mixture of sounds depends on conscious effort and prior learning, and called the process behind this *schema-based* segregation and integration [49]. From an engineering viewpoint, the schema-based process often corresponds to knowledge-based or top-down processing.

Here, we distinguish between the ideas of knowledge-based and top-down processing. The words bottom-up and top-down signify the direction of the processing between the levels of abstraction. In bottom-up processing, low-level features are transformed to a higher level to form larger-scale or more abstract entities, whereas in top-down processing, entities at a high-abstraction level predict, verify, or control lower-level description. On the other hand, knowledge-based means that the process uses some prior information. That is, both bottom-up and top-down processes can be knowledge-based in the sense that they can use prior information stored in the system. To clarify this, in this chapter we will use the term 'internal model' to refer to such prior information.

Many researchers have addressed the integration of bottom-up and top-down processing modules. Regarding auditory scene analysis tasks, Lesser et al. built a sophisticated audio signal understanding system based on the blackboard architecture [407]. The blackboard architecture was employed to integrate and control the different processing modules; for example, the system creates high-level hypotheses which are then used for tuning the front-end signal processing parameters.

Godsmark et al. proposed a blackboard model for computational auditory scene analysis [215]. It was designed to accommodate various grouping cues including pitch proximity, timbral similarity, and source-specific knowledge such as metre and melodic phrases. The grouping mechanisms interact in a context-sensitive and retroactive[4] manner on a blackboard containing eight levels of abstraction. For example, the model allows high-level predictions about metre and melody to influence the primitive organization at lower levels of the blackboard.

Ellis presented a prediction-driven architecture for computational auditory scene analysis [171]. His motivation included the detection of non-tonal events; if we use harmonicity to group frequency components as reviewed above, non-harmonic sounds such as drums, noises, and clicks will not be recognized as one sound. His idea was to generate both noise component and periodic component hypotheses and then to look at the acoustic features in order to either justify or reject the hypotheses. His system was able to simulate old-plus-new heuristics [49],[5] sequential integration of successive tone events into a stream [49], and auditory restoration phenomenon, which means that a partially masked sound event tends to be restored based on the elements before and after the masking event [662]. Recently, the idea has been further extended based on a quantitative formulation [539].

For bottom-up and top-down integration it is essential to formulate a quantitative method that clarifies what is computed or optimized by the algorithm. In the following sections in this chapter, we review probabilistic approaches that form a quantitative base for integration. It should be noted, however, that sometimes the rules can still be effective even in a probabilistic framework.

10.3 Probabilistic Models for Music Scene Analysis

Recently, researchers have proposed probabilistic models for music scene analysis. This section begins by introducing the Bayesian network, which is a flexible tool for obtaining probabilistic inferences. Then, its applications to music scene analysis are reviewed.

10.3.1 Posterior Estimation by the Bayesian Network

A Bayesian network is a directed acyclic graph (DAG) where the nodes correspond to random variables and the links between the nodes encode probabilistic dependences between corresponding random variables [509], [205].

A random variable corresponds to an event to be modelled. A directed link, represented by an arrow, shows the direction of the probabilistic dependency. The origin of the arrow is called the parent and the end point is called the child. Each link can encode conditional probabilities, which are the probabilities of

---

[4]In a retroactive system, the interpretation of previous material can be affected by what happens afterwards.

[5]Old-plus-new heuristics loosely refers to the principle that, whenever possible, a change in the signal is interpreted as a continuation of the previously played sounds plus new sound elements.

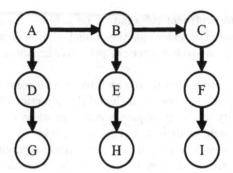

Fig. 10.7. An example Bayesian network.

child events given the parent events. The word acyclic means that there is no route from any node that returns to the original node as long as the links are followed in the designated direction. Such a graph can be singly or multiply connected. Singly connected means there is only one path between any two nodes in the graph. Otherwise the graph is referred to as multiply connected. When a graph is directed and singly connected, it is called a tree if none of the nodes has more than one parent.

The objective of considering such a network is to calculate posteriors efficiently after some of the random variables have been fixed or observed. The absence of a link between two nodes means that there is no direct relationship between the corresponding random variables. Even if these nodes are linked indirectly (i.e., via other nodes), they are independent when at least one node existing between the two is fixed. Posterior calculation on the Bayesian network takes advantage of this property.

Figure 10.7 shows an example of a Bayesian network. If the network is singly connected as in the figure, the posterior calculation is straightforward. As an example, assume we wish to find the posterior probabilities induced at node B. Letting D_B^- represent the data contained in the tree rooted at B and D_B^+ for the data contained in the rest of the network, we have

$$P(B) = P(B|D_B^+, D_B^-). \tag{10.9}$$

Based on the singly connected structure of the Bayesian network, we assume the independence of random variables given B, namely

$$P(D_B^+, D_B^-|B) = P(D_B^+|B)P(D_B^-|B), \tag{10.10}$$

and then, according to Bayes's theorem, we have

$$P(B|D_B^+, D_B^-) = \alpha P(D_B^-|B)P(B|D_B^+), \tag{10.11}$$

where the constant $\alpha = 1/P(D_B^-, D_B^+)$ can be determined so that the left-hand side of (10.11) is normalized.

Substituting as $\lambda(B) = P(D_B^-|B)$ and $\pi(B) = P(B|D_B^+)$, Equation (10.11) can be written as

$$P(B) = \alpha\lambda(B)\pi(B). \tag{10.12}$$

Now we want to obtain $\lambda(B)$ and $\pi(B)$. First, we consider $\lambda(B)$. Denoting the data encoded in the tree rooted at the k-th child of B by D^{k-}, we have

$$\lambda(B) = P(D_B^-|B) \tag{10.13}$$

$$= \beta \prod_k P(D^{k-}|B) \tag{10.14}$$

$$= \beta \prod_k \lambda_k(B), \tag{10.15}$$

where β is a normalization constant that can be chosen so that the sum $\sum_j \lambda(b_j)$ equals 1. Here b_j denote the different values of B. In the last equation, we defined $\lambda_k(B) = P(D^{k-}|B)$. In a singly connected graph, there is no direct link between children, and therefore given a fixed parent node, the children are independent. This makes it possible to factor the children. If we assume E is the k-th child and e_i are different values of E, we have

$$P(D^{k-}|B)=\sum_i P(D_E^-|B, e_i)P(e_i|B) \tag{10.16}$$

$$=\sum_i P(D_E^-|e_i)P(e_i|B) \tag{10.17}$$

$$=\sum_i \lambda(e_i)P(e_i|B), \tag{10.18}$$

and therefore, using conditional probabilities of children given the parent such as $P(e_i|b_j)$ and (10.15), we obtain λ's node by node. Here, $P(e_i|b_j)$ is to be obtained from a statistical model or data, that is learned by or provided to the system.

Now we consider $\pi(B)$. We have

$$\pi(B)=P(B|D_B^+) \tag{10.19}$$

$$=\sum_i P(B|a_i, D_B^+) \, P(a_i|D_B^+) \tag{10.20}$$

$$=\sum_i P(B|a_i) \, P(a_i|D_B^+) \tag{10.21}$$

$$=\sum_i P(B|a_i) \left\{ \gamma\pi(a_i) \prod_m \lambda_m(a_i) \right\}, \tag{10.22}$$

where m is a suffix enumerating the siblings of B except for B, a_i are different values of A, and γ is a normalization constant. The last equation follows from

$$P(A|D_B^+) = \frac{P(A)}{P(D_A^-|A)}. \tag{10.23}$$

The term in parenthesis in (10.22) is already calculated when $P(A)$ is calculated. This means that π can also be calculated node by node, using the conditional probabilities of children given the parent such as $P(b_j|a_i)$.

Now we have shown that $\lambda(X)$ can be derived from λ(children of X) and $\pi(X)$ from π(parent of X), using conditional probabilities P(child|parent) between two adjacent nodes. This allows us to calculate posteriors using (10.12). The calculation of $\lambda(\cdot)$ and $\pi(\cdot)$ can be viewed as the propagation of diagnostic and causal support for X, respectively.

In the above discussion, we assumed that the network is singly connected. If the network is multiply connected, the algorithm presented above will not properly terminate due to the loops of the probability propagation paths. Various methods have been developed to obtain the posteriors for such cases, including the junction tree algorithm and approximative methods using Monte Carlo sampling and variational methods [396].

The junction tree algorithm utilizes a tree whose nodes are a group of nodes in the original graph. First, the original graph is converted to an undirected graph by operations called moralization and triangulation. Then, a junction tree is constructed by substituting the cliques[6] in the converted graph with a node.

In the junction tree, posterior probabilities can be factored using a function defined for each node called a clique potential, and calculated with the probability propagation process designed for undirected trees.

10.3.2 Bayesian Networks Applied to Music Signals

Here, we review how the Bayesian networks were applied to a music scene analysis task. The first example is a processing model called Organized Processing Toward Integrated Music Scene Analysis (OPTIMA) [332]. The input of the model is assumed to be monaural music signals. The output is a music scene description, that is, a hierarchical representation of musical events such as frequency components, notes and chords. As shown in Fig. 10.8, the model consists of three blocks: (A) a pre-processing block, (B) a main processing block, and (C) internal models.

In the pre-processing block, frequency analysis is performed and a sound spectrogram is obtained. Then, frequency components are extracted. An example of the power transition of a frequency component is shown in Fig. 10.9. With complicated spectrum patterns, it is difficult to recognize the onset and offset times solely based on bottom-up information. Thus the system creates several terminal point candidates for each extracted component.

Rosenthal's rhythm recognition method [547] and Desain's time quantization method [140] are used to obtain rhythm information for the precise extraction of frequency components and recognition of the onset/offset time.

---

[6]A clique of a graph is its maximal complete subgraph. A complete graph is a graph in which each pair of nodes is connected by a link.

Main Processes   Internal Models

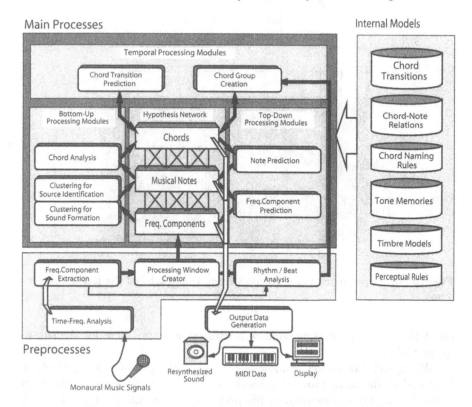

Fig. 10.8. An overview of the OPTIMA processing model.

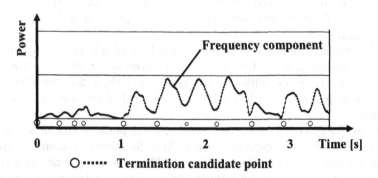

Fig. 10.9. A frequency component and its terminal point candidates.

Based on the integration of the beat probabilities and termination probabilities of terminal point candidates, the candidates are determined as being continuous or terminated, and consequently *processing windows* are formed. Here a processing window is a group of frequency components with similar

processing windows

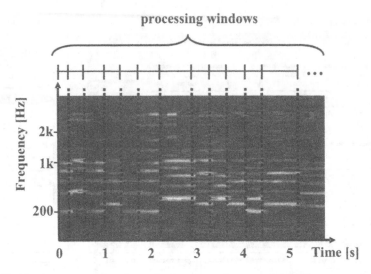

Fig. 10.10. A spectrogram of a polyphonic music excerpt and the processing windows.

onset times. The processing window is utilized as a time base for the subsequent main processes.

When each processing window is created in the pre-processing block, it is passed to the main processing block. The main block involves a Bayesian network. Exactly as discussed in the previous section, the Bayesian network has three layers: component, note, and chord levels. The chord level nodes are connected in time as the time proceeds. Each node in the network is a random variable that encodes multiple hypotheses. That is, the model holds hypotheses of the external acoustic events as a probability distribution in a hierarchical space.

The Bayesian network is actually built by multiple processing modules. The modules are classified into two types: those for creating the nodes and providing initial probabilities to the nodes, and those for providing conditional probabilities to the links. The former is called creators, and the latter predictors.

There are two creators in OPTIMA: a note hypothesis creator and a chord hypothesis creator. As described above, first, frequency component hypotheses and processing windows are created. Then the note hypothesis creator generates the hypotheses by referring to perceptual rules such as harmonic mistuning and onset asynchrony as described in the previous section. The creator also consults timbre models for a timbre discrimination analysis to identify the sound source of each note. A chord hypothesis creator generates the chord hypotheses when note hypotheses are given. This creator refers to chord naming rules.

Table 10.3. Examples of the chord-note relation knowledge. The conditional probabilities P(note|chord) were obtained by statistical analysis of printed music.

Note	Given chord (5 examples)				
	A	A7	Am	Am7	Adim
A	.983 ± .023	.852 ± .074	1.00 ± .000	.933 ± .081	.781 ± .143
A#	.000 ± .000	.023 ± .031	.030 ± .058	.000 ± .000	.188 ± .135
B	.150 ± .064	.364 ± .101	.091 ± .098	.182 ± .132	.094 ± .101
C	.008 ± .016	.023 ± .031	.848 ± .122	1.00 ± .000	.656 ± .165
C#	.850 ± .064	.818 ± .081	.000 ± .000	.030 ± .058	.031 ± .060
D	.025 ± .028	.057 ± .048	.182 ± .132	.394 ± .167	.031 ± .060
D#	.067 ± .045	.023 ± .031	.030 ± .058	.000 ± .000	.406 ± .170
E	.842 ± .065	.545 ± .104	.909 ± .098	.879 ± .111	.031 ± .060
F	.017 ± .023	.023 ± .031	.000 ± .000	.030 ± .058	.094 ± .101
F#	.125 ± .059	.148 ± .074	.000 ± .000	.061 ± .081	.469 ± .173
G	.025 ± .028	.773 ± .088	.121 ± .111	.939 ± .081	.062 ± .084
G#	.075 ± .047	.045 ± .044	.000 ± .000	.000 ± .000	.156 ± .126

$\pm$: 95% reliable range

There are three predictors in OPTIMA that provide the conditional probabilities of a child, given a parent, to the links. A frequency component predictor calculates P(component|note). A note predictor evaluates P(note|chord). A chord transition predictor provides chord transition probabilities P(chord|chord). These processing modules use six types of internal models.

The *chord transition model* holds statistical information on chord progressions, under the tri-gram assumption (see Chapter 5 for further details on N-gram models). This dictionary is based on a statistical analysis of 206 traditional songs (all Western tonal music). It is used by a chord transition predictor.

The *chord-note relation* model is used by a note predictor. The model involves a database that stores the probabilities of notes that can be played in a given chord. This information is also obtained by a statistical analysis of the 206 songs. Part of the stored data is shown in Table 10.3.

The *chord naming rules* are based on music theory, and are used by a chord hypothesis creator to recognize chords when the hypotheses of played notes are given.

The *tone memory* stores instances of frequency component data of a single note played by various musical instruments such as a clarinet, flute, piano, trumpet, or violin at different degrees of loudness (forte, medium, piano), frequency range, and durations. This memory is used by a frequency component predictor.

The *timbre models* are formed in the feature space of the timbre. An eleven-dimensional feature space was created using principal component analysis, and

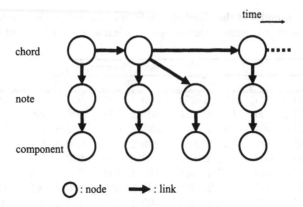

Fig. 10.11. An example of created Bayesian networks.

each of the above-mentioned musical instruments was modelled by a Gaussian distribution in the feature space. This model is used by a note hypothesis creator to provide the initial probabilities of sound sources (i.e., instrument names).

Finally, the *perceptual rules* are used by a note hypothesis creator to create note hypotheses from the frequency component hypotheses. The rules involve the clustering algorithm described in the previous section [334].

Each time a frequency component hypothesis is created in a processing window, the hypothesis creators create the node instances and links. When a link is created, the predictors provide conditional probabilities. Then, a probability propagation series is automatically performed and posterior probabilities at that time are calculated. As shown in Fig. 10.11, multiple successive note-level nodes can be linked to a single chord level node based on rhythm information extracted during the pre-processing stage. Thus, the instance of the network structure grows dynamically as the input signal arrives. Although the higher-level hypotheses (e.g. chords) are created based on lower-level information (e.g. notes), the higher-level information is still useful for the lower-level hypotheses, because it reflects and integrates a larger-scale context and information from the corresponding internal models.

In the OPTIMA model, temporal transitions are solely represented at the chord level. This simplifies the network structure to a tree, but generally this is insufficient because note-level temporal dependency is not considered.

The second example we review here is a music stream network [331], in which note-level temporal transitions were introduced. In this example, the Bayesian network was employed for sound source identification for ensemble music.[7]

---

[7]The work toward a musical sound source identification is discussed in Chapter 6 in this book.

Consider two musical notes n_k, n_{k-1} (k denotes the order of the onset times of these notes, n_{k-1} preceding n_k). An 'impedance' measure $Z(n_k, n_{k-1})$ is defined as

$$Z(n_k, n_{k-1}) = W \sum_i \left\{ - w_i \log P_i(n_k, n_{k-1}) \right\} , \qquad (10.24)$$

where i is a suffix that enumerates different terms of Z, P_i is the conditional probability of the occurrence of the transition from n_{k-1} to n_k in a given musical context, and w_i (> 0) is the weight for each term. Z can be viewed as a weighted sum of the amount of information, $-\log P_i$, delivered by the transition from n_{k-1} to n_k. Thus, Z reflects the infrequency of the transition for these two notes. Then, a 'music stream' [447] is formed as a sequence of musical notes that gives a local minimum of Z.

The term W is a time window that is defined as

$$W(\delta t) = \exp \left(\frac{\delta t}{\tau} \right) , \qquad (10.25)$$

where δt is the difference between the onset times of these two notes, and τ is a time constant. Unlike ordinary time windows, W becomes greater as δt increases. This loosely corresponds to the proximity rule of auditory stream organization as described in [49].

In this example, the following three Z factors are considered: (P_1) the transition probabilities of musical intervals, (P_2) the transition probabilities of timbres, and (P_3) the transition probabilities of musical roles.

The first factor is the musical interval probability. In tonal music, the musical intervals of note transitions do not appear equally often; some intervals are more frequent than others. Thus the pitch transition probability in a melody can be utilized as P_1 in (10.24). The probabilities P_1 were obtained from 397 melodies extracted from 196 pop scores and 201 jazz scores, where the total number of note transitions was 62,689. Figure 10.12 shows the estimated probabilities. The analysis was made only for the principal melodies and may not be precisely valid for the other melodies such as bass lines or parts arranged for polyphonic instruments such as the piano. For simplicity, however, probabilities shown in Fig. 10.12 for P_1 were used for all cases.

The second factor is the timbre transition probability. It is reasonable to suppose that a sequence of notes tends to be composed of notes that have similar timbres. To incorporate this tendency, a distance measure was defined between the timbres of two notes, so as to estimate the probability that two notes a certain distance apart would appear sequentially in a musical stream. These probabilities form P_2 in (10.24).

The distance between timbres is defined as the Euclidean distance between the timbre vectors in a timbre space. A timbre space can be spanned in several ways. In the experiment described in [331], each axis of the space corresponds to a musical instrument name, and a timbre vector is composed of correlation

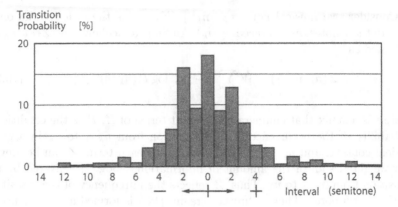

Fig. 10.12. Probabilities of musical intervals.

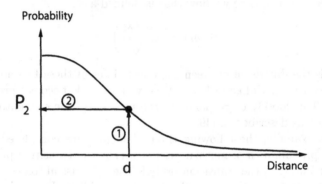

Fig. 10.13. Conversion from distance to probability. The calculated distance d is converted to P_2, which is the probability of the appearance of the distance in a sequence of musical notes, by using a histogram of distances. The histogram is normalized so that the histogram values add up to one.

values between the input signal and each of the template signals of musical instruments stored in advance. Then, the distances between the timbre vectors of successive notes in a sequence are translated into probabilities using a normalized histogram as explained in Fig. 10.13. This histogram models the distribution of the timbre vectors for successive notes.

The third factor is musical role consistency. In ensemble music, a sequence of notes can be regarded as carrying a musical role such as a principal melody or a bass line. To introduce such musical semantics, the probability P_3 is introduced:

$$P_3 = ar + b, \qquad (10.26)$$

where a and b are constants, and r is the rate of the highest (or lowest) notes in the music stream under consideration. Equation (10.26) represents a musical heuristic that the music stream formed by the highest (lowest) notes tends to continue to flow to the highest (lowest) note.

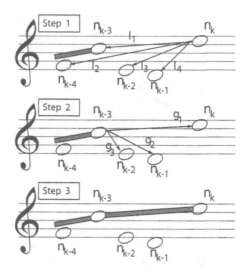

Fig. 10.14. A procedure for creating music stream networks. See text for details.

Using (10.24), the networks that correspond to sequences of musical notes are built by a procedure that is illustrated in Fig. 10.14:

1. When a new node n_k has just been created, the system first chooses the link that gives the minimum Z value (l_1) among the candidate links $(l_1, \cdots, l_4)$.
2. The system then evaluates Z values for the link candidates $(g_1, \cdots, g_3)$ from the selected node (n_{k-3}), to choose the link with the minimum Z value (g_1).
3. If g_1 and l_1 are identical, the link composes a music stream. If a music stream from n_{k-3} has already been formed in a direction other than g_1, the stream is cut; the direction of the music stream is changed to $g_1 (= l_1)$.

Thus the networks are built by connecting nodes that give the locally minimum Z value.

Once the network has been built, then it can be considered as a Bayesian network and the posterior probabilities of sound sources are calculated. An example of the system in operation is shown in Figure 10.15. The input here is a monaural recording of a real ensemble performance of 'Auld Lang Syne', a Scottish folk song, arranged in thee parts and performed by a violin, a flute, and a piano. Figure 10.15 displays the recognized music streams as well as the status of nodes for the beginning part of the song. The bars in each node indicate the probabilities at the node (not normalized). The links between the nodes are the extracted music streams. It is shown that each part is correctly recognized as the music stream. The thickness of the link line corresponds to its Z value given by (10.24); a thick line represents a link with a low Z value.

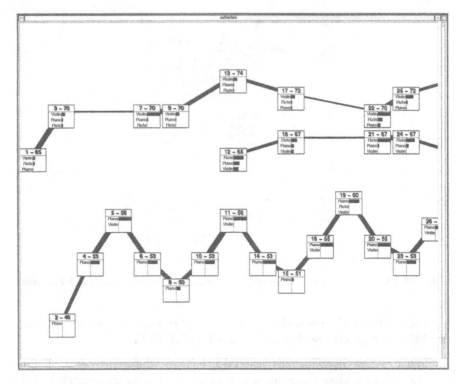

Fig. 10.15. An example of a created music stream network.

10.4 Conclusion: From Grouping to Generative Estimation

As shown throughout this chapter, early work on computational auditory scene analysis was rather directly inspired by psychological and physiological findings. In particular, frequency component grouping rules were intensively investigated. Various systems have been designed to simulate simultaneous, sequential, and schema-based grouping and integration.

Fortunately, probabilistic approaches have become computationally feasible in recent years [101], [672], [208]. For example, as discussed in Chapter 7, the Bayesian approach has been successfully applied to fundamental frequency estimation for music signals [212]. Although the first trials of the Bayesian approach for music scene analysis were simplified in terms of model structure and parameter exploration [332], the approach has been greatly extended in recent years. For example, Sterian developed a music transcription system based on a probabilistic framework [611]. The system tracks frequency components with a Kalman filter, which is equivalent to Bayesian estimation, and then it maintains multiple hypotheses for note formation. Cemgil proposed a generative model for music transcription [74], as described in Chapter 7.

We anticipate that it will be possible to explain certain auditory restoration phenomena in terms of Bayesian inference. When this is proven to be the case, the probabilistic inference approach to music scene analysis may be viewed as a newer version of the ASA-inspired approach.

11

Music Scene Description

Masataka Goto

National Institute of Advanced Industrial Science and Technology (AIST).
1-1-1 Umezono, Tsukuba, Ibaraki 305-8568, Japan. m.goto@aist.go.jp

11.1 Introduction

This chapter introduces a research approach called *'music scene description'* [232], [225], [228], where the goal is to build a computer system that can understand musical audio signals at the level of untrained human listeners without trying to extract every musical note from music. People listening to music can easily hum the melody, clap hands in time to the musical beat, notice a phrase being repeated, and find chorus sections. The brain mechanisms underlying these abilities, however, are not yet well understood. In addition, it has been difficult to implement these abilities on a computer system, although a system with them is useful in various applications such as music information retrieval, music production/editing, and music interfaces. It is therefore an important challenge to build a music scene description system that can understand complex real-world music signals like those recorded on commercially distributed compact discs (CDs).

Music scene description differs from two popular approaches to deal with music signals, sound source separation and traditional automatic music transcription (in the narrow sense[1]). Although these technologies are valuable from an engineering viewpoint, neither separation nor transcription is necessary or sufficient for understanding music.

- *It is possible to understand music without sound source separation.*
  The fact that human listeners understand various properties of audio signals is not necessarily evidence that the human auditory system extracts the audio signal of each individual source. Even if a mixture of two components cannot be separated, it can be understood from their salient features that the mixture includes them. In fact, from the viewpoint of auditory

---

[1]The term 'automatic music transcription' in this chapter refers to a traditional approach of transcribing all musical notes as a score, while the term 'automatic music transcription' in this book has a broader meaning including the music scene description as described in Chapter 1 of this volume.

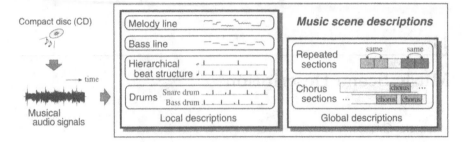

Fig. 11.1. Music scene descriptions.

psychology, it has been pointed out that human listeners do not perform sound source separation: perceptual sound source segregation is different from signal-level separation. For example, Bregman noted that 'there is evidence that the human brain does not completely separate sounds' [50]. The approach of developing methods for monaural or binaural sound source separation might deal with a hard problem which is not solved by any mechanism in this world (not solved even by the human brain).

- *It is possible to understand music without complete music transcription.*
  Music transcription, identifying the names (symbols) of musical notes, is a difficult skill mastered only by trained musicians. As pointed out by Goto [239], [240], [232] and Scheirer [565], untrained listeners understand music to some extent without mentally representing audio signals as musical scores. For example, as known from the observation that a listener who cannot identify the name and constituent notes of a chord can nevertheless feel the harmony and chord changes, a chord is perceived as combined whole sounds (tone colour) without reducing it to its constituent notes (like reductionism). Furthermore, even if it is possible to derive separated signals and musical notes, it would still be difficult to obtain high-level music descriptions like melody lines and chorus sections.

The music scene description approach therefore emphasizes methods that can obtain a certain description of a music scene from sound mixtures of various musical instruments in a musical piece. Here, it is important to discuss what constitutes an appropriate *description* of music signals. Since various levels of abstraction for the description are possible, it is necessary to consider which level is an appropriate first step towards the ultimate description in human brains. Goto [232], [228] proposed the following three viewpoints:

- An intuitive description that can be easily obtained by untrained listeners.
- A basic description that trained musicians can use as a basis for higher-level music understanding.
- A useful description facilitating the development of various practical applications.

According to these viewpoints, the following local and global descriptions (Fig. 11.1) have been proposed for Western music:

1. *Melody and bass lines*
   Melody and bass lines represent the temporal trajectory of the melody and bass. The melody is a series of single tones and is heard more distinctly than the rest. The bass is a series of single tones and is the lowest frequency part in polyphonic music. Note that a melody or bass line here is not represented as a series of musical notes; it is a continuous representation of fundamental frequency (F0, perceived as pitch) and power transitions. Only music with distinct melody and bass lines is dealt with for this description.

2. *Hierarchical beat structure*
   Hierarchical beat structure represents the fundamental temporal structure of music and comprises the quarter-note (beat) and measure levels—i.e., the positions of quarter-note beats and bar lines (corresponding to the metrical levels of 'beat' and 'bar' in Fig. 4.1, p. 106).

3. *Drums*
   Drums represent onset times of principal drum sounds, such as bass and snare drums. Their temporal patterns form drum patterns. Only music with drum sounds is dealt with for this description.

4. *Chorus sections and repeated sections*
   Chorus sections represent the most representative, uplifting, and prominent thematic sections in the structure of a musical piece (especially in popular music). Since chorus sections are usually repeated, they are represented as a list of the start and end points of every chorus section. Repeated sections represent the repetition of temporal regions with various lengths. Only music with distinct repeated choruses, such as popular music, is dealt with for the description of chorus sections, while any music can be dealt with for the description of repeated sections.

The idea behind these descriptions came from introspective observation of how untrained listeners listen to music. The following sections introduce methods for producing these descriptions from music signals such as CD recordings, which contain simultaneous sounds of various instruments (with or without drum sounds). In general, these methods deal with monaural audio signals because stereo signals on CDs can be easily converted to monaural signals by averaging the left and right channels. While methods depending on stereo information [24] can have better performance than methods dealing with monaural signals, such stereo-based methods cannot be applied to monaural signals. Methods assuming monaural signals, on the other hand, can be applied to stereo signals and be considered essential to music understanding since human listeners have no difficulty understanding the above descriptions even from monaural signals.

11.2 Estimating Melody and Bass Lines

The estimation of melody and bass lines is important because the melody forms the core of Western music and is very influential in the identity of a musical piece, while the bass is closely related to the tonality (see Chapter 1). These lines are fundamental to the perception of music by both musically trained and untrained listeners. They are also useful in various applications such as automatic music indexing for information retrieval (e.g., searching for a song by singing a melody), computer participation in live human performances, musical performance analysis of outstanding recorded performances, and automatic production of accompaniment tracks for karaoke using CDs.

It is difficult to estimate the fundamental frequency (F0) of melody and bass lines in monaural sound mixtures from CD recordings. Most previous F0 estimation methods cannot be applied to this estimation because they require that the input audio signal contain just a single-pitch sound with aperiodic noise or that the number of simultaneous sounds be known beforehand. The main reason F0 estimation in sound mixtures is difficult is that, in the time-frequency domain, the frequency components of one sound often overlap the frequency components of simultaneous sounds. In popular music, for example, part of the voice's harmonic structure is often overlapped by harmonics (overtone partials) of the keyboard instrument or guitar, by higher harmonics of the bass guitar, and by noisy inharmonic frequency components of the snare drum. A simple method for locally tracing a frequency component is therefore neither reliable nor stable. Moreover, F0 estimation methods relying on the existence of the F0s frequency component (the frequency component corresponding to the F0) not only cannot handle the *missing fundamental*, but are also unreliable when the F0s frequency component is smeared by the harmonics of simultaneous sounds.

F0 estimation of melody and bass lines in CD recordings was first achieved in 1999 by Goto [232], [222], [228]. Goto proposed a real-time method called *PreFEst* (P̲re̲dominant-F̲0 E̲st̲imation method) which estimates the melody and bass lines in monaural sound mixtures. Unlike previous F0 estimation methods, PreFEst does not assume the number of sound sources, locally trace frequency components, or even rely on the existence of the F0s frequency component. PreFEst basically estimates the F0 of the most predominant harmonic structure—the most predominant F0 corresponding to the melody or bass line—within an intentionally limited frequency range of the input mixture. It simultaneously takes into consideration all possibilities for the F0 and treats the input mixture as if it contained all possible harmonic structures with different weights (amplitudes). To enable the application of statistical methods, the input frequency components are represented as a probability density function (pdf), called an *observed pdf*. The point is that the method regards the observed pdf as a weighted mixture of harmonic-structure tone

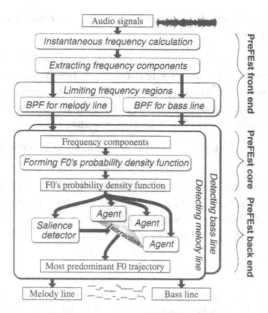

Fig. 11.2. Overview of PreFEst (<u>Pre</u>dominant-<u>F</u>0 <u>Est</u>imation method) for estimating melody and bass lines in CD recordings. In this figure, BPF denotes bandpass filtering.

models (represented by pdfs) of all possible F0s. It simultaneously estimates both their weights corresponding to the relative dominance of every possible harmonic structure and the shape of the tone models by maximum *a posteriori* probability (MAP) estimation (see Chapter 2, p. 40 for an introduction to MAP estimation methods) considering their prior distribution. It then considers the maximum-weight model as the most predominant harmonic structure and obtains its F0. The method also considers the F0s temporal continuity by using a multiple-agent architecture.

The following sections first explain the PreFEst method in detail and then introduce other methods for estimating the melody line developed by Paiva, Mendes, and Cardoso [494], [493], Marolt [435], [436], and Eggink and Brown [169], and a method for estimating the bass line developed by Hainsworth and Macleod [264]. Figure 11.2 shows an overview of PreFEst. PreFEst consists of three components, the *PreFEst front end* for frequency analysis, the *PreFEst core* to estimate the predominant F0, and the *PreFEst back end* to evaluate the temporal continuity of the F0. Since the melody line tends to have the most predominant harmonic structure in middle and high-frequency regions, and the bass line tends to have the most predominant harmonic structure in a low-frequency region, the F0s of the melody and bass lines can be estimated by applying the PreFEst core with appropriate frequency-range limitation.

11.2.1 PreFEst Front End: Forming the Observed Probability Density Functions

The PreFEst front end first uses a multirate filterbank to obtain adequate time-frequency resolution under a real-time constraint. By using an instantaneous frequency-related measure [84], [7], [338] for the existence of frequency components, it then extracts frequency components $\Psi^{(t)}(\nu)$ from the short-time Fourier transform (STFT) $X(\nu, t)$ of a signal

$$\Psi^{(t)}(\nu) = \begin{cases} |X(\nu, t)| & \text{if } \nu \text{ has a frequency component,} \\ 0 & \text{otherwise,} \end{cases} \tag{11.1}$$

where t is the time measured in units of frame-shifts (10 ms), and ν is the log-scale frequency denoted in units of *cents* (a musical-interval measurement). Frequency f_{Hz} in Hertz is converted to frequency f_{cent} in cents so that there are 100 cents to a tempered semitone and 1200 to an octave:

$$f_{\text{cent}} = 1200 \log_2 \frac{f_{\text{Hz}}}{440 \times 2^{\frac{3}{12}-5}}. \tag{11.2}$$

To obtain two sets of bandpass-filtered frequency components, one for the melody line (261.6–4186 Hz) and the other for the bass line (32.7–261.6 Hz) [228],[2] the PreFEst front end uses bandpass filters (BPFs) whose frequency response is $\text{BPF}_u(\nu)$ where u denotes the melody line ($u = $ 'melody') or the bass line ($u = $ 'bass line'). Each set of the bandpass-filtered components is finally represented as an *observed pdf* $\text{p}_\Psi^{(t)}(\nu)$

$$\text{p}_\Psi^{(t)}(\nu) = \frac{\text{BPF}_u(\nu)\,\Psi^{(t)}(\nu)}{\int_{-\infty}^{\infty} \text{BPF}_u(\eta)\,\Psi^{(t)}(\eta)\,d\eta}. \tag{11.3}$$

11.2.2 PreFEst Core: Estimating the F0s Probability Density Function

For each melody or bass line set of filtered frequency components represented as an observed pdf $\text{p}_\Psi^{(t)}(\nu)$, the PreFEst core forms a probability density function of the F0, called the *F0s pdf*, $\text{p}_{F0}^{(t)}(\nu_0)$, where ν_0 is the log-scale fundamental frequency in cents. The PreFEst core considers each observed pdf to have been generated from a weighted-mixture model of the tone models of all possible F0s; the tone model is the pdf corresponding to a typical harmonic structure and indicates where the harmonics (overtone partials) of the F0 tend to occur (Fig. 11.3). Because the weights of tone models represent the relative dominance of every possible harmonic structure, these weights can be regarded as the F0s pdf: the more dominant a tone model is in the mixture, the higher the probability of the F0 of its model.

---

[2]The method finds the F0 whose harmonics are most predominant in those limited frequency ranges. In other words, whether the F0 is within each limited range or not, PreFEst tries to estimate the F0 which is supported by predominant harmonic frequency components within that range.

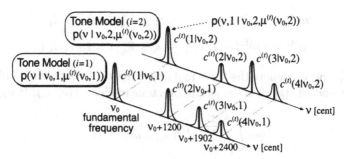

Fig. 11.3. Model parameters of multiple adaptive tone models $p(\nu|\nu_0, i, \boldsymbol{\mu}^{(t)}(\nu_0, i))$.

Weighted-Mixture Model of Adaptive Tone Models

To deal with diversity of the harmonic structure, the PreFEst core can use several types of harmonic-structure tone models. The pdf of the i-th tone model for each F0 ν_0 is denoted by $p(\nu|\nu_0, i, \boldsymbol{\mu}^{(t)}(\nu_0, i))$ (see Fig. 11.3), where the model parameter $\boldsymbol{\mu}^{(t)}(\nu_0, i)$ represents the shape of the tone model. The number of tone models is I_u (that is, $i = 1, \ldots, I_u$), where u denotes the melody line ($u = $ 'melody') or the bass line ($u = $ 'bass line'). Each tone model is defined by

$$p(\nu|\nu_0, i, \boldsymbol{\mu}^{(t)}(\nu_0, i)) = \sum_{m=1}^{M_u} p(\nu, m|\nu_0, i, \boldsymbol{\mu}^{(t)}(\nu_0, i)), \qquad (11.4)$$

$$p(\nu, m|\nu_0, i, \boldsymbol{\mu}^{(t)}(\nu_0, i)) = c^{(t)}(m|\nu_0, i)\, \mathcal{N}\big(\nu; \nu_0 + 1200 \log_2 m, \sigma_u^2\big), (11.5)$$

$$\boldsymbol{\mu}^{(t)}(\nu_0, i) = \{c^{(t)}(m|\nu_0, i) \mid m = 1, \ldots, M_u\}, \qquad (11.6)$$

where M_u is the number of harmonics considered, σ_u^2 is the variance of the Gaussian distribution $\mathcal{N}\big(\nu; \nu_0, \sigma_u^2\big)$ (see (2.16), p. 29 for a definition), and $c^{(t)}(m|\nu_0, i)$ determines the relative amplitude of the m-th harmonic component (the shape of the tone model) and satisfies

$$\sum_{m=1}^{M_u} c^{(t)}(m|\nu_0, i) = 1. \qquad (11.7)$$

In short, this tone model places a weighted Gaussian distribution at the position of each harmonic component.

The PreFEst core then considers the observed pdf $p_\Psi^{(t)}(\nu)$ to have been generated from the following model $p(\nu|\boldsymbol{\theta}^{(t)})$, which is a weighted mixture of all possible tone models $p(\nu|\nu_0, i, \boldsymbol{\mu}^{(t)}(\nu_0, i))$:

$$p(\nu|\boldsymbol{\theta}^{(t)}) = \int_{F_u^l}^{F_u^h} \sum_{i=1}^{I_u} w^{(t)}(\nu_0, i)\, p(\nu|\nu_0, i, \boldsymbol{\mu}^{(t)}(\nu_0, i))\, d\nu_0, \qquad (11.8)$$

$$\boldsymbol{\theta}^{(t)} = \{w^{(t)}, \boldsymbol{\mu}^{(t)}\}, \qquad (11.9)$$

$$w^{(t)} = \{w^{(t)}(\nu_0, i) \mid F_u^l \leq \nu_0 \leq F_u^h, i = 1, \ldots, I_u\}, \qquad (11.10)$$

$$\boldsymbol{\mu}^{(t)} = \{\mu^{(t)}(\nu_0, i) \mid F_u^l \leq \nu_0 \leq F_u^h, i = 1, \ldots, I_u\}, \qquad (11.11)$$

where F_u^l and F_u^h denote the lower and upper limits of the possible (allowable) F0 range and $w^{(t)}(\nu_0, i)$ is the weight of a tone model $p(\nu|\nu_0, i, \boldsymbol{\mu}^{(t)}(\nu_0, i))$ that satisfies

$$\int_{F_u^l}^{F_u^h} \sum_{i=1}^{I_u} w^{(t)}(\nu_0, i) \, d\nu_0 = 1. \qquad (11.12)$$

Because the number of sound sources cannot be known *a priori*, it is important to simultaneously take into consideration all F0 possibilities as expressed in (11.8). If it is possible to estimate the model parameter $\boldsymbol{\theta}^{(t)}$ such that the observed pdf $p_\Psi^{(t)}(\nu)$ is likely to have been generated from the model $p(\nu|\boldsymbol{\theta}^{(t)})$, the weight $w^{(t)}(\nu_0, i)$ can be interpreted as the F0s pdf $p_{F0}^{(t)}(\nu_0)$:

$$p_{F0}^{(t)}(\nu_0) = \sum_{i=1}^{I_u} w^{(t)}(\nu_0, i) \qquad (F_u^l \leq \nu_0 \leq F_u^h). \qquad (11.13)$$

Introducing a Prior Distribution

To use prior knowledge about F0 estimates and the tone model shapes, a prior distribution $p_{0u}(\boldsymbol{\theta}^{(t)})$ of $\boldsymbol{\theta}^{(t)}$ is defined as follows:

$$p_{0u}(\boldsymbol{\theta}^{(t)}) = p_{0u}(w^{(t)}) \, p_{0u}(\boldsymbol{\mu}^{(t)}), \qquad (11.14)$$

$$p_{0u}(w^{(t)}) = \frac{1}{Z_w} e^{-\beta_{wu}^{(t)} \, D_w(w_{0u}^{(t)}; w^{(t)})}, \qquad (11.15)$$

$$p_{0u}(\boldsymbol{\mu}^{(t)}) = \frac{1}{Z_\mu} e^{-\int_{F_u^l}^{F_u^h} \sum_{i=1}^{I_u} \beta_{\mu u}^{(t)}(\nu_0, i) \, D_\mu(\mu_{0u}^{(t)}(\nu_0, i); \mu^{(t)}(\nu_0, i)) \, d\nu_0}. \qquad (11.16)$$

Here, $p_{0u}(w^{(t)})$ and $p_{0u}(\boldsymbol{\mu}^{(t)})$ are unimodal distributions: $p_{0u}(w^{(t)})$ takes its maximum value at $w_{0u}^{(t)}(\nu_0, i)$ and $p_{0u}(\boldsymbol{\mu}^{(t)})$ takes its maximum value at $\mu_{0u}^{(t)}(\nu_0, i)$ $(= \{c_{0u}^{(t)}(m|\nu_0, i) \mid m = 1, \ldots, M_u\})$, where $w_{0u}^{(t)}(\nu_0, i)$ and $\mu_{0u}^{(t)}(\nu_0, i)$ are the most probable parameters. Figure 11.4 shows two examples of the most probable tone model shape parameters, $\mu_{0u}^{(t)}(\nu_0, i)$, used in Goto's implementation. Z_w and Z_μ are normalization factors, and $\beta_{wu}^{(t)}$ and $\beta_{\mu u}^{(t)}(\nu_0, i)$ are parameters determining how much emphasis is put on the maximum value. The prior distribution is not informative (i.e., it is uniform) when $\beta_{wu}^{(t)}$ and $\beta_{\mu u}^{(t)}(\nu_0, i)$ are 0, corresponding to the case when no prior knowledge is available. In practice, however, $\beta_{\mu u}^{(t)}(\nu_0, i)$ should not be 0 and a prior distribution of the tone model shapes should be provided. This is because if the prior distribution of the tone model shapes is not used, there are too many

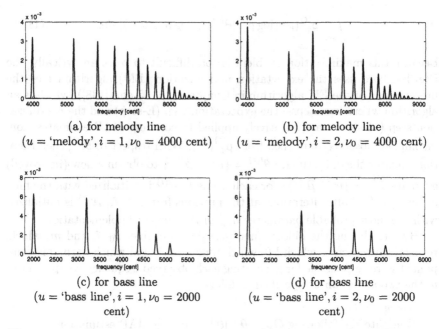

(a) for melody line
($u = $ 'melody', $i = 1, \nu_0 = 4000$ cent)

(b) for melody line
($u = $ 'melody', $i = 2, \nu_0 = 4000$ cent)

(c) for bass line
($u = $ 'bass line', $i = 1, \nu_0 = 2000$ cent)

(d) for bass line
($u = $ 'bass line', $i = 2, \nu_0 = 2000$ cent)

Fig. 11.4. Examples of prior distribution of the tone model shapes $\mathsf{p}(\nu|\nu_0, i, \boldsymbol{\mu}_{0u}^{(t)}(\nu_0, i))$.

degrees of freedom in their shapes. Without the prior distribution, unrealistic tone model shapes, such as a shape having only one salient component at frequency of the fourth harmonic component, could be estimated. In (11.15) and (11.16), $\mathsf{D}_{\boldsymbol{w}}(\boldsymbol{w}_{0u}^{(t)}; \boldsymbol{w}^{(t)})$ and $\mathsf{D}_{\boldsymbol{\mu}}(\boldsymbol{\mu}_{0u}^{(t)}(\nu_0, i); \boldsymbol{\mu}^{(t)}(\nu_0, i))$ are the following Kullback–Leibler information:

$$\mathsf{D}_{\boldsymbol{w}}(\boldsymbol{w}_{0u}^{(t)}; \boldsymbol{w}^{(t)}) = \int_{\mathrm{F}_u^l}^{\mathrm{F}_u^h} \sum_{i=1}^{I_u} w_{0u}^{(t)}(\nu_0, i) \log \frac{w_{0u}^{(t)}(\nu_0, i)}{w^{(t)}(\nu_0, i)} d\nu_0, \qquad (11.17)$$

$$\mathsf{D}_{\boldsymbol{\mu}}(\boldsymbol{\mu}_{0u}^{(t)}(\nu_0, i); \boldsymbol{\mu}^{(t)}(\nu_0, i)) = \sum_{m=1}^{M_u} c_{0u}^{(t)}(m|\nu_0, i) \log \frac{c_{0u}^{(t)}(m|\nu_0, i)}{c^{(t)}(m|\nu_0, i)}. \qquad (11.18)$$

These prior distributions were originally introduced for the sake of analytical tractability of the expectation maximization (EM) algorithm to obtain intuitive (11.25) and (11.26).

MAP Estimation Using the EM Algorithm

The problem to be solved is to estimate the model parameter $\boldsymbol{\theta}^{(t)}$, taking into account the prior distribution $\mathsf{p}_{0u}(\boldsymbol{\theta}^{(t)})$, when $\mathsf{p}_{\Psi}^{(t)}(\nu)$ is observed. The MAP estimator of $\boldsymbol{\theta}^{(t)}$ is obtained by maximizing

$$\int_{-\infty}^{\infty} \mathsf{p}_{\psi}^{(t)}(\nu) \left(\log \mathsf{p}(\nu|\boldsymbol{\theta}^{(t)}) + \log \mathsf{p}_{0u}(\boldsymbol{\theta}^{(t)}) \right) d\nu. \tag{11.19}$$

Because this maximization problem is too difficult to solve analytically, the PreFEst core uses the expectation maximization (EM) algorithm (see the presentation of the EM algorithm in Chapter 2, p. 35 and [138]), which is an algorithm where two steps—the expectation step (E-step) and the maximization step (M-step)—are iteratively applied to compute MAP estimates from incomplete observed data (i.e., from $\mathsf{p}_{\psi}^{(t)}(\nu)$). With respect to $\boldsymbol{\theta}^{(t)}$, each iteration updates the old estimate $\boldsymbol{\theta}'^{(t)} = \{w'^{(t)}, \boldsymbol{\mu}'^{(t)}\}$ to obtain a new (improved) estimate $\widehat{\boldsymbol{\theta}}^{(t)} = \{\widehat{w}^{(t)}, \widehat{\boldsymbol{\mu}}^{(t)}\}$. For each frame t, $w'^{(t)}$ is initialized with the final estimate $\widehat{w}^{(t-1)}$ after iterations at the previous frame $t-1$; $\boldsymbol{\mu}'^{(t)}$ is initialized with the most probable parameter $\boldsymbol{\mu}_{0u}^{(t)}$ in the current implementation.

By introducing the hidden (unobservable) variables ν_0, i, and m, which, respectively, describe which F0, which tone model, and which harmonic component were responsible for generating each observed frequency component at ν, the two steps can be specified as follows:

1. E-step:
   Compute the following $Q_{\mathrm{MAP}}(\boldsymbol{\theta}^{(t)}|\boldsymbol{\theta}'^{(t)})$ for the MAP estimation:

$$Q_{\mathrm{MAP}}(\boldsymbol{\theta}^{(t)}|\boldsymbol{\theta}'^{(t)}) = Q(\boldsymbol{\theta}^{(t)}|\boldsymbol{\theta}'^{(t)}) + \log \mathsf{p}_{0u}(\boldsymbol{\theta}^{(t)}), \tag{11.20}$$

$$Q(\boldsymbol{\theta}^{(t)}|\boldsymbol{\theta}'^{(t)}) = \int_{-\infty}^{\infty} \mathsf{p}_{\psi}^{(t)}(\nu) \mathbb{E}_{\nu_0,i,m}[\log \mathsf{p}(\nu, \nu_0, i, m|\boldsymbol{\theta}^{(t)}) \mid \nu, \boldsymbol{\theta}'^{(t)}] \, d\nu, \tag{11.21}$$

   where $Q(\boldsymbol{\theta}^{(t)}|\boldsymbol{\theta}'^{(t)})$ is the conditional expectation of the mean log-likelihood for the maximum likelihood estimation. $\mathbb{E}_{\nu_0,i,m}[a|b]$ denotes the conditional expectation of a with respect to the hidden variables ν_0, i, and m, with the probability distribution determined by condition b.

2. M-step:
   Maximize $Q_{\mathrm{MAP}}(\boldsymbol{\theta}^{(t)}|\boldsymbol{\theta}'^{(t)})$ as a function of $\boldsymbol{\theta}^{(t)}$ to obtain an updated (improved) estimate $\widehat{\boldsymbol{\theta}}^{(t)}$:

$$\widehat{\boldsymbol{\theta}}^{(t)} = \operatorname*{argmax}_{\boldsymbol{\theta}^{(t)}} Q_{\mathrm{MAP}}(\boldsymbol{\theta}^{(t)}|\boldsymbol{\theta}'^{(t)}). \tag{11.22}$$

In the E-step, $Q(\boldsymbol{\theta}^{(t)}|\boldsymbol{\theta}'^{(t)})$ is expressed as

$$Q(\boldsymbol{\theta}^{(t)}|\boldsymbol{\theta}'^{(t)}) = \int_{-\infty}^{\infty} \int_{\mathrm{F}_u^l}^{\mathrm{F}_u^h} \sum_{i=1}^{I_u} \sum_{m=1}^{M_u} \mathsf{p}_{\psi}^{(t)}(\nu)$$

$$\times \mathsf{p}(\nu_0, i, m|\nu, \boldsymbol{\theta}'^{(t)}) \log \mathsf{p}(\nu, \nu_0, i, m|\boldsymbol{\theta}^{(t)}) \, d\nu_0 d\nu, \tag{11.23}$$

where the complete-data log-likelihood is given by

$$\log \mathsf{p}(\nu, \nu_0, i, m | \boldsymbol{\theta}^{(t)}) = \log(w^{(t)}(\nu_0, i)\, \mathsf{p}(\nu, m | \nu_0, i, \boldsymbol{\mu}^{(t)}(\nu_0, i))). \qquad (11.24)$$

Regarding the M-step, (11.22) is a conditional problem of variation, where the conditions are given by (11.7) and (11.12). This problem can be solved by using Euler–Lagrange differential equations with Lagrange multipliers [222], [228] and the following new parameter estimates are obtained:

$$\widehat{w}^{(t)}(\nu_0, i) = \frac{\widehat{w}_{\mathrm{ML}}^{(t)}(\nu_0, i) + \beta_{wu}^{(t)} w_{0u}^{(t)}(\nu_0, i)}{1 + \beta_{wu}^{(t)}}, \qquad (11.25)$$

$$\widehat{c}^{(t)}(m | \nu_0, i) = \frac{\widehat{w}_{\mathrm{ML}}^{(t)}(\nu_0, i)\, \widehat{c}_{\mathrm{ML}}^{(t)}(m | \nu_0, i) + \beta_{\mu u}^{(t)}(\nu_0, i) c_{0u}^{(t)}(m | \nu_0, i)}{\widehat{w}_{\mathrm{ML}}^{(t)}(\nu_0, i) + \beta_{\mu u}^{(t)}(\nu_0, i)}, \qquad (11.26)$$

where $\widehat{w}_{\mathrm{ML}}^{(t)}(\nu_0, i)$ and $\widehat{c}_{\mathrm{ML}}^{(t)}(m | \nu_0, i)$ are, when the noninformative prior distribution ($\beta_{wu}^{(t)} = 0$ and $\beta_{\mu u}^{(t)}(\nu_0, i) = 0$) is given, the following maximum likelihood estimates:

$$\widehat{w}_{\mathrm{ML}}^{(t)}(\nu_0, i) = \int_{-\infty}^{\infty} \mathsf{p}_{\Psi}^{(t)}(\nu) \frac{w'^{(t)}(\nu_0, i)\, \mathsf{p}(\nu | \nu_0, i, \boldsymbol{\mu}'^{(t)}(\nu_0, i))}{\int_{\mathrm{F}_u^l}^{\mathrm{F}_u^h} \sum_{k=1}^{I_u} w'^{(t)}(\eta, k)\, \mathsf{p}(\nu | \eta, k, \boldsymbol{\mu}'^{(t)}(\eta, k))\, d\eta}\, d\nu, \qquad (11.27)$$

$$\widehat{c}_{\mathrm{ML}}^{(t)}(m | \nu_0, i) = \frac{1}{\widehat{w}_{\mathrm{ML}}^{(t)}(\nu_0, i)}$$
$$\times \int_{-\infty}^{\infty} \mathsf{p}_{\Psi}^{(t)}(\nu) \frac{w'^{(t)}(\nu_0, i)\, \mathsf{p}(\nu, m | \nu_0, i, \boldsymbol{\mu}'^{(t)}(\nu_0, i))}{\int_{\mathrm{F}_u^l}^{\mathrm{F}_u^h} \sum_{k=1}^{I_u} w'^{(t)}(\eta, k)\, \mathsf{p}(\nu | \eta, k, \boldsymbol{\mu}'^{(t)}(\eta, k))\, d\eta}\, d\nu. \qquad (11.28)$$

After the above iterative computation of (11.25) and (11.26),[3] the F0s pdf $\mathsf{p}_{F0}^{(t)}(\nu_0)$ can be obtained from $w^{(t)}(\nu_0, i)$ according to (11.13). The tone model shape $c^{(t)}(m | \nu_0, i)$, which is the relative amplitude of each harmonic component of all types of tone models $\mathsf{p}(\nu | \nu_0, i, \boldsymbol{\mu}^{(t)}(\nu_0, i))$, can also be obtained.

11.2.3 PreFEst Back End: Sequential F0 Tracking by Multiple-Agent Architecture

A simple way to identify the most predominant F0 is to find the frequency that maximizes the F0s pdf. This result is not always stable, however, because peaks corresponding to the F0s of simultaneous sounds sometimes compete in the F0s pdf for a moment and are transiently selected, one after another, as the maximum.

---

[3] In implementing the PreFEst core, this iterative computation is simple enough to perform only (11.25), (11.26), (11.27), and (11.28).

338   Masataka Goto

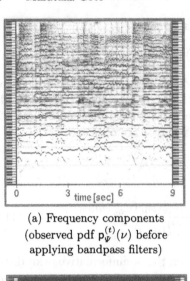

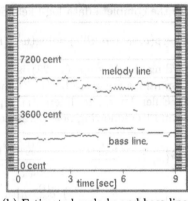

(a) Frequency components
(observed pdf $p_\Psi^{(t)}(\nu)$ before
applying bandpass filters)

(b) Estimated melody and bass lines
(the most dominant and stable
F0 trajectory in each $p_{F0}^{(t)}(\nu_0)$)

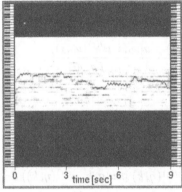

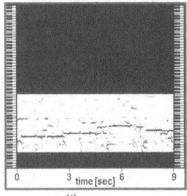

(c) F0s pdf ($p_{F0}^{(t)}(\nu_0)$) for estimating
the melody line in (b)

(d) F0s pdf ($p_{F0}^{(t)}(\nu_0)$) for estimating the
bass line in (b)

Fig. 11.5. Audio-synchronized real-time graphics output for a popular music excerpt with drum sounds: (a) frequency components, (b) the corresponding melody and bass lines estimated (final output), (c) the corresponding F0s pdf obtained when estimating the melody line, and (d) the corresponding F0s pdf obtained when estimating the bass line. These interlocking windows have the same vertical axis of log-scale frequency.

The PreFEst back end therefore considers the global temporal continuity of the F0 by using a multiple-agent architecture in which agents track different temporal trajectories of the F0 [228]. Each agent starts tracking from each salient peak in the F0s pdf, keeps tracking as long as it is temporally continued, and stops tracking when its next peak cannot be found for a while. The final F0 output is determined on the basis of the most dominant and stable F0 trajectory. Figure 11.5 shows an example of the final output.

11.2.4 Other Methods

While the PreFEst method resulted from pioneering research regarding melody and bass estimation and weighted-mixture modelling for F0 estimation, many issues still need to be resolved. For example, if an application requires MIDI-level note sequences of the melody line, the F0 trajectory should be segmented and organized into notes. Note that the PreFEst method does not deal with the problem of detecting the absence of melody and bass lines: it simply outputs the predominant F0 for every frame. In addition, since the melody and bass lines are generated from a process that is statistically biased rather than random—i.e., their transitions are musically appropriate this bias can also be incorporated into their estimation. This section introduces other recent approaches [494], [493], [435], [436], [169] that deal with these issues in describing polyphonic audio signals.

Paiva, Mendes, and Cardoso [494], [493] proposed a method of obtaining the melody note sequence by using a model of the human auditory system [595] as a frequency-analysis front end and applying MIDI-level note tracking, segmentation, and elimination techniques. Although the techniques used differ from the PreFEst method, the basic idea that 'the melody generally clearly stands out of the background' is the same as the basic PreFEst concept that the F0 of the most predominant harmonic structure is considered the melody. The advantage of this method is that MIDI-level note sequences of the melody line are generated, while the output of PreFEst is a simple temporal trajectory of the F0. The method first estimates predominant F0 candidates by using correlograms (see Chapter 8) that represent the periodicities in a cochleagram (auditory nerve responses of an ear model). It then forms the temporal trajectories of F0 candidates: it quantizes their frequencies to the closest MIDI note numbers and then tracks them according to their frequency proximity, where only one-semitone transition is considered continuous. After this tracking, F0 trajectories are segmented into MIDI-level note candidates by finding a sufficiently long trajectory having the same note number and by dividing it at clear local minima of its amplitude envelope. Because there still remain many inappropriate notes, it eliminates notes whose amplitude is too low, whose duration is too short, or which have harmonically related F0s and almost same onset and offset times. Finally, the melody note sequence is obtained by selecting the most predominant notes according to heuristic rules. Since simultaneous notes are not allowed, the method eliminates simultaneous notes that are less dominant and not in a middle frequency range.

Marolt [435], [436] proposed a method of estimating the melody line by representing it as a set of short vocal fragments of F0 trajectories. This method is based on the PreFEst method with some modifications: it uses the PreFEst core to estimate predominant F0 candidates, but uses a spectral modelling synthesis (SMS) front end that performs the sinusoidal modelling and analysis (see Chapters 1 and 3) instead of the PreFEst front end. The advantage of this method is that the F0 candidates are tracked and grouped into melodic fragments (reasonably segmented signal regions that exhibit strong and stable F0) and these fragments are then clustered into the melody line. The method first tracks temporal trajectories of the F0 candidates (salient peaks) to form the melodic fragments by using a salient peak tracking approach similar to the PreFEst back end (though it does not use multiple agents). Because the fragments belong to not only the melody (lead vocal), but also to different parts of the accompaniment, they are clustered to find the melody cluster by using Gaussian mixture models (GMMs) according to their five properties:

- Dominance (average weight of a tone model estimated by the EM algorithm),
- Pitch (centroid of the F0s within the fragment),
- Loudness (average loudness of harmonics belonging to the fragment),
- Pitch stability (average change of F0s during the fragment), and
- Onset steepness (steepness of overall loudness change during the first 50 ms of the fragment).

Eggink and Brown [169] proposed a method of estimating the melody line with the emphasis on using various knowledge sources, such as knowledge about instrument pitch ranges and interval transitions, to choose the most likely succession of F0s as the melody line. Unlike other methods, this method is specialized for a classical sonata or concerto, where a solo melody instrument can span the whole pitch range, ranging from the low tones of a cello to a high-pitched flute, so the frequency range limitation used in the PreFEst method is not feasible. In addition, because the solo instrument does not always have the most predominant F0, additional knowledge sources are necessary to extract the melody line. The main advantage of this method is the leverage provided by knowledge sources, including local knowledge about an instrument recognition module and temporal knowledge about tone durations and interval transitions, which are integrated in a probabilistic search. Those sources can both help to choose the correct F0 among multiple concurrent F0 candidates and to determine sections where the solo instrument is actually present. The knowledge sources consist of two categories, local knowledge and temporal knowledge. The local knowledge concerning F0 candidates obtained by picking peaks in the spectrum includes

- F0 strength (the stronger the spectral peak, the higher its likelihood of being the melody),
- Instrument-dependent F0 likelihood (the likelihood values of an F0 candidate in terms of its frequency and the pitch range of each solo instrument, which are evaluated by counting the frequency of its F0 occurrence in different standard MIDI files), and
- Instrument likelihood (the likelihood values of an F0 candidate being produced by each solo instrument, which are evaluated by the instrument recognition module).

The instrument recognition module uses trained Gaussian classifiers of the frequency and power of the first ten harmonic components, their deltas, and their delta-deltas, which are taken from the spectrum for each F0 candidate. On the other hand, the temporal knowledge concerning tone candidates obtained by connecting F0 candidates includes

- Instrument-dependent interval likelihood (the likelihood values of an interval transition between succession tones, which are evaluated by counting the frequency of its interval occurrence in different standard MIDI files), and
- Relative tone usage (measures related to tone durations between successive tones, which are used to penalize overlapped tones).

These knowledge sources are combined to find the most likely 'path' of the melody through the space of all F0 candidates in time. Since the melody path occasionally follows the accompaniment, additional postprocessing is done to eliminate sections where the solo instrument is actually silent.

While the above methods deal with the melody line, Hainsworth and Macleod [264] proposed a method of obtaining the bass note sequence by maintaining multiple hypotheses. The method first extracts the onset times of bass notes by picking peaks of a smoothed temporal envelope of a total power below 200 Hz. It then generates hypotheses regarding the F0 of each extracted note; the F0 and amplitude of each hypothesis are estimated by fitting a quadratic polynomial to a large amplitude peak and subtracting it from the spectrum. The first four harmonic components of those hypotheses are tracked over time by using a comb-filter-like analysis. Finally, the method selects the most likely hypothesis for each onset on the basis of its duration and the amplitude of harmonic components and further tidies up these hypotheses by removing inappropriate overlaps and relatively low amplitude notes.

11.3 Estimating Beat Structure

Beat tracking (including measure or bar line estimation) is defined as the process of organizing musical audio signals into a hierarchical beat structure (including beat and measure levels). It is also an important initial step in the computational modelling of music understanding because the beat is fundamental, for both trained and untrained listeners, to the perception of Western music. As described in Section 11.7.2 and Section 4.1, p. 101, there are many applications such as music-synchronized computer graphics, stage lighting control, video/audio synchronization, and human–computer improvisation in live ensembles.

Various methods for estimating the beat structure are described in detail in Chapter 4. Here, the synergy between the estimation of the hierarchical beat structure, drum patterns, and chord changes is briefly discussed. This synergy is exploited in a real-time beat-tracking system developed by Goto and

Muraoka [235], [220], [221]. The estimation of the hierarchical beat structure, especially the measure (bar line) level, requires the use of musical knowledge about drum patterns and chord changes; on the other hand, drum patterns and chord changes are difficult to estimate without referring to the beat structure of the beat level (quarter note level). The system addresses this issue by leveraging the integration of top-down and bottom-up processes (Fig. 11.6) under the assumption that the time signature of an input song is 4/4. The system first obtains multiple possible hypotheses of provisional beat times (quarter-note-level beat structure) on the basis of onset times without using musical knowledge about drum patterns and chord changes. Because the onset times of the sounds of bass drum and snare drum can be detected by a bottom-up frequency analysis described in Section 5.2.3, p. 137, the system makes use of the provisional beat times as top-down information to form the detected onset times into drum patterns whose grid is aligned with the beat times. The system also makes use of the provisional beat times to detect chord changes in a frequency spectrum without identifying musical notes or chords by name. The frequency spectrum is sliced into strips at the beat times and the dominant frequencies of each strip are estimated by using a histogram of frequency components in the strip [240]. Chords are considered to be changed when the dominant frequencies change between adjacent strips. After the drum patterns and chord changes are obtained, the higher-level beat structure, such as the measure level, can be estimated by using musical knowledge regarding them.

11.4 Estimating Drums

The detection of the onset times of drum sounds is important because the basic rhythms of popular music pieces including drum sounds are mainly characterized by drum performances. As described in Section 11.7.1, there are many applications such as rhythm-based music information retrieval and genre classification.

Various methods for detecting drum sounds are described in detail in Chapter 5.

11.5 Estimating Chorus Sections and Repeated Sections

Chorus ('hook' or 'refrain') sections of popular music are the most representative, uplifting, and prominent thematic sections in the music structure of a song, and human listeners can easily understand where the chorus sections are because these sections are the most repeated and memorable portions of a song. Automatic detection of chorus sections is essential for the computational modelling of music understanding and is useful in various practical applications. In music browsers or music retrieval systems, it enables a listener to quickly preview a chorus section as a 'music thumbnail' (a musical equivalent

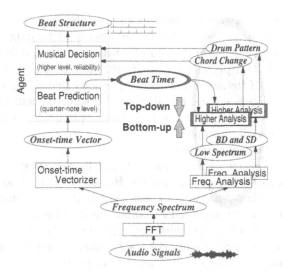

Fig. 11.6. Synergy between the estimation of the hierarchical beat structure, drum patterns, and chord changes. Drum patterns and chord changes are obtained, at 'Higher Analysis' in the figure, by using provisional beat times as top-down information. The hierarchical beat structure is then estimated, at 'Musical Decision' in the figure, by using the drum patterns and chord changes. A drum pattern is represented by the temporal pattern of a bass drum (BD) and a snare drum (SD).

of an image thumbnail) to find a desired song. It can also provide novel music listening interfaces for end users as described in Section 11.7.3.

To detect chorus sections, typical approaches do not rely on prior information regarding acoustic features unique to choruses but focus on the fact that chorus sections are usually the most repeated sections of a song. They thus adopt the following basic strategy: detect similar sections that repeat within a musical piece (such as a repeating phrase) and output those that appear most often. On entering the 2000s, this strategy has led to methods for extracting a single segment from several chorus sections by detecting a repeated section of a designated length as the most representative part of a musical piece [417], [27], [103]; methods for segmenting music, discovering repeated structures, or summarizing a musical piece through bottom-up analyses without assuming the output segment length [110], [111], [512], [516], [23], [195], [104], [82], [664], [420]; and a method for exhaustively detecting all chorus sections by determining the start and end points of every chorus section [224].

Although this basic strategy of finding sections that repeat most often is simple and effective, it is difficult for a computer to judge repetition because it is rare for repeated sections to be exactly the same. The following summarizes the main problems that must be addressed in finding music repetition and determining chorus sections.

Problem 1: Extracting acoustic features and calculating their similarity

Whether a section is a repetition of another must be judged on the basis of the similarity between the acoustic features obtained from each frame or section. In this process, the similarity must be high between acoustic features even if the accompaniment or melody line changes somewhat in the repeated section (e.g., the absence of accompaniment on bass and/or drums after repetition). That is, it is necessary to use features that capture useful and invariant properties.

Problem 2: Finding repeated sections

A pair of repeated sections can be found by detecting contiguous temporal regions having high similarity. However, the criterion establishing how high similarity must be to indicate repetition depends on the song. For a song in which repeated accompaniment phrases appear very often, for example, only a section with very high similarity should be considered the chorus section repetition. For a song containing a chorus section with accompaniments changed after repetition, on the other hand, a section with somewhat lower similarity can be considered the chorus section repetition. This criterion can be easily set for a small number of specific songs by manual means. For a large open song set, however, the criterion should be automatically modified based on the song being processed.

Problem 3: Grouping repeated sections

Even if many pairs of repeated sections with various lengths are obtained, it is not obvious how many times and where a section is repeated. It is therefore necessary to organize repeated sections that have common sections into a group. Both ends (the start and end points) of repeated sections must also be estimated by examining the mutual relationships among various repeated sections. For example, given a song having the structure (A B C B C C), the long repetition corresponding to (B C) would be obtained by a simple repetition search. Both ends of the C section in (B C) could be inferred, however, from the information obtained regarding the final repetition of C in this structure.

Problem 4: Detecting modulated repetition

Because the acoustic features of a section generally undergo a significant change after modulation (key change; see Section 1.1, p. 7), similarity with the section before modulation is low, making it difficult to judge repetition. The detection of modulated repetition is important since modulation sometimes occurs in chorus repetitions, especially in the latter half of a song.[4]

Problem 5: Selecting chorus sections

Because various levels of repetition can be found in a musical piece, it is necessary to select a group of repeated sections corresponding to chorus

---

[4]Masataka Goto's survey of Japan's popular music hit chart (top 20 singles ranked weekly from 2000 to 2003) showed that modulation occurred in chorus repetitions in 152 songs (10.3%) out of 1481.

sections. A simple selection of the most repeated sections is not always appropriate though. For example, another section such as verse A is occasionally repeated more often than chorus sections.

Regarding the above repetition-based methods, the following sections mainly describe a method called *RefraiD* (Refrain Detection method) [224] and briefly introduce techniques used in the other methods in each relevant section and Section 11.5.6. Since the RefraiD method addresses all of the above problems and detects all chorus sections in a popular music song regardless of whether a key change occurs, it is suitable for music scene description. Figure 11.7 shows the process flow of the RefraiD method. First, a 12-dimensional feature vector called a *chroma vector*, which is robust with respect to changes of accompaniments, is extracted from each frame of an input audio signal and then the similarity between these vectors is calculated *(solution to Problem 1)*. Each element of the chroma vector corresponds to one of the 12 pitch classes (C, C#, D, ..., B) and is the sum of the magnitude spectrum at frequencies of its pitch class over six octaves. Pairs of repeated sections are then listed (found) using an adaptive repetition-judgement criterion which is configured by an automatic threshold selection method based on a discriminant criterion *(solution to Problem 2)*. To organize common repeated sections into groups and to identify both ends of each section, the pairs of repeated sections are integrated (grouped) by analysing their relationships over the whole song *(solution to Problem 3)*. Because each element of a chroma vector corresponds to a different pitch class, a before-modulation chroma vector is close to the after-modulation chorus vector whose elements are shifted (exchanged) by the pitch difference of the key change. By considering 12 kinds of shift (pitch differences), 12 sets of the similarity between non-shifted and shifted chroma vectors are then calculated, pairs of repeated sections from those sets are listed, and all of them are integrated *(solution to Problem 4)*. Finally, the *chorus measure*, which is the possibility of being chorus sections for each group, is evaluated *(solution to Problem 5)*, and the group of chorus sections with the highest chorus measure as well as other groups of repeated sections are output (Fig. 11.8).

11.5.1 Extracting Acoustic Features and Calculating Their Similarity

The following acoustic features, which capture pitch and timbral features of audio signals in different ways, were used in various methods: chroma vectors [224], [27], [110], [111], mel-frequency cepstral coefficients (MFCC) [417], [103], [23], [195], [104], (dimension-reduced) spectral coefficients [103], [195], [104], [82], [664], pitch representations using F0 estimation or constant-Q filterbanks [110], [111], [82], [420], and dynamic features obtained by supervised learning [512], [516].

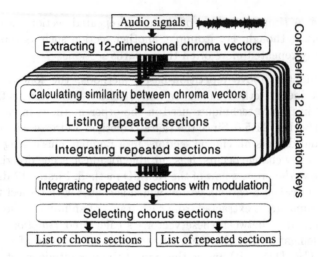

Fig. 11.7. Overview of RefraiD (Refrain Detection method) for detecting all chorus sections with their start and end points while considering modulations (key changes).

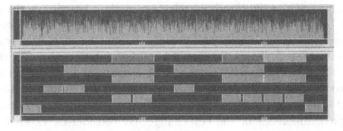

Fig. 11.8. An example of chorus sections and repeated sections detected by the RefraiD method. The horizontal axis is the time axis (in seconds) covering the entire song. The upper window shows the power. On each row in the lower window, coloured sections indicate similar (repeated) sections. The top row shows the list of the detected chorus sections, which were correct for this song (RWC-MDB-P-2001 No. 18 of the RWC Music Database [229], [227]) and the last of which was modulated. The bottom five rows show the list of various repeated sections (only the five longest repeated sections are shown). For example, the second row from the top indicates the structural repetition of 'verse A ⇒ verse B ⇒ chorus'; the bottom row with two short coloured sections indicates the similarity between the 'intro' and 'ending'.

Pitch Feature: Chroma Vector

The chroma vector is a perceptually motivated feature vector using the concept of *chroma* in Shepard's helix representation of musical pitch perception [584]. According to Shepard [584], the perception of pitch with respect to a musical context can be graphically represented by using a continually cyclic helix that has two dimensions, *chroma* and *height*, as shown at the right of Fig. 11.9. Chroma refers to the position of a musical pitch within an octave that corresponds to a cycle of the helix; i.e., it refers to the position on the circumference of the helix seen from directly above. On the other hand, height refers to the vertical position of the helix seen from the side (the position of an octave).

Figure 11.9 shows an overview of calculating the chroma vector used in the RefraiD method [224]. This represents magnitude distribution on the chroma that is discretized into twelve pitch classes within an octave. The 12-dimensional chroma vector $\mathbf{v}(t)$ is extracted from the magnitude spectrum, $\Psi_p(\nu, t)$ at the log-scale frequency ν at time t, calculated by using the short-time Fourier transform (STFT). Each element of $\mathbf{v}(t)$ corresponds to a pitch class c ($c = 1, 2, \ldots, 12$) in the equal temperament and is represented as $v_c(t)$:

$$v_c(t) = \sum_{h=\mathrm{Oct_L}}^{\mathrm{Oct_H}} \int_{-\infty}^{\infty} \mathrm{BPF}_{c,h}(\nu) \, \Psi_p(\nu, t) \, d\nu. \tag{11.29}$$

The $\mathrm{BPF}_{c,h}(\nu)$ is a bandpass filter that passes the signal at the log-scale centre frequency $F_{c,h}$ (in cents) of pitch class c (chroma) in octave position h (height), where

$$F_{c,h} = 1200h + 100(c - 1). \tag{11.30}$$

The $\mathrm{BPF}_{c,h}(\nu)$ is defined using a Hanning window as follows:

$$\mathrm{BPF}_{c,h}(\nu) = \frac{1}{2} \left(1 - \cos \frac{2\pi(\nu - (F_{c,h} - 100))}{200} \right), \quad \nu \in [0, 200]. \tag{11.31}$$

This filter is applied to octaves from $\mathrm{Oct_L}$ to $\mathrm{Oct_H}$. In Goto's implementation [224], an STFT with a 256 ms Hanning window[5] shifted by 80 ms is calculated for audio signals sampled at 16 kHz, and the $\mathrm{Oct_L}$ and $\mathrm{Oct_H}$ are respectively 3 and 8, covering six octaves (130 Hz to 8 kHz).

There are variations in how the chroma vector is calculated. For example, Bartsch and Wakefield [27] developed a technique where each STFT bin of the log-magnitude spectrum is mapped directly to the most appropriate pitch class, and Dannenberg and Hu [110], [111] also used this technique. A similar continuous concept was called the chroma spectrum [655].

There are several advantages to using the chroma vector. Because it captures the overall harmony (pitch-class distribution), it can be similar even if accompaniments or melody lines are changed to some degree after repetition. In fact, the chroma vector is effective for identifying chord names [201], [678], [679], [583], [684]. The chroma vector also enables modulated repetition to be detected as described in Section 11.5.4.

Timbral Feature: MFCC and Dynamic Features

While the chroma vectors capture pitch-related content, the MFCCs (see Section 2.1.3, p. 25 for a presentation of MFCCs) typically used in speech

---

[5]The window length is determined to obtain good frequency resolution in a low-frequency region.

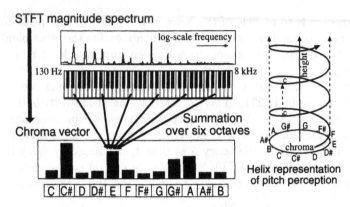

Fig. 11.9. Overview of calculating a 12-dimensional chroma vector. The magnitude at six different octaves is summed into just one octave which is divided into 12 log-spaced divisions corresponding to pitch classes. Shepard's helix representation of musical pitch perception [584] is shown at the right.

recognition capture spectral content and general pitch range, and are useful for finding timbral or 'texture' repetitions. Dynamic features [512], [516] are more adaptive spectral features that are designed for music structure discovery through a supervised learning method. Those features are selected from the spectral coefficients of a filterbank output by maximizing the mutual information between the selected features and hand-labelled music structures. The dynamic features are beneficial in that they reduce the size of the results when calculating similarity (i.e., the size of the similarity matrix described in Section 11.5.1) because the frame shift can be longer (e.g., 1 s) than for other features.

Calculating Similarity

Given a feature vector such as the chroma vector or MFCC at every frame, the next step is to calculate the similarity between feature vectors. Various distance or similarity measures, such as the Euclidean distance and the cosine angle (inner product), can be used for this. Before calculating the similarity, feature vectors are usually normalized, for example, to a mean of zero and a standard deviation of one or to a maximum element of one.

In the RefraiD method [224], the similarity $r(t, l)$ between the feature vectors (chroma vectors) $\mathbf{v}(t)$ and $\mathbf{v}(t-l)$ is defined as

$$r(t, l) = 1 - \frac{1}{\sqrt{12}} \left| \frac{\mathbf{v}(t)}{\max_c v_c(t)} - \frac{\mathbf{v}(t-l)}{\max_c v_c(t-l)} \right|, \quad (11.32)$$

where l is the lag and $v_c(t)$ is an element of $\mathbf{v}(t)$ (11.29). Since the denominator $\sqrt{12}$ is the length of the diagonal line of a 12-dimensional hypercube with edge length 1, $r(t, l)$ satisfies $0 \leq r(t, l) \leq 1$.

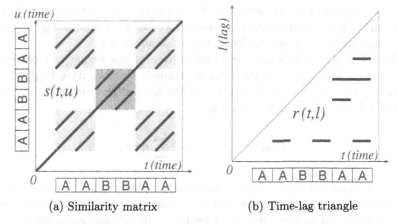

(a) Similarity matrix   (b) Time-lag triangle

Fig. 11.10. An idealized example of a similarity matrix and time-lag triangle drawn from the same feature vectors of a musical piece consisting of four 'A' sections and two 'B' sections. The diagonal line segments in the similarity matrix or horizontal line segments in the time-lag triangle, which represent similar sections, appear when short-time pitch features like chroma vectors are used.

For the given 36-dimensional feature vectors of a constant-Q filterbank output with centre frequencies at 36 tempered semitones in 3 octaves, Lu, Wang, and Zhang [420] introduced an original distance measure that emphasizes melody similarity and suppresses timbre similarity. This measure does not depend on the norm of the difference between the 36-dimensional feature vectors, but on the structure of it. It considers how the peak intervals in the difference conform to harmonic relationships such as perfect fifth and octave.

11.5.2 Finding Repeated Sections

By using the same similarity measure $r(t, l)$, two equivalent representations can be obtained: a *similarity matrix* [103], [110], [111], [195], [104], [664] and a *time-lag triangle* (or time-lag matrix) [224], [27], [516], [420], as shown in Fig. 11.10. For the similarity matrix, the similarity $s(t, u)$ between feature vectors $\mathbf{v}(t)$ and $\mathbf{v}(u)$,

$$s(t, u) = r(t, t - u), \qquad (11.33)$$

is drawn within a square in the two-dimensional $(t\text{-}u)$ space.[6] For the time-lag triangle, the similarity $r(t, l)$ between feature vectors $\mathbf{v}(t)$ and $\mathbf{v}(t - l)$ is drawn within a right-angled isosceles triangle in the two-dimensional time-lag $(t\text{-}l)$ space. If a nearly constant tempo can be assumed, each pair of similar sections is represented by two non-central diagonal line segments in the

---

[6]As described in Section 4.6, p. 112, the similarity matrix can also be used to examine rhythmic structure.

similarity matrix or a horizontal line segment in the time-lag triangle. Because the actual $r(t, l)$ obtained from a musical piece is noisy and ambiguous, it is not a straightforward task to detect these line segments.

The RefraiD method [224] finds all horizontal line segments (contiguous regions with high $r(t, l)$) in the time-lag triangle by evaluating $R_{all}(t, l)$, the possibility of containing line segments at the lag l at the current time t (e.g., at the end of a song[7]) as follows (Fig. 11.11):[8]

$$R_{all}(t, l) = \frac{1}{t - l + 1} \sum_{\tau = l}^{t} r(\tau, l). \tag{11.34}$$

Before this calculation, $r(t, l)$ is normalized by subtracting a local mean value while removing noise and emphasizing horizontal lines. In more detail, given each point $r(t, l)$ in the time-lag triangle, six-directional local mean values along the right, left, upper, lower, upper right, and lower left directions starting from the point $r(t, l)$ are calculated, and the maximum and minimum are obtained. If the local mean along the right or left direction takes the maximum, $r(t, l)$ is considered part of a horizontal line and emphasized by subtracting the minimum from $r(t, l)$. Otherwise, $r(t, l)$ is considered noise and suppressed by subtracting the maximum from $r(t, l)$; noise tends to appear as lines along the upper, lower, upper right, and lower left directions.

The method then picks up each peak in $R_{all}(t, l)$ along the lag l after smoothing $R_{all}(t, l)$ with a moving average filter along the lag and removing a global drift (bias) caused by cumulative noise in $r(t, l)$[9] from $R_{all}(t, l)$. The method next selects only high peaks above a threshold to search the line segments. Because this threshold is closely related to the repetition-judgement criterion which should be adjusted for each song, an automatic threshold selection method based on a discriminant criterion [491] is used. When dichotomizing the peak heights into two classes by a threshold, the optimal threshold is obtained by maximizing the discriminant criterion measure defined by the following between-class variance:

$$\sigma_B^2 = \omega_1 \omega_2 (\mu_1 - \mu_2)^2, \tag{11.35}$$

where ω_1 and ω_2 are the probabilities of class occurrence (number of peaks in each class/total number of peaks), and μ_1 and μ_2 are the means of the peak heights in each class.

---

[7]$R_{all}(t, l)$ is evaluated along with the real-time audio input for a real-time system based on RefraiD. On the other hand, it is evaluated at the end of a song for a non-real-time off-line analysis.

[8]This can be considered the Hough transform where only horizontal lines are detected: the parameter (voting) space $R_{all}(t, l)$ is therefore simply one dimensional along l.

[9]Because the similarity $r(\tau, l)$ is noisy, its sum $R_{all}(t, l)$ tends to be biased: the longer the summation period for $R_{all}(t, l)$, the higher the summation result by (11.34).

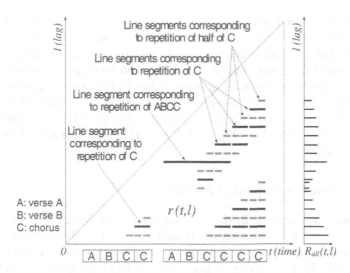

Fig. 11.11. A sketch of line segments, the similarity $r(t, l)$ in the time-lag triangle, and the possibility $R_{all}(t, l)$ of containing line segments at lag l.

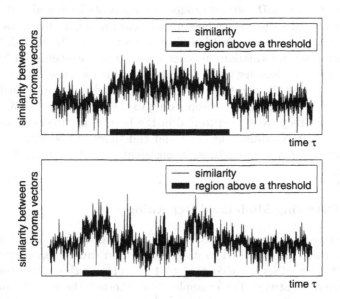

Fig. 11.12. Examples of the similarity $r(\tau, l_1)$ at high-peak lags l_1. The bottom horizontal bars indicate the regions above an automatically adjusted threshold, which means they correspond to line segments.

For each picked-up high peak with lag l_1, the line segments are finally searched on the one-dimensional function $r(\tau, l_1)$ ($l_1 \leq \tau \leq t$). After smoothing $r(\tau, l_1)$ using a moving average filter, the method obtains line segments on which the smoothed $r(\tau, l_1)$ is above a threshold (Fig. 11.12). This threshold is also adjusted through the automatic threshold selection method.

Instead of using the similarity matrix and time-lag triangle, there are other approaches that do not explicitly find repeated sections. To segment music, represent music as a succession of states (labels), and obtain a music thumbnail or summary, these approaches segment and label (i.e., categorize) contiguous frames (feature vectors) by using clustering techniques [417] or ergodic hidden Markov models (HMMs) [417], [512], [516] (HMMs are introduced on p. 63 of this volume).

11.5.3 Grouping Repeated Sections

Since each line segment in the time-lag triangle indicates just a pair of repeated sections, it is necessary to organize into a group the line segments that have common sections—i.e., overlap in time. When a section is repeated N times ($N \geq 3$), the number of line segments to be grouped together should theoretically be $N(N-1)/2$ if all of them are found in the time-lag triangle.

Aiming to exhaustively detect all the repeated (chorus) sections appearing in a song, the RefraiD method groups line segments having almost the same section while redetecting some missing (hidden) line segments not found in the bottom-up detection process (described in Section 11.5.2) through top-down processing using information on other detected line segments. In Fig. 11.11, for example, two line segments corresponding to the repetition of the first and third C and the repetition of the second and fourth C, which overlap with the long line segment corresponding to the repetition of ABCC, can be found even if they were hard to find in the bottom-up process. The method also appropriately adjusts the start and end times of line segments in each group because they are sometimes inconsistent in the bottom-up line segment detection.

11.5.4 Detecting Modulated Repetition

The processes described above do not deal with modulation (key change), but they can easily be extended to it. A modulation can be represented by the pitch difference of its key change, ζ $(0, 1, \ldots, 11)$, which denotes the number of tempered semitones. For example, $\zeta = 9$ means the modulation of nine semitones upward or the modulation of three semitones downward. One of the advantages of the 12-dimensional chroma vector $\mathbf{v}(t)$ is that a transposition amount ζ of the modulation can naturally correspond to the amount by which its 12 elements are shifted (rotated). When $\mathbf{v}(t)$ is the chroma vector of a certain performance and $\mathbf{v}(t)'$ is the chroma vector of the performance that is modulated by ζ semitones upward from the original performance, they tend to satisfy

$$\mathbf{v}(t) \approx \mathbf{S}^{\zeta} \mathbf{v}(t)^{\mathsf{T}}, \tag{11.36}$$

where **S** is a 12-by-12 shift matrix defined by

$$
\mathbf{S} = \begin{pmatrix}
0 & 1 & 0 & \cdots\cdots & 0 \\
0 & 0 & 1 & 0 & \cdots & 0 \\
\vdots & & \ddots & \ddots & \ddots & \vdots \\
0 & \cdots\cdots & & 0 & 1 & 0 \\
0 & \cdots\cdots\cdots & & & 0 & 1 \\
1 & 0 & \cdots\cdots\cdots & & & 0
\end{pmatrix}. \tag{11.37}
$$

To detect modulated repetition by using this feature of chroma vectors and considering 12 destination keys, the RefraiD method [224] calculates 12 kinds of extended similarity as follows:

$$
r_\varsigma(t, l) = 1 - \frac{1}{\sqrt{12}} \left| \frac{\mathbf{S}^\varsigma \mathbf{v}(t)}{\max_c v_c(t)} - \frac{\mathbf{v}(t-l)}{\max_c v_c(t-l)} \right|. \tag{11.38}
$$

Starting from each $r_\varsigma(t, l)$, the processes of finding and grouping the repeated sections are performed again. Non-modulated and modulated repeated sections are then grouped if they share the same section.

11.5.5 Selecting Chorus Sections

A group corresponding to the chorus sections is finally selected from groups of repeated sections (line segments). In general, a group that has many and long repeated sections tends to be the chorus sections. In addition to this property, the RefraiD method evaluates the *chorus measure*, which is the possibility of being chorus sections for each group, by considering the following three heuristic rules with a focus on popular music:

1. The length of the chorus has an appropriate, allowed range (7.7 to 40 s in Goto's implementation).
2. When there is a repeated section that is long enough to likely correspond to the repetition of a long section like (verse A ⇒ verse B ⇒ chorus) × 2, the chorus section is likely to be at the end of that repeated section.
3. Because a chorus section tends to have two half-length repeated subsections within its section, a section having those subsections is likely to be the chorus section.

The group that maximizes the chorus measure is finally selected as the chorus sections.

11.5.6 Other Methods

Since the above sections mainly describe the RefraiD method [224] with the focus on detecting all chorus sections, this section briefly introduces other methods [417], [27], [103], [110], [111], [512], [516], [23], [195], [104], [82], [664],

[420] that aim at music thumbnailing, music segmentation, structure discovery, or music summarization.

Several methods for detecting the most representative part of a song for use as a music thumbnail have been studied. Logan and Chu [417] developed a method using clustering techniques and hidden Markov models (HMMs) to categorize short segments (1 s) in terms of their acoustic features, where the most frequent category is then regarded as a chorus. Bartsch and Wakefield [27] developed a method that calculates the similarity between acoustic features of beat-length segments obtained by beat tracking and finds the given-length segment with the highest similarity averaged over its segment. Cooper and Foote [103] developed a method that calculates a similarity matrix of acoustic features of short frames (100 ms) and finds the given-length segment with the highest similarity between it and the whole song. Note that these methods assume that the output segment length is given and do not identify both ends of a repeated section.

Music segmentation or structure discovery methods where the output segment length is not assumed have also been studied. Dannenberg and Hu [110], [111] developed a structure discovery method of clustering pairs of similar segments obtained by several techniques such as efficient dynamic programming or iterative greedy algorithms. This method finds, groups, and removes similar pairs from the beginning to group all the pairs. Peeters, La Burthe, and Rodet [512], [516] developed a supervised learning method of modelling dynamic features and studied two structure discovery approaches: the sequence approach of obtaining repetitions of patterns and the state approach of obtaining a succession of states. The dynamic features are selected from the spectrum of a filterbank output by maximizing the mutual information between the selected features and hand-labelled music structures. Aucouturier and Sandler [23] developed two methods for finding repeated patterns in a succession of states (texture labels) obtained by HMMs. They used two image processing techniques, the kernel convolution and Hough transform, to detect line segments in the similarity matrix between the states. Foote and Cooper [195], [104] developed a method of segmenting music by correlating a kernel along the diagonal of the similarity matrix, and clustering the obtained segments on the basis of the self-similarity of their statistics. Chai and Vercoe [82] developed a method of detecting segment repetitions by using dynamic programming, clustering the obtained segments, and labelling the segments based on heuristic rules such as the rule of first labelling the most frequent segments, removing them, and repeating the labelling process. Wellhausen and Crysandt [664] studied the similarity matrix of spectral envelope features defined in the MPEG-7 descriptors and a technique of detecting non-central diagonal line segments. Lu, Wang, and Zhang [420] developed a method of analysing all repeated sections by using a structure-based distance measure that emphasizes pitch similarity over timbral similarity. Their method also estimates the tempo of a song and discriminates between vocal and instrumental sections to facilitate music structure analysis.

11.6 Evaluation Issues

To evaluate automatic music scene description methods, it is necessary to label musical pieces in an adequate-size music database with their correct descriptions (metadata). This labelling task is time consuming and troublesome. More seriously, there was no available common music database with correct metadata since most musical pieces used by researchers are generally copyrighted and cannot be shared by other researchers.

But since 2000, a copyright-cleared music database, called the RWC (Real World Computing) Music Database [229], [230], [227], was developed and has been available to researchers as a common foundation for research. It contains six original collections: the Popular Music Database (100 pieces), Royalty-Free Music Database (15 pieces), Classical Music Database (50 pieces), Jazz Music Database (50 pieces), Music Genre Database (100 pieces), and Musical Instrument Sound Database (50 instruments). For all 315 musical pieces, audio signals, standard MIDI files, and text files of lyrics were prepared. For the 50 instruments, individual sounds at half-tone intervals were captured. This database has been distributed to researchers around the world and has already been widely used. For musical instrument sounds, there are other databases released for public use: the McGill University Master Samples [487] and the University of Iowa Musical Instrument Samples [198]. Musical pieces licensed under a Creative Commons license can also be used for evaluation purposes.

To establish benchmarks (evaluation frameworks) for music scene description by labelling copyright-cleared musical pieces with correct descriptions, a multipurpose music-scene labelling editor (metadata editor) was also developed [225]. It enables a user to hand-label a musical piece with music scene descriptions shown in Fig. 11.1. The editor can deal with both audio files and standard MIDI files and supports interactive audio/MIDI playback while editing. Along a wave or MIDI piano-roll display it shows subwindows in which any selected descriptions can be displayed and edited. To facilitate the support of various descriptions, its architecture is based on a plug-in system in which an external module for editing each description is installed as plug-in software. As a first step, the RefraiD method was evaluated by using the chorus section metadata for 100 songs of the RWC Music Database: Popular Music (80 of the 100 songs were correctly detected) [224].

11.7 Applications of Music Scene Description

Music scene description methods that can deal with real-world audio signals of musical pieces sampled from CD recordings have various practical applications such as music information retrieval, music-synchronized computer graphics, and music listening stations. The following sections introduce these applications.

Fig. 11.13. Virtual dancer 'Cindy'.

11.7.1 Music Information Retrieval

Music scene description contributes to content-based music information retrieval since it can provide various acoustical metadata (annotations) of musical pieces. For example, the automatic melody estimation described in Section 11.2 is useful for *query by humming (QBH)* [323], [207], [604], [484], [603], [507], [605], [585], [301], [109] which enables a user to retrieve a musical piece by humming or singing its melody: a QBH database consisting of audio signals of musical pieces can be indexed using their melody lines. Moreover, the description of chorus sections (Section 11.5) can increase the efficiency and precision of QBH by enabling a QBH system to match a query with only the chorus sections.

Temporal or rhythmic descriptions such as beat structure, tempo, and drums (Sections 11.3 and 11.4) are also useful for retrieving musical pieces on the basis of rhythm and tempo. Indexing musical pieces using drum descriptions, for example, will enable a user to retrieve music by voice percussion or beat boxing (verbalized expression of drum sounds by voice) [479], [326].

In addition, various music scene descriptions facilitate the computation of similarity between musical pieces. Similarity measures based on music scene descriptions enable a user to use musical pieces themselves as the search key to retrieve a musical piece having a similar feeling. These measures can also be used to automatically classify musical pieces into genres or music styles.

11.7.2 Music-Synchronized Computer Graphics

Because the beat tracking described in Section 11.3 and Chapter 4 can be used to automate the time-consuming tasks that must be done to synchronize events with music, there are various applications. In fact, Goto and Muraoka [235], [220], [221] developed a real-time system that displays virtual dancers and several graphic objects whose motions and positions change in time to beats (Fig. 11.13). This system has several dance sequences, each for a different mood of dance motions. While a user selects a dance sequence manually, the timing of each motion in the selected sequence is determined automatically on the basis of the beat-tracking results. Such a computer graphics system is suitable for live stage, TV program, and karaoke uses.

Beat tracking also facilitates the automatic synchronization of computer-controlled stage lighting with the beats in a musical performance. Various properties of lighting—such as colour, brightness, and direction—can be changed in time to music. In the above virtual dancer system, this was simulated on a computer graphics display with virtual dancers.

11.7.3 Music Listening Station

The automatic chorus section detection described in Section 11.5 enables new music-playback interfaces that facilitate content-based manual browsing of entire songs. As an application of the RefraiD method, Goto [226] developed a music listening station for trial listening, called SmartMusicKIOSK. Customers in music stores often search out the chorus or 'hook' of a song by repeatedly pressing the fast-forward button, rather than passively listening to the music. This activity is not well supported by current technology. SmartMusicKIOSK provides the following two functions to facilitate an active listening experience by eliminating the hassle of manually searching for the chorus and making it easier for a listener to find desired parts of a song:

1. *'Jump to chorus' function: automatic jumping to the beginning of sections relevant to a song's structure*
    Functions are provided enabling automatic jumping to sections that will be of interest to listeners. These functions are 'jump to chorus (NEXT CHORUS button)', 'jump to previous section in song (PREV SECTION button)', and 'jump to next section in song (NEXT SECTION button)', and they can be invoked by pushing the buttons shown above in parentheses (in the lower window of Fig. 11.14). With these functions, a listener can directly jump to and listen to chorus sections, or jump to the previous or next repeated section of the song.
2. *'Music map' function: visualization of song contents*
    A function is provided to enable the contents of a song to be visualized to help the listener decide where to jump next. Specifically, this function provides a visual representation of the song's structure consisting of chorus sections and repeated sections, as shown in the upper window of Fig. 11.14. While examining this display, the listener can use the automatic jump buttons, the usual fast-forward/rewind buttons, or a playback slider to move to any point of interest in the song.

This interface, which enables a listener to look for a section of interest by interactively changing the playback position, is useful not only for trial listening but also for more general purposes in selecting and using music. While entire songs of no interest to a listener can be skipped on conventional music-playback interfaces, SmartMusicKIOSK is the first interface that allows the listener to easily skip sections of no interest even within a song.

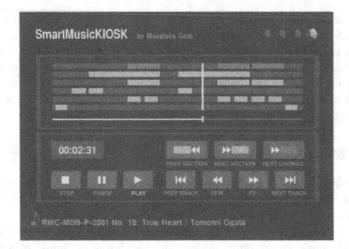

Fig. 11.14. SmartMusicKIOSK screen display. The lower window provides content-based controls allowing a listener to skim rapidly through music as well as common playback controls. The upper window provides a graphical overview of the music structure (results of automatic chorus section detection using RWC-MDB-P-2001 No. 18 of the RWC Music Database [229], [227]). The horizontal axis of the upper window is the time axis covering the entire song; the top row shows chorus sections, the five lower rows show repeated sections, and the bottom horizontal bar is a playback slider.

11.8 Conclusion

This chapter has described the music scene description research approach towards developing a system that understands real-world musical audio signals without deriving musical scores or separating signals. This approach is important from an academic viewpoint because it explores what is essential for understanding audio signals in a human-like fashion. The ideas and techniques are expected to be extended to not only music signals but also general audio signals including music, speech, environmental sounds, and mixtures of them. Traditional speech recognition frameworks have been developed for dealing with only monophonic speech signals or a single-pitch sound with background noise, which should be removed or suppressed without considering their relationship. Research on understanding musical audio signals is a good starting point for creating a new framework for understanding general audio signals, because music is polyphonic, temporally structured, and complex, yet still well organized. In particular, relationships between various simultaneous or successive sounds are important and unique to music. This chapter, as well as other chapters in this book, will contribute to such a general framework.

The music scene description approach is also important from industrial or application viewpoints since end users can now easily 'rip' audio signals from CDs, compress and store them on a personal computer, load a huge number of songs onto a portable music player, and listen to them anywhere and anytime. These users want to retrieve and listen to their favourite music or a portion of a musical piece in a convenient and flexible way. Reflecting these demands, the target of processing has expanded from the internal content of individual musical pieces to entire musical pieces and even sets of musical pieces [233]. While the primary target of music scene description is the internal content of a piece, the obtained descriptions are useful for dealing with sets of musical pieces as described in Section 11.7.1. The more accurate and detailed we can make the obtained music scene descriptions, the more advanced and intelligent music applications and interfaces will become.

Although various methods for detecting melody and bass lines, tracking beats, detecting drums, and finding chorus sections have been developed and successful results have been achieved to some extent, there is much room for improving these methods and developing new ones. For example, in general each method has been researched independently and implemented separately. An integrated method exploiting the relationships between these descriptions will be a promising next step. Other music scene descriptions apart from those described in this chapter should also be investigated in the future. Ten years ago it was considered too difficult for a computer to obtain most of the music scene descriptions described here, but today we can obtain them with a certain accuracy. I look forward to experiencing further advances in the next ten years.

12

Singing Transcription

Matti Ryynänen

Institute of Signal Processing, Tampere University of Technology
Korkeakoulunkatu 1, 33720 Tampere, Finland
Matti.Ryynanen@tut.fi

12.1 Introduction

Singing refers to the act of producing musical sounds with the human voice, and *singing transcription* refers to the automatic conversion of a recorded singing signal into a parametric representation (e.g., a MIDI file) by applying signal-processing methods. Singing transcription is an important topic in computational music-content analysis since it is the most natural way of human-computer interaction in the musical sense: even a musically untrained subject is usually able to hum the melody of a piece. This chapter introduces the singing transcription problem and presents an overview of the main approaches to solve it, including the current state-of-the-art singing transcription systems.

During the last ten years, the rapid growth of digital music databases has challenged researchers to develop natural user interfaces for accessing them by using the singing voice. Consequently, most of the research on singing transcription has been conducted in the context of *query-by-humming* systems where singing transcription acts as a front end. After converting a singing signal into a notated query, music pieces corresponding to the query can be retrieved from the database. However, singing transcription enables a wide range of other applications as well, including singing-input functionalities in applications such as computer games or singing tutors, automatic tools for annotating large corpora of singing, audio editor applications for professional music production, and naturally, applications that convert singing signals into musical scores.

12.1.1 Problem Formulation and the Scope of This Chapter

The singing transcription problem is here formulated as follows.

> Given the acoustic waveform of a single-voice singing performance, produce a sequence of notes and rests which is melodically and rhythmically as close to the performance as possible.

The sequence of notes and rests forms a parametric representation that is considered as the transcription of the singing signal. A *note* has an identifiable pitch, a beginning (onset) and an ending (offset) time, and a duration determined by these. *Rests* are silent moments in music. The *fundamental frequency (F0)* of a note is measured in Hertz (Hz) units and corresponds to the pitch of a note (see Chapter 1 for the definitions of pitch and F0). The ratio of the fundamental frequencies of two notes is referred to as an *interval*. In particular, an interval with the ratio 1 : 2 is called an *octave*, which is divided into twelve notes in Western music. This leads to the F0 ratio $1 : 2^{1/12}$ between adjacent notes, which is called a *semitone*.

A *note pitch label* identifies the standard F0 of a note either with a note name (e.g., A4) or with an integer *MIDI note number*. The latter is defined for a note with a fundamental frequency F_0 by

$$\text{MIDI note number} = 69 + 12 \log_2 \left(\frac{F_0}{440 \, \text{Hz}} \right), \qquad (12.1)$$

where 69 and 440 Hz correspond to the MIDI note number and the fundamental frequency of the note A4, respectively. The term *absolute tuning* refers to a standard tuning where the note A4 has a F0 value of 440 Hz, and the F0s of all the other notes are related to it. Equation (12.1) provides a musically convenient way of representing an arbitrary F0 value in semitone units by omitting the rounding to an integer. The term *note labelling* refers to the assignment of a pitch label for a note (e.g., an integer MIDI note number). The term *note segmentation* refers to a process where the onset and offset time of a note are determined.

The biggest difficulty in singing transcription lies in the conversion of a continuous F0 curve into note pitch labels. Simply rounding F0 estimates into MIDI note numbers produces poor results since the F0 curve may exhibit large deviations from the nominal note pitches. Singing performances are typically far from perfect and usually there is no straightforward one-to-one correspondence between a F0 curve and a transcribed note. In addition, it cannot be assumed that singing is performed in absolute tuning. In this respect, singing transcription is particularly challenging as compared to the transcription of other musical instruments which usually produce more stable F0 curves for each note. The relationship between singing sounds and note pitches is discussed in more detail in Section 12.2.

This chapter concentrates on the quantization of note pitches rather than their onset times and durations, although the common musical notation expresses also durations with discrete labels. In this sense, the aim here is to produce a piano-roll or MIDI-type representation of singing rather than the actual score of the performance. There exist algorithms for the time quantization of notes for producing score-type representations, such as work by Cemgil et al. [75], [77], which can be used to post-process the transcription result. See Chapter 4 for a discussion on time quantization.

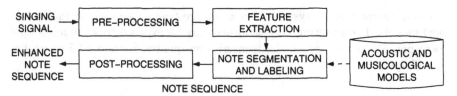

Fig. 12.1. A flow chart of the basic steps in singing transcription.

The scope of this chapter is restricted to the transcription of monophonic singing performances in which only one note is sounding at a time. The presence of other instruments is not allowed. The singing style is not restricted but singing with lyrics, syllables, or humming is considered, and the 'quality' of the performance or the recorded signal is not limited in any way. It should be noticed that we do not try to transcribe lyrics but only the note pitch labels and their non-quantized timing information.

There are some research areas that are closely related to singing transcription. These include the analysis, synthesis, and transformations of singing voice [44], [337], the detection of singing-voice segments in polyphonic music [33], melody extraction from polyphonic music [435], [228], and *music information retrieval* (MIR) systems in general [455].

Currently, some transcription software are available on the Internet which are more or less suitable also for singing signals. These include the MAMI C++ library[1] for singing transcription [136] and commercial applications such as Autoscore, Digital Ear, Solo Explorer, and Akoff Music Composer. Free software demonstrations can be downloaded to give an idea of the current capabilities of software-based singing transcription. Also, there exist auto-tuning applications which correct inaccuracies in singing pitch instead of producing actual transcriptions. These are widely used in professional music production.

12.1.2 Organization of This Chapter

This chapter is organized as follows. Section 12.2 introduces singing-voice production by humans and considers the relationship between singing sounds and the corresponding musical notations. The rest of the chapter follows the basic steps of singing transcription systems shown in Fig. 12.1.

Section 12.3 discusses the pre-processing of singing signals and the measurement of acoustic features (such as the F0) from them. The pre-processing aims at making the feature extraction more robust against additive noise and the formant structure of singing sounds. Section 12.4 considers the conversion of acoustic features into a sequence of notes and rests. This involves the segmentation and labelling of notes, for which there exist two main approaches: the segmentation and labelling can be performed either jointly or as separate

---

[1]Available at http://www.ipem.ugent.be/MAMI

steps in cascade. Some systems also apply pre-trained acoustic or musicological models. The resulting note sequence may be exposed to some additional post-processing to improve the transcription quality. Section 12.5 concludes the chapter.

12.2 Singing Signals

The human vocal organ produces several types of voice sounds, such as speech, laughing, or whispering, in addition to singing. In contrast with the other voice sounds, singing exhibits musically meaningful information, and here we assume that singing sounds can be associated with notes in musical transcriptions.

In this section, we study the act of singing in order to understand the target signals of our transcription systems. First, we consider how the vocal organ produces singing sounds and controls the acoustical properties of them. Second, we discuss how humans associate singing sounds with notes and what are the problematic properties of singing signals from the transcription point of view. For an extensive study on the singing voice and its properties, the reader is referred to the work of Sundberg in [613], [614].

The following discussion concentrates on voiced sounds, i.e., sounds with a distinguishable pitch, although singing sounds can be considered as more or less modified speech sounds and include also unvoiced and transient-type sounds. These affect the note segmentation in singing. For a more extensive review of speech sounds, please refer to [536], [321].

12.2.1 Production of Singing Sounds

Singing sounds are produced by the human vocal organ, which consists of three basic units: (i) the respiratory system, (ii) the vocal folds, and (iii) the vocal tract [614]. The sound production process is as follows. First, the respiratory system creates an overpressure of air in the lungs, called the *subglottic pressure*, which results in an air flow through the vocal folds. The vocal folds start to vibrate and chop the air flow into a sequence of quasi-periodic air pulses, thus producing a sound with a measurable fundamental frequency. The sequence of air pulses is called the *voice source* and the process of sound generation via the vocal fold vibration is referred to as *phonation*. At the final stage, the voice source passes through the vocal tract, which modifies the spectral shape and determines the timbre of the voice sound. This stage is referred to as *articulation* and it controls the production of different speech sounds and lyrics in singing.

Figure 12.2 shows a block diagram of the singing-sound production process. The vocal-organ units control various acoustic properties of singing sounds, including their fundamental frequency, timbre, and loudness. These are discussed in more detail in the following.

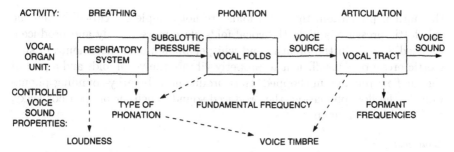

Fig. 12.2. The stages of the singing-sound production. The dashed arrows indicate the acoustic properties controlled by the three vocal organ units. (Modified from Sundberg [613, p. 10], used by permission.)

Fundamental frequency

The fundamental frequency of a singing sound is primarily controlled by the vocal folds. The term *phonation frequency* refers to the vibration frequency of the vocal folds and, during singing sounds, this is the fundamental frequency of the generated tone. The vibration is mainly controlled with the musculature of the vocal folds and to some extent with the amount of the subglottic pressure so that the greater the subglottic pressure, the higher the phonation frequency. In addition to the fundamental frequency, the generated tone includes frequency components called the overtone partials at the integer multiples of the fundamental frequency. In an idealized case, the amplitudes of the overtone partials can be expected to decrease by about 12 dB per octave [613, p. 64].

The phonation frequencies may range from around 100 Hz for male singers to over 1 kHz for female singers. The term *register* refers to a range of phonation frequencies where the singing sounds are produced in a similar manner, thus producing similar timbral characteristics. Basically, there exist two registers for male voices (normal and falsetto) and three registers for female voices (chest, middle, and head).

Timbre

The vocal tract acts as the most important controller of the singing sound timbre at the articulation stage. It functions as a resonating filter which emphasizes certain frequencies called the *formant frequencies*. These depend on the configuration of the articulators, including the jaw, the tongue, and the lips. In voiced sounds, the two lowest formants contribute to the identification of the vowel and the higher formants to the personal voice timbre.

In addition, singing sound timbre is affected by the amount of subglottic pressure and the tension of the vocal folds, resulting in different types of phonation. The phonation types include pressed, normal, flow, breathy, and whisper phonation, given in decreasing order of subglottic pressure. In the normal phonation, the vocal folds are not completely closed. When the subglottic pressure is high, the vocal folds close more rapidly and produce a 'pressed' sound. In the breathy and whisper phonations, the amount of subglottic pressure is insufficient to properly vibrate the vocal folds and a part of the air flow remains unchopped, thus producing a breathy-sounding phonation. In the flow phonation, the produced sound is neither pressed nor leaky, which is ideal for singing.

Loudness

The loudness of singing is influenced at several stages of voice production. The loudness is mainly controlled with the amount of subglottic pressure and the type of phonation. Although the subglottic pressure is smaller in the flow phonation than in the normal phonation, the sound pressure level of the voice source is maximal in the flow phonation.

Trained singers can modify their vocal tract configuration so that the phonation frequency and a formant frequency match in order to create a louder sound. This is known as the *singer's formant*. In the both extremes of the register, the control over loudness is more restricted, especially for untrained singers.

12.2.2 Singing Sounds and Notes

The phonation frequency enables the association of a singing sound with a musical note. If a singer could perform each note with a stable phonation frequency corresponding to the note pitch, the singing transcription task would be readily accomplished. However this is not the case, because singing performances possess both intentional and unintentional deviations from the nominal note pitches. Deliberate phonation-frequency deviations are commonly used to enhance the expressiveness of singing performances, whereas the unintended deviations are mostly due to the lack of voice training.

Figure 12.3 shows the fundamental frequency curve and the loudness curve of the note A4 performed by a female singer. Listening to the note performance, one unambiguously associates a note with A4, although the phonation-frequency curve is not stable during the note. To address this perceptual association of phonation frequencies with notes, the following discussion considers issues such as vibrato, tremolo, glissando, legato, and out of tune singing.

Vibrato and tremolo

The term *vibrato* refers to the modulation of the phonation frequency during a performed note. Vibrato can be characterized with the rate and the depth of the modulation. The rate typically varies between 4–7 Hz [425], and the depth between 0.3–1 semitones. In [530], the depth of vibrato was measured in ten recordings performed by professional singers and was reported to range from

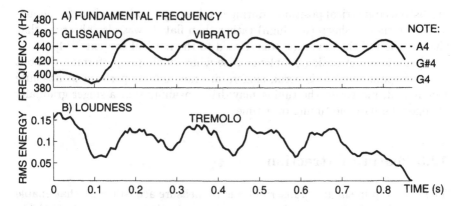

Fig. 12.3. A) The fundamental frequency curve and B) the loudness curve measured from a note A4 (440 Hz) performed by a female singer. The F0 curve clearly shows glissando and vibrato, whereas the loudness curve shows tremolo. Notice the correlation between these two.

0.34 to 1.23 semitones. Experimental results with human listeners indicate that the mean of the fundamental frequency during a vibrato note is close to the perceived pitch of the note [589].

The fluctuation of loudness is usually referred to as *tremolo*. In Fig. 12.3, the phonation-frequency and loudness curves imply vibrato and tremolo effect during the performed note, respectively. The vibrato rate is approximately 6.5 Hz and the vibrato depth varies from half a semitone to slightly over one semitone. The figure also shows a high correlation between vibrato and tremolo.

Glissando and legato

The term *glissando* refers to a phonation frequency slide to the note pitch. Glissando is usually employed at the beginning of long notes: these often begin flat (too low) and the phonation frequency is matched to the note pitch during the first 200 ms of a note [570]. Figure 12.3 shows a clear glissando in which the note begins over two semitones flat. Singing in legato means that consecutive notes are tied together, for example by using glissandi to change the note pitch.

Singing out of tune

Singing out of tune refers to a situation where a note pitch differs annoyingly from the tuning of the other notes within a performance. According to the listening test results reported by Sundberg in [614], the mean phonation frequency may deviate ±0.07 semitones from the nominal note pitch for notes which were generally judged to be in tune. However, mean phonation-frequency deviation larger than ±0.2 semitones can be acceptable at unstressed metrical positions, during tragic song mood, or when the phonation frequency is sharp (too high) rather than flat. It was also observed that the direction of the deviation was related to the musical context [614].

Other types of tuning problems arise when singing is performed without an accompaniment. In general, singers are not able to perform in absolute tuning and, moreover, the tuning may drift, meaning that a singer gradually changes the baseline tuning over time.

12.3 Feature Extraction

The first step in singing transcription is to measure acoustic cues that enable the detection of notes in singing signals. The acoustic cues are here referred to as *features*, and the term *feature extraction* refers to the measurement process. The most salient features include the fundamental frequency and features related to note segmentation. Features are commonly extracted frame-by-frame from acoustic waveforms and a pre-processing stage may precede this to facilitate the extraction process.

12.3.1 Pre-Processing

Pre-processing of the input signals aims at facilitating the feature extraction process. If singing has been recorded in a noisy environment or with low-quality equipment, the signal may include additive background noise which needs to be attenuated. The noise reduction problem has been studied for decades in speech processing and, to some extent, the same methods can be applied in singing transcription. A comprehensive study of noise reduction can be found in [638]. An issue related to noise reduction is audio restoration that aims at improving the quality of old music recordings, for which an interested reader is referred to [214].

The two main steps in noise reduction are the estimation of the noise component in the signal and its subsequent suppression. For example, Pollastri estimated the background noise spectrum during the time segments where singing activity was not detected [528]. The noise component was then suppressed using linear power-spectral subtraction.

In addition to noise reduction, spectral whitening can be applied at the pre-processing stage to flatten strong formants in the signal spectrum. This aims at facilitating the measurement of fundamental frequency, since it is most salient in the voice source signal before the vocal tract influence. In [627], inverse filtering with a warped linear prediction (WLP) model was applied to perform spectral whitening. The idea of WLP is the same as for ordinary linear prediction, but implemented on a frequency scale resembling that of human hearing.

12.3.2 Fundamental Frequency Estimation

Fundamental frequency is the most important feature in singing transcription systems. F0 estimation has been widely studied in speech processing, and most of the state-of-the-art singing transcription systems use algorithms originally designed for speech signals. The following survey introduces a few useful F0 estimation algorithms for singing transcription. For a more thorough review of different methods, see [289], [290], [618], [134].

Time-Domain Methods

Autocorrelation-based F0 estimators are widely used in singing transcription. The idea of these is to measure the amount of correlation between a time-domain signal and its time-shifted version. For periodic signals, the autocorrelation function has local maxima at time shifts that equal the fundamental period and its multiples.

Given a sampled time-domain signal $s(k)$ and a frame length W, the short-time autocorrelation function $r_t(\tau)$ at time t is defined as

$$r_t(\tau) = \sum_{k=t}^{t+W-1} s(k)s(k+\tau) \,, \qquad (12.2)$$

where τ is called the lag. Figure 12.4 shows $r_t(\tau)$ as calculated for a frame of a singing signal. The function $r_t(\tau)$ peaks at lags which correspond to the multiples of the fundamental period. A fundamental frequency estimate is obtained by dividing the sampling rate of the signal with the smallest non-zero lag value for which $r_t(\tau)$ reaches a value above a chosen threshold.

The autocorrelation method is straightforward to implement and easy to use. A drawback is, however, that the method is sensitive to formants in signal spectrum and therefore tends to make octave errors. Spectral whitening makes the autocorrelation more robust in this respect. Autocorrelation has been employed in the singing transcription systems of Ghias et al. [207], Shih et al. [587], [586], [588], and Wang et al. [659], for example.

YIN algorithm

The YIN algorithm for F0 estimation was proposed by de Cheveigné and Kawahara [135]. It resembles the idea of autocorrelation but introduces certain improvements that make it more convenient to use. The YIN algorithm was successfully used for singing transcription by Viitaniemi et al. [643] and Ryynänen and Klapuri [558]. Given that $s(k)$ is a discrete time-domain signal with sampling rate f_s, the YIN algorithm produces a F0 estimate as follows.

1. Calculate the squared difference function $d_t(\tau)$ where τ is the lag:

$$d_t(\tau) = \sum_{k=t}^{t+W-1} (s(k) - s(k+\tau))^2 \,. \qquad (12.3)$$

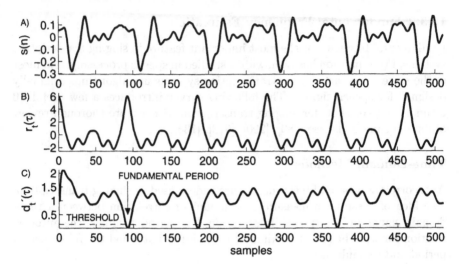

Fig. 12.4. A) An excerpt of a singing signal with sampling rate 44.1 kHz. B) The ACF $r_t(\tau)$ calculated using (12.2). C) The cumulative-mean-normalized difference function $d'_t(\tau)$ calculated using (12.4) (YIN algorithm). See text for details.

2. Derive the cumulative-mean-normalized difference function $d'_t(\tau)$ from $d_t(\tau)$:

$$d'_t(\tau) = \begin{cases} 1 \, , & \tau = 0 \\ d_t(\tau) \, / \, [(1/\tau) \sum_{j=1}^{\tau} d_t(j)] \, , & \text{otherwise.} \end{cases} \tag{12.4}$$

3. Search for the smallest τ for which a local minimum of $d'_t(\tau)$ is smaller than a given absolute threshold value κ. If no such value can be found, search for the global minimum of $d'_t(\tau)$ instead. Denote the found lag value with τ'.
4. Interpolate the $d'_t(\tau)$ function values at abscissas $\{\tau' - 1, \tau', \tau' + 1\}$ with a second-order polynomial.
5. Search for the minimum of the polynomial in the continuous range $(\tau' - 1, \tau' + 1)$ and denote the corresponding lag value with $\hat{\tau}$. The estimated fundamental frequency is then $f_s/\hat{\tau}$.

The normalization in (12.4) makes $d'_t(\tau)$ independent of the absolute signal level and removes a spurious local minimum around the lag zero, thus making the subsequent thresholding in Step 3 more convenient to implement. The threshold value $\kappa = 0.15$ was used in [558]. The interpolation in Step 4 increases the numerical accuracy of the F0 estimate.

Panel C in Fig. 12.4 shows the function $d'_t(\tau)$ calculated from the signal in panel A, and the dashed line indicates the threshold value $\kappa = 0.15$. The first local minimum below the threshold is found at $\tau' = 92$ and then refined to $\hat{\tau} = 92.37$ by the interpolation. For sampling rate $f_s = 44.1$ kHz, this yields a F0 estimate $f_s/\hat{\tau} \approx 477$ Hz. Compared to the ACF in panel B, $r_t(\tau)$ and $d'_t(\tau)$ exhibit maxima and minima at the same lag values. To conclude

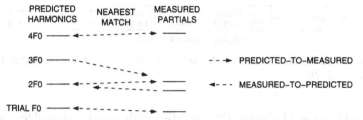

Fig. 12.5. The two-way mismatch error calculation is based on combining the predicted-to-measured and measured-to-predicted errors. The two-way matching makes the method robust for both 'too low' and 'too high' octave errors. (After Maher and Beauchamp [428].)

with, normalization and thresholding are important in autocorrelation-based methods in general, and the YIN algorithm proposes one useful alternative for this.

Frequency-Domain Methods

Cepstrum-based F0 estimators originated in speech processing (see Chapter 2 for the definition of the cepstrum). The fundamental period of a signal can be found from the cepstrum in a way similar to the ACF-based methods. Cepstrum-based F0 estimators perform well for speech sounds with strong formant structures, and they can also be used for singing transcription. However, these F0 estimators are rather sensitive to noise, since the logarithm operation involved in the cepstrum calculation raises the level of the noise floor in relation to harmonic components in the spectrum. A cepstrum-based F0 estimator was used in the query-by-humming system of Liu et al. [411], for example.

Maher and Beauchamp proposed a F0 estimation method for music signals which was based on a two-way mismatch procedure [428]. The method searches for the most likely fundamental frequency among a set of trial F0 values. The amplitudes and the frequencies of prominent sinusoids are first measured from the short-time spectrum of a signal frame. Given a trial F0, the measured partials are compared to the predicted harmonics of the trial F0 (measured-to-predicted error) and vice versa (predicted-to-measured error), thus producing two error measures for the F0 estimate in question. Figure 12.5 illustrates the matching process. The two error measures are then combined to derive a total error measure. The trial F0 with the smallest total error is considered as the best F0 estimate within the frame. Details of the algorithm can be found in [428]. The algorithm is relatively straightforward to implement, and it has been used to separate duet signals [427] and to analyse the melody line in polyphonic recordings [519].

Peak-picking F0-estimation algorithms search for peaks in the magnitude spectrum and group them to derive an F0 estimate. In the systems of Pollastri

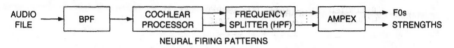

Fig. 12.6. A block diagram of the auditory model-based F0 estimator used by Clarisse et al. [92].

[528] and Haus and Pollastri [278], an F0 estimate was determined by searching for the most prominent peak in a spectrum between 80 Hz and 800 Hz for which at least two clear harmonics were found. Another peak-detection scheme was applied by Zhu and Kankanhalli in [689], where a set of the highest peaks in the spectrum were detected and a rule-based peak-grouping algorithm was employed to obtain an F0 estimate.

Auditory Model-Based Methods

Clarisse et al. used an auditory model-based F0 estimator in their singing transcription system [92]. Figure 12.6 shows the block diagram of the method. The algorithm was originally designed for speech signals and its detailed explanation can be found in [637]. In brief, the algorithm works as follows.

An audio file is first bandpass filtered to model the frequency response of the outer and the middle ear. The signal is then fed into a cochlear processor, where a filterbank models the frequency selectivity of the inner ear. The signal at each band is processed by a hair-cell model in which the signal is half-wave rectified (i.e., negative signal-sample values are set to zero) and its dynamic range is compressed. The output of each hair-cell model is followed by a lowpass filter which extracts the envelope of the signal in the channel.

The envelope signals in different channels are processed by a pitch extraction module called AMPEX. It performs pseudo-autocorrelation analysis and sums the results across channels to obtain a global pseudo-autocorrelation function. The major peaks in the function are searched for to obtain a set of pitch candidates and their corresponding strengths. The final pitch estimate is chosen by analysing the continuity of the pitch estimates in the analysis frame and its surrounding frames. If the strength of the final pitch estimate exceeds a threshold value, the frame is considered to be voiced. Later, De Mulder et al. concluded that the AMPEX algorithm was unsuitable for the F0 estimation of very high-pitched sounds and proposed another F0 estimation algorithm working in parallel with AMPEX [137].

A drawback of the auditory model-based algorithms is that they are computationally quite demanding. As an exception to this, Tolonen and Karjalainen have proposed an auditory model-based F0 estimator which is computationally efficient [627]. The method is described in Chapter 8 but, to our knowledge, it has not yet been applied to singing transcription.

12.3.3 Features for Note Segmentation

Singing transcription requires the temporal segmentation of notes, i.e., the determination of where notes begin and end in time. This is usually done based on two types of features: those that indicate silence, noisy, or voiced segments in audio signals; and those that imply the beginning or the ending of a note. The first type of features include signal energy, the degree of voicing, and zero crossing rate, and the second type of features include phenomenal and metrical accents (to be explained below).

Energy

Features related to the signal energy are widely used for note segmentation, with the assumption that the signal level reflects the loudness of the target singing voice. Segments where the signal energy exceeds a given threshold value are often considered to be notes and the other segments are treated as silence or background noise. Energy-related measures are straightforward to use and work well if the notes are separated with more quiet regions. However, this is not the case in most signals; usually there are also legato-type note transitions. Therefore, robust note segmentation cannot be based on energy measures alone.

One of the most often used energy measures is the root-mean-square (RMS) energy (see Chapter 6 for the RMS definition). This feature has been applied by Haus and Pollastri [278], Pollastri [528], McNab et al. [454], [455], and Shih et al. [587], [586], [588], for example. Different variants of energy calculations have been applied in note segmentation, including the systems by Clarisse et al. [92], Liu et al. [411], and Orio and Sette [489].

Voicing

A more reliable feature for note segmentation is the *degree of voicing* of a signal frame. Voiced frames possess clear periodicity, whereas unvoiced frames can represent transient noise with a great amount of signal energy, or just silence. Commonly, the voicing determination is embedded within the F0 estimation algorithms. In autocorrelation-based F0 estimators, for example, the degree of voicing is straightforwardly given by the ACF function value at the lag corresponds to the estimated fundamental period, divided by the value at lag zero. If the ratio of these two does not exceed a given threshold, the frame is considered to be unvoiced.

When the YIN algorithm is used for F0 estimation, the degree of voicing can be directly derived from the value $d'_t(\hat{\tau})$ of the cumulative-mean-normalized difference function defined in (12.4). The $d'_t(\hat{\tau})$ value itself describes the amount of non-periodicity in F0 measurement. To obtain a voicing value, we have to map the $d'_t(\hat{\tau})$ value to a voicing feature for example as

$$\nu(\hat{\tau}) = 1 - d'_t(\hat{\tau}) . \tag{12.5}$$

The voiced/unvoiced decision can also be based on the zero crossing rate (ZCR) together with an energy measure, as done in [528], [421]. ZCR loosely describes the brightness of the sound. High ZCR values imply transient or noisy segments since these tend to have lots of energy at high frequencies. ZCR is defined in Chapter 6.

Accents

Features indicating note beginnings include the degree of phenomenal accent and the degree of metrical accent as a function of time. Phenomenal accents refer to moments that have perceptual emphasis in music signals, whereas the metrical accent corresponds to the underlying pulse of a music performance [404].

Klapuri proposed a method for estimating both of these accents in music signals [349]. In his system, a (phenomenal) *accent signal* is measured based on the amount of spectral change as a function of time. Briefly, the signal power is first measured at 36 sub-bands. At each band, the power of the signal is computed and then log-compressed and smoothed over time by lowpass filtering. The power envelopes are then subjected to time differentiation and half-wave rectification to measure changes in intensity. Finally, the resulting signals are summed across the bands to produce the accent signal. The *metrical accent signal* is constructed by inferring regularities in the accent signal and by estimating the temporal pulse of the music performance. The method of Klapuri can be used as a tempo estimator and is capable of following the tempo changes during music performances. For further discussion on beat-tracking and musical metre analysis, see Chapter 4.

Ryynänen and Klapuri used the accent signal and the metrical accent signal as features in a singing transcription system [558]. The accent signal indicated the singing note beginnings whereas the metrical accent signal was used to predict possible note beginnings according to the estimated tempo of the performance.

12.3.4 Summary of the Features and an Example

Table 12.1 summarizes the features discussed in this section. Figure 12.7 shows a selection of features extracted from a short singing excerpt containing five notes performed by a professional female singer. From top to bottom, the panels in the figure show the recorded singing waveform, fundamental frequency estimates, the degree of voicing, RMS energy, the accent signal, and the zero crossing rate. The panel with F0 estimates also shows a manual transcription of the notes in the performance (B4, A4, A4, F♯4, and A4). The F0 estimates and the degree of voicing were obtained using the YIN algorithm and (12.5).

At the beginning of the excerpt there is silence, which can be observed from the RMS and the degree of voicing. When the first note begins, the accent signal has a clear peak. Interestingly, the second note begins with an

Table 12.1. Acoustic features for singing transcription.

Feature	Usage	Methods	Comments
F0	note labelling	autocorrelation	well-motivated & understood
		YIN	related to the ACF
		cepstrum	robust against formants, sensitive to noise
		two-way mismatch	rather robust, intuitive
		auditory model-based	computationally complex
Energy	segmentation	RMS + variants	reflects the loudness
Voicing	segmentation	ACF value	side-product of F0 estimation
		ZCR	reflects the brightness
Accents	segmentation	(phenomenal) accent	indicates note onsets
		metrical accent	predicts note onsets

almost two-semitones flat glissando. Simply by following the F0 estimates, the beginning of this note would be easily interpreted as note G4 ($\approx$ 390 Hz). In a manual transcription, however, the choice of a single note A4 is obvious. The last note includes strong vibrato and tremolo as well as a glissando in the beginning of the note. Although the accent signal indicates note beginnings quite reliably, especially strong vibrato causes false peaks to the accent signal, as can be seen during the last note. The zero crossing rate has a peak at time 2.8 seconds caused by the consonant /s/.

As the example shows, it is not always obvious what notes should be transcribed when given the extracted features, and this makes the singing transcription task rather difficult. Conversion of the frame-level features into note pitch labels is considered in the next section.

12.4 Converting Features into Note Sequences

The greatest challenge in singing transcription lies in the conversion of frame-level features into note sequences. Singing performances exhibit inaccuracies both in pitch and timing, usually as a consequence of expressive singing or the lack of voice training. Therefore, a singing transcription system may not assume that the notes are performed in an absolute tuning, that the phonation frequency is stable during the notes, or that the notes are performed in a rhythmically accurate manner.

The features-to-notes conversion involves note segmentation and labelling. These two steps can be performed either (i) in a cascade or (ii) jointly. The former approach is here referred to as the *segment-and-label approach* where singing recordings are first segmented into notes and rests, and each note segment is then assigned a pitch label. The joint segmentation and labelling of notes usually applies statistical models of note events and is here

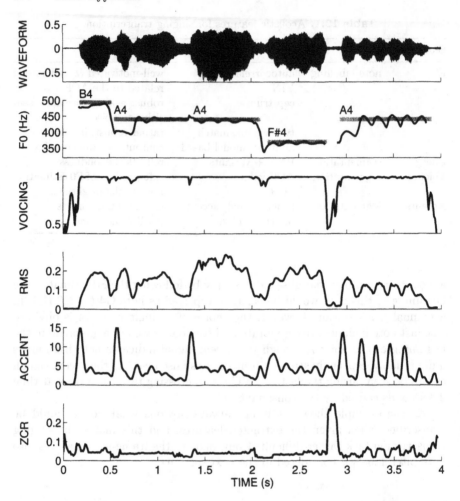

Fig. 12.7. Selected features extracted from an excerpt performed by a professional female singer. The panels show (in top-to-bottom order) the singing recording waveform F0 estimates, the degree of voicing, RMS energy, the accent signal, and the zero crossing rate. The excerpt was manually transcribed to contain five notes, B4, A4, A4, F♯4, and A4, and they are shown in the panel of fundamental frequencies. See text for details.

referred to as the *statistical modelling approach*. Both of these approaches may apply musical rules or musical context to obtain a better transcription result. At an optional post-processing stage, the transcription can be enhanced, for example by adding a parametric representation of an estimated expression to the transcription.

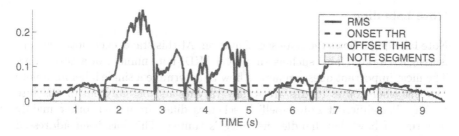

Fig. 12.8. Note segmentation based on RMS-energy thresholding.

12.4.1 Segment-and-Label Approach

As mentioned above, the idea of the segment-and-label approach is to first segment an audio signal into notes and rests and then to assign a pitch label to each note segment. This approach has been taken in the singing transcription systems of McNab et al. [454], [455], Haus and Pollastri [278], Pollastri [528], Lu et al. [421], Wang et al. [659], and Clarisse et al. [92].

Note Segmentation

Note segmentation is usually based on features related to the signal energy or voicing, or on abrupt changes in fundamental frequency. The methods usually apply fixed or adaptive thresholds to decide whether a note boundary has occurred or not.

McNab et al. used the RMS energy for note segmentation [454]. When the signal energy exceeded a given threshold, the method interpreted this as a note onset, and similarly, when the signal energy dropped below another threshold, a note offset was inferred. In addition, the segment had to be at least 100 ms long to be accepted. Figure 12.8 shows the measured RMS contour of ten singing notes performed with a /da/ syllable. The onset and offset thresholds are set to 50% and 30% of the overall RMS energy as in [454]. The note offsets are accurately found, whereas for example the first note onset is detected at 1.1 s although it should be at 0.5 s.

As an alternative for the energy-based note segmentation, McNab et al. used a scheme where slowly varying regions of F0 estimates were grouped. A note boundary was detected when an F0 estimate deviated more than half a semitone from the average of the previous estimates during the segment.

Pollastri used both the signal energy and F0 changes for note segmentation, in addition to ZCR, which was used to make voiced/unvoiced decisions [528]. Clarisse et al. performed note segmentation based on adaptive adjustment of threshold values for signal energy. Later, the system of Clarisse et al. was improved by adding a number of decision rules for note segmentation, so as to handle notes with legato, vibrato, and tremolo [137].

Note Labelling

Note labelling follows the note segmentation. At this stage, each note segment is assigned a pitch label such as an integer MIDI note number or a note name. The most important question here is how to determine a single label for a note segment where F0 estimates are widely varying and possibly out of absolute tuning. The different note-labelling schemes differ from each other mainly in terms of how they handle the singer's tuning. This has been addressed using three different assumptions:

1. The singer performs in absolute tuning (note A4 corresponds to 440 Hz);
2. The singer has a relative tuning in mind and makes consistent deviations from the absolute tuning;
3. The singer allows the baseline tuning to drift during a performance.

Corresponding to the three assumptions, the transcription systems either perform no tuning, estimate a constant tuning, or perform time-adaptive tuning, respectively. Three systems applying these three approaches are briefly introduced in the following.

No tuning

The system of Clarisse et al. aimed at transcribing notes as precisely as possible according to singing pitch instead of trying to infer the intention of the singer [92]. The F0 for a note segment was calculated as the arithmetic mean of the F0 estimates (Hz) in the central part of the note segment. The F0s which differed more than 10% from the mean were discarded and the mean was recalculated. The note pitch label was then obtained by rounding the resulting mean F0 into an integer MIDI note number.

Constant tuning

Haus and Pollastri assumed that the performed notes differ a constant amount in semitones from the absolute tuning [278]. Figure 12.9 shows a block diagram of their pitch-labelling process. Given a note segment, the F0 estimates within the segment were first 3-point median filtered to remove F0 outliers. Then a group of four contiguous frames with similar F0 values constituted a block. At the block level, legato with note pitch change was detected when pitch between adjacent blocks had changed more than 0.8 semitones. In the case of a detected legato, the note segment was divided into two new segments. Otherwise, the adjacent blocks constituted a note event for which the F0 was calculated as the arithmetic mean of the F0 in the blocks and represented as an unrounded MIDI note number. This process was repeated for each note segment, resulting in note segments with unrounded MIDI note labels.

To determine the deviation from the absolute tuning, the authors calculated a histogram of distances from the unrounded MIDI note numbers to their nearest integer numbers. The highest peak in the histogram then indicated the

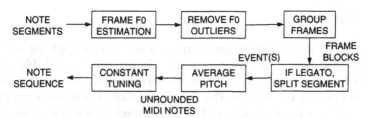

Fig. 12.9. Pitch-labelling process of Haus and Pollastri [278].

most-often occurring offset from the absolute tuning and, subsequently, every note segment label was shifted by this offset, thus minimizing the rounding error. Finally, the shifted MIDI note labels were rounded to integers, thus obtaining the note pitch labels.

Time-adaptive tuning

McNab et al. considered a labelling technique which continuously adapted to the singer's tuning during a performance [454]. For the first note segment, a histogram of the F0 estimates in the segment was calculated and the most prominent peak in the histogram was rounded to a MIDI note number. For the next note segment, however, the histogram was shifted by the amount of the rounding done in labelling the previous note. In other words, the tuning was constantly adjusted according to the rounding errors made while labelling the preceding notes. This approach, however, tends to adjust tuning too much at a time. In the worst case, the maximum rounding error in each note labelling can be half a semitone. Thus the reference tuning might have changed one octave (twelve semitones) after labelling 24 notes.

12.4.2 Statistical Modelling of Notes

The statistical approach discussed in this section applies pre-trained *acoustic models* of note events and performs the note segmentation and labelling jointly. This has certain advantages in the singing transcription problem in particular. First, it is theoretically advantageous to optimize the note segments and their labels jointly instead of doing these as two consecutive steps. Secondly, statistical models allow representing uncertain information and learning from examples, thus providing a more convenient way to deal with the highly varying acoustic data.

Statistical note models are derived by measuring low-level acoustic features during note events and by training a parametric statistical model to describe the behaviour of the features during the events. The models can then be used to calculate the likelihoods of different notes in previously unseen material. The note sequence which maximizes the overall posterior probability according to the note models is considered to be the transcribed note sequence.

Hidden Markov models

In practice, all the current singing transcription systems applying statistical note modelling have been based on *hidden Markov models (HMMs)*. HMMs are widely used to model phonemes and words in automatic speech recognition and, therefore, the theory of HMMs is well established. The following presentation introduces the basic idea of HMMs and their usage in note modelling. For a more formal consideration of HMMs, see Chapter 2. A good tutorial on HMMs can be found for example in [534].

A hidden Markov model is a state machine which models stochastic signal sources. The HMM is mathematically formulated by the following parameters:

1. A *set of states* $\Theta = \{e_1, e_2, \ldots, e_E\}$ within the model where E is the number of states.
2. *State-transition probabilities* $P(\theta_t = e_j \mid \theta_{t-1} = e_i)$, i.e., the conditional probabilities that state e_i is followed by state e_j within a random state sequence $\theta_{1:t}$. In particular, a direct transition from e_i to e_j is not possible if $P(\theta_t = e_j \mid \theta_{t-1} = e_i) = 0$.
3. The *observation likelihood distributions* $p(\mathbf{x}_t \mid \theta_t = e_j)$, i.e., the likelihoods that an observation vector $\mathbf{x}$ is emitted from state $e_j \in \Theta$. Here the observation vector consists of the acoustic features extracted in frame t.
4. The initial probabilities $P(\theta_1 = e_i)$ and the final state probabilities $P(\theta_T = e_i)$, i.e., the probabilities that a state sequence begins or ends to state e_i.

The state-transition probabilities, the observation likelihood distributions, and the initial and the final state probabilities can be learned from acoustic data either with supervised or unsupervised training techniques, depending on whether the underlying state sequence $\theta_{1:t}$ is known a priori or not. In the case of singing transcription, the training data consists of the frame-level features extracted from singing recordings, and the training is usually performed in an unsupervised manner with the Baum–Welch algorithm, which is described for example in [534].

Once the HMM parameters have been learned, the state sequence which maximizes the posterior probability of the observed data can be estimated using the *Viterbi algorithm* [654]. See Chapter 2 for a description of the Viterbi algorithm. An alternative state-sequence estimation scheme is the *token-passing algorithm*, which is designed for finding the most probable path through a network of connected HMMs. For the algorithm details, see [685]. Briefly, the token-passing algorithm propagates *tokens* which represent alternative paths through the network. Inside a HMM, each state contributes to the weight of the tokens by the observation likelihoods and the state-transition probabilities. When a token is emitted out of an HMM, the algorithm identifies the boundary between the connected HMMs and appends it to a list of boundaries for the purpose of backtracking the most probable path after the analysis. The tokens are further weighted with predefined transition probabilities between different HMMs. Eventually, the most probable path is defined by the token with the maximum posterior probability and the corresponding boundaries between different HMMs in the network.

Note-Event Model of Ryynänen and Klapuri

Ryynänen and Klapuri proposed a singing transcription system based on modelling note events with HMMs [557], [558]. The system performs note segmentation and labelling simultaneously and also utilizes the musical context in singing performances. The note-event model and its usage are described in the following, whereas the use of musical context is considered later in Section 12.4.3.

Note-event model

Note events are described with a three-state left-to-right hidden Markov model where $P(\theta_t = e_j \mid \theta_{t-1} = e_i) \neq 0$ only when $j = i$ or $j = i + 1$. The state e_i in the model represents the typical acoustic characteristics of the ith temporal segment of a performed singing note. The model uses three features: fundamental frequency estimates (represented as unrounded MIDI note numbers), the degree of voicing, and the accent signal. The features are extracted as explained in Section 12.3. Different notes are represented with a separate HMM for each MIDI note $n = 36, \ldots, 79$. For note n, the features in frame t form the observation vector $\mathbf{x}_t$ where the difference between the fundamental frequency estimate and note n is used instead of the F0 estimate directly. This is referred to as the *pitch difference* ΔF_0:

$$\Delta F_0 = F_0 - n . \tag{12.6}$$

The use of pitch difference facilitates the training of the model. Usually, there is a limited amount of training data available, at least for each possible singing note. Due the pitch-difference feature, it is possible to train only one set of note HMM parameters with greater amount of data, and the same parameters can be used to represent all the different MIDI notes.

The state-transition probabilities and the observation likelihood distributions were estimated from an acoustic database containing audio material performed by eleven non-professional singers. The singers were accompanied by MIDI representations of the melodies which the singers heard through headphones while performing. Only the performed melodies were recorded, and later, the reference accompaniments were synchronized with the performances. The reference notes were used to determine note boundaries in the training material, and the Baum–Welch algorithm was then used to learn the note HMM parameters.

Figure 12.10 illustrates the trained note HMM. The HMM states are shown on top of the figure where the three states are referred to as the attack,

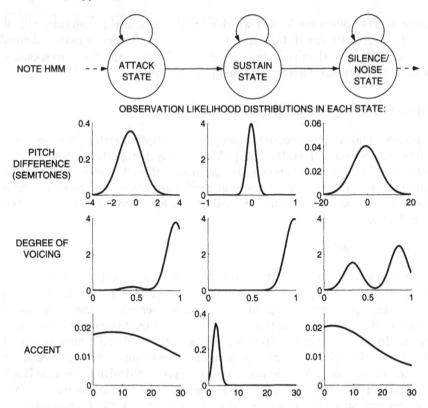

Fig. 12.10. A note HMM with three states: an attack, a sustain, and a silence/noise state. The arrows show the possible state-transitions. The observation likelihood distributions are shown below each state for the features used (pitch difference, the degree of voicing, and accent).

the sustain, and the silence/noise state, each of which corresponds to a time segment within note events. The arrows show the possible transitions between the states. The observation likelihood distributions shown below each state express the typical behaviour of these features during the different segments within note events. It is a little surprising that unsupervised learning leads to such an intuitive interpretation of the three states of the note model:

Attack: The singing pitch may vary about a few semitones from the actual note pitch. Since the mean of the pitch-difference distribution is at -0.5 semitones, this implies that notes usually begin slightly flat. Accent value distribution is widely spread, which indicates the presence of large accent values during note attacks.

Sustain: The variance of pitch difference is much smaller than in the attack state. Most of the F0 estimates stay within ± 0.2 semitone distance from the nominal pitch of the note. In addition, the frames during the sustain state are mostly voiced (i.e., the F0 estimation can be reliably performed) and the accent values are small.

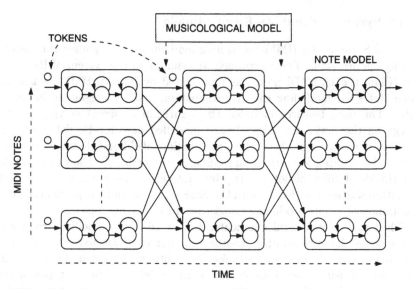

Fig. 12.11. Combination of the note models and the musicological model.

Silence/noise: The F0 estimates are almost random valued, since most of the frames are unvoiced (notice the high likelihood of voicing values near 0.3).

The authors did not consider modelling the rests, since the note model itself includes a state for silent or noisy segments. Thus the time instants where the silent/noise state is visited can be considered as rests.

Using the note event model

The HMMs of different MIDI notes are joined into a note model network illustrated in Figure 12.11. The probability of a transition from one note to another is determined by a musicological model (explained in Section 12.4.3). Notice that the figure shows the note models at successive time instants although there actually exists only one note model per MIDI note. Singing melodies can be transcribed by finding the most probable path through the network according to the probabilities given by the note HMMs and the musicological model. The authors used the token-passing algorithm for this purpose.

The transcription system of Ryynänen and Klapuri was evaluated using 57 singing melodies from the same database that was used for training the note models (the training signals were not included in the evaluation set). An error rate below 10% was achieved when the transcriptions were compared to the reference MIDI notes. Simply rounding F0 estimates to MIDI note numbers produced an error rate of about 20%. The use of note models reduced the error rate to 13%, and including also the musicological model decreased the error rate below 10%. A similar note model network was later applied by the authors to the transcription of polyphonic music from various genres [559].

Other Systems with Note-Event Models

Orio and Sette used a HMM-based note modelling technique for transcribing singing queries [489]. Their approach was similar to the system of Ryynänen and Klapuri: each MIDI note was represented by its own HMM, which had three states for modelling the attack, sustain, and rest segments in singing notes. The used features included the logarithm of signal energy, spectral energy on the first harmonics of each modelled note for F0 detection, and the first derivatives of these. The observation likelihood distributions were derived by making statistical analysis of a set of labelled audio examples. The HMMs of different notes were joined into a note network, and the Viterbi algorithm was used to decide both the note segments and the pitch note labels simultaneously, thus producing transcriptions of singing queries.

Their preliminary results indicated that most of the labelling errors were due to glissandi between notes. Another common error was to insert additional note boundaries during note attacks. The authors discussed the interesting possibility of using several attack states for modelling different types of note beginnings and an enhanced sustain state with two additional states modelling slight detunings upwards and downwards from the note pitch. However, these ideas were not implemented in the reported system.

Viitaniemi et al. proposed an HMM for singing transcription, where the HMM states corresponded to different notes [643]. The observation likelihood distributions consisted of the F0 value distributions for each note and the state-transitions were controlled with a musicological model. The optimal state sequence was found with the Viterbi algorithm in order produce note labels at each time instant. In the optimal state sequence, transitions between different states were interpreted as note boundaries. The system achieved error rates around 13% for the database that was used by Ryynänen and Klapuri in [558].

Shih et al. used a three-state HMM to model note events [587]. The features included mel-frequency cepstral coefficients (MFCC), one energy measure, and a pitch ratio. The pitch ratio was calculated as

$$\log(F_0) - \log(F_0^{\text{ref}}) , \qquad (12.7)$$

where F_0 (Hz) denotes a fundamental frequency estimate and F_0^{ref} a reference F0. Two different definitions for the reference F0 were tested: F_0^{ref} was defined as the arithmetic mean of the F0 values (i) in the first detected note segment or, alternatively, (ii) in the immediately preceding note segment. In the former case, they trained 15 note models describing one octave of a major scale upwards and downwards from the first detected note. In the latter case, they trained models for one and two semitone intervals upwards and downwards with respect to the previous note. In addition, they defined a model for segmenting the first note. Once the first note of a performance is segmented, the mean F0 is calculated as the reference F0. Depending on the selected set of HMMs, the reference F0 was then either kept fixed or updated after each detected note, leading to the constant and the time-adaptive tuning, respectively (see Section 12.4.1). The time-adaptive tuning was reported to work better for which approximately 80% of notes were correctly labelled.

Shih et al. soon modified their system to transcribe singing in two consecutive stages of note segmentation and labelling [588]. The note segmentation was performed with two three-state HMMs, one of which modelled note segments and the other rests. The note and rest segments were then found using the Viterbi algorithm. After a note segment was detected, it was labelled with respect to its previous note by using pitch interval in semitones and the standard deviation of F0 estimates within the note segment. These attribute values were modelled with a Gaussian mixture model (GMM) and the interval label was decided according to the most likely GMM. Some minor improvements in transcription quality were achieved compared to their previous system. In addition, the improved system operated in real time.

12.4.3 Utilizing Musical Context

So far, we have discussed singing transcription without any prior knowledge about the musical relevance of individual notes or their relationships. *Musical context* plays an important role in singing: some notes and note sequences are much more probable than others when the whole performance is treated as a musical entity. This section considers how to exploit the knowledge about musical context in singing transcription.

Musical Key

The tonality of a singing performance is largely determined by its *musical key*, which characterizes the basic note *scale* used within the performance (see Chapter 1 for a discussion of tonal music). The first note of the scale is called a *tonic note*, which also gives the name for the key. For example, the C major scale consists of the notes C, D, E, F, G, A, and B (the white piano keys), where the intervals between the notes in semitones are 2, 2, 1, 2, 2, and 1, respectively. The natural A minor scale consists of the notes A, B, C, D, E, F, and G, where 2, 1, 2, 2, 1, and 2 are the corresponding intervals. Since C major and natural A minor scales actually contain the same notes, the corresponding keys (C major and A minor) are referred to as *relative keys* (or the relative-key pair). In general, the keys are relative if they correspond to a major and a natural minor scale with the same notes.

The term *pitch class* stems from the octave-equivalence of notes. That is, notes separated by an octave (or several octaves) belong to the same pitch class, so that for example C is the pitch class of C3, C4, and C5. Krumhansl reported the occurrence frequencies of different pitch classes with respect to the tonic note of a piece, measured from a large amount of classical music [377]. The distributions are shown in Fig. 12.12, where the pitch-class names are listed for the relative keys C major and A minor. As can be observed,

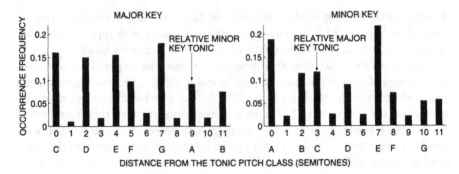

Fig. 12.12. Pitch class occurrence frequencies in major and minor keys with respect to the tonic pitch class (after Krumhansl [377, p. 67]). As an example, the pitch class names are listed below the figure axes for the relative keys C major and A minor.

pitch classes such as C, E, and A often occur in both of these keys, and the pitch classes which belong to the scale of the key are much more frequent than the others. To conclude, given the key of a singing performance, we may determine the probabilities of different pitch classes from a musical point of view and use this knowledge to solve ambiguous note labellings.

A musical key estimation method

Viitaniemi et al. proposed a key estimation method which produces the probabilities of different keys given a sequence of F0 estimates [643]. The required prior knowledge consists of the occurrence probabilities of different notes n given a key k, $P(n \mid k)$. As only the pitch class of a note is assumed to affect the mentioned probability, distributions such as those shown in Fig. 12.12 can be used. Then the probability of a key given a note is obtained using Bayes's formula:

$$P(k \mid n) = \frac{P(n \mid k)P(k)}{P(n)} \ . \tag{12.8}$$

Further, the authors used singing performances as training data to estimate the probabilities of different F0 values given a note n, i.e., $P(F_0 \mid n)$, which was modelled with a GMM. This was to represent singing pitch deviation from the nominal note pitch, and the F0 estimation errors. Then the probability of a key given an F0 estimate was calculated as

$$P(k|F_0) = \frac{P(F_0 \mid k)P(k)}{P(F_0)} \overset{\text{(assumpt.)}}{=} \sum_{\text{all notes } n} P(F_0 \mid n)P(n \mid k)\frac{P(k)}{P(F_0)} \ ,$$

where it was assumed that the probability of an F0 estimate is independent of the key when the note is given, i.e., $P(F_0 \mid n, k) = P(F_0 \mid n)$. The probability of key k given a sequence of F0 estimates, $O = (F_0(1), \ldots, F_0(T))$, was then calculated as

$$P(k \mid O) = \prod_{t=1}^{T} P(k \mid F_0(t)) .$$ (12.9)

The method was reported to find the correct relative-key pair in 86% of singing performances [643]. The estimated key probabilities were applied as a musicological model to favour certain notes in singing transcription. At time t, the probability of note n_t given the sequence of F0 estimates up to time t was defined as

$$P(n_t \mid O) = \sum_{k=1}^{24} P(n_t \mid k) P(k \mid O) ,$$ (12.10)

where the index k sums over all the 24 major and minor keys. It should be noted that the above equation gives the probability of note n at time t from the viewpoint of the musicological model alone, without any special emphasis to the most recent observation $F_0(t)$.

Note N-Grams

Note N-grams formulate the idea that the probability of a note depends on the previous $N-1$ notes. The probability of a note n_t given the previous notes $n_{1:t-1}$ is here denoted by $P(n_t \mid n_{1:t-1})$. N-grams are based on $(N-1)$-th order Markov assumption which can be written as

$$P(n_t \mid n_{1:t-1}) = P(n_t \mid n_{t-N+1:t-1}) .$$ (12.11)

The note N-gram probabilities are estimated from databases containing note sequences. In practice, the databases do not include all the possible note sequences, and therefore a method is usually employed to estimate also the probabilities of the non-occurring note sequences. In general, this process is called *smoothing*. For different smoothing methods, see [321].

Ryynänen and Klapuri applied note N-gram probabilities for $N \in \{2,3\}$ under a given key to control the transitions between different note HMMs [558] (see Fig. 12.11). The probability of note n_t at time t was defined as $P(n_t \mid n_{\mathrm{prev}}, k)$, where k denotes the key and $n_{\mathrm{prev}} = n_{t-N+1:t-1}$ is used to denote the $N - 1$ previous notes for convenience. The probabilities were estimated by counting the occurrences of different note sequences in a large database of monophonic MIDI files and by smoothing them with the Witten–Bell discounting algorithm [673].

The estimated note N-gram probabilities were applied in singing transcription as follows. First, the major and the minor keys $\hat{k}_{\mathrm{maj}}$, $\hat{k}_{\mathrm{min}}$ of the most probable relative key-pair were determined from the singing performance using the key estimation method of Viitaniemi et al. [643]. Then the probability of moving to note n_t was obtained by

$$P(n_t \mid n_{\mathrm{prev}}) = \frac{P(n_t \mid n_{\mathrm{prev}}, \hat{k}_{\mathrm{maj}}) + P(n_t \mid n_{\mathrm{prev}}, \hat{k}_{\mathrm{min}})}{2} .$$ (12.12)

If the key information was not available, the probability was given by

$$P(n_t|n_{\text{prev}}) = \frac{1}{24} \sum_{k=1}^{24} P(n_t|n_{\text{prev}}, k) , \qquad (12.13)$$

that is, assuming all the major and minor keys equally probable. If the musicological model was completely disabled, equal transition probabilities were used for all note transitions.

Simulation experiments with the described musicological model showed that the use of note N-grams without the key information did not improve transcription results compared to the situation where the musicological model was completely disabled. When using the key information with the note N-grams, however, the error rate was reduced from 13% to under 10%, where the note bigrams ($N = 2$) worked slightly better than the note trigrams ($N = 3$) [558], [557].

Metrical Context

Metrical context refers to the temporal arrangement of notes with respect to the musical tempo of the performance. Both the note beginnings and their durations are usually related to the tempo. As an example of the note durations, given a tempo of 100 beats per minute, the duration of quarter notes is around $60\,\text{s} / 100 \approx 0.6\,\text{s}$, a half of this for eighth notes, two times the duration of quarter notes for the half notes, and so forth. The note beginnings, on the other hand, are usually located at the positions of the beats of an underlying pulse of the performance (metrically strong positions). For a further discussion on musical metre analysis, see Chapter 4. The singing notes, however, are not necessarily performed at a constant tempo due to the expressive nature of singing performances.

Viitaniemi et al. tested the explicit modelling of note durations within an HMM [643]. They weighted the transitions between HMM states (corresponding to different notes) by note-duration likelihoods estimated from a large amount of MIDI files. However, it was reported that the duration modelling did not improve transcription results. Ryynänen and Klapuri estimated the tempo during singing performances and derived a metrical accent function based on the estimated tempo [558]. The metrical accent function was used as an acoustic feature in the note event model in order to predict note beginnings in singing performances. However, the advantage of using the metrical accent was found to be insignificant compared to the increased complexity of the transcription system.

Summary

To conclude, the utilization of musical context can significantly improve the accuracy of singing transcription systems, although the techniques for musical context modelling are still under development. If the musical key or the tempo of a performance is known in advance, the quality of the transcription is very likely to be improved. Use of the tonal context for pitch labelling appears to be more important than the use of the metrical context for note segmentation. However, the metrical context becomes essential if in addition the note durations are quantized, as described in Chapter 4.

12.4.4 Post-Processing and Encoding Expression

Note-sequence post-processing can be used to encode the expressive aspects of a singing performance into the transcription, or to perform time quantization. The term *time quantization* refers to the process of associating note onsets and durations with a discrete temporal grid. Time quantization is needed, for example, in singing-to-score types of applications. A reader interested in time quantization is referred to the work of Cemgil et al. [75], [77], and to Chapter 4 in this volume.

The focus in this section is on the expression encoding. Singing performances are often quite expressive in nature, and note sequences are insufficient representations of singing performances in this respect. If singing transcription is used in an application where the transcribed note sequence is resynthesized, the plain note sequence sounds rather dull. Therefore, the transcriptions for this type of applications should contain additional data representing the expressive aspects of singing performances, such as glissandi, vibrato, tremolo, and loudness changes during the notes. In addition, traditional music notation provides some limited number of expression symbols, such as loudness indicators (*pp, p, mf, ff*). These can be derived from the estimated parameters of expression and added to the score.

Vibrato is a very important characteristic of singing sounds. Research on the computational analysis of vibrato has addressed the detection, estimation, or synthesis of vibrato in singing signals [284], [461], [337]. Herrera and Bonada detected vibrato in the F0 trajectory during a note as follows [284]. Given a note segment, the global mean of the fundamental frequency trajectory during the note is first removed to obtain a zero-mean F0 trajectory. The trajectory is then analysed in half-overlapping windows of length 0.37 s. For each window, the FFT of the windowed F0 trajectory is computed and prominent spectral peaks are subjected to parabolic interpolation. If a prominent peak around 5–6 Hz is found in the spectrum, the frequency and the amplitude of the peak are interpreted as the vibrato rate and the depth in the analysed window, respectively. The method can be straightforwardly used to detect tremolo, too, by replacing the fundamental frequencies with framewise energy measures. Once the vibrato parameters have been detected during a note, they can be encoded as pitch-bend MIDI messages in the MIDI representation of the performance.

The loudness of a note over its duration can be directly measured with framewise energies in the note segment. Each MIDI note-on message includes a velocity parameter which indicates the loudness of the note. To decide the velocity value of an entire note, the RMS energies of voiced frames can be averaged and mapped to a velocity value, for example.

The note event models which consider different types of time segments during the singing notes allow different measurements to be made within the different segments of transcribed notes. The amount of glissando can be measured during the attack segment, vibrato and tremolo during the sustain segment, and so forth. The estimation and encoding of expression parameters improves the quality of resynthesized singing transcriptions considerably. Some examples on expression encoding can be found at http://www.cs.tut.fi/sgn/arg/matti/demos/monomel.

12.5 Summary and Discussion

In this chapter, we have considered the automatic conversion of monophonic singing performances into musical transcriptions and covered the basic methodology for solving this problem. In the current singing transcription systems, frame-level features are first extracted from a singing waveform and the features are then segmented into note events and assigned pitch labels. In the literature, the latter two operations have been performed either consecutively (the segment-and-label approach) or jointly with statistical methods.

In both of these approaches, it is important to also consider the musical context, since music transcription in general is based on the perception of musical entities at many levels rather than on the local evidence of individual notes only. The musicological modelling has been addressed using key estimation, note N-grams, and metrical context where especially the use of tonal context has brought improvements in terms of transcription accuracy.

Singing transcription can be accomplished quite satisfactorily with the state-of-the-art methods, and these are already applicable in real-world situations. However, the quality of singing transcription is still far from perfect as compared to the superior quality of hand-made transcriptions. To design better methods, we need a deeper understanding of the perception of singing sounds, and of the influence of the musical context on this. Further development of both the acoustic and the musicological models is likely to improve the transcription quality. This inevitably leads to the development of complex computational models and provides interesting challenges for the future.

References

1. S. Abdallah and M. Plumbley. Unsupervised onset detection: A probabilistic approach using ICA and a hidden Markov classifier. In *Cambridge Music Processing Colloquium*, Cambridge, UK, 2003.
2. S.A. Abdallah. *Towards Music Perception by Redundancy Reduction and Unsupervised Learning in Probabilistic Models*. PhD thesis, Department of Electronic Engineering, King's College London, 2002.
3. S.A. Abdallah and M.D. Plumbley. If the independent components of natural images are edges, what are the independent components of natural sounds? In *International Conference on Independent Component Analysis and Signal Separation*, pp. 534–539, San Diego, USA, 2001.
4. S.A. Abdallah and M.D. Plumbley. An independent component analysis approach to automatic music transcription. In *Audio Engineering Society 114th Convention*, Amsterdam, Netherlands, March 2003.
5. S.A. Abdallah and M.D. Plumbley. Polyphonic transcription by non-negative sparse coding of power spectra. In *International Conference on Music Information Retrieval*, pp. 318–325, Barcelona, Spain, October 2004.
6. M. Abe and S. Ando. Application of loudness/pitch/timbre decomposition operators to auditory scene analysis. In *IEEE International Conference on Acoustics, Speech, and Signal Processing*, pp. 2646–2649, Atlanta, USA, 1996.
7. T. Abe, T. Kobayashi, and S. Imai. Robust pitch estimation with harmonics enhancement in noisy environments based on instantaneous frequency. In *International Conference on Spoken Language Processing*, pp. 1277–1280, Philadelphia, USA, 1996.
8. M. Abramowitz and I.A. Stegun. *Handbook of Mathematical Functions*. Dover Publications, New York, USA, 1970.
9. K. Achan, S. Roweis, and B. Frey. Probabilistic inference of speech signals from phaseless spectrograms. In *Neural Information Processing Systems*, pp. 1393–1400, Vancouver, Canada, 2003.
10. G. Agostini. Sistemi informatici per il riconoscimento automatico di timbriche musicali in segnali monofonici mediante tecniche di statistica multivariata. Technical report, Departimento di Elettronica e Informazione, Facoltá di Ingegneria, Politecnico di Milano, 1999. Tesi di Laurea.

11. G. Agostini, M. Longari, and E. Pollastri. Musical instrument timbres classification with spectral features. In *IEEE Fourth Workshop on Multimedia Signal Processing*, pp. 97–102, Cannes, France, 2001.

12. G. Agostini, M. Longari, and E. Pollastri. Musical instrument timbres classification with spectral features. *EURASIP Journal of Applied Signal Processing*, 1:5–14, 2003.

13. W.A. Ainsworth, editor. *Advances in Speech, Hearing and Language Processing*. JAI Press, Greenwich, Connecticut, 1996.

14. P.E. Allen and R.B. Dannenberg. Tracking musical beats in real time. In *International Computer Music Conference*, Glasgow, Scotland, 1990.

15. M. Alonso, M. Badeau, B. David, and G. Richard. Musical tempo estimation using noise subspace projections. In *IEEE Workshop on Applications of Signal Processing to Audio and Acoustics*, New Paltz, USA, October 2003.

16. M. Alonso, B. David, and G. Richard. Tempo and beat estimation of musical signals. In *International Conference on Music Information Retrieval*, Barcelona, Spain, October 2004.

17. B. Alpert. A class of bases in l^2 for the sparse representation of integral operators. *SIAM Journal of Mathematical Analysis*, 24(1):246–262, 1993.

18. B.D.O. Anderson and J.B. Moore. *Optimal Filtering*. Prentice Hall, Englewood Cliffs, USA, 1979.

19. C. Andrieu, M. Davy, and A. Doucet. Efficient particle filtering for jump Markov Systems: Application to time-varying autoregressions. *IEEE Transactions on Signal Processing*, 51(7):1762–1769, July 2003.

20. C. Andrieu, P.M. Djuric, and A. Doucet. Model selection by MCMC computation. *Signal Processing*, 81(1):19–37, January 2001.

21. C. Andrieu and A. Doucet. Joint Bayesian detection and estimation of noisy sinusoids via reversible jump MCMC. *IEEE Transactions on Signal Processing*, 47(10):2667–2676, 1999.

22. M.S. Arulampalam, S. Maskell, N. Gordon, and T. Clapp. A tutorial on particle filters for online nonlinear/non-Gaussian Bayesian tracking. *IEEE Transactions on Signal Processing*, 50(2):174–88, February 2002.

23. J.-J. Aucouturier and M. Sandler. Finding repeating patterns in acoustic musical signals: Applications for audio thumbnailing. In *AES International Conference on Virtual, Synthetic and Entertainment Audio*, pp. 412–421, Espoo, Finland, June 2002.

24. C. Avendano. Frequency-domain source identification and manipulation in stereo mixes for enhancement, suppression and re-panning applications. In *IEEE Workshop on Applications of Signal Processing to Audio and Acoustics*, New Paltz, USA, 2003.

25. F.R. Bach and M.I. Jordan. Discriminative training of hidden Markov models for multiple pitch tracking. In *IEEE International Conference on Audio, Speech and Signal Processing*, Philadelphia, USA, March 2005.

26. Y. Bar-Shalom. Tracking methods in a multitarget environment. *IEEE Transactions on Automatic Control*, 23(4):618–626, August 1978.

27. M.A. Bartsch and G.H. Wakefield. To catch a chorus: Using chroma-based representations for audio thumbnailing. In *IEEE Workshop on Applications of Signal Processing to Audio and Acoustics*, pp. 15–18, New Paltz, USA, 2001.

28. A.J. Bell and T.J. Sejnowski. An information-maximization approach to blind separation and blind deconvolution. *Neural Computation*, 7:1129–1159, 1995.

29. S.D. Bella and I. Peretz. Music agnosias: Selective impairments of music recognition after brain damage. *Journal of New Music Research*, 28(3):209–216, 1999.

30. J.P. Bello, L. Daudet, S. Abdallah, C. Duxbury, M. Davies, and M.B. Sandler. A tutorial on onset detection in music signals. *IEEE Transactions on Speech and Audio Processing*, 2005. (in press).

31. J.P. Bello and M. Sandler. Phase-based note onset detection for music signals. In *IEEE International Conference on Acoustics, Speech, and Signal Processing*, Hong Kong, China.

32. L. Benaroya, F. Bimbot, L. McDonagh, and R. Gribonval. Non-negative sparse representation for Wiener based source separation with a single sensor. In *IEEE International Conference on Acoustics, Speech, and Signal Processing*, Hong Kong, China, 2003.

33. A.L. Berenzweig and D.P.W. Ellis. Locating singing voice segments within music signals. In *IEEE Workshop on Applications of Signal Processing to Audio and Acoustics*, pp. 119–122, New Paltz, USA, October 2001.

34. J. Berger, R. Coifman, and M. Goldberg. Removing noise from music using local trigonometric bases and wavelet packets. *Journal of the Audio Engineering Society*, 42(10):808–818, 1994.

35. D.P. Bertsekas. *Nonlinear Programming*. Athena Scientific, Belmont, Massachusetts, 1995.

36. E. Bigand. Contributions of music to research on human auditory cognition. In S. McAdams and E. Bigand, editors, *Thinking in Sound: The Cognitive Psychology of Human Audition*, pp. 231–277. Oxford University Press, 1993.

37. J.A. Bilmes. *Timing is of the Essence: Perceptual and Computational Techniques for Representing, Learning, and Reproducing Expressive Timing in Percussive Rhythm*. Master's thesis, Massachusetts Institute of Technology, September 1993.

38. J.A. Bilmes. A gentle tutorial of the EM algorithm and its application to parameter estimation for Gaussian mixture and hidden Markov models. Technical report, International Computer Science Institute, Berkeley, USA, 1998.

39. E. Bingham and H. Mannila. Random projection in dimensionality reduction: Applications to image and text data. In *The Seventh ACM SIGKDD International Conference on Knowledge Discovery and Data Mining*, pp. 245–250, San Francisco, California, 2001.

40. C.M. Bishop. *Neural Networks for Pattern Recognition*. Oxford University Press, Oxford, England, 1995.

41. S.S. Blackman and R. Popoli. *Design and Analysis of Modern Tracking Systems*. Artech House, 1999.

42. A. Blum and P. Langley. Selection of relevant features and examples in machine learning. *Artificial Intelligence*, 97(1–2):245–271, 1997.

43. T. Blumensath and M. Davies. Unsupervised learning of sparse and shift-invariant decompositions of polyphonic music. In *IEEE International Conference on Acoustics, Speech, and Signal Processing*, Montreal, Canada, 2004.

44. J. Bonada and A. Loscos. Sample-based singing voice synthesizer by spectral concatenation. In *Stockholm Music Acoustics Conference*, Stockholm, Sweden, 2003.

45. I. Borg and P. Groenen. *Modern Multidimensional Scaling: Theory and Applications*. Springer, 1997.

46. G.C. Bowker and S.L. Star. *Sorting Things Out: Classification and Its Consequences.* MIT Press, Cambridge, USA, 1999.
47. G.E.P. Box and D.R. Cox. An analysis of transformations. *Journal of the Royal Statistical Society: Series B*, 26:296–311, 1964.
48. R. Bracewell. *Fourier Transform and Its Applications.* McGraw-Hill, 1999.
49. A.S. Bregman. *Auditory Scene Analysis.* MIT Press, Cambridge, USA, 1990.
50. A.S. Bregman. Constraints on computational models of auditory scene analysis, as derived from human perception. *Journal of the Acoustical Society of Japan (E)*, 16(3):133–136, May 1995.
51. L. Breiman. Random forests. *Machine Learning*, 45:5–32, 2001.
52. G.J. Brown. *Computational Auditory Scene Analysis: A Representational Approach.* PhD thesis, University of Sheffield, Sheffield, UK, 1992.
53. G.J. Brown and M. Cooke. Perceptual grouping of musical sounds: A computational model. *Journal of New Music Research*, 23(1):107–132, 1994.
54. J.C. Brown. Musical fundamental frequency tracking using a pattern recognition method. *Journal of the Acoustical Society of America*, 92(3):1394–1402, 1992.
55. J.C. Brown. Determination of the meter of musical scores by autocorrelation. *Journal of the Acoustical Society of America*, 94(4):1953–1957, October 1993.
56. J.C. Brown. Computer identification of musical instruments using pattern recognition with cepstral coefficients as features. *Journal of the Acoustical Society of America*, 105:1933–1941, 1999.
57. J.C. Brown, O. Houix, and S. McAdams. Feature dependence in the automatic identification of musical woodwind instruments. *Journal of the Acoustical Society of America*, 109:1064–1072, 2001.
58. J.C. Brown and P. Smaragdis. Independent component analysis for automatic note extraction from musical trills. *Journal of the Acoustical Society of America*, 115, May 2004.
59. C.J.C. Burges. A tutorial on support vector machines for pattern recognition. *Data Mining and Knowledge Discovery*, 2(2):121–167, 1998.
60. E.M. Burns. Intervals, scales, and tuning. In Deutsch [144], pp. 215–264.
61. C.S. Burrus. Multiband least square FIR filter design. *IEEE Transactions on Signal Processing*, 43(2):412–421, February 1995.
62. P. Cano, M. Koppenberger, S. Le Groux, J. Ricard, P. Herrera, and N. Wack. Nearest-neighbor automatic sound annotation with a wordnet taxonomy. *Journal of Intelligent Information Systems*, 24:99–111, 2005.
63. J.-F. Cardoso. Source separation using higher order moments. In *IEEE International Conference on Acoustics, Speech, and Signal Processing*, pp. 2109–2112, Glasgow, UK, 1989.
64. J.-F. Cardoso. Multidimensional independent component analysis. In *IEEE International Conference on Acoustics, Speech, and Signal Processing*, Seattle, USA, 1998.
65. J.-F. Cardoso. High-order contrasts for independent component analysis. *Neural Computation*, 11(1), 1999.
66. P. Cariani. Recurrent timing nets for auditory scene analysis. In *International Joint Conference on Neural Networks*, Portland, Oregon, July 2003.
67. P.A. Cariani and B. Delgutte. Neural correlates of the pitch of complex tones. I. Pitch and pitch salience. II. Pitch shift, pitch ambiguity, phase invariance, pitch circularity, rate pitch, and the dominance region for pitch. *Journal of Neurophysiology*, 76(3):1698–1734, 1996.

68. R.P. Carlyon. Temporal pitch mechanisms in acoustic and electric hearing. *Journal of the Acoustical Society of America*, 112(2):621–633, 2002.

69. F. Carreras, M. Leman, and M. Lesaffre. Automatic description of musical signals using schema-based chord decomposition. *Journal of New Music Research*, 28(4):310–331, 1999.

70. M.A. Casey. *Auditory Group Theory with Applications to Statistical Basis Methods for Structured Audio*. PhD thesis, Massachusetts Institute of Technology, 1998.

71. M.A. Casey. General sound classification and similarity in MPEG-7. *Organized Sound*, 6:153–164, 2001.

72. M.A. Casey. MPEG-7 sound-recognition tools. *IEEE Transactions on Circuits and Systems for Video Technology*, 11(6), June 2001.

73. M.A. Casey and A. Westner. Separation of mixed audio sources by independent subspace analysis. In *International Computer Music Conference*, Berlin, Germany, 2000.

74. A. Cemgil. *Bayesian Music Transcription*. PhD thesis, Nijmegen University, 2004.

75. A.T. Cemgil, P. Desain, and B. Kappen. Rhythm quantization for transcription. *Computer Music Journal*, 24(2):60–76, 2000.

76. A.T. Cemgil and B. Kappen. Tempo tracking and rhythm quantization by sequential Monte Carlo. In *Neural Information Processing Systems*, Vancouver, British Columbia, Canada, 2001.

77. A.T. Cemgil and B. Kappen. Monte Carlo methods for tempo tracking and rhythm quantization. *Journal of Artificial Intelligence Research*, 18:45–81, 2003.

78. A.T. Cemgil, B. Kappen, and D. Barber. A generative model for music transcription. *IEEE Transactions on Speech and Audio Processing*, 13(6), 2005.

79. A.T. Cemgil, B. Kappen, P. Desain, and H. Honing. On tempo tracking: Tempogram representation and Kalman filtering. *Journal of New Music Research*, 28(4):259–273, 2001.

80. C. Chafe and D. Jaffe. Source separation and note identification in polyphonic music. In *IEEE International Conference on Acoustics, Speech, and Signal Processing*, pp. 1289–1292, Atlanta, USA, 1986.

81. C. Chafe, J. Kashima, B. Mont-Reynaud, and J. Smith. Techniques for note identification in polyphonic music. In *International Computer Music Conference*, pp. 399–405, Vancouver, Canada, 1985.

82. W. Chai and B. Vercoe. Structural analysis of musical signals for indexing and thumbnailing. In *ACM/IEEE Joint Conference on Digital Libraries*, pp. 27–34, Texas, USA, 2003.

83. G. Charbonneau. Timbre and the perceptual effects of three types of data reduction. *Organized Sound*, 5:10–19, 1981.

84. F.J. Charpentier. Pitch detection using the short-term phase spectrum. In *IEEE International Conference on Acoustics, Speech, and Signal Processing*, pp. 113–116, Tokyo, Japan, 1986.

85. G. Chechik, A. Globerson, M.J. Anderson, E.D. Young, I. Nelken, and N. Tishby. Group redundancy measures reveal redundancy reduction in the auditory pathway. In *Neural Information Processing Systems*, Vancouver, British Columbia, Canada, 2001.

86. C.E. Cherry. Some experiments on the recognition of speech, with one and with two ears. *Journal of the Acoustical Society of America*, 25(5):975–979, 1953.

87. N. Chétry, M. Davies, and M. Sandler. Identification of monophonic instrument recordings using k-means and support vector machines. In *Digital Music Research Network Conference*, Glasgow, England, 2005.

88. N. Chétry, M. Davies, and M Sandler. Musical instrument identification using lsf and k-means. In *Audio Engineering Society 118th Convention*, Barcelona, Spain, 2005.

89. C.K. Chui. *Wavelets: A Tutorial in Theory and Applications*. Academic Press, 1992.

90. K.W. Church and W.A. Gale. A comparison of the enhanced Good-Turing and deleted estimation methods for estimating probabilities of English speech. *Computer Speech and Language*, 5(1):19–54, 1991.

91. G. Ciuperca, A. Ridolfi, and J. Idier. Penalized maximum likelihood estimator for normal mixtures. *Scandinavian Journal of Statistics*, 30(1):45–59, March 2003.

92. L.P. Clarisse, J.P. Martens, M. Lesaffre, B. De Baets, H. De Meyer, and M. Leman. An auditory model based transcriber of singing sequences. In *International Conference on Music Information Retrieval*, Paris, France, October 2002.

93. E.F. Clarke. Rhythm and timing in music. In Deutsch [144], pp. 473–500.

94. D. Cliff. Hang the DJ: Automatic sequencing and seamless mixing of dance-music tracks. Technical Report HPL-2000-104, HP Laboratories Bristol, August 2000.

95. N. Collins. A comparison of sound onset detection algorithms with emphasis on psychoacoustically motivated detection functions. In *Audio Engineering Society 118th Convention*, Barcelona, Spain, May 2005.

96. P. Comon. Independent component analysis - a new concept? *Signal Processing*, 36(3):287–314, April 1994.

97. M. Cooke. *Modelling Auditory Processing and Organisation*. PhD thesis, University of Sheffield, Sheffield, UK, 1991.

98. M. Cooke and D.P.W. Ellis. The auditory organization of speech and other sources in listeners and computational models. *Speech Communication*, 35(3–4):141–177, 2001.

99. M. Cooke, P. Green, L. Josifovski, and A. Vizinho. Robust automatic speech recognition with missing and unreliable acoustic data. *Speech Communication*, 34:267–285, 2001.

100. M.P. Cooke. *Modeling Auditory Processing and Organisation*. Cambridge University Press, Cambridge, UK, 1993.

101. G. Cooper. The computational complexity of probabilistic inference using Bayesian belief networks. *Artificial Intelligence*, 42(2–3):393–405, 1990.

102. G. Cooper and L.B. Meyer. *The Rhythmic Structure of Music*. University of Chicago Press, 1960.

103. M. Cooper and J. Foote. Automatic music summarization via similarity analysis. In *International Conference on Music Information Retrieval*, pp. 81–85, Paris, France, October 2002.

104. M. Cooper and J. Foote. Summarizing popular music via structural similarity analysis. In *IEEE Workshop on Applications of Signal Processing to Audio and Acoustics*, pp. 127–130, New Paltz, USA, 2003.

105. P. Cosi, G. De Poli, and G. Lauzzana. Auditory modelling and self-organizing neural networks for timbre classification. *Journal of New Music Research*, 21:71–98, 1994.

106. H. Cottereau, J.-M. Piasco, C. Doncarli, and M. Davy. Two approaches for the estimation of time-varying amplitude multichirp signals. In *IEEE International Conference on Acoustics, Speech, and Signal Processing*, Hong Kong, China, April 2003.

107. M.S. Crouse, R.D. Nowak, and R.G. Baraniuk. Wavelet-based signal processing using hidden Markov models. *IEEE Transactions on Signal Processing*, 46:886–902, April 1998. Special issue on filter banks.

108. R. Cusack and R.P. Carlyon. Auditory perceptual organization inside and outside the laboratory. In J.G. Neuhoff, editor, *Ecological Psychoacoustics*, pp. 15–48. Elsevier Academic Press, 2004.

109. R.B. Dannenberg, W.P. Birmingham, G. Tzanetakis, C. Meek, N. Hu, and B. Pardo. The MUSART testbed for query-by-humming evaluation. In *International Conference on Music Information Retrieval*, pp. 41–47, Baltimore, USA, October 2003.

110. R.B. Dannenberg and N. Hu. Discovering musical structure in audio recordings. In *International Conference on Music And Artificial Intelligence*, pp. 43–57, Edinburgh, Scotland, UK, September 2002.

111. R.B. Dannenberg and N. Hu. Pattern discovery techniques for music audio. *Journal of New Music Research*, 32(2):153–163, 2003.

112. T. Dasu and T. Johnson. *Exploratory data mining and data cleaning*. John Wiley & Sons, second edition, 2003.

113. I. Daubechies. *Ten Lectures on Wavelets*. SIAM, Philadelphia, PA, 1992.

114. L. Daudet, S. Molla, and B. Torrésani. Towards a hybrid audio coder. In Jianping Li, editor, *Wavelets and Its Applications, Proceedings of a Conference Held in Chongqing (China)*. World Scientific Publishing Company, 2004.

115. L. Daudet and B. Torrésani. Hybrid representations for audiophonic signal encoding. *Signal Processing*, 82(11):1595–1617, 2002. Special issue on image and video coding beyond standards.

116. Laurent Daudet. Sparse and structured decompositions of signals with the molecular matching pursuit. *IEEE Transactions on Acoustics, Speech, and Signal Processing*, 2004. (in press).

117. W.B. Davenport and W.L. Root. *An Introduction to the Theory of Random Signals and Noise*. IEEE Press, New York, 1987.

118. M.E. Davies and M.D. Plumbley. Causal tempo tracking of audio. In *International Conference on Music Information Retrieval*, pp. 164–169, 2004.

119. G. Davis, S. Mallat, and M. Avellaneda. Adaptive greedy approximations. *Constructive Approximation*, 13:57–98, 1997.

120. M. Davy. Bayesian separation of harmonic sources. In *Joint Statistical Meeting of the American Statistial Association*, Minneapolis, USA, August 2005.

121. M. Davy, C. Doncarli, and J.Y. Tourneret. Classification of chirp signals using hierarchical Bayesian learning and MCMC methods. *IEEE Transactions on Signal Processing*, 50(2):377–388, 2002.

122. M. Davy and S. Godsill. Bayesian harmonic models for musical signal analysis. In *Seventh Valencia International meeting Bayesian statistics 7*, Tenerife, Spain, June 2002.

123. M. Davy and S. Godsill. Detection of abrupt spectral changes using support vector machines: An application to audio signal segmentation. In *IEEE International Conference on Acoustics, Speech, and Signal Processing*, Volume 2, pp. 1313–1316, Orlando, USA, May 2002.

124. M. Davy, S. Godsill, and J. Idier. Bayesian analysis of polyphonic western tonal music. *Journal of the Acoustical Society of America*, 2005. (in press).

125. M. Davy and S.J. Godsill. Audio information retrieval: A bibliographical study. Technical Report CUED/F-INFENG/TR.429, Department of Engineering, University of Cambridge, February 2002.

126. M. Davy and S.J. Godsill. Bayesian harmonic models for musical pitch estimation and analysis. Technical Report CUED/F-INFENG/TR.431, Department of Engineering, University of Cambridge, Cambridge, UK, April 2002.

127. M. Davy and J. Idier. Fast MCMC computations for the estimation of sparse processes from noisy observations. In *IEEE International Conference on Acoustics, Speech, and Signal Processing*, Montreal, Canada, May 2004.

128. E. de Boer and H.R. de Jongh. On cochlear encoding: Potentials and limitations of the reverse-correlation technique. *Journal of the Acoustical Society of America*, 63(1):115–135, 1978.

129. A. de Cheveigné. Separation of concurrent harmonic sounds: Fundamental frequency estimation and a time-domain cancellation model for auditory processing. *Journal of the Acoustical Society of America*, 93(6):3271–3290, 1993.

130. A. de Cheveigné. Concurrent vowel identification. III. A neural model of harmonic interference cancellation. *Journal of the Acoustical Society of America*, 101(5):2857–2865, 1997.

131. A. de Cheveigné. Cancellation model of pitch perception. *Journal of the Acoustical Society of America*, 103(3):1261–1271, 1998.

132. A. de Cheveigné. Pitch perception models. In Plack and Oxenham [522].

133. A. de Cheveigné and H. Kawahara. Multiple period estimation and pitch perception model. *Speech Communication*, 27:175–185, 1999.

134. A. de Cheveigné and H. Kawahara. Comparative evaluation of F0 estimation algorithms. In *7th European Conf. Speech Communication and Technology*, Aalborg, Denmark, 2001.

135. A. de Cheveigné and H. Kawahara. YIN, a fundamental frequency estimator for speech and music. *Journal of the Acoustical Society of America*, 111(4):1917–1930, 2002.

136. T. De Mulder, J.P. Martens, M. Lesaffre, M. Leman, B. De Baets, and H. De Meyer. An auditory model based transcriber of vocal queries. In *International Conference on Music Information Retrieval*, Baltimore, USA, 2003.

137. T. De Mulder, J.P. Martens, M. Lesaffre, M. Leman, B. De Baets, and H. De Meyer. Recent improvements of an auditory model based front-end for the transcription of vocal queries. In *IEEE International Conference on Acoustics, Speech, and Signal Processing*, Volume 4, pp. 257–260, Montreal, Canada, 2004.

138. A.P. Dempster, N.M. Laird, and D.B. Rubin. Maximum likelihood from incomplete data via the EM algorithm. *Journal of the Royal Statistical Society: Series B*, 39(1):1–38, 1977.

139. P. Depalle, G. Garcia, and X. Rodet. Tracking of partials for additive sound synthesis using hidden Markov model. In *IEEE International Conference on Acoustics, Speech, and Signal Processing*, Volume 1, pp. 225–228, Minneapolis, USA, 1993.

140. P. Desain and H. Honing. The quantization of musical time: A connectionist approach. *Computer Music Journal*, 13(3):56–66, 1989.

141. P. Desain and H. Honing. Computational models of beat induction: The rule-based approach. *Journal of New Music Research*, 28(1):29–42, 1999.

142. P. Desain and H. Honing. The formation of rhythmic categories and metric priming. *Perception*, 32(3):241–365, 2003.
143. F. Desobry, M. Davy, and C. Doncarli. An online kernel change detection algorithm. *IEEE Transactions on Signal Processing*, 2005. (in press).
144. D. Deutsch, editor. *The Psychology of Music*. Academic Press, San Diego, California, 2nd edition, 1999.
145. T.G. Dietterich. Approximate statistical tests for comparing supervised classification learning algorithms. *Neural Computation*, 10:1895–1923, 1998.
146. T.G. Dietterich. Machine learning for sequential data: A review. In T. Caelli, A. Amin, R.P.W. Duin, M. Kamel, and de Ridder D., editors, *Structural, Syntactic, and Statistical Pattern Recognition*, Volume 2396 of *Lecture Notes in Computer Science*, pp. 15–30. Springer-Verlag, 2002.
147. C. Dittmar and C. Uhle. Further steps towards drum transcription of polyphonic music. In *Audio Engineering Society 116th Convention*, Berlin, Germany, May 2004.
148. S. Dixon. Automatic extraction of tempo and beat from expressive performaces. *Journal of New Music Research*, 30(1):39–58, 2001.
149. S. Dixon, F. Gouyon, and G. Widmer. Towards characterisation of music via rhythmic patterns. In *International Conference on Music Information Retrieval*, pp. 509–516, Barcelona, Spain, 2004.
150. M. Dörfler. *Gabor Analysis for a Class of Signals called Music*. PhD thesis, NuHAG, University of Vienna, 2002.
151. D. Dorran and R. Lawlor. An efficient audio time-scale modification algorithm for use in a sub-band implementation. In *International Conference on Digital Audio Effects*, pp. 339–343, 2003.
152. A. Doucet, N. de Freitas, and N. Gordon. *Sequential Monte Carlo Methods in Practice*. Springer, New York, USA, 2001.
153. B. Doval and X. Rodet. Estimation of fundamental frequency of musical sound signals. In *IEEE International Conference on Acoustics, Speech, and Signal Processing*, pp. 3657–3660, Toronto, Canada, 1991.
154. J.S. Downie, J. Futrelle, and D. Tcheng. The international music information systems evaluation laboratory: Governance, access and security. In *International Conference on Music Information Retrieval*, Barcelona, Spain, 2004.
155. C. Drake, A. Penel, and E. Bigand. Tapping in time with mechanical and expressively performed music. *Music Perception*, 18(1):1–23, 2000.
156. J. Drish. Obtaining calibrated probability estimates from support vector machines. Technical report, University of California, San Diego, California, USA, June 2001.
157. S. Dubnov. Extracting sound objects by independent subspace analysis. In *22nd International Audio Engineering Society Conference*, Espoo, Finland, June 2002.
158. C. Dubois and M. Davy. Harmonic tracking in spectrograms. *IEEE Transactions on Signal Processing*, 2005. Submitted.
159. C. Dubois and M. Davy. Harmonic tracking using sequential Monte Carlo. In *IEEE Statistical Signal Processing Workshop*, Bordeaux, France, July 2005.
160. C. Dubois and M. Davy. Suivi de trajectoires temps-fréquence par filtrage particulaire. In *20st GRETSI colloquium*, Louvain-La-Neuve, Belgium, September 2005. In French.
161. R.O. Duda, P.E. Hart, and D.G. Stork. *Pattern Classification*. Wiley, New York, USA, second edition, 2001.

162. C. Duxbury, J.P. Bello, M. Davies, and M. Sandler. A combined phase and amplitude based approach to onset detection for audio segmentation. In *4th European Workshop on Image Analysis for Multimedia Interactive Services*, London, UK, 2003.

163. C. Duxbury, J.P. Bello, M. Davies, and M. Sandler. Complex domain onset detection for musical signals. In *International Conference on Digital Audio Effects*, London, UK, 2003.

164. C. Duxbury, M. Sandler, and M.E. Davies. A hybrid approach to musical note onset detection. In *International Conference on Digital Audio Effects*, Hamburg, Germany, September 2002.

165. D. Eck. A positive-evidence model for classifying rhythmical patterns. Technical Report IDSIA-09-00, Dalle Molle Institute for Artificial Intelligence, 2000.

166. D. Eck. A network of relaxation oscillators that finds downbeats in rhythms. Technical Report IDSIA-06-01, Dalle Molle Institute for Artificial Intelligence, 2001.

167. J. Eggink and G.J. Brown. Application of missing feature theory to the recognition of musical instruments in polyphonic audio. In *International Conference on Music Information Retrieval*, pp. 125–131, Baltimore, USA, 2003.

168. J. Eggink and G.J. Brown. Application of missing feature theory to the recognition of musical instruments in polyphonic audio. In *IEEE International Conference on Acoustics, Speech, and Signal Processing*, pp. 553–556, Hong Kong, China, 2003.

169. J. Eggink and G.J. Brown. Extracting melody lines from complex audio. In *International Conference on Music Information Retrieval*, pp. 84–91, Barcelona, Spain, October 2004.

170. J. Eggink and G.J. Brown. Instrument recognition in accompanied sonatas and concertos. In *IEEE International Conference on Acoustics, Speech, and Signal Processing*, pp. 217–220, Montreal, Canada, 2004.

171. D.P.W. Ellis. *Prediction-Driven Computational Auditory Scene Analysis*. PhD thesis, Massachusetts Institute of Technology, 1996.

172. D.P.W. Ellis. Using knowledge to organize sound: The prediction-driven approach to computational auditory scene analysis, and its application to speech/nonspeech mixtures. *Speech Communication*, 27:281–298, 1999.

173. D.P.W. Ellis and D.F. Rosenthal. Mid-level representations for computational auditory scene analysis. In *International Joint Conference on Artificial Intelligence*, Montreal, Quebec, 1995.

174. A. Eronen. Automatic musical instrument recognition. Master's thesis, Tampere University of Technology, 2001.

175. A. Eronen. Musical instrument recognition using ICA-based transform of features and discriminatively trained HMMs. In *Seventh International Symposium on Signal Processing and its Applications*, pp. 133–136, Paris, France, 2003.

176. A. Eronen and A. Klapuri. Musical instrument recognition using cepstral coefficients and temporal features. In *IEEE International Conference on Acoustics, Speech, and Signal Processing*, pp. 753–756, Istanbul, Turkey, 2000.

177. S. Essid, G. Richard, and B. David. Musical instrument recognition based on class pairwise feature selection. In *International Conference on Music Information Retrieval*, Barcelona, Spain, 2004.

178. S. Essid, G. Richard, and B. David. Musical instrument recognition on solo performance. In *European Signal Processing Conference*, Vienna, Austria, 2004.

179. S. Essid, G. Richard, and B. David. Inferring efficient hierarchical taxonomies for mir tasks: Application to musical instruments. In *6th International Conference on Music Information Retrieval*, London, UK, 2005.

180. S. Essid, G. Richard, and B. David. Instrument recognition in polyphonic music. In *IEEE International Conference on Audio, Speech and Signal Processing*, Philadelphia, USA, 2005.

181. A.M. Fanelli, G. Castellano, and C.A. Buscicchio. A modular neuro-fuzzy network for musical instruments classification. In J. Kittler and F. Roli, editors, *Proceedings of the First International Workshop on Multiple Classifier Systems*, Volume 1857 of *Lecture Notes In Computer Science*, pp. 372–382. Springer, 2000.

182. FastICA package for MATLAB. http://www.cis.hut.fi/projects/ica/fastica/.

183. U. Fayyad and K.B. Irani. Multi-interval discretisation of continuous-valued attributes for classification learning. In *13th International Joint Conference on Artificial Intelligence*, pp. 1022–1027, Berlin, Heidelberg, 1993.

184. H.G. Feichtinger and T. Strohmer. *Gabor Analysis and Algorithms: Theory and Applications*. Birkhauser, 1998.

185. H.G. Feichtinger and T. Strohmer, editors. *Advances in Gabor analysis*. Birkhauser, Boston, 2003.

186. D. FitzGerald. *Automatic Drum Transcription and Source Separation*. PhD thesis, Dublin Institute of Technology, 2004.

187. D. FitzGerald, E. Coyle, and B. Lawlor. Sub-band independent subspace analysis for drum transcription. In *International Conference on Digital Audio Effects*, Hamburg, Germany, 2002.

188. D. FitzGerald, E. Coyle, and B. Lawlor. Prior subspace analysis for drum transcription. In *Audio Engineering Society 114th Convention*, Amsterdam, Netherlands, March 2003.

189. D. FitzGerald, R. Lawlor, and E. Coyle. Drum transcription in the presence of pitched instruments using prior subspace analysis. In *Irish Signals & Systems Conference 2003*, Limerick, Ireland, July 2003.

190. D. FitzGerald, R. Lawlor, and E. Coyle. Drum transcription using automatic grouping of events and prior subspace analysis. In *4th European Workshop on Image Analysis for Multimedia Interactive Services*, pp. 306–309, 2003.

191. P. Flandrin. *Time-Frequency/Time-Scale Analysis*. Academic Press, 1999.

192. H. Fletcher. Auditory patterns. *Reviews of Modern Physics*, 12:47–65, 1940.

193. N.H. Fletcher and T.D. Rossing. *The Physics of Musical Instruments*. Springer, Berlin, Germany, second edition, 1998.

194. J. Foote and S. Uchihashi. The beat spectrum: A new approach to rhythm analysis. In *IEEE International Conference on Multimedia and Expo*, 2001.

195. J.T. Foote and M.L. Cooper. Media segmentation using self-similarity decomposition. In *SPIE Storage and Retrieval for Media Databases 2003*, Volume 5021, pp. 167–175, 2003.

196. Y. Freund and R.E. Schapire. Experiments with a new boosting algorithm. In *International Conference on Machine Learning*, Bari, Italy, 1996.

197. J.H. Friedman and J.W. Tukey. A projection pursuit algorithm for exploratory data analysis. *IEEE Transactions on Computers*, 23:881–890, 1974.

198. Lawrence Fritts. University of Iowa Musical Instrument Samples. http://theremin.music.uiowa.edu/MIS.html.

199. I. Fujinaga and K. MacMillan. Realtime recognition of orchestral instruments. In *International Computer Music Conference*, Berlin, Germany, 2000.

200. I. Fujinaga, S. Moore, and D.S. Sullivan. Implementation of exemplar-based learning model for music cognition. In *5th Int. Conf. on Music Perception and Cognition*, Seoul, Korea, 1998.

201. Takuya Fujishima. Realtime chord recognition of musical sound: A system using common lisp music. In *International Computer Music Conference*, pp. 464–467, Beijing, China, 1999.

202. K. Fukunaga. *Introduction to Statistical Pattern Recognition*. Academic Press, second edition, 1990.

203. S. Furui and M.M. Sondhi, editors. *Advances in Speech Signal Processing*. Marcel Dekker, New York, USA, 1991.

204. T. Gautama and M.M. Van Hulle. Separation of acoustic signals using self-organizing neural networks. In *IEEE Neural Network for Signal Processing Workshop*, Madison, USA, 1999.

205. Z. Ghahramani. An introduction to hidden Markov models and Bayesian networks. *International Journal of Pattern Recognition and Artificial Intelligence*, 15(1):9–42, 2001.

206. M.A. Ghezzo. *Music Theory, Ear Training, Rhythm, Solfége, and Dictation: A Comprehensive Course*. The University of Alabama Press, Alabama, 1980.

207. A. Ghias, J. Logan, D. Chamberlin, and B.C. Smith. Query by humming: Musical information retrieval in an audio database. In *ACM Multimedia Conference*, pp. 231–236, San Fransisco, California, November 1995.

208. W.R. Gilks, S. Richardson, and D.J. Spiegelhalter, editors. *Markov Chain Monte Carlo in Practice*. Chapman & Hall, London, UK, 1996.

209. O. Gillet and G. Richard. Automatic transcription of drum loops. In *IEEE International Conference on Acoustics, Speech, and Signal Processing*, Montreal, Canada, May 2004.

210. O. Gillet and G. Richard. Drum loops retrieval from spoken queries. *Journal of Intelligent Information Systems*, 2005.

211. O.K. Gillet and G. Richard. Automatic labeling of tabla signals. In *International Conference on Music Information Retrieval*, Baltimore, USA, October 2003.

212. S. Godsill and M. Davy. Bayesian harmonic models for musical pitch estimation and analysis. In *IEEE International Conference on Acoustics, Speech, and Signal Processing*, Orlando, USA, 2002.

213. S.J. Godsill and M. Davy. Bayesian computational models for inharmonicity in musical instruments. In *IEEE Workshop on Applications of Signal Processing to Audio and Acoustics*, New Paltz, USA, October 2005.

214. S.J. Godsill and P.J.W. Rayner. *Digital Audio Restoration: A Statistical Model-Based Approach*. Springer, Berlin, Germany, September 1998.

215. D. Godsmark and G.J. Brown. A blackboard architecture for computational auditory scene analysis. *Speech Communication*, 27(3):351–366, 1999.

216. B. Gold and N. Morgan. *Speech and Audio Signal Processing: Processing and Perception of Speech and Music*. John Wiley & Sons, 2000.

217. D.E. Goldberg. *Genetic Algorithms in Search, Optimization and Machine Learning*. Addison Wesley, 1989.

218. I.J. Good. The population frequencies of species and the estimation of population parameters. *Biometrika*, 16(40):16–264, 1953.

219. M. Goodwin. Residual modeling in music analysis-synthesis. In *IEEE International Conference on Acoustics, Speech, and Signal Processing*, pp. 1005–1008, Atlanta, USA, 1996.

220. M. Goto. *A Study of Real-Time Beat Tracking for Musical Audio Signals*. PhD thesis, Waseda University, 1998.

221. M. Goto. An audio-based real-time beat tracking system for music with or without drum-sounds. *Journal of New Music Research*, 30(2):159–171, 2001.

222. M. Goto. A predominant-F0 estimation method for CD recordings: MAP estimation using EM algorithm for adaptive tone models. In *IEEE International Conference on Acoustics, Speech, and Signal Processing*, Volume 5, pp. 3365–3368, Salt Lake City, USA, May 2001.

223. M. Goto. A predominant-F0 estimation method for real-world musical audio signals: Map estimation for incorporating prior knowledge about F0s and tone models. In *Proc. Workshop on Consistent and Reliable Acoustic Cues for Sound Analysis*, Aalborg, Denmark, 2001.

224. M. Goto. A chorus-section detecting method for musical audio signals. In *IEEE International Conference on Acoustics, Speech, and Signal Processing*, Volume 5, pp. 437–440, Hong Kong, China, April 2003.

225. M. Goto. Music scene description project: Toward audio-based real-time music understanding. In *International Conference on Music Information Retrieval*, pp. 231–232, Baltimore, USA, October 2003.

226. M. Goto. SmartMusicKIOSK: Music listening station with chorus-search function. In *ACM Symposium on User Interface Software and Technology*, pp. 31–40, Vancouver, British Columbia, Canada, 2003.

227. M. Goto. Development of the RWC music database. In *the 18th International Congress on Acoustics*, Volume 1, pp. 553–556, Kyoto, Japan, 2004.

228. M. Goto. A real-time music scene description system: Predominant-F0 estimation for detecting melody and bass lines in real-world audio signals. *Speech Communication*, 43(4):311–329, 2004.

229. M. Goto, H. Hashiguchi, T. Nishimura, and R. Oka. RWC music database: Popular, classical, and jazz music databases. In *International Conference on Music Information Retrieval*, pp. 287–288, Paris, France, October 2002.

230. M. Goto, H. Hashiguchi, T. Nishimura, and R. Oka. RWC music database: Music genre database and musical instrument sound database. In *International Conference on Music Information Retrieval*, pp. 229–230, Baltimore, USA, October 2003.

231. M. Goto and S. Hayamizu. A real-time music scene description system: Detecting melody and bass lines in audio signals. In *International Joint Conference on Artificial Intelligence*, pp. 31–40, Stockholm, Sweden, 1999.

232. M. Goto and S. Hayamizu. A real-time music scene description system: Detecting melody and bass lines in audio signals. In *Working Notes of the IJCAI-99 Workshop on Computational Auditory Scene Analysis*, pp. 31–40, Stockholm, Sweden, 1999.

233. M. Goto and K. Hirata. Invited review "Recent studies on music information processing". *Acoustical Science and Technology* (edited by the Acoustical Society of Japan), 25(6):419–425, November 2004.

234. M. Goto and H. Muraoka. Issues in evaluating beat tracking systems. In *Working Notes of the IJCAI-97 Workshop on Issues in AI and Music*, pp. 9–17, August 1997.

235. M. Goto and Y. Muraoka. A beat tracking system for acoustic signals of music. In *ACM International Conference on Multimedia*, pp. 365–372, San Fransisco, California, October 1994.

236. M. Goto and Y. Muraoka. A sound source separation system for percussion instruments. *Transactions of the Institute of Electronics, Information and Communication Engineers D-II*, J77-D-II(5):901–911, May 1994. (in Japanese)

237. M. Goto and Y. Muraoka. Music understanding at the beat level: Real-time beat tracking for audio signals. In *International Joint Conference on Artificial Intelligence*, pp. 68–75, Montreal, Quebec, 1995.

238. M. Goto and Y. Muraoka. A real-time beat tracking system for audio signals. In *International Computer Music Conference*, Tokyo, Japan, 1995.

239. M. Goto and Y. Muraoka. Real-time rhythm tracking for drumless audio signals: Chord change detection for musical decisions. In *Working Notes of the IJCAI-97 Workshop on Computational Auditory Scene Analysis*, pp. 135–144, Nagoya, Japan, 1997.

240. M. Goto and Y. Muraoka. Real-time beat tracking for drumless audio signals: Chord change detection for musical decisions. *Speech Communication*, 27(3–4):311–335, 1999.

241. M. Goto, M. Tabuchi, and Y. Muraoka. An automatic transcription system for percussion instruments. In *the 46th National Convention of Information Processing Society of Japan*, 7Q-2, Tokyo, Japan, March 1993. (in Japanese)

242. F. Gouyon. *Towards Automatic Rhythm Description of Musical Audio Signals: Representations, Computational Models and Applications*. Master's thesis, UPF, Barcelona, 2003.

243. F. Gouyon and S. Dixon. A review of automatic rhythm description systems. *Computer Music Journal*, 29(1), 2005.

244. F. Gouyon, L. Fabig, and J. Bonada. Rhythmic expressiveness transformations of audio recordings: swing modifications. In *International Conference on Digital Audio Effects*, 2003.

245. F. Gouyon and P. Herrera. Exploration of techniques for automatic labeling of audio drum tracks' instruments. In *MOSART Workshop on Current Research Directions in Computer Music*, Barcelona, Spain, 2001.

246. F. Gouyon and P. Herrera. Determination of the meter of musical audio signals: Seeking recurrences in beat segment descriptors. In *Audio Engineering Society 114th Convention*, Amsterdam, Netherlands, March 2003.

247. F. Gouyon, P. Herrera, and P. Cano. Pulse-dependent analysis of percussive music. In *22nd International Audio Engineering Society Conference*, Espoo, Finland, 2002.

248. F. Gouyon, A. Klapuri, S. Dixon, M. Alonso, G. Tzanetakis, C. Uhle, and P. Cano. An experimental comparison of audio tempo induction algorithms. *IEEE Transactions on Speech and Audio Processing*, 2005. (in press).

249. F. Gouyon and B. Meudic. Towards rhythmic content processing of musical signals: Fostering complementary approaches. *Journal of New Music Research*, 32(1), 2003.

250. F. Gouyon, F. Pachet, and O. Delerue. On the use of zero-crossing rate for an application of classification of percussive sounds. In *International Conference on Digital Audio Effects*, Verona, Italy, December 2000.

251. P.J. Green. On use of the EM algorithm for penalized likelihood estimation. *Journal of the Royal Statistical Society: Series B*, 52:443–452, 1990.

252. P.J. Green. Penalized likelihood. In *Encyclopaedia of Statistical Sciences*, Volume 3, pp. 578–586, 1999.

253. J.M. Grey. Multidimensional perceptual scaling of musical timbres. *Journal of the Acoustical Society of America*, 61:1270–1277, 1977.

254. J.M. Grey and J.A. Moorer. Perceptual evaluations of synthesized musical instrument tones. *Journal of the Acoustical Society of America*, 62:454–462, 1977.

255. R. Gribonval. *Approximations Non-linéaires pour l'Analyse des Signaux Sonores*. PhD thesis, Université de Paris IX Dauphine, 1999.

256. R. Gribonval and E. Bacry. Harmonic decomposition of audio signals with matching pursuit. *IEEE Transactions on Signal Processing*, 51(1):101–111, 2003.

257. R. Gribonval, E. Bacry, S. Mallat, Ph. Depalle, and X. Rodet. Analysis of sound signals with high resolution matching pursuit. In *IEEE Symposium on Time-Frequency and Time-Scale Analysis*, pp. 125–128, Paris, France, June 1996.

258. R. Gribonval, L. Benaroya, E. Vincent, and C. Févotte. Proposals for performance measurement in source separation. In *the 4th International Symposium on Independent Component Analysis and Blind Signal Separation*, Nara, Japan, 2003.

259. D. Griffin and J. Lim. Signal estimation from modified short-time fourier transform. *IEEE Transactions on Acoustics, Speech, and Signal Processing*, 32:236–242, 1984.

260. T.D. Griffiths, J.D. Warren, S.K. Scott, I. Nelken, and A.J. King. Cortical processing of complex sound: A way forward? *Trends in Neuroscience*, 27:181–185, 1977.

261. K. Gröchenig. *Foundations of Time-frequency Analysis*. Birkhäuser, Boston, MA, 2001.

262. I. Guyon and A. Elisseeff. An introduction to variable and feature selection. *Journal of Machine Learning Research*, 3:1157–1182, 2003.

263. S.W. Hainsworth. *Techniques for the Automated Analysis of Musical Audio*. PhD thesis, Department of Engineering, University of Cambridge, 2004.

264. S.W. Hainsworth and M.D. Macleod. Automatic bass line transcription from polyphonic music. In *International Computer Music Conference*, pp. 431–434, Havana, Cuba, 2001.

265. S.W. Hainsworth and M.D. Macleod. Onset detection in musical audio signals. In *International Computer Music Conference*, Singapore, 2003.

266. S.W. Hainsworth and M.D. Macleod. Particle filtering applied to musical tempo tracking. *Journal of Applied Signal Processing*, 15:2385–2395, 2004.

267. M.A. Hall. *Correlation-Based Feature Selection for Machine Learning*. PhD thesis, Department of Computer Science, University of Waikato, Hamilton, New Zealand, 1998.

268. M.A. Hall. Correlation-based feature selection for discrete and numeric class machine learning. In *Seventeeth International Conference on Machine Learning*, Stanford, CA, USA, 2000.

269. K.N. Hamdy, A. Ali, and A.H. Tewfik. Low bit rate high quality audio coding with combined harmonic and wavelet representations. In *IEEE International Conference on Acoustics, Speech, and Signal Processing*, Volume 2, pp. 1045–1048, Atlanta, USA, 1996.

270. S. Handel. *Listening: An Introduction to the Perception of Auditory Events*. MIT Press, 1989.

271. S. Handel. Timbre perception and auditory object identification. In Moore [474], pp. 425–460.

272. A. Härmä, M. Karjalainen, L. Savioja, V. Välimäki, U. K. Laine, and J. Huopaniemi. Frequency-warped signal processing for audio applications. *Journal of the Audio Engineering Society*, 48(11):1011–1031, 2000. URL: www.acoustics.hut.fi/software/warp/.

273. A. Härmä and K. Palomäki. HUTear—A free Matlab toolbox for modeling of auditory system. In *Matlab DSP conference*, pp. 96–99, Espoo, Finland, 1999. URL: www.acoustics.hut.fi/software/HUTear/.

274. F.J. Harris. On the use of windows for harmonic analysis with the discrete Fourier transform. *Proceedings of the IEEE*, 66(1):51–83, January 1978.

275. W.M. Hartmann. Pitch, periodicity, and auditory organization. *Journal of the Acoustical Society of America*, 100(6):3491–3502, 1996.

276. W.M. Hartmann. *Signals, Sound, and Sensation.* Springer, New York, 1998.

277. W.M. Hartmann, S. McAdams, and B.K. Smith. Hearing a mistuned harmonic in an otherwise periodic complex tone. *Journal of the Acoustical Society of America*, 88(4):1712–1724, 1990.

278. G. Haus and E. Pollastri. An audio front end for query-by-humming systems. In *2nd Annual International Symposium on Music Information Retrieval*, Bloomington, Indiana, USA, 2001.

279. S. Haykin. *Neural Networks: A Comprehensive Foundation.* Prentice Hall, New Jersey, 1999.

280. D.P. Hedges. *Taking Notes: The History, Practice, and Innovation of Musical Dictation in English and American Aural Skills Pedagogy.* PhD thesis, School of Music, Indiana University, 1999.

281. T. Heittola. *Automatic Classification of Music Signals.* Master's thesis, Tampere University of Technology, December 2003.

282. M. Helén and T. Virtanen. Separation of drums from polyphonic music using non-negtive matrix factorization and support vector machine. In *European Signal Processing Conference*, Antalya, Turkey, 2005.

283. P. Herrera, X. Amatriain, E. Batlle, and X. Serra. Towards instrument segmentation for music content description: A critical review of instrument classification techniques. In *International Conference on Music Information Retrieval*, Plymouth, Massachusetts, USA, 2000.

284. P. Herrera and J. Bonada. Vibrato extraction and parametrization in the spectral modeling synthesis framework. In *First COST-G6 Workshop on Digital Audio Effects*, Barcelona, Spain, November 1998.

285. P. Herrera, A. Dehamel, and F. Gouyon. Automatic labeling of unpitched percussion sounds. In *Audio Engineering Society 114th Convention*, Amsterdam, Netherlands, March 2003.

286. P. Herrera, G. Peeters, and S. Dubnov. Automatic classification of musical instrument sounds. *Journal of New Music Research*, 32:3–21, 2003.

287. P. Herrera, A. Yeterian, and F. Gouyon. Automatic classification of drum sounds: A comparison of feature selection methods and classification techniques. In *International Conference on Music and Artificial Intelligence*, pp. 69–80, Edinburgh, Scotland, UK, September 2002.

288. P. Herrera, A. Yeterian, and F. Gouyon. Automatic classification of drum sounds: a comparison of feature selection methods and classification techniques. In C. Anagnostopoulou, M. Ferrand, and A. Smaill, editors, *Proceedings of the Second International Conference on Music and Artificial Intelligence*, Volume 2445 of *Lecture Notes In Computer Science*, pp. 69–80. Springer, 2002.

289. W.J. Hess. *Pitch Determination of Speech Signals*. Springer, Berlin Heidelberg, 1983.

290. W.J. Hess. Pitch and voicing determination. In Furui and Sondhi [203], pp. 3–48.

291. M.J. Hewitt and R. Meddis. An evaluation of eight computer models of mammalian inner hair-cell function. *Journal of the Acoustical Society of America*, 90(2):904–917, 1991.

292. T.W. Hilands and S.C.A. Thomopoulos. Nonlinear filtering methods for harmonic retrieval and model order selection in Gaussian and non-Gaussian noise. *IEEE Transactions on Signal Processing*, 45(4):982–995, April 1997.

293. J. Holland. *Practical Percussion: A Guide to the Instruments and Their Sources*. Oxford University Press, 2001.

294. H. Honing. From time to time: The representation of timing and tempo. *Computer Music Journal*, 25(3):50–61, 2001.

295. R.A. Horn and C.R. Johnson. *Topics in Matrix Analysis*. Cambridge University Press, Cambridge, UK, 1994.

296. E. v. Hornbostel and C. Sachs. The classification of musical instruments. *Galpin Society Journal*, pp. 3–29, 1961.

297. A.J.M. Houtsma. Pitch perception. In Moore [474], pp. 267–295.

298. A.J.M. Houtsma and J.L. Goldstein. The central origin of the pitch of complex tones: Evidence from musical interval recognition. *Journal of the Acoustical Society of America*, 51(2):520–529, 1972.

299. P. Hoyer. Non-negative sparse coding. In *IEEE Workshop on Networks for Signal Processing XII*, Martigny, Switzerland, 2002.

300. P.O. Hoyer. Non-negative matrix factorization with sparseness constraints. *Journal of Machine Learning Research*, 5:1457–1469, 2004.

301. N. Hu and R.B. Dannenberg. A comparison of melodic database retrieval techniques using sung queries. In *ACM/IEEE Joint Conference on Digital Libraries*, pp. 301–307, Oregon, USA, 2002.

302. A. Hyvärinen. Fast and robust fixed-point algorithms for independent component analysis. *IEEE Transactions on Neural Networks*, 10(3):626–634, 1999.

303. A. Hyvärinen and P. Hoyer. Emergence of phase and shift invariant features by decomposition of natural images into independent feature subspaces. *Neural Computation*, 12(7):1705–1720, 2000.

304. A. Hyvärinen, J. Karhunen, and E. Oja. *Independent Component Analysis*. John Wiley & Sons, 2001.

305. Vamtech Enterprises Inc. *Drumtrax 3.0*. Buffalo, New York, USA, 1999.

306. H. Indefrey, W. Hess, and G. Seeser. Design and evaluation of double-transform pitch determination algorithms with nonlinear distortion in the frequency domain—preliminary results. In *IEEE International Conference on Acoustics, Speech, and Signal Processing*, pp. 415–418, Tampa, Florida, 1985.

307. International Organization for Standardization. *ISO/IEC 15938-4:2002 Information Technology – Multimedia Content Description Interface – Part 4: Audio*. International Organization for Standardization, Geneva, Switzerland, 2002.

308. R. Irizarry. Local harmonic estimation in musical sound signals. *Journal of the American Statistical Association*, 96(454):357–367, June 2001.

309. R. Irizarry. Weighted estimation of harmonic components in a musical sound signal. *Journal of Time Series Analysis*, 23(1):29–48, 2002.

310. R.A. Irizarry. *Statistics and Music: Fitting a Local Harmonic Model to Musical Sound Signals.* PhD thesis, University of California, Berkeley, 1998.

311. F. Jaillet. *Représentations et Traitement Temps-fréquence des Signaux Audionumériques pour des Applications de Design Sonore.* PhD thesis, LATP and Université de Provence, Marseille, 2005.

312. F. Jaillet and B. Torrésani. Time-frequency jigsaw puzzle: Adaptive multiwindow and multilayered gabor expansions. Technical report, LATP and Université de Provence, Marseille, 2004. (submitted)

313. A.K. Jain, R.P.W. Duin, and J. Mao. Statistical pattern recognition: A review. *IEEE Transactions on Pattern Analysis and Machine Intelligence,* 22:4–37, 2000.

314. G.-J. Jang and T.-W. Lee. A maximum likelihood approach to single channel source separation. *Journal of Machine Learning Research,* 23:1365–1392, 2003.

315. H. Järveläinen, V. Välimäki, and M. Karjalainen. Audibility of the timbral effects of inharmonicity in stringed instrument tones. *Acoustics Research Letters Online,* 2(3):79–84, 2001.

316. F. Jelinek. *Statistical Methods for Speech Recognition.* MIT Press, Cambridge, Massachusetts, 1997.

317. F. Jelinek and R.L. Mercer. Interpolated estimation of Markov source parameters from sparse data. In *Internation Workshop on Pattern Recognition in Practice,* Amsterdam, The Netherlands, May 1980.

318. K. Jensen and T.H. Andersen. Beat estimation on the beat. In *IEEE Workshop on Applications of Signal Processing to Audio and Acoustics,* New Paltz, USA, 2003.

319. K. Jensen and J. Arnspang. Binary decission tree classification of musical sounds. In *International Computer Music Conference,* Beijing, China, 1999.

320. I.T. Jolliffe. *Principal Component Analysis.* Springer-Verlag, New York, USA, 1986.

321. D. Jurafsky and J.H. Martin. *Speech and Language Processing.* Prentice Hall, New Jersey, USA, 2000.

322. C. Kaernbach and L. Demany. Psychophysical evidence against the autocorrelation theory of auditory temporal processing. *Journal of the Acoustical Society of America,* 104(4):2298–2306, 1998.

323. T. Kageyama, K. Mochizuki, and Y. Takashima. Melody retrieval with humming. In *International Computer Music Conference,* pp. 349–351, Tokyo, Japan, 1993.

324. H. Kameoka, T. Nishimoto, and S. Sagayama. Separation of harmonic structures based on tied Gaussian mixture model and information criterion for concurrent sounds. In *IEEE International Conference on Acoustics, Speech, and Signal Processing,* Montreal, Canada, 2004.

325. I. Kaminskyj and T. Czaszejko. Automatic recognition of isolated monophonic musical instrument sounds using KNNC. *Journal of Intelligent Information Systems,* 24:199–221, 2005.

326. A. Kapur, M. Benning, and G. Tzanetakis. Query-by-beat-boxing: Music retrieval for the DJ. In *International Conference on Music Information Retrieval,* pp. 170–177, Barcelona, Spain, October 2004.

327. M. Karjalainen and T. Tolonen. Multi-pitch and periodicity analysis model for sound separation and auditory scene analysis. In *IEEE International Conference on Acoustics, Speech, and Signal Processing,* Phoenix, USA, 1999.

328. M. Karjalainen and T. Tolonen. Separation of speech signals using iterative multipitch analysis and prediction. In *6th European Conf. Speech Communication and Technology*, pp. 2187–2190, Budapest, Hungary, 1999.

329. M. Kartomi. *On Concepts and Classification of Musical Instruments*. The University of Chicago Press, 1990.

330. K. Kashino and H. Murase. A music stream segregation system based on adaptive multi-agents. In *International Joint Conference on Artificial Intelligence*, pp. 1126–1136, Nagoya, Japan, 1997.

331. K. Kashino and H. Murase. A sound source identification system for ensemble music based on template adaptation and music stream extraction. *Speech Communication*, 27:337–349, 1999.

332. K. Kashino, K. Nakadai, T. Kinoshita, and H. Tanaka. Organisation of hierarchical perceptual sounds: Music scene analysis with autonomous processing modules and a quantitative information integration mechanism. In *International Joint Conference on Artificial Intelligence*, pp. 158–164, Montreal, Quebec, 1995.

333. K. Kashino and H. Tanaka. A sound source separation system with the ability of automatic tone modeling. In *International Computer Music Conference*, pp. 248–255, Tokyo, Japan, 1993.

334. K. Kashino and H. Tanaka. A computational model of segregation of two frequency components: Evaluation and integration of multiple cues. *Electronics and Communications in Japan*, 77(7):35–47, 1994.

335. H. Katayose and S. Inokuchi. An intelligent transcription system. In *First International Conference on Music Perception and Cognition*, pp. 95–98, 1989.

336. S.M. Katz. Estimation of probabilities from sparse data for the language model component of a speech recogniser. *IEEE Transactions on Acoustics, Speech, and Signal Processing*, 35(3):400–401, March 1987.

337. H. Kawahara, Y. Hirachi, M. Morise, and H. Banno. Procedure "senza vibrato": A key component for morphing singing. In *ISCA Tutorial and Research Workshop on Statistical and Perceptual Audio Processing*, Jeju, Korea, October 2004.

338. H. Kawahara, H. Katayose, A. de Cheveigné, and R.D. Patterson. Fixed point analysis of frequency to instantaneous frequency mapping for accurate estimation of F0 and periodicity. In *6th European Conf. Speech Communication and Technology*, pp. 2781–2784, Budapest, Hungary, 1999.

339. S.M. Kay. *Fundamentals of Statistical Signal Processing: Estimation Theory*. Prentice Hall, 1993.

340. R.A. Kendall. The role of acoustic signal partitions in listener categorization of musical phrases. *Music Perception*, 4:185–214, 1986.

341. H.-G. Kim, J.J. Burred, and T. Sikora. How efficient is MPEG-7 for general sound recognition? In *25th International Audio Engineering Society Conference Metadata for Audio*, London, UK, 2004.

342. H.-G. Kim, N. Moreau, and T. Sikora. *MPEG-7 Audio and Beyond: Audio Content Indexing and Retrieval*. Wiley, 2005.

343. T. Kinoshita, S. Sakai, and H. Tanaka. Musical sound source identification based on frequency component adaptation. In *International Joint Conference on Artificial Intelligence*, pp. 18–24, Stockholm, Sweden, 1999.

344. G. Kitagawa. Monte Carlo filter and smoother for non-Gaussian nonlinear state space models. *Journal of Computational and Graphical Statistics*, 5(1):1–25, 1996.

345. T. Kitahara, M. Goto, and H.G. Okuno. Musical instrument identification based on F0-dependent multivariate normal distribution. In *IEEE International Conference on Acoustics, Speech, and Signal Processing*, pp. 421–424, Hong Kong, China, 2003.

346. T. Kitahara, M. Goto, and H.G. Okuno. Category-level identificatioon of non-registered musical instrument sounds. In *IEEE International Conference on Acoustics, Speech, and Signal Processing*, pp. 253–256, Montreal, Canada, 2004.

347. A. Klapuri. Sound onset detection by applying psychoacoustic knowledge. In *IEEE International Conference on Acoustics, Speech, and Signal Processing*, Phoenix, USA, 1999.

348. A. Klapuri. Musical meter estimation and music transcription. In *Cambridge Music Processing Colloquium*, pp. 40–45, 2003.

349. A. Klapuri, A Eronen, and J. Astola. Analysis of the meter of acoustic musical signals. *IEEE Transactions on Speech and Audio Processing*, 14(1), 2006.

350. A. Klapuri, T. Virtanen, A. Eronen, and J. Seppänen. Automatic transcription of musical recordings. In *Workshop on Consistent and Reliable Acoustic Cues for Sound Analysis*, Aalborg, Denmark, 2001.

351. A.P. Klapuri. Multiple fundamental frequency estimation based on harmonicity and spectral smoothness. *IEEE Transactions on Speech and Audio Processing*, 11(6):804–815, 2003.

352. A.P. Klapuri. Automatic music transcription as we know it today. *Journal of New Music Research*, 33(3), 2004.

353. A.P. Klapuri. *Signal Processing Methods for the Automatic Transcription of Music*. PhD thesis, Tampere University of Technology, 2004.

354. A.P. Klapuri. A perceptually motivated multiple-F0 estimation method for polyphonic music signals. In *IEEE Workshop on Applications of Signal Processing to Audio and Acoustics*, New Paltz, USA, 2005.

355. W.B. Kleijn and K.K. Paliwal, editors. *Speech Coding and Synthesis*. Elsevier Academic Press, Amsterdam, 1995.

356. T. Kohonen. Emergence of invariant-feature detectors in the adaptive-subspace self-organizing map. *Biological Cybernetics*, 75:281–291, 1996.

357. B. Kostek. Application of learning algorithms to musical sound analysis. In *Audio Engineering Society 97th Convention*, San Francisco, USA, 1994.

358. B. Kostek. Feature extraction methods for the intelligent processing of musical sounds. In *Audio Engineering Society 99th Convention*, New York, NY, 1995.

359. B. Kostek. Soft computing-based recognition of musical sounds. In L. Polkowski and A. Skowron, editors, *Rough Sets in Knowledge Discovery*, pp. 193–213. Physica-Verlag, Heidelberg, 1998.

360. B. Kostek. *Soft Computing in Acoustics: Applications of Neural Networks, Fuzzy Logic and Rough Sets to Musical Acoustics*. Physica-Verlag, Heidelberg, 1999.

361. B. Kostek. Application of soft computing to automatic music information retrieval. *Journal of the American Society for Information Science and Technology*, 55:1108–1116, 2004.

362. B. Kostek. Musical instrument recognition and duet analysis employing music information retrieval techniques. In *Proceedings of the IEEE*, Volume 92, pp. 712–729, 2004.

363. B. Kostek and A. Czyzewski. An approach to the automatic classification of musical sounds. In *Audio Engineering Society 108th Convention*, Paris, France, 2000.

364. B. Kostek and A. Czyzewski. Representing musical instrument sounds for their automatic classification. *Journal of the Audio Engineering Society*, 49:768–785, 2001.

365. B. Kostek, M. Dziubinski, and P. Zwan. Further developments of methods for searching optimum musical and rhythmic feature vectors. In *Audio Engineering Society 21st International Conference*, St. Petersburg, Russia, 2002.

366. B. Kostek and R. Krolikowski. Application of artificial neural networks to the recognition of musical sounds. *Archives of Acoustics*, 22:27–50, 1997.

367. B. Kostek, P. Szczuko, P. Zwan, and P Dalka. Processing of musical data employing rough sets and artificial neural networks. In *Transactions on Rough Sets III*, pp. 112–133, 2005.

368. B. Kostek and A. Wieczorkowska. Study of parameter relations in musical instrument patterns. In *Audio Engineering Society 100th Convention*, Copenhangen, Denmark, 1996.

369. B. Kostek and A. Wieczorkowska. Parametric representation of musical sounds. *Archives of Acoustics*, 22:3–26, 1997.

370. B. Kostek and P. Zwan. Wavelet-based automatic recognition of musical instrument classes. In *142nd Meeting of the Acoustical Society of America*, Melville, New York, 2001.

371. B. Kostek, P. Zwan, and M. Dziubinski. Statistical analysis of musical sound features derived from wavelet representation. In *Audio Engineering Society 112nd Convention*, Munich, Germany, 2002.

372. J.R. Koza. *Genetic Programming: On the Programming of Computers by Means of Natural Selection*. MIT Press, 1992.

373. K. Kreutz-Delgado, J.F. Murray, B.D. Rao, K. Engan, T. Lee, and T.J. Sejnowski. Dictionary learning algorithms for sparse representation. *Neural Computation*, 15:349–396, 2003.

374. J. Krimphoff, S. McAdams, and S. Winsberg. Caractérisation du timbre des sons complexes. II: Analyses acoustiques et quantification psychophysique. *Journal de Physique*, 4:625–628, 1994.

375. A. Krishna and T. Sreenivas. Music instrument recognition: From isolated notes to solo phrases. In *IEEE International Conference on Acoustics, Speech, and Signal Processing*, Montreal, Canada, 2004.

376. S. Krstulović, R. Gribonval, P. Leveau, and L. Daudet. A comparison of two extensions of the matching pursuit algorithm for the harmonic decomposition of sounds. In *IEEE Workshop on Applications of Signal Processing to Audio and Acoustics*, New Paltz, USA, 2005.

377. C. Krumhansl. *Cognitive Foundations of Musical Pitch*. Oxford University Press, 1990.

378. C.L. Krumhansl. Why is musical timbre so hard to understand? In S. Nielzenand and O. Olsson, editors, *Structure and Perception of Electroacoustic Sound and Music*, pp. 43–53. Elsevier Academic Press, Amsterdam, 1989.

379. L. Kuncheva. *Combining Pattern Classifiers: Methods and Algorithms*. Wiley, 2004.

380. N. Kunieda, T. Shimamura, and J. Suzuki. Robust method of measurement of fundamental frequency by ACLOS—autocorrelation of log spectrum. In *IEEE International Conference on Acoustics, Speech, and Signal Processing*, pp. 232–235, Atlanta, USA, 1996.

381. J. Tin-Yau Kwok. Moderating the outputs of support vector machine classifiers. *IEEE Transactions on Neural Networks*, 10(5):1018–1031, September 1999.
382. T.I. Laakso, V. Välimäki, M. Karjalainen, and U.K. Laine. Splitting the unit delay: Tools for fractional delay filter design. *IEEE Signal Processing Magazine*, 13(1):30–60, 1996.
383. M. Lagrange, S. Marchand, and J.B. Rault. Using linear prediction to enhance the tracking of partials. In *IEEE International Conference on Acoustics, Speech, and Signal Processing*, Montreal, Canada, 2004.
384. M. Lahat, R. Niederjohn, and D.A. Krubsack. Spectral autocorrelation method for measurement of the fundamental frequency of noise-corrupted speech. *IEEE Transactions on Acoustics, Speech, and Signal Processing*, 6:741–750, June 1987.
385. S. Lakatos. A common perceptual space for harmonic and percussive timbres. *Perception and Psychophysics*, 62:1426–1439, 1994.
386. T.L. Lam. *Beat Tracking*. Master's thesis, Department of Engineering, University of Cambridge, 2003.
387. D. Lang and N. de Frietas. Beat tracking the graphical model way. In *Neural Information Processing Systems*, Vancouver, Canada, 2004.
388. K. Lange. *Numerical Analysis for Statisticians*. Springer, New York, USA, 1999.
389. E.W. Large. *Dynamic Representation of Musical Structure*. PhD thesis, Ohio State Univ., 1994.
390. E.W. Large. Beat tracking with a nonlinear oscillator. In *International Joint Conference on Artificial Intelligence*, pp. 24–31, Stockholm, Sweden, 1995.
391. E.W. Large and J.F. Kolen. Resonance and the perception of musical meter. *Connection Science*, 6(1):177–208, 1994.
392. J. Laroche. Estimating tempo, swing and beat locations in audio recordings. In *IEEE Workshop on Applications of Signal Processing to Audio and Acoustics*, pp. 135–138, 2001.
393. J. Laroche. Efficient tempo and beat tracking in audio recordings. *Journal of the Audio Engineering Society*, 51(4):226–233, April 2003.
394. J. Laroche, Y. Stylianou, and E. Moulines. HNS: Speech modification based on a harmonic + noise model. In *IEEE International Conference on Acoustics, Speech, and Signal Processing*, Volume 2, pp. 550–553, Minneapolis, USA, April 1993.
395. H. Laurent and C. Doncarli. Stationarity index for abrupt changes detection in the time-frequency plane. *IEEE Signal Processing Letters*, 5(2):43–45, 1998.
396. S. Lauritzen and D. Spiegelhalter. Local computations with probabilities on graphical structures and their application to expert systems. *Journal of the Royal Statistical Society: Series B*, 50(2):157–224, 1988.
397. C.L. Lawson and R.J. Hanson. *Solving Least Squares Problems*. Prentice Hall, Englewood Cliffs, New Jersey, 1974.
398. C.S. Lee. The perception of metrical structure: Experimental evidence and a model. In P. Howell, R. West, and I. Cross, editors, *Representing Musical Structure*. Academic Press, London, 1991.
399. D.D. Lee and H.S. Seung. Learning the parts of objects by non-negative matrix factorization. *Nature*, 401:788–791, October 1999.
400. D.D. Lee and H.S. Seung. Algorithms for non-negative matrix factorization. In *Neural Information Processing Systems*, pp. 556–562, Denver, USA, 2001.

401. T. Lee and R. Orglmeister. A contextual blind separation of delayed and convolved sources. In *IEEE International Conference on Acoustics, Speech, and Signal Processing*, pp. 1199–1202, Munich, Germany, 1997.
402. M. Leman. *Music and Schema Theory.* Springer, Heidelberg, 1995.
403. P. Lepain. Polyphonic pitch extraction from musical signals. *Journal of New Music Research*, 28(4):296–309, 1999.
404. F. Lerdahl and R. Jackendoff. *A Generative Theory of Tonal Music.* MIT Press, 1983.
405. M. Lesaffre, M. Leman, B. De Baets, and J.-P. Martens. Methodological considerations concerning manual annotation of musical audio in function of algorithm development. In *International Conference on Music Information Retrieval*, Barcelona, Spain, 2004.
406. M. Lesaffre, M. Leman, K. Tanghe, B. De Baets, H. De Meyer, and J.P. Martens. User-dependent taxonomy of musical features as a conceptual framework for musical audio-mining technology. In *Stockholm Music Acoustics Conference*, Stockholm, Sweden, 2003.
407. V. Lesser, S.H. Nawab, I. Gallastegi, and F. Klassner. IPUS: An architecture for integrated signal processing and signal interpretation in complex environments. In *11th National Conference on Artificial Intelligence*, pp. 249–255, 1993.
408. S. Levine. *Audio Representations for Data Compression and Compressed Domain Processing.* PhD thesis, Center for Computer Research in Music and Acoustics, Stanford University, 1998.
409. J.C.R. Licklider. A duplex theory of pitch perception. *Experientia*, 7:128–133, 1951. Reproduced in Schubert, E.D. (Ed.): *Psychological Acoustics* (Benchmark papers in acoustics, vol. 13), Stroudsburg, Pennsylvania: Dowden, hutchinson & Ross, Inc., pp. 155–160.
410. T.M. Little and F.J. Hills. *Statistical Methods in Agricultural Research.* University of California Press, 1972.
411. B. Liu, Y. Wu, and Y. Li. Linear hidden Markov model for music information retrieval based on humming. In *IEEE International Conference on Acoustics, Speech, and Signal Processing*, Volume 5, pp. 533–536, Hong Kong, China, 2003.
412. M. Liu and C. Wan. Feature selection for automatic classification of musical instrument sounds. In *ACM/IEEE Joint Conference on Digital Libraries*, Roanoke, VA, USA, 2001.
413. A. Livshin, G. Peeters, and X. Rodet. Studies and improvements in automatic classification of musical sound samples. In *International Computer Music Conference*, Singapore, 2001.
414. A. Livshin and X. Rodet. The importance of cross database evaluation in sound classification. In *International Conference on Music Information Retrieval*, Baltimore, USA, 2003.
415. A. Livshin and X. Rodet. Instrument recognition beyond separate notes: Indexing continuous recordings. In *International Computer Music Conference*, Miami, Florida, USA, 2004.
416. A. Livshin and X. Rodet. Musical instrument identification in continuous recordings. In *International Conference on Digital Audio Effects*, Naples, Italy, 2004.
417. B. Logan and S. Chu. Music summarization using key phrases. In *IEEE International Conference on Acoustics, Speech, and Signal Processing*, Volume 2, pp. 749–752, Istanbul, Turkey, 2000.

418. H.C. Longuet-Higgins and C.S. Lee. The perception of musical rhythms. *Perception*, 11(2):115–128, 1982.
419. M.A. Loureiro, H.B. de Paula, and H.C. Yehia. Timbre classification of a single musical instrument. In *International Conference on Music Information Retrieval*, Barcelona, Spain, 2004.
420. L. Lu, M. Wang, and H.-J. Zhang. Repeating pattern discovery and structure analysis from acoustic music data. In *ACM SIGMM International Workshop on Multimedia Information Retrieval*, pp. 275–282, 2004.
421. L. Lu, H. You, and H.-J. Zhang. A new approach to query by humming in music retrieval. In *IEEE International Conference on Multimedia and Expo*, pp. 776–779, Tokyo, Japan, 2001.
422. R.F. Lyon. Computational models of neural auditory processing. In *IEEE International Conference on Acoustics, Speech, and Signal Processing*, pp. 36.1.1–36.1.4, San Diego, California, 1984.
423. A. Madevska-Bogdanova and D. Nikolic. A geometrical modification of SVM outputs for pattern recognition. In *22nd International Conference on Information Technology Interfaces*, pp. 165–170, Pula, Croatia, June 2000.
424. A. Madevska-Bogdanova and D. Nikolic. A new approach of modifying SVM outputs. In *International Joint Conference on Neural Networks*, Volume 6, pp. 395–398, Como, Italy, July 2000.
425. R. Maher and J. Beauchamp. An investigation of vocal vibrato for synthesis. *Applied Acoustics*, 30:219–245, 1990.
426. R.C. Maher. *An Approach for the Separation of Voices in Composite Music Signals*. PhD thesis, Univ. of Illinois, Urbana, 1989.
427. R.C. Maher. Evaluation of a method for separating digitized duet signals. *Journal of the Audio Engineering Society*, 38(12):956–979, 1990.
428. R.C. Maher and J.W. Beauchamp. Fundamental frequency estimation of musical signals using a two-way mismatch procedure. *Journal of the Acoustical Society of America*, 95(4):2254–2263, April 1994.
429. S. Mallat. *A Wavelet Tour of Signal Processing*. Academic Press, 1998.
430. S. Mallat and Z. Zhang. Matching pursuits with time-frequency dictionaries. *IEEE Transactions on Signal Processing*, 41:3397–3415, 1993.
431. H.S. Malvar. *Signal Processing with Lapped Transforms*. Artech House, Norwood, MA, 1992.
432. B.S. Manjunath, P. Salembier, and T. Sikora. *Introduction to MPEG-7: Multimedia Content Description Language*. John Wiley & Sons, 2002.
433. M. Marolt. SONIC: Transcription of polyphonic piano music with neural networks. In *MOSART Workshop on Current Research Directions in Computer Music*, Barcelona, Spain, November 2001.
434. M. Marolt. A connectionist approach to transcription of polyphonic piano music. *IEEE Transactions on Multimedia*, 6(3):439–449, 2004. URL: lgm.fri.uni-lj.si/SONIC.
435. M. Marolt. Gaussian mixture models for extraction of melodic lines from audio recordings. In *International Conference on Music Information Retrieval*, Barcelona, Spain, October 2004.
436. M. Marolt. On finding melodic lines in audio recordings. In *International Conference on Digital Audio Effects*, Naples, Italy, 2004.
437. J. Marques and P.J. Moreno. A study of musical instrument classification using Gaussian mixture models and support vector machines. Technical Report CRL 99/4, Compaq, 1999.

438. D. Marr. *Vision.* W.H. Freeman and Company, 1982.

439. K.D. Martin. Automatic transcription of simple polyphonic music: Robust front end processing. Technical Report 399, MIT Media Laboratory Perceptual Computing Section, 1996.

440. K.D. Martin. A blackboard system for automatic transcription of simple polyphonic music. Technical Report 385, MIT Media Laboratory Perceptual Computing Section, 1996.

441. K.D. Martin. Toward automatic sound source recognition: Identifying musical instruments. In *Computational Hearing Conference of NATO Advanced Study Institute*, Tuscany, Italy, 1998.

442. K.D. Martin. *Sound-Source Recognition: A Theory and Computational Model.* PhD thesis, Massachusetts Institute of Technology, 1999.

443. K.D. Martin and Y.E. Kim. Musical instrument identification: A pattern-recognition approach. In *136th meeting of the Acoustical Society of America*, Norfolk, Virginia, USA, 1998.

444. P. Masri and A. Bateman. Improved modelling of attack transients in music analysis-resynthesis. In *International Computer Music Conference*, pp. 100–103, Hong Kong, China, August 1996.

445. O. Mayor. An adaptive real-time beat tracking system for polyphonic pieces of audio using multiple hypotheses. In *MOSART Workshop on Current Research Directions in Computer Music*, 2001.

446. S. McAdams, J.W. Beauchamp, and S. Meneguzzi. Discrimination of musical instrument sounds resynthesized with simplified spectrotemporal parameters. *Journal of the Acoustical Society of America*, 105:882–897, 1999.

447. S. McAdams and A.S. Bregman. Hearing musical streams. *Computer Music Journal*, 3(4):26–43, 1979.

448. S. McAdams, S. Winsberg, G. de Soete, and J. Krimphoff. Perceptual scaling of synthesized musical timbres: Common dimensions, specificities, and latent subject classes. *Psychological Research*, 58:177–192, 1995.

449. R.J. McAulay and Th.F. Quatieri. Speech analysis/synthesis based on a sinusoidal representation. *IEEE Transactions on Acoustics, Speech, and Signal Processing*, 34:744–754, 1986.

450. J.D. McAuley. *Perception of Time as Phase: Towards an Adaptive-Oscillator Model of Rhythmic Pattern Processing.* PhD thesis, Computer Science and Cognitive Science, Univ. Indiana, July 1995.

451. M.E. McIntyre, R.T. Schumacher, and J. Woodhouse. On the oscillations of musical instruments. *Journal of the Acoustical Society of America*, 74(5):1325–1345, 1983.

452. M.F. McKinney and D. Moelants. Extracting the perceptual tempo from music. In *International Conference on Music Information Retrieval*, Barcelona, Spain, October 2004.

453. G.J. McLachlan. *Discriminant Analysis and Statistical Pattern Recognition.* John Wiley & Sons, 1992.

454. R.J. McNab, L.A. Smith, and I.H. Witten. Signal processing for melody transcription. In *19th Australasian Computer Science Conference*, Melbourne, Australia, February 1996.

455. R.J. McNab, L.A. Smith, I.H. Witten, and C.L. Henderson. Tune retrieval in the multimedia library. *Multimedia Tools and Applications*, 10(2):113–132, April 2000.

456. R. Meddis. Simulation of mechanical to neural transduction in the auditory receptor. *Journal of the Acoustical Society of America*, 79(3):702–711, 1986.

457. R. Meddis and M.J. Hewitt. Virtual pitch and phase sensitivity of a computer model of the auditory periphery. I: Pitch identification. II: Phase sensitivity. *Journal of the Acoustical Society of America*, 89(6):2866–2894, 1991.

458. R. Meddis and M.J. Hewitt. Modeling the identification of concurrent vowels with different fundamental frequencies. *Journal of the Acoustical Society of America*, 91(1):233–245, 1992.

459. R. Meddis and L. O'Mard. A unitary model of pitch perception. *Journal of the Acoustical Society of America*, 102(3):1811–1820, 1997.

460. D.K. Mellinger. *Event Formation and Separation of Musical Sound*. PhD thesis, Stanford University, Stanford, USA, 1991.

461. Y. Meron and K. Hirose. Synthesis of vibrato singing. In *IEEE International Conference on Acoustics, Speech, and Signal Processing*, Volume 2, pp. 745–748, Istanbul, Turkey, 2000.

462. The MIDI Manufacturers Association. *The Complete MIDI 1.0 Detailed Specification*, second edition, 1996. Website: www.midi.org.

463. I. Mierswa and K. Morik. Automatic feature extraction for classifying audio data. *Machine Learning*, 58:127–149, 2005.

464. B.O. Miller, D.L. Scarborough, and J.A. Jones. On the perception of meter. In M. Balaben, K. Ebeioglu, and O. Laske, editors, *Understanding Music With AI*, chapter 18, pp. 428–47. MIT Press, 1992.

465. MIREX: Annual Music Information Retrieval Evaluation eXchange, 2005. URL: www.music-ir.org/ mirexwiki/ index.php/ MIREX_2005.

466. O.M.E. Mitchell, C.A. Ross, and G.H. Yates. Signal processing for a cocktail party effect. *Journal of the Acoustical Society of America*, 50(2):656–660, 1971.

467. T.M. Mitchell. *Machine Learning*. McGraw-Hill, 1997.

468. S. Molla and B. Torrésani. Determining local transientness of audio signals. *IEEE Signal Processing Letters*, 11(7):625–628, 2004.

469. S. Molla and B. Torrésani. A hybrid scheme for encoding audio signals using hidden Markov models of waveforms. *Applied and Computational Harmonic Analysis*, 2005. (in press)

470. B. Mont-Reynaud. Problem-solving strategies in a music transcription system. In *9th International Joint Conference on Artificial Intelligence*, pp. 916–918, Los Angeles, USA, 1985.

471. T.K. Moon. The expectation-maximization algorithm. *IEEE Signal Processing Magazine*, 13(6):47–60, November 1996.

472. T.K. Moon and W.C. Stirling. *Mathematical Methods and Algorithms for Signal Processing*. Prentice Hall, Upper Saddle River, NJ, 2000.

473. B.C.J. Moore. Frequency analysis and masking. In Moore [474], pp. 161–205.

474. B.C.J. Moore, editor. *Hearing—Handbook of Perception and Cognition*. Academic Press, San Diego, California, 2nd edition, 1995.

475. B.C.J. Moore. *An Introduction to the Psychology of Hearing*. Academic Press, 4th edition, 1997.

476. B.C.J. Moore, B.R. Glasberg, and R.W. Peters. Thresholds for hearing mistuned partials as separate tones in harmonic complexes. *Journal of the Acoustical Society of America*, 80(2):479–483, 1986.

477. J.A. Moorer. *On the Segmentation and Analysis of Continuous Musical Sound by Digital Computer*. PhD thesis, Dept. of Music, Stanford University, 1975. Distributed as Dept. of Music report No. STAN-M-3.

478. J.A. Moorer. On the transcription of musical sound by computer. *Computer Music Journal*, 1(4):32–38, 1977.

479. T. Nakano, J. Ogata, M. Goto, and Y. Hiraga. A drum pattern retrieval method by voice percussion. In *International Conference on Music Information Retrieval*, pp. 550–553, Barcelona, Spain, October 2004.

480. T. Nakatani. *Computational Auditory Scene Analysis Based on Residue-Driven Architecture and Its Application to Mixed Speech Recognition*. PhD thesis, Kyoto University, Kyoto, Japan, 2002.

481. T. Nakatani, G.H. Okuno, and T. Kawabata. Residue-driven architecture for computational auditory scene analysis. In *International Joint Conference on Artificial Intelligence*, pp. 165–172, Montreal, Quebec, 1995.

482. A. Nehorai and B. Porat. Adaptive comb filtering for harmonic signal enhancement. *IEEE Transactions on Acoustics, Speech, and Signal Processing*, 34(5), 1986.

483. T. Niihara and S. Inokuchi. Transcription of sung song. In *IEEE International Conference on Acoustics, Speech, and Signal Processing*, pp. 1277–1280, Tokyo, Japan, 1986.

484. T. Nishimura, H. Hashiguchi, J. Takita, J. Xin Zhang, M. Goto, and R. Oka. Music signal spotting retrieval by a humming query using start frame feature dependent continuous dynamic programming. In *2nd Annual International Symposium on Music Information Retrieval*, pp. 211–218, Bloomington, Indiana, USA, October 2001.

485. H. Okuno and D Rosenthal, editors. *Computational Auditory Scene Analysis*. Lawrence Erlbaum Associates, Publishers, 1997.

486. B.A. Olshausen and D.F. Field. Sparse coding with an overcomplete basis set: A strategy employed by v1? *Vision Research*, 37:3311–3325, 1997.

487. F. Opolko and J. Wapnick. McGill University Master Samples. Technical report, McGill University, Montreal, Canada, 1987.

488. I.F.O Orife. *Riddim: A Rhythm Analysis and Decomposition Tool Based on Independent Subspace Analysis*. Master's thesis, Dartmouth College, Hanover, New Hampshire, USA, May 2001.

489. N. Orio and M.S. Sette. A HMM-based pitch tracker for audio queries. In *International Conference on Music Information Retrieval*, pp. 249–250, Baltimore, USA, 2003.

490. J. Osborne. Notes on the use of data transformations. *Practical Assessment, Research and Evaluation*, 8(6), 2002.

491. N. Otsu. A threshold selection method from gray-level histograms. *IEEE Transactions on Systems, Man, and Cybernetics*, SMC-9(1):62–66, 1979.

492. P. Paatero and U. Tapper. Positive matrix factorization: A non-negative factor model with optimal utilization of error estimates of data values. *Environmetrics*, 5:111–126, 1994.

493. R.P. Paiva, T. Mendes, and A. Cardoso. An auditory model based approach for melody detection in polyphonic musical recordings. In *International Symposium on Computer Music Modeling and Retrieval*, May 2004.

494. R.P. Paiva, T. Mendes, and A. Cardoso. A methodology for detection of melody in polyphonic musical signals. In *Audio Engineering Society 116th Convention*, Berlin, Germany, May 2004.

495. T.H. Park. *Towards Automatic Musical Instrument Timbre Recognition*. PhD thesis, Princeton University, 2004.

496. T.H. Park and P. Cook. Nearest centroid error clustering for radial/elliptical basis function neural networks in timbre classification. In *International Computer Music Conference*, pp. 833–866, Barcelona, Spain, 1998.

497. R. Parncutt. A perceptual model of pulse salience and metrical accent in musical rhythms. *Music Perception*, 11(4):409–464, 1994.

498. T.W. Parsons. Separation of speech from interfering speech by means of harmonic selection. *Journal of the Acoustical Society of America*, 60(4), 1976.

499. R.D. Patterson. Auditory filter shapes derived with noise stimuli. *Journal of the Acoustical Society of America*, 59(3):640–654, 1976.

500. R.D. Patterson. Auditory images: How complex sounds are represented in the auditory system. *Journal of the Acoustical Society of Japan (E)*, 21(4):183–190, 2000.

501. R.D. Patterson and M.H. Allerhand. Time-domain modeling of peripheral auditory processing: A modular architecture and a software platform. *Journal of the Acoustical Society of America*, 98(4):1890–1894, 1995. URL: http://www.mrc-cbu.cam.ac.uk/cnbh/web2002/ bodyframes/AIM.htm.

502. R.D. Patterson and J. Holdsworth. A functional model of neural activity patterns and auditory images. In Ainsworth [13], pp. 551–567.

503. J. Paulus and A. Klapuri. Measuring the similarity of rhythmic patterns. In *International Conference on Music Information Retrieval*, Paris, France, 2002.

504. J. Paulus and A. Klapuri. Model-based event labeling in the transcription of percussive audio signals. In M. Davies, editor, *International Conference on Digital Audio Effects*, pp. 73–77, London, UK, September 2003.

505. J. Paulus and T. Virtanen. Drum transcription with non-negative spectrogram factorisation. In *European Signal Processing Conference*, Antalya, Turkey, September 2005.

506. J.K. Paulus and A.P. Klapuri. Conventional and periodic N-grams in the transcription of drum sequences. In *IEEE International Conference on Multimedia and Expo*, Volume 2, pp. 737–740, Baltimore, Maryland, USA, July 2003.

507. Steffen Pauws. CubyHum: A fully operational query by humming system. In *International Conference on Music Information Retrieval*, pp. 187–196, Paris, France, October 2002.

508. Z. Pawlak. Rough set elements. In L. Polkowski and A. Skowron, editors, *Rough Sets in Knowledge Discovery*, pp. 10–30. Physica-Verlag, Heidelberg, 1998.

509. J. Pearl. *Probabilistic Reasoning in Intelligent Systems: Networks of Plausible Inference*. Morgan Kaufmann Publishers, 1988.

510. G. Peeters. Automatic classification of large musical instrument databases using hierarchical classifiers with inertia ratio maximization. In *Audio Engineering Society 115th Convention*, New York, NY, USA, 2003.

511. G. Peeters. A large set of audio features for sound description (similarity and classification) in the CUIDADO project. Technical report, IRCAM, Paris, France, April 2004.

512. G. Peeters, A. La Burthe, and X. Rodet. Toward automatic music audio summary generation from signal analysis. In *International Conference on Music Information Retrieval*, pp. 94–100, Paris, France, October 2002.

513. G. Peeters, S. McAdams, and P. Herrera. Instrument sound description in the context of MPEG-7. In *International Computer Music Conference*, pp. 166–169, Berlin, Germany, 2000.

514. G. Peeters and X. Rodet. Automatically selecting signal descriptors for sound classification. In *International Computer Music Conference*, Göteborg, Sweden, September 2002.

515. G. Peeters and X. Rodet. Hierarchical Gaussian tree with inertia ratio maximization for the classification of large musical instrument databases. In *International Conference on Digital Audio Effects*, London, UK, 2003.

516. G. Peeters and X. Rodet. Signal-based music structure discovery for music audio summary generation. In *International Computer Music Conference*, pp. 15–22, Singapore, 2003.

517. I. Peretz. Music perception and recognition. In B. Rapp, editor, *The Handbook of Cognitive Neuropsychology*, pp. 519–540. Hove: Psychology Press, 2001.

518. I. Peretz and M. Coltheart. Modularity of music processing. *Nature Neuroscience*, 6(7), 2003.

519. G. Peterschmitt, E. Gómez, and P. Herrera. Pitch-based solo location. In *MOSART Workshop on Current Research Directions in Computer Music*, Barcelona, Spain, 2001.

520. M. Piszczalski. *A Computational Model of Music Transcription*. PhD thesis, Univ. of Michigan, Ann Arbor, 1986.

521. M. Piszczalski and B.A. Galler. Automatic music transcription. *Computer Music Journal*, 1(4):24–31, 1977.

522. C.J. Plack, A.J. Oxenham, R.R. Fay and A.N. Popper, editors. *Pitch*. Springer, New York, 2005.

523. C.J. Plack and R.P. Carlyon. Loudness perception and intensity coding. In Moore [474], pp. 123–160.

524. J.C. Platt. Probabilistic outputs for support vector machines and comparisons to regularized likelihood methods. In A.J. Smola, P. Bartlett, B. Schölkopf, and D. Schuurmans, editors, *Advances in Large Margin Classifiers*. MIT Press, 1999.

525. M.D. Plumbley. Conditions for non-negative independent component analysis. *IEEE Signal Processing Letters*, 9(6), 2002.

526. M.D. Plumbley and E. Oja. A 'non-negative PCA' algorithm for independent component analysis. *IEEE Transactions on Neural Networks*, 15(1):66–67, 2004.

527. H.F. Pollard and E.V. Janson. A tristimulus method for the specification of musical timbre. *Acustica*, 51:162–171, 1982.

528. E. Pollastri. A pitch tracking system dedicated to process singing voice for musical retrieval. In *IEEE International Conference on Multimedia and Expo*, Volume 1, pp. 341–344, Lusanne, Switzerland, 2002.

529. D-J. Povel and P. Essens. Perception of musical patterns. *Music Perception*, 2(4):411–440, Summer 1985.

530. E. Prame. Vibrato extent and intonation in professional Western lyric singing. *Journal of the Acoustical Society of America*, 102(1):616–621, July 1997.

531. W.H. Press, S.A. Teukolsky, W.T. Vetterling, and B.P. Flannery. *Numerical Recipes in C/C++; The Art of Scientific Computing*. Cambridge University Press, Cambridge, UK, 2002.

532. H. Purnhagen and N. Meine. HILN: The MPEG-4 parametric audio coding tools. In *the IEEE International Symposium on Circuits and Systems (ISCAS 2000)*, Geneva, Switzerland, 2000.

533. J.R. Quinlan. *C4.5: Programs for Machine Learning*. Morgan Kaufmann Publishers, San Francisco, CA, USA, 1993.

534. L.R. Rabiner. A tutorial on hidden Markov models and selected applications in speech recognition. *Proc. of IEEE*, 77(2):257–289, February 1989.

535. L.R. Rabiner, M.J. Cheng, A.E. Rosenberg, and C.A. McGonegal. A comparative performance study of several pitch detection algorithms. *IEEE Transactions on Acoustics, Speech, and Signal Processing*, 24(5):399–418, 1976.

536. L.R. Rabiner and B.-H. Juang. *Fundamentals of Speech Recognition*. Prentice Hall, New Jersey, 1993.

537. C. Raphael. Automated rhythm transcription. In *2nd Annual International Symposium on Music Information Retrieval*, Bloomington, Indiana, USA, 2001.

538. C. Raphael. A probabilistic expert system for automatic musical accompaniment. *Journal of Computational and Graphical Statistics*, 10(3):486–512, 2001.

539. M. Reyes-Gomez, N. Jojic, and D. Ellis. Deformable spectrograms. In *10th International Workshop on Artificial Intelligence and Statistics*, pp. 285–292, Barbados, 2005.

540. E. Rich and K. Knight. *Artificial Intelligence*. McGraw-Hill, New York, 1991.

541. S. Richardson and P.J. Green. On Bayesian analysis of mixtures with an unknown number of components. *Journal of the Royal Statistical Society: Series B*, 59(4):731–792, 1997.

542. C. Roads. Research in music and artificial intelligence. *ACM Computing Surveys*, 17(2):163–190, 1985.

543. C. Roads. *The Computer Music Tutorial*. MIT Press, Cambridge, USA, 1996.

544. C.P. Robert and G. Casella. *Monte Carlo Statistical Methods*. Springer, New York, USA, 2000.

545. A. Ron and Z. Shen. Frames and stable bases for shift-invariant subspaces of $L^2(R^d)$. *Canadian Journal of Mathematics*, 47:1051–1094, 1995.

546. D. Rosenthal. Emulation of human rhythm perception. *Computer Music Journal*, 16(1):64–72, Spring 1992.

547. D.F. Rosenthal. *Machine Rhythm: Computer Emulation of Human Rhythm Perception*. PhD thesis, Massachusetts Institute of Technology, 1992.

548. J. Rosenthal. *A First Look at Rigorous Probability Theory*. World Scientific Publishing Company, 2000.

549. M.J. Ross, H.L. Shaffer, A. Cohen, R. Freudberg, and H.J. Manley. Average magnitude difference function pitch extractor. *IEEE Transactions on Acoustics, Speech, and Signal Processing*, 22:353–362, 1974.

550. T.D. Rossing. *The Science of Sound*. Addison Wesley, second edition, 1990.

551. T.D. Rossing. *Science of Percussion Instruments*. World Scientific Publishing Company, 2000.

552. C. Röver, F. Klefenz, and C. Weihs. *Identification of Musical Instruments by Means of the Hough-Transformation*. Springer, 2005.

553. R. Rowe. *Machine Musicianship*. MIT Press, Cambridge, Massachusetts, 2001.

554. S. Roweis. One microphone source separation. In T.K. Leen, T.G. Dietterich, and V. Tresp, editors, *Neural Information Processing Systems*, pp. 793–799, Denver, USA, 2000.

555. D.E. Rumelhart, G.E. Hinton, and R.J. Williams. Learning internal representations by error propagation. In D.E. Rumelhart, J.L. McClelland, and the PDP Research Group, editors, *Parallel Distributed Processing: Explorations in the Microstructure of Cognition*, pp. 318–362. MIT Press, Cambridge, MA, 1986.

556. S. Rüping. A simple method for estimating conditional probabilities for SVMs. Technical report, CS Department, Dortmund University, Dortmund, Germany, December 2004.

557. M. Ryynänen. *Probabilistic Modelling of Note Events in the Transcription of Monophonic Melodies.* Master's thesis, Tampere University of Technology, March 2004.
558. M. Ryynänen and A. Klapuri. Modelling of note events for singing transcription. In *ISCA Tutorial and Research Workshop on Statistical and Perceptual Audio Processing,* Jeju, Korea, October 2004.
559. M. Ryynänen and A. Klapuri. Polyphonic music transcription using note event modeling. In *IEEE Workshop on Applications of Signal Processing to Audio and Acoustics,* New Paltz, USA, 2005.
560. G.J. Sandell and W.L. Martens. Perceptual evaluation of principal-component-based synthesis of musical timbres. *Journal of the Acoustical Society of America,* 43:1013–1028, 1995.
561. V. Sandvold, F. Gouyon, and P. Herrera. Percussion classification in polyphonic audio recordings using localized sound models. In *International Conference on Music Information Retrieval,* Barcelona, Spain, October 2004.
562. T.J. Santner, B.J. Williams, and W.I. Notz. *The Design and Analysis of Computer Experiments.* Berlin: Springer Verlag, 2003.
563. L.K. Saul, F. Sha, and D.D. Lee. Statistical signal processing with nonnegativity constraints. In *EuroSpeech,* Geneva, Switzerland, 2003.
564. E.D. Scheirer. Tempo and beat analysis of acoustical musical signals. *Journal of the Acoustical Society of America,* 103(1):588–601, January 1998.
565. E.D. Scheirer. *Music-Listening Systems.* PhD thesis, Massachusetts Institute of Technology, 2000.
566. E.D. Scheirer. About this business of metadata. In *International Conference on Music Information Retrieval,* pp. 252–254, 2002.
567. W.A. Schloss. *On the Automatic Transcription of Percussive Music – from Acoustic Signal to High-Level Analysis.* PhD thesis, Center for Computer Research in Music and Acoustics, Stanford University, Stanford, California, USA, May 1985.
568. B. Schölkopf and A. Smola. *Learning with Kernels.* MIT Press, Cambridge, USA, 2002.
569. E. Schuijers, W. Oomen, B. den Brinker, and J. Breebaart. Advances in parametric coding for high quality audio. In *Audio Engineering Society 114th Convention,* Amsterdam, Netherlands, 2003.
570. C.E. Seashore. *Psychology of Music.* Dover Publications, 1967.
571. E. Selfridge-Field. *Beyond MIDI: The Handbook of Musical Codes.* MIT Press, Cambridge, Massachusetts, 1997.
572. J. Seppänen. *Computational Models of Musical Meter Estimation.* Master's thesis, Tampere University of Technology, Tampere, Finland, November 2001.
573. J. Seppänen. Tatum grid analysis of musical signals. In *IEEE Workshop on Applications of Signal Processing to Audio and Acoustics,* pp. 131–134, New Paltz, USA, 2001.
574. X. Serra. *A System for Sound Analysis/Transformation/Synthesis Based on a Deterministic Plus Stochastic Decomposition.* PhD thesis, Stanford University, 1990.
575. X. Serra. Musical sound modeling with sinusoids plus noise. In C. Roads, S. Pope, A. Picialli, and G. De Poli, editors, *Musical Signal Processing.* Swets & Zeitlinger, 1997.

576. X. Serra and J.O. Smith. Spectral modeling synthesis: A sound analysis/synthesis system based on a deterministic plus stochastic decomposition. *Computer Music Journal*, 14(4):12–24, Winter 1990.

577. W.A. Sethares, R.D. Morris, and J.C. Sethares. Beat tracking of musical performances using low-level audio features. *IEEE Transactions on Speech and Audio Processing*, 13(2):1063–1076, 2005.

578. W.A. Sethares and T.A. Staley. Meter and periodicity in musical performance. *Journal of New Music Research*, 30(2), June 2001.

579. F. Sha and L.K. Saul. Real-time pitch determination of one or more voices by nonnegative matrix factorization. In *Neural Information Processing Systems*, Vancouver, Canada, 2004.

580. G. Shafer. *A Mathematical Theory of Evidence*. Princeton University Press, 1976.

581. R.V. Shannon, F.-G. Zeng, V. Kamath, J. Wygonski, and M. Ekelid. Speech recognition with primarily temporal cues. *Science*, 270(5234):303–304, 1995.

582. J.M. Shapiro. Embedded image coding using zerotrees of wavelet coefficients. *IEEE Transactions on Signal Processing*, 41(12):3445–3462, 1993.

583. A. Sheh and D.P.W. Ellis. Chord segmentation and recognition using EM-trained hidden Markov models. In *International Conference on Music Information Retrieval*, pp. 183–189, Baltimore, USA, October 2003.

584. R.N. Shepard. Circularity in judgments of relative pitch. *Journal of the Acoustical Society of America*, 36(12):2346–2353, 1964.

585. J. Shifrin, B. Pardo, C. Meek, and W. Birmingham. HMM-based musical query retrieval. In *ACM/IEEE Joint Conference on Digital Libraries*, pp. 295–300, Oregon, USA, 2002.

586. H. Shih, S.S. Narayanan, and C.-C.J. Kuo. A statistical multidimensional humming transcription using phone level hidden Markov models for query by humming systems. In *IEEE International Conference on Multimedia and Expo*, Volume 1, pp. 61–64, Baltimore, Maryland, USA, 2003.

587. H.-H. Shih, S.S. Narayanan, and C.-C.J. Kuo. An HMM-based approach to humming transcription. In *IEEE International Conference on Multimedia and Expo*, Lusanne, Switzerland, 2002.

588. H.-H. Shih, S.S. Narayanan, and C.-C.J. Kuo. Multidimensional humming transcription using a statistical approach for query by humming systems. In *IEEE International Conference on Acoustics, Speech, and Signal Processing*, Volume 5, pp. 541–544, Hong Kong, China, 2003.

589. J.I. Shonle and K.E. Horan. The pitch of vibrato tones. *Journal of the Acoustical Society of America*, 67(1):246–252, January 1980.

590. J. Sillanpää, A. Klapuri, J. Seppänen, and T. Virtanen. Recognition of acoustic noise mixtures by combined bottom-up and top-down processing. In *European Signal Processing Conference*, 2000.

591. M. Slaney. An efficient implementation of the Patterson Holdsworth auditory filter bank. Technical Report 35, Perception Group, Advanced Technology Group, Apple Computer, 1993.

592. M. Slaney. A critique of pure audition. In *International Joint Conference on Artificial Intelligence*, pp. 13–18, Montreal, Quebec, 1995.

593. M. Slaney. Mixtures of probability experts for audio retrieval and indexing. In *IEEE International Conference on Multimedia and Expo*, Lusanne, Switzerland, 2002.

594. M. Slaney and R.F. Lyon. A perceptual pitch detector. In *IEEE International Conference on Acoustics, Speech, and Signal Processing*, pp. 357–360, Albuquerque, New Mexico, 1990.

595. M. Slaney and R.F. Lyon. On the importance of time—a temporal representation of sound. In M. Cooke, S. Beet, and M. Crawford, editors, *Visual Representations of Speech Signals*, pp. 95–116. John Wiley & Sons, 1993.

596. D. Sleator and D. Temperley. Melisma music analyser code., 2001. http://www.link.cs.cmu.edu/music-analysis/.

597. D. Slezak, P. Synak, A. Wieczorkowska, and J. Wróblewski. KDD-based approach to musical instrument sound recognition. In M.-S. Hacid, Z.W. Ras, D.A. Zighed, and Y. Kodratoff, editors, *International Symposium on Methodologies for Intelligent Systems*, Volume 2366 of *Lecture Notes in Artificial Intelligence*, pp. 28–36. Springer, 2002.

598. P. Smaragdis. *Redundancy Reduction for Computational Audition, a Unifying Approach.* PhD thesis, Massachusetts Institute of Technology, 2001.

599. P. Smaragdis. Discovering auditory objects through non-negativity constraints. In *ISCA Tutorial and Research Workshop on Statistical and Perceptual Audio Processing*, Jeju, Korea, 2004.

600. P. Smaragdis and J.C. Brown. Non-negative matrix factorization for polyphonic music transcription. In *IEEE Workshop on Applications of Signal Processing to Audio and Acoustics*, New Paltz, USA, 2003.

601. L.M. Smith and P. Kovesi. A continuous time-frequency approach to representing rhythmic strata. In *4th Int. Conf. on Music Perception and Cognition*, Montreal, Canada, 1996.

602. B. Snyder. *Music and Memory.* MIT Press, Cambridge, Massachusetts, 2000.

603. J. Song, S.Y. Bae, and K. Yoon. Mid-level music melody representation of polyphonic audio for query-by-humming system. In *International Conference on Music Information Retrieval*, pp. 133–139, Paris, France, October 2002.

604. T. Sonoda, M. Goto, and Y. Muraoka. A WWW-based melody retrieval system. In *International Computer Music Conference*, pp. 349–352, Michigan, USA, 1998.

605. T. Sonoda, T. Ikenaga, K. Shimizu, and Y. Muraoka. The design method of a melody retrieval system on parallelized computers. In *International Conference on Web Delivering of Music*, pp. 66–73, Darmstadt, Germany, December 2002.

606. A. Srinivasan, D. Sullivan, and I. Fujinaga. Recognition of isolated instrument tones by conservatory students. In *7th Int. Conf. on Music Perception and Cognition*, pp. 720–723, Sydney, Australia, 2002.

607. M.J. Steedman. The perception of musical rhythm and metre. *Perception*, 6(5):555–569, 1977.

608. D. Van Steelant, K. Tanghe, S. Degroeve, M. Baets, B. De Leman, and J.-P. Martens. Classification of percussive sounds using support vector machines. In *Machine Learning Conference of Belgium and The Netherlands*, Brussels, Belgium, January 2004.

609. A. Sterian, M.H. Simoni, and G.H. Wakefield. Model-based musical transcription. In *International Computer Music Conference*, Beijing, China, 1999.

610. A. Sterian and G.H. Wakefield. Music transcription systems: From sound to symbol. In *Workshop on AI and Music*, 2000.

611. A.D. Sterian. *Model-Based Segmentation of Time-Frequency Images for Musical Transcription.* PhD thesis, MusEn Project, University of Michigan, Ann Arbor, 1999.

612. J.V. Stone, J. Porrill, C. Buchel, and K. Friston. Spatial, temporal, and spatiotemporal independent component analysis of fMRI data. In *the 18th Leeds Statistical Research Workshop on Spatial-Temporal Modelling and its Applications*, 1999.

613. J. Sundberg. *The Science of the Singing Voice*. Northern Illinois University Press, 1987.

614. J. Sundberg. The perception of singing. In D. Deutsch, editor, *The Psychology of Music*, pp. 171–214. Academic Press, 1999.

615. P. Szczuko, P. Dalka, M. Dabrowski, and B. Kostek. MPEG-7-based low-level descriptor effectiveness in the automatic musical sound classification. In *Audio Engineering Society 116th Convention*, Berlin, Germany, 2004.

616. J. Tabrikian, S. Dubnov, and Y. Dickalov. Maximum a posteriori probability pitch tracking in noisy environments using harmonic model. *IEEE Transactions on Speech and Audio Processing*, 12(1):76–87, 2004.

617. H. Takeda, T. Nishimoto, and S. Sagayama. Rhythm and tempo recognition of musical performance from a probabilistic approach. In *International Conference on Music Information Retrieval*, Barcelona, Spain, October 2004.

618. D. Talkin. A robust algorithm for pitch tracking. In Kleijn and Paliwal [355], pp. 495–517.

619. A.S. Tanguiane. *Artificial Perception and Music Recognition*. Springer, Berlin Heidelberg, 1993.

620. T. Tarvainen. *Automatic Drum Track Transcription from Polyphonic Music*. Master's thesis, Department of Computer Science, University of Helsinki, Helsinki, Finland, May 2004.

621. D. Temperley. *The Cognition of Basic Musical Structures*. MIT Press, Cambridge, Massachusetts, 2001.

622. D. Temperley and D. Sleator. Modeling meter and harmony: A preference-rule approach. *Computer Music Journal*, 23(1):10–27, Spring 1999.

623. M. Tervaniemi and K. Hugdahl. Lateralization of auditory-cortex functions. *Brain Research Reviews*, 43(3):231–46, 2003.

624. D. Thompson, editor. *Concise Oxford English Dictionary*. Clarendon Press, 9 edition, 1995.

625. H.D. Thornburg, R.J. Leistikow, and J. Berger. Melody retrieval and musical onset detection from the STFT. *IEEE Transactions on Speech and Audio Processing*, 2005. (submitted)

626. P. Toiviainen. An interactive MIDI accompanist. *Computer Music Journal*, 22(4):63–75, Winter 1998.

627. T. Tolonen and M. Karjalainen. A computationally efficient multipitch analysis model. *IEEE Transactions on Speech and Audio Processing*, 8(6):708–716, 2000.

628. T. Tolonen, V. Välimäki, and M. Karjalainen. Evaluation of modern sound synthesis methods. Technical Report 48, Helsinki University of Technology, March 1998.

629. I. Tsochantaridis, T. Hofmann, T. Joachims, and Y. Altun. Support vector machine learning for interdependent and structured output spaces. In *International Conference on Machine Learning*, Banff, Canada, 2004.

630. G. Tzanetakis. Song-specific bootstrapping of singing voice structure. In *IEEE International Conference on Multimedia and Expo*, Sorrento (Naples), Italy, 2004.

631. G. Tzanetakis and P. Cook. Musical genre classification of audio signals. *IEEE Transactions on Speech and Audio Processing*, 10(5):293–302, July 2002.
632. G. Tzanetakis, G. Essl, and P. Cook. Automatic musical genre classification of audio signals. In *2nd Annual International Symposium on Music Information Retrieval*, 2001.
633. G. Tzanetakis, G. Essl, and P. Cook. Human perception and computer extraction of beat strength. In *International Conference on Digital Audio Effects*, pp. 257–261, 2002.
634. C. Uhle, C. Dittmar, and T. Sporer. Extraction of drum tracks from polyphonic music using independent subspace analysis. In *the 4th International Symposium on Independent Component Analysis and Blind Signal Separation*, Nara, Japan, 2003.
635. C. Uhle and J. Herre. Estimation of tempo, micro time and time signature from percussive music. In *International Conference on Digital Audio Effects*, London, UK, 2003.
636. M. Unoki and M. Akagi. A method of signal extraction from noisy signal based on auditory scene analysis. *Speech Communication*, 27(3):261–279, 1999.
637. L. Van Immerseel and J.P. Martens. Pitch and voiced/unvoiced determination with an auditory model. *Journal of the Acoustical Society of America*, 91(6):3511–3526, June 1992.
638. S.V. Vaseghi. *Advanced Digital Signal Processing and Noise Reduction*. John Wiley & Sons, 1996.
639. S.V. Vaseghi. *Advanced Digital Signal Processing and Noise Reduction*, pp. 270–290. Wiley, 2nd edition, July 2000.
640. R. Ventura-Miravet, F. Murtagh, and J. Ming. Pattern recognition of musical instruments using hidden Markov models. In *Stockholm Music Acoustics Conference*, Stockholm, Sweden, 2003.
641. T. Verma and T. Meng. Extending spectral modeling synthesis with transient modeling synthesis. *Computer Music Journal*, 24(2):47–59, 2000.
642. M. Vetterli and J. Kovacevic. *Wavelets and Subband Coding*. Prentice Hall, Englewood Cliffs, NJ, USA, 1995.
643. T. Viitaniemi, A. Klapuri, and A. Eronen. A probabilistic model for the transcription of single-voice melodies. In *2003 Finnish Signal Processing Symposium*, pp. 59–63, Tampere, Finland, May 2003.
644. E. Vincent. *Modéles d'Instruments pour la Séparation de Sources et la Transcription d'Enregistrements Musicaux*. PhD thesis, IRCAM and Université Pierre et Marie Curie, Paris, 2004.
645. E. Vincent. Musical source separation using time-frequency source priors. *IEEE Transactions on Acoustics, Speech, and Signal Processing*, 2005. special issue on Statistical and Perceptual Audio Processing, to appear.
646. E. Vincent and M.D. Plumbley. A prototype system for object coding of musical audio. In *IEEE Workshop on Applications of Signal Processing to Audio and Acoustics*, New Paltz, USA, October 2005.
647. E. Vincent and X. Rodet. Instrument identification in solo and ensemble music using independent subspace analysis. In *International Conference on Music Information Retrieval*, Barcelona, Spain, 2004.
648. E. Vincent and X. Rodet. Music transcription with ISA and HMM. In *the 5th International Symposium on Independent Component Analysis and Blind Signal Separation*, 2004.

649. T. Virtanen. *Audio Signal Modeling with Sinusoids Plus Noise*. Technical report, Tampere University of Technology, Department of Information Technology, 2000. Master's thesis.

650. T. Virtanen. Sound source separation using sparse coding with temporal continuity objective. In *International Computer Music Conference*, Singapore, 2003.

651. T. Virtanen. Separation of sound sources by convolutive sparse coding. In *ISCA Tutorial and Research Workshop on Statistical and Perceptual Audio Processing*, Jeju, Korea, 2004.

652. T. Virtanen and A. Klapuri. Separation of harmonic sound sources using sinusoidal modeling. In *IEEE International Conference on Acoustics, Speech, and Signal Processing*, Volume 2, pp. 765–768, Istanbul, Turkey, 2000.

653. T. Virtanen and A. Klapuri. Separation of harmonic sounds using linear models for the overtone series. In *IEEE International Conference on Acoustics, Speech, and Signal Processing*, Orlando, USA, 2002.

654. A. Viterbi. Error bounds for convolutional codes and an asymptotically optimum decoding algorithm. *IEEE Transactions on Information Theory*, 13(2):260–269, April 1967.

655. G.H. Wakefield. Mathematical representation of joint time-chroma distributions. In *SPIE Conference on Advanced Signal Processing Algorithms, Architectures, and Implementations*, pp. 637–645, 1999.

656. P.J. Walmsley. *Signal Separation of Musical Instruments: Simulation-Based Methods for Musical Signal Decomposition and Transcription*. PhD thesis, Department of Engineering, University of Cambridge, September 2000.

657. P.J. Walmsley, S.J. Godsill, and P.J.W. Rayner. Multidimensional optimisation of harmonic signals. In *European Signal Processing Conference*, Island of Rhodes, Greece, September 1998.

658. P.J. Walmsley, S.J. Godsill, and P.J.W. Rayner. Polyphonic pitch tracking using joint Bayesian estimation of multiple frame parameters. In *IEEE Workshop on Applications of Signal Processing to Audio and Acoustics*, New Paltz, USA, October 1999.

659. C. Wang, R. Lyu, and Y. Chiang. A robust singing melody tracker using adaptive round semitones (ARS). In *3rd International Symposium on Image and Signal Processing and Analysis*, pp. 549–554, Rome, Italy, 2003.

660. Y. Wang. A beat-pattern based error concealment scheme for music delivery with burst packet loss. In *IEEE International Conference on Multimedia and Expo*, 2001.

661. Y. Wang and M. Vilermo. A compressed domain beat detector using mp3 audio bitstreams. In *ACM International Multimedia Conference*, Ottawa, Canada, 2001.

662. R.M. Warren. Perceptual restoration of missing speech sounds. *Science*, 167:392–393, 1970.

663. M. Weintraub. A computational model for separating two simultaneous talkers. In *IEEE International Conference on Acoustics, Speech, and Signal Processing*, pp. 81–84, Tokyo, Japan, 1986.

664. J. Wellhausen and H. Crysandt. Temporal audio segmentation using MPEG-7 descriptors. In *SPIE Storage and Retrieval for Media Databases 2003*, Volume 5021, pp. 380–387, 2003.

665. D. Wessel. Timbre space as a musical control structure. *Computer Music Journal*, 3:45–52, 1979.

666. H. White. Connectionist nonparametric regression: Multilayer feedforward networks can learn arbitrary mappings. *Neural Networks*, 3:535–549, 1990.

667. M.V. Wickerhauser. *Adapted Wavelet Analysis from Theory to Software*. AK Peters, Boston, MA, USA, 1994.

668. A. Wieczorkowska. Classification of musical instrument sounds using decision trees. In *8th International Symposium on Sound Engineering and Mastering*, pp. 225–230, 1999.

669. A. Wieczorkowska. Rough sets as a tool for audio signal classification. In Z.W. Ras and A. Skowron, editors, *Foundations of Intelligent Systems: Proceedings of the 11th International Symposium on Foundations of Intelligent Systems*, pp. 367–375. Springer, 1999.

670. A. Wieczorkowska, J. Wróblewski, P. Synak, and D. Slezak. Application of temporal descriptors to musical instrument sound recognition. *Journal of Intelligent Information Systems*, 21:71–93, 2003.

671. A. Wieczorkowska and J.M. Zytkow. Analysis of feature dependencies in sound description. *Journal of Intelligent Information Systems*, 20:285–302, 2003.

672. W. Wiegerinck and D. Barber. Mean field theory based on belief networks for approximate inference. In *8th International Conference on Artificial Neural Networks*, pp. 499–504, Skövde, Sweden, 1998.

673. I.H. Witten and T.C. Bell. The zero-frequency problem: Estimating the probabilities of novel events in adaptive text compression. *IEEE Transactions on Information Theory*, 37(4):1085–1094, 1991.

674. I.H. Witten and E. Frank. *Data Mining: Practical Machine Learning Tools and Techniques*. Amsterdam: Elsevier-Morgan Kaufmann, second edition, 2005.

675. E. Wold, T. Blum, D. Keislar, and J. Wheaton. Classification, search and retrieval of audio. In B. Furth, editor, *Handbook of Multimedia Computing*, pp. 207–226. CRC Press, 1999.

676. P.J. Wolfe, S.J. Godsill, and W.J. Ng. Bayesian variable selection and regularisation for time-frequency surface estimation. *Journal of the Royal Statistical Society: Series B*, 66(3):575–589, August 2004.

677. M. Wu, D. Wang, and G.J. Brown. A multipitch tracking algorithm for noisy speech. *IEEE Transactions on Speech and Audio Processing*, 11(3):229–241, 2003. URL: www.cse.ohio-state.edu/ dwang/pnl/ software.html.

678. H. Yamada, M. Goto, H. Saruwatari, and K. Shikano. Multi-timbre chord classification for musical audio signals. In *2002 Autumn Meeting of the Acoustical Society of Japan*, pp. 641–642, Akita, Japan, September 2002. (in Japanese)

679. H. Yamada, M. Goto, H. Saruwatari, and K. Shikano. Multi-timbre chord classification method for musical audio signals: Application to musical pieces. In *the 2003 Spring Meeting of the Acoustical Society of Japan*, pp. 835–836, Tokyo, Japan, March 2003. (in Japanese)

680. G.K. Yates. Cochlear structure and function. In Moore [474], pp. 41–74.

681. C. Yeh and A. Röbel. A new score function for joint evaluation of multiple F0 hypotheses. In *International Conference on Digital Audio Effects*, Naples, Italy, October 2004.

682. K. Yoshii, M. Goto, and H.G. Okuno. Automatic drum sound description for real-world music using template adaptation and matching methods. In *International Conference on Music Information Retrieval*, Barcelona, Spain, October 2004.

683. K. Yoshii, M. Goto, and H.G. Okuno. Drum sound identification for polyphonic music using template adaptation and matching methods. In *ISCA Tutorial and Research Workshop on Statistical and Perceptual Audio Processing*, Jeju, Korea, October 2004.

684. T. Yoshioka, T. Kitahara, K. Komatani, T. Ogata, and H.G. Okuno. Automatic chord transcription with concurrent recognition of chord symbols and boundaries. In *International Conference on Music Information Retrieval*, pp. 100–105, Barcelona, Spain, October 2004.

685. S.J. Young, N.H. Russell, and J.H.S. Thornton. Token passing: A simple conceptual model for connected speech recognition systems. Technical report, Department of Engineering, University of Cambridge, July 1989.

686. B. Zadrozny and C. Elkan. Obtaining calibrated probability estimates from decision trees and naive Bayesian classifiers. In *International Conference on Machine Learning*, pp. 202–209, Williamstown, Massachusetts, USA, June 2001.

687. R.J. Zatorre, P. Belin, and V.B. Penhune. Structure and function of auditory cortex: Music and speech. *TRENDS in Cognitive Sciences*, 6(1):37–46, 2002.

688. T. Zhang. Instrument classification in polyphonic music based on timbre analysis. In *SPIE, Internet Multimedia Management Systems II*, pp. 136–147, 2001.

689. Y. Zhu and M.S. Kankanhalli. Robust and efficient pitch tracking for query-by-humming. In *2003 Joint Conference of the Fourth International Conference on Information, Communications and Signal Processing, 2003 and the Fourth Pacific Rim Conference on Multimedia*, Volume 3, pp. 1586–1590, Singapore, December 2003.

690. M. Zibulski and Y. Zeevi. Analysis of multiwindow Gabor-type schemes by frame methods. *Applied and Computational Harmonic Analysis*, 4(2):188–212, 1997.

691. A. Zils. *Extraction de Descripteurs Musicaux: Une Approche Volutionniste*. PhD thesis, Sony CSL Paris and Laboratoire d'Informatique de l'universit Paris 6, 2004.

692. A. Zils and F. Pachet. Automatic extraction of music descriptors from acoustic signals using eds. In *Audio Engineering Society 116th Convention*, Berlin, Germany, 2004.

693. A. Zils, F. Pachet, O. Delerue, and F. Gouyon. Automatic extraction of drum tracks from polyphonic music signals. In *International Conference on Web Delivering of Music*, Darmstadt, Germany, December 2002.

694. E. Zwicker and H. Fastl. *Psychoacoustics: Facts and Models*. Springer, 1999.

Index

Accent, 107, 112, 374
Acoustic features, *see* Feature extraction
Acoustic model, *see* Source model
Adaptive modelling, *see* Source model adaptation
Adaptive oscillator, *see* Oscillator
Additive synthesis, *see* Sinusoidal model
Aerophone, 167, 231
Amplitude envelope, 176
Amplitude evolution model, 208
  damped amplitudes, 208, 218
  exponential decay model, 207, 219
Amplitude modulation
  as clue for component fusion, 306
  tremolo, *see* Tremolo
Amplitude prior, 212
Analysis by synthesis, 140
Analysis frame, 23, 66, 68
Analytic mode of listening, 6
Annotation of music, 17, 355
Arrhythmic, 105
Articulation, 9, 193
  in singing, 364
Artificial neural network, *see* Neural network
Attack time, 177
Attack-point, 134
Audio coding, 3, 70, 88, 91
Auditory cortex, 12
  left/right asymmetry, 12
Auditory filters, 235
  bandwidth, 236
  centre frequency, 236

frequency response, 237
impulse response, 237
Auditory model, 6, 14, 234, 372
  of Cooke, 246
  of Ellis, 245
  of Meddis and Hewitt, 241
  toolboxes, 238
Auditory nerve, 234, 235, 237, 238, 242
Auditory scene analysis, 7, 246, 247, 299, 306
Auditory system, 65, 234, 300
  auditory cortex, 234
  central auditory processing, 235, 241, 242, 265
  cochlea, *see* Cochlea
  peripheral hearing, 14, 234
Autocorrelation, 241, 243, 245
  autocorrelation function, 233
  calculation via FFT, 256
  enhanced, 288
  generalized ACF, 252, 256
  of a frequency-domain signal, 233
  summary ACF, *see* Summary autocorrelation
  use for beat tracking, 112
  use for F0 estimation, 232, 369
Automatic music transcription, *see* Transcription
Automatic threshold selection, 350
Autoregressive model, 39, 175, 211
Autoregressive noise, 211
Average magnitude difference function, 251

Bandwise energy features, 135
Bar line, 10, 106
Bar line estimation, 124, 159, 341
Basilar membrane, 234, 235
Basis function, 269
  amplitude, *see* Gain function
  frequency domain, 143, 146, 149, 273
  time domain, 271
Bass line, 13, 321, 329
Bass line transcription, 330
Bassoon, 167
Bayes's rule, 38, 40
Bayesian
  classification, 62
  estimation, 40
  model, 203
  network, 313
Beat, 10, 105, 134, 329, 341
Beat induction
  by humans, *see* Metre perception
  computational, *see* Beat tracking
Beat tracking, 6, 101, 341
Beating, 239, 256
Bell, 167
Binning, 154
Blackboard architecture, 246, 247, 312
Blind source separation, *see* Source
  separation
Bottom-up processing, 16, 307, 312,
  342, 352
Box-Cox power transform, 179
Brain damage, 11
Brain imaging, *see* Neuroimaging
Brass instrument, 167
Bugle, 167

Cancellation filter, 196, 250
Canonical discriminant analysis, 180
Celesta, 167
Cello, 167, 209, 340
Cent, 332
Central auditory processing, *see*
  Auditory system
Central limit theorem, 274
Cepstral coefficients, 25, 26
  MFCC, *see* Mel-frequency cepstral
  coefficients
Cepstrum, 25, 253
Cepstrum pitch detection, 253, 264, 371

Channel selection
  in auditory model, 244, 249, 251
Choral music, 109, 125
Chord, 9, 13
  chord change detection, 117, 328, 341
  notation, 3
  recognition, 319
  root detection, 246, 254, 263
Chordophone, 167, 231
Chorus measure, 345, 353
Chorus sections, 329
  estimation, 342
Chroma and height, 346
Chroma vector, 345, 347
Clarinet, 96, 166, 167, 176, 177, 206,
  207, 319
Classical music, 7, 112, 125, 128, 340,
  355
Classification, 52, 137, 139, 163, 184
  combining, 193
  ensemble, 184
  flat, 184
  generalization, 169, 189
  genre, 13, 112, 356
  hierarchical, 184, 192
  unsupervised, 53, 133, 164
Clustering, 133, 151, 159, 164, 193, 286,
  307, 311, 340, 352, 354
Cochlea, 164, 234
  cochlear filters, *see* Auditory filters
  computational models, 234
Cocktail party problem, 300
Comb filter, 115, 243, 257
Combining classifiers, 193
Common musical notation, 3, 4
  expression symbols, 389
Complete family, 72
Complete transcription, 3
Compression
  in auditory model, 238, 239, 248
  νth-law compression, 255
Computational auditory scene analysis,
  see Auditory scene analysis
Concurrent vowel identification, 249
Conditional probability density
  function, 31
Conjugate prior, 47
Contextual information, 198
  in auditory scene analysis, 313

in instrument classification, 166
in singing, 385
short-term acoustic context, 154
tonal context, 320, 385
Continuous-length evaluation measure,
125
Contrabass, 167, 198
Convolution, 22, 25, 289
time-frequency, 25
Correlation-based feature selection, 182
Correlogram, 227, 242, 245, 339
Covariance matrix, 29, 30
empirical, 53, 61
Cramér–Rao bound, 34
Creative Commons initiative, 170
Critical band, 236, 274
Critical-band scale, 236
Cross-validation, 170

Damped amplitudes, see Amplitude
evolution model
Damping factor, 208
Dance music, 102, 125, 160
ballroom dance, 117
Data representation, see Mid-level data
representation
Data-driven processing, 16
Databases, 16
of musical instrument sounds, 169,
260, 355
RWC Music Database, 355
DCT-IV transform, 75
decibel, 8, 294
Decision tree, 138, 186
Decorrelation, 275
nonlinear, 275
Delta features, 175
Delta-MFCC, see Mel-frequency
cepstral coefficients
Description of music signal, 328
Deterministic approach to signal
modelling, 67
Diatonic scale, 271, 293
Dictation, see Human transcription
Dictionary of waveforms, see Waveform
dictionary
Dilation, 71
Directed acyclic graph, 120, 313
Discrete cosine transform, 75

Discrete Fourier transform, see Fourier
transform, 273
Discrete wavelet transform, see Wavelet
transform
Discriminant analysis, 61, 185
Dissonance, 9, 287
Distribution, see Probability density
function
Divergence, 282
Double bass, see Contrabass
Drums, see Percussion
Dual frame, 78
Duplex theory of pitch perception, 241
Duration
of a note, 362
perceived duration, 8
Dyadic grid, 74
Dynamic model
finite, discrete state space, 63
Dynamic range
of hearing, 238, 273
of inner hair cells, 238

Ear training, 12
Eardrum, 234
Eigenvalue, 54, 180
Eigenvector, 54
Electric guitar, see Guitar
Elementary waveform, 70
EM algorithm, see Expectation
maximization
Empirical average, 29
Empirical probability density function,
82
Empirical risk, 56
English horn, 167
Enhanced summary ACF, see Summary
autocorrelation function
Ensemble of classifiers, 184
Equal-tempered scale, 9, 224
Equivalent rectangular bandwidth, 236
Estimation,
bias, 33, 41
covariance, 33
unbiased estimator, 34
see also Fundamental frequency
estimation, 21
Estimation theory, 33
Euclidean distance, 282

Evaluation methodology, 170
Expectation, 29
Expectation maximization, 35, 221, 336
Exponential decay, *see* Amplitude
    evolution model
Expression, 193
  estimation and encoding, 389

FastICA algorithm, 275
Feature extraction, 66, 135, 171, 247,
    270, 287, 307, 344, 345, 368
  delta, 175
  sliding window, 155
Feature scale transformation, 178
Feature selection, 137, 181
Feature transformation, 191
Fisher ratio, 61
Flat classifier, 184
Flute, 205, 206, 209, 319, 323, 340
Folk music, 125, 323
Formant, 253, 294, 365
Fourier transform, 21, 22
  continuous, 22
  discrete, 22
  inverse, 22
Fractional-delay filter, 250
Frame, *see* Analysis frame
Frame (family of vectors), 77
  dual, 78
  frame inversion, 78
  frame representation, 78
  Gabor frame, 78
  hybrid system, 81
  multiple Gabor frames, 79
  overcomplete system, 78
  quilted frame, 81
French horn, 166, 167
Frequency evolution model, 209
Frequency grid, 219, 224
Frequency modulation
  as clue for component fusion, 306
  vibrato, *see* Vibrato
Frequency proposal distribution, 216
Frequency warping, 254
Frobenius norm, 280
Front end, 244
Full-wave νth-law compression, 255
Fundamental frequency estimation
  basic principles, 232

multiple F0 estimation, 203, 229, 248,
    287
  of melody and bass lines, 330
  single F0 estimation, 232, 369
  typical errors, 233, 253
Fundamental frequency pdf, 332
Fundamental frequency, term definition,
    8

g-prior, 212
Gabor frame, 78
Gabor function, 78
Gabor representation, 24, 209
Gabor waveforms, 71
Gain function, 143, 269
Gamma prior, 211
Gammatone filter, 237, 241, 247, 255,
    306
  efficient implementation, 238
Gaussian distribution, 28
  Gaussianity of variable, 274
  generalized, 286
  truncated, 213
Gaussian mixture model, 35, 37, 139,
    191, 385, 386
Gaussian noise, 279, 282
Generalization in classification, 169, 189
Generalized autocorrelation, *see*
    Autocorrelation
Generative model, 62, 112, 122, 204,
    277, 324
Generative signal, speech and music, 16
Genre classification, 13, 112, 356
Gestalt psychology, 300
Glissando, 389
  in singing, 367
Glockenspiel, 232
Greedy algorithm, 84, 89, 225, 354
Grid, temporal, *see* Temporal grid
Grouping, *see* Rhythmic grouping
Guitar, 79, 166, 167, 177, 227
  electric, 198, 200, 209
  onset detection, 108
  sound separation, 270, 290
  transcription, 301

Hair cell, *see* Inner hair cell, Outer hair
    cell
Half-wave rectification, 239, 256, 258

Harmonic matching pursuit, 92
Harmonic partial, 107, 231, 291
Harmonic pattern matching, 233
Harmonic selection, 257
Harmonic sound, 231
Harmonic trajectory, 225
Harmonicity, as clue for component fusion, 306
Harmonics and Individual Lines plus Noise, 69
Harmony, 9
Harp, 167
Harpsichord, 167
    transcription, 282
Heisenberg–Gabor inequality, 24
Hidden Markov model, 63, 121, 122, 139, 191, 286, 380
Hierarchical beat structure, see Metre
Hierarchical classifier, 184, 192
Higher-order statistics, 275, 287
History of automatic music transcription, 6, 301
Human auditory system, see Auditory system
Human transcription, 5, 12
Human–computer interaction, 341
Human-computer interaction, 5, 361
Hybrid representation, 67
Hybrid system, see Frame (family of vectors)
Hyperparameter, 47, 62

i.i.d., 32
Idiophone, 131, 145, 167
Ill-posed problem, 38
Importance probability density function, 49
Importance sampling, 49, 223
Independence, see Statistical independence
Independent component analysis, 109, 143, 274, 305
    multidimensional, 276, 277
    non-negative, 278
    spatiotemporal, 277
Independent subspace analysis, 144, 276, 285
    sub-band, 145

inertia ratio maximization using feature space projection, 183
Information gain, 182
Information retrieval, see Music information retrieval
Inharmonicity, 175, 206, 222
    in string instruments, 231
    inharmonicity factor, 232
    model for piano, 206, 222, 232
Inner ear, 234
Inner hair cell, 234, 235
    models, 238, 241, 246, 247, 255
Inner lines in music, 6, 13
Inner product, 71
Inner product space, 71
Instance-based classification, 138, 185
    k-NN, see k-nearest neighbours
Instantaneous frequency, 246, 332
Instrument
    sample databases, see Databases
    sounds, see Musical sounds
Instrument classification, 7, 65, 163
    in humans, 164
    percussion, see Percussion sound recognition
Instrument families, 167
Integration of information, 16, 154, 313, 383
Inter-onset interval, 8, 115
Intermediate data representation, see Mid-level data representation
Internal model, 15
Interval, 12, 321, 340, 341, 362
Invariant feature extraction, 276, 277
Inverse Fourier transform, see Fourier transform
Inverse gamma distribution, 47
Iterative F0 estimation and cancellation, 196, 250, 254, 259, 263

JADE algorithm, 275
Jazz music, 105, 125, 127, 321, 355
Joint estimation of multiple F0s, 214, 251, 252, 263
Joint probability density function, 31
Junction tree algorithm, 316

k-nearest neighbours, 60, 185, 191
Kalimba, 167

Kalman filter, 51, 52, 117, 121, 221–223, 226, 324
Kernel, 57
  Gaussian, 57
  positive definite, 57
Kullback–Leibler divergence, 282
  symmetric, 287
Kullback–Leibler information, 335
Kurtosis, 275
  spectal, 136

Language model, 15
Laplace approximation, 225
Laplacian distribution, 280
Latent variable, 35
Law of large numbers, 30
Lazy learning, 185
Least-squares, 285
Leave-one-out cross-validation, 170
Lebesgue measure, 28
Legato, 9
  in singing, 367
Level adaptation in auditory model, 239, 246
Level compression, see Compression
Likelihood, 210, 219, 332
Likelihood function, 32
  degenerate, 33, 38
  penalized, 38, 41, 57
Linear discriminant analysis, 186, 191
Linear interpolation, 253
Linear prediction, 39
Linear programming, 286
Local cosine basis, 73
Localized source model, 142
Locally harmonic sources, 69
Log-Gaussian distribution, 224
Loss function, 55
Loudness, 8, 172
  of instrument sounds, 319
  of melody vs. accompaniment, 340
  of singing, 366

Mallet percussion instrument, 232
Marginal MMSE, 214
Marginal probability density function, 31
Marimba, 167, 232
Markov chain, see N-gram model

Markov chain Monte Carlo, 43, 117, 121, 215
Markov tree, 90
Masking, 236
Matching pursuit, 84
Matrix diagonalization, 54
Matrix factorization, see Non-negative matrix factorization
Maximum a posteriori estimation, 40, 121, 214, 279, 331
Maximum likelihood estimation, 33, 35
Mbira, 167
McNemar's test, 171
Mean, 29, 30
  empirical, 54, 61
Mean square error of an estimator, 34
Measure
  musical measure, see Bar line
  musical measure estimation, see Bar line estimation
Mechanical-to-neural transduction, 238
Mel frequency cepstral coefficients, 26
Mel frequency scale, 26
Mel-frequency cepstral coefficients, 63, 135, 174
  delta-MFCC, 175
mel-frequency cepstral coefficients, 270
Mel-frequency scale, 173
Mel-scale filterbank, 26
mel-scale filterbank, 173
Melodic phrase, see Phrase
Melody, 9, 12, 13, 329
  perceptual coherence, 15
  segregation of melodic lines, 247
  transcription, see Predominant F0 estimation
Membranophone, 131, 145, 147, 167
Memory for music, 11, 13
Message passing, 221
Metadata, see Annotation
Metre, 10, 105, 312, 329, 341
Metre analysis, 101, 134, 341, 388
Metre perception, 10, 11, 102
MFCC, see Mel frequency cepstral coefficients
Mid-level data representation, 12, 13, 65, 244, 248, 251, 256, 264
  desirable qualities, 14
  hybrid, see Hybrid representation

Middle ear, 234
MIDI, 3, 4, 9, 101, 102
MIDI note number, 362
Minimum mean square error estimation, 41, 214
Missing feature theory, 197
Mixing matrix, 270
Mixture-of-experts approach, 193
Model adaptation, source, *see* Source model adaptation
Model selection, 39
Model, signal, *see* Signal model
Modified discrete cosine transform, 75
   MDCT basis, 73
Modularity, 11
Modulation, 71
   amplitude, *see* Amplitude modulation
   frequency, *see* Frequency modulation
Modulation (musical key change), 344
Molecular matching pursuit, 94
Moment order r, 30
Monophonic signal, 5
Monte Carlo, 41
MPEG-7, 136, 167, 172, 270, 354
Multi-class classification, 55
Multi-layer perceptron, 187, 191
Multidimensional scaling, 168, 179
Multiple F0 (non)stationary model, *see* Signal model
Multiple F0 estimation, *see* Fundamental frequency estimation
Multiplicative update rule, 283
Multiresolution analysis, 74
Multiwavelet, 77
Music cognition, 11
   impaired cognition, 11
Music information retrieval, 5, 102, 170, 327, 356, 363
Music listening station, 357
Music map, 357
Music perception, 5, 103, 327
Music scene analysis, 299
Music scene description, 327, 328
Music structure, *see* Structure
Music structure analysis, *see* Structure analysis
Music thumbnail, 342
Music transcription, *see* Transcription
Music-playback interface, 357

Music-synchronized computer graphics, 356
Musical context, *see* Contextual information
Musical instrument, *see* Instrument
Musical instrument classification, *see* Instrument classification
Musical key, 9, 385
   change, 344
   estimation, 386
Musical metre, *see* Metre
Musical scale, *see* Scale
Musical sounds
   percussive, 107, 131
   pitched, 107, 167, 231
Musicological modelling, 15, 153–155
   melodic continuity, 331
   musical key, 385
   of periodic patterns, 157
   of rhythmic patterns, 159
   short-term context modelling, 154
   with N-grams, *see* N-gram model
Mutual information, 274

N-fold cross-validation, 170
N-gram model, 155
   for chord sequences, 319
   of melody, 321, 387
   of percussion sequences, 155, 156
   periodic N-gram, 157, 158
Neural firing probability, 237, 238
Neural impulse, 234, 238, 242
Neural network, 187, 191
   MLP, 187, 191
   time-delay neural network, 247
Neural spike, *see* Neural impulse
Neuroimaging, 11
Neurophysiology of music cognition, 11
Noise, 32
   autoregressive, 211
   Gaussian, 279, 282
   Poisson, 282
   white, 32, 39
Noise robustness in F0 estimation, 251, 253, 264
Noisy sum-of-sines model, *see* Signal model
Non-negative matrix deconvolution, 293

Non-negative matrix factorization, 148, 282

Norm, 71

Normal distribution, *see* Gaussian distribution

Notation, *see* Common musical notation

Note, 3, 362
  MIDI note number, 362

Note birth move, 216

Note death move, 216

Note labelling in singing transcription, 362, 378

Note model, *see* Source model

Note segmentation, *see* Temporal segmentation

Note update move, 216

Nuisance parameters, 220

Nyquist frequency, 23

Oboe, 167, 209

Observation density, *see* Likelihood

Octave equivalence, 9

Odd-to-even ratio, 175

Offset (a)synchrony, 306

Onset, 101, 260, 262
  and beat, 106
  of percussive sounds, 108
  of pitched sounds, 108

Onset (a)synchrony, 306

Onset detection, 102, 107, 134, 195, 287

Organ of Corti, 234

Orthogonal matching pursuit, 85

Orthonormal basis, 72

Oscillator, 113
  adaptive, 113, 247
  comb filter, 115, 244, 257
  oscillator net, 247

Outer ear, 234

Outer hair cell, 234

Overcomplete system, *see* Frame (family of vectors)

Overlap-add, 288

Overtone partial, *see* Harmonic partial

Parametric model of signal, *see* Signal model

Parseval formula, 73

Partial de-tuning, 206, 306

Partial transcription, 3

Particle filter, 50, 117, 121
  Rao–Blackwellized, 222

Peak selection in autocorrelation function, 251

Penalized likelihood, *see* Likelihood function

Perception
  of metre, *see* Metre perception
  of music, *see* Music perception
  of pitch, *see* Pitch perception

Perceptual attributes of sounds, 8

Perceptual categorization, 9

Perceptual sound vs. physical sound, 302

Percussion notation, 3

Percussion sound recognition, 133, 137, 174, 184
  clustering and labelling, 159

Percussion transcription, 6, 7, 131, 329, 342
  pattern recognition-based, 133
  separation-based, 142

Periodicity
  in the frequency domain, 233
  in the time domain, 232

Peripheral hearing, *see* Auditory system

Phase generation, 288

Phenomenal accent, *see* Accent

Phonation, 364
  frequency, 365
  types, 365

Phoneme, 16

Phrase, 10

Phrasing, 193

Physical sound vs. perceptual sound, 302

Piano, 24, 167, 203, 206, 207, 209, 319, 323
  identification in music, 198
  inharmonicity, *see* Inharmonicity
  keyboard, 9
  onset detection, 108
  transcription, 245, 247

Piano roll, 4, 220

Pitch, 8, 107
  of noise signals, 229
  perception, *see* Pitch perception
  tonal encoding, *see* Tonal encoding
  zoo of pitch effects, 229

Pitch class, 9, 345, 385
Pitch label of a note, 362
Pitch perception, 229, 242
Pitch perception model, 234, 241
    autocorrelation model, 241, 242
    duplex theory, 241
    shortcomings, 248, 262
    unitary model, 242, 252
Pitched musical sounds, 107, 167, 231
Pizzicato, 192
Poisson noise, 282
Poisson prior, 214
Polyphonic signal, 5
Popular music, 10, 125, 128, 131, 137,
    151, 329, 330, 342, 344, 345, 353,
    355
Predominant F0 estimation, 198, 260,
    330
PreFEst method, 330
Principal component analysis, 54, 137,
    139, 144, 151, 152, 155, 179, 181,
    275
Prior, 40
    conjugate, 47
    sequential, 217, 220, 221, 225
Prior distribution, 40
Prior selection, 211
Prior subspace analysis, 146, 150
    input generated priors, 151
    non-negative, 148
Probabilistic model, 31
Probability density function, 28
    conditional, 31
    empirical, 82
    Gaussian, see Gaussian distribution
    importance, 49
    inverse gamma, 47
    joint, 31
    Laplacian, 280
    marginal, 31
    normal, see Gaussian distribution
    prior, 40
    proposal, 44
    uniform, 28
Projected steepest descent algorithm,
    281
Proposal distribution, 44
Pseudoinverse, 147
Psychoacoustics, 8, 173, 235, 308

Pulse, see Beat; Metre
Pulse Code Modulation (PCM), 65

Quantization, 9
    of F0 values, 213, 362
    of onset times, 101, 106, 118, 389
    of signal sample values, 65
Query by humming, 356, 361
Quilted frame, see Frame (family of
    vectors)

Random variable, 28
    expectation, see Expectation
    i.i.d., see i.i.d.
    independent, see Statistical indepen-
        dence
Random walk, 208
Rayleigh quotient, 61
Recognition
    of percussion sounds, see Percussion
        sound recognition
    of pitched sounds, see Instrument
        classification
Rectification, see Half-wave rectification
Redundancy, 78
Reed instrument, 167
RefraiD method, 345
Register (in singing), 365
Regular grid, 75
Regularized risk, 57, 58
Repeated sections, see Structure
Reproducing kernel Hilbert space, 57
Resolvability, 258
Rest (in music), 362
Rhythm, 10, 105
    analysis, see Metre analysis
Rhythmic grouping, 10, 105
Rhythmic pattern modelling, 159
Risk, 55
    empirical, 56
    regularized, 57, 58
Rock music, 105, 112, 125, 131, 151
Root-mean-square level, 172
Rotation matrix, 218
Royalty-free music, 355
RWC Music Database, 355

Salience
    of an onset, 107

of F0 candidate, 257
Sampling, 65
Saxophone, 167, 198, 207, 256, 259
Scale, 10, 385
    equal-tempered, 9, 224
Scale tone, 10, 385
Scale transformation, 178
Scaling function, 74
Scene analysis, 299, see also Auditory
    scene analysis
Schema-based segregation, 312
Score, see Common musical notation
Scoretime, 105
Segmentation, see Temporal segmenta-
    tion
Semitone, 291, 292, 332, 362
Sequential dependency, see N-gram
    model
Sequential Monte Carlo, 117
Sequential prior, 217, 220, 221, 225
Shannon theorem, 23
Short-time Fourier transform, 23
Signal model, 66, 204
    multiple F0 non-stationary, 209
    noisy sum-of-sines, 204
    parametric, 67, 70
    single F0 non-stationary, 207
    single F0 stationary, 204
    sinusoidal, see Sinusoidal model
    sum-of-sines model, 204
    weighted mixture of tone models, 332
Signal space, 71
Significance map, 88
Significance tree, 89
Similarity matrix, 349
Simplex algorithm, 286
Singer's formant, 366
Singing
    acoustic characteristics, 364
    production of singing sounds, 364
    singing out of tune, 367
Singing transcription
    applications, 361
    expression encoding, 389
    problem formulation, 361
    segment-and-label approach, 377
    statistical approach, 379
Single F0 (non)stationary model, see
    Signal model

Singular value decomposition, 277
Sinusoidal model, 14, 68, 225, 246
    other variants, see Signal model
    single sinusoid, 31
Sinusoids + transients + noise model,
    70
Sliding window method, 155
SmartMusicKIOSK, 357
Sound production mechanism
    in musical instruments, 167
    in singing, 364
Sound source separation, see Source
    separation
Source model, 15
    percussion sounds, 149
    pitched sounds, 319
    statistical note model, 379
    tone model, 319, 333
Source model adaptation, 140, 142, 197,
    333
    in time domain, 140
    in time-frequency domain, 141
Source separation, 15, 65, 142, 143, 195,
    249, 267, 304, 305, 327
Sparse coding, 143, 278, 285
    non-negative, 153, 281, 285
Sparse expansion, 82
Sparse representation, 67
Spatial information, 267, 306
Spectral features
    spectral centroid, 136
    spectral flatness, 173
    spectral flux, 174
    spectral irregularity, 174
    spectral kurtosis, 136
    spectral rolloff, 174
    spectral shape, 135
    spectral skewness, 136
    spectral spread, 136
Spectral model synthesis, 69
Spectral organization, see Auditory
    scene analysis
Spectral smoothness, 207, 240
Spectral whitening, 239, 248, 252, 255,
    263
Spectrogram, 23, 24, 65
Spectrogram factorization, 143, 268
Spectrum envelope, 207

Speech
    recognition, 15, 16, 63, 156, 307, 358
    speech separation, 249
    speech signals, 16
Squared difference function, 250
Staccato, 9
Statistical independence, 31, 274
Statistical significance tests, 170
Stiffness of vibrating strings, 232
Stochastic signal components, 69
Strobed temporal integration, 244
Structure (of a musical work), 10, 329
Structure analysis, 10
    by humans, 13
    computational, 342
Structured approximation, 87
Structured audio coding, 3
Student–Fisher t-test, 170
Style detection, see Genre classification
Sub-band coding, 74
Sub-beat structure, 103, 106, 117
Subglottic pressure, 364
Sum-of-sines model, 204
Summary autocorrelation function, 242, 245, 248
    enhanced summary ACF, 253
Super-beat structure, 103, 106
Supervised classification, 55, 164
Support vector machine, 57, 138, 191
    output moderating, 154
Swing, 105
Symbolic representation, 9, 102, 107, 302
Synchrony strand, 247
Synthesis of separated sources, 288

Tactus, 10, 105
Tatum, 106, 134
    analysis, 116, 124, 134, 159
Taxonomy, 164
    of musical instruments, 167
Template matching, 139
Tempo, 6, 101, 105
    estimation, 103, 118, 354, 374
    variation, 105, 106, 388
Tempogram, 121
Temporal centroid, 136, 178
Temporal grid, 134
Temporal segmentation, 12, 134

of instrument sounds, 177
of singing notes, 362, 373, 377
Threshold of hearing, 273
Thumbnail, 342
Tick, see Tatum
Timbre, 8, 231
    acoustic correlates, 168
    acoustic features, see Feature extraction
    F0 dependency, 189
    of singing sounds, 365
    perceived similarity, 168
    use for auditory organization, 306
Timbre space, 168
Time quantization, see Quantization
Time–frequency molecule, 87
Time-frequency atom, 70
Time-frequency covariance, 24
Time-frequency jigsaw puzzle, 85
Time-frequency lattice, 209
Time-frequency representation, 13, 23, 25, 65
Time-lag triangle, 349
Time-persistence, 88
Timing deviations, 105, 375
Timpani, 167
Token-passing algorithm, 380
Tonal encoding of pitch, 10, 11
Tonal music, 10, 304, 330, 385
Tone model, see Source model
Tonic note, 10
Top-down processing, 16, 312, 342, 352
Transcription
    by humans, 5, 12
    complete vs. partial, 3
    designing transcription system, 11
    state of the art, 7
    subtopics, 5, 11
    trends and approaches, 6, 301
Transient, 70
Transientness index, 89
Translation, 71
Tremolo, 178, 389
    in singing, 367
Tristimulus, 176
Trombone, 166, 167
Trumpet, 167, 176, 198, 209, 245, 254, 319
Tuba, 167

Tuning, 9
  absolute, 362
  drift of, 368
  in note labelling, 378
  singing out of tune, 367
Two-way mismatch, 288, 371

Ukulele, 167
Unbiased estimator, see Estimation
Uniform distribution, 28
Unitary model, see Pitch perception
    model
Unmixing matrix, 147, 275
Unsupervised
  classification, 53, 133, 164
  clustering, see Clustering
  learning, 7, 16, 267, 380, 382

Validation, 169, 170
Vibraphone, 232
Vibrating bar, 232
Vibrato, 178, 389
  in singing, 366
  rate and depth estimation, 389
Viola, 167
Violin, 9, 166, 167, 177, 207, 209, 231,
    254, 259, 319, 323
  onset detection, 109
Viterbi algorithm, 63, 122, 225, 380

Vocal organ, 364
Voice source, 364
Voicing, degree of, 373

Waveform dictionary, 67, 81
Waveform representation, 13, 72
Wavelet, 71
Wavelet basis, 73
Wavelet transform, 74
Weighted-mixture model, see Signal
    model
Well-tempered scale, see Equal-
    tempered scale
Western music, 3, 5, 9, 10, 231, 304
White noise, 32, 39
Whitening, see Spectral whitening
Wiener–Khintchine theorem, 252
Wigner–Ville representation, 25
Window function, 23, 75, 208
Windowing, 23
Woodwind instrument, 167
Written music, see Common musical
    notation

Xylophone, 167, 232

Zero crossing rate, 136, 174, 182
Zero tree, 87